THE PRINCIPLES OF ELECTROMAGNETIC THEORY AND OF RELATIVITY

MARIE-ANTOINETTE TONNELAT

THE PRINCIPLES OF ELECTROMAGNETIC THEORY AND OF RELATIVITY

D. REIDEL PUBLISHING COMPANY / DORDRECHT-HOLLAND

LES PRINCIPES DE LA THÉORIE

ÉLECTROMAGNÉTIQUE ET DE LA RELATIVITÉ

First published 1959, by Masson et Cie, Paris

Translated by Arthur J. Knodel and revised by the author

ISBN-13: 978-94-010-3552-1 e-ISBN: 978-94-010-3550-7

DOI: 10.1007/978-94-010-3550-7

1966

Softcover reprint of the hardcover 1st edition 1959

CONTENTS

PREFACE 1

PART ONE / ELECTROMAGNETIC THEORY

INTRODUCTION 9

CHAPTER I / ELECTROSTATICS 12

1. The experimental laws. Coulomb's law 12
2. The general laws of electrostatics 13
3. The first law. Gauss' theorem 14
4. Applications. The electric field on the surface of a conductor. Electrostatic pressure 15
5. The second law. Definition of the potential 17
6. The solutions of Laplace's and Poisson's equations 18
7. Poisson's equation and boundary conditions 20
8. Applications 23
9. Dielectrics 26
10. Diclcctrics and dipoles 28
11. Polarization and displacement 29

CHAPTER II / MAGNETOSTATICS 34

1. Permanent states. The experimental law of Biot and Savart 34
2. The general laws of magnetism 35
3. Magnetic dipoles 39
4. Magnetic media 42
5. The magnetic moment of a layer. Magnetic permeability and susceptibility 44

CHAPTER III / ELECTROMAGNETISM 48

A. Electromagnetic Induction. Displacement Current 49

1. Faraday's experimental law 49
2. Conduction current. Displacement current 50

B. Maxwell's Equations 53
3. Systems of units 53
4. Basic relations 54
5. The potential 55
6. Equations of propagation. Retarded potentials 55
C. Electromagnetic Energy. Energy Flux 58
7. Electric and magnetic energy densities 58
8. Poynting vector and Poynting's theorem 60
D. Electromagnetic Waves 61
9. Equations for the propagation of fields 61
10. Plane waves 63
11. Wave-trains 67
12. Spherical waves 70
E. Electromagnetic Equations Valid for Non-Magnetic Bodies in Slow Motion 73
13. Application of Maxwell's theory to moving media 73
14. Motion of a conductor or an insulator in an electric field 74
15. Displacement of a conductor or an insulator in a magnetic field 76
16. Hertz' and Lorentz' hypotheses 78

CHAPTER IV / SOURCES OF THE ELECTROMAGNETIC FIELD. LORENTZ' THEORY 83

1. "Microscopic" fields and potentials connected with an electron 84
2. "Structure" of the Lorentz electron 87
3. Potentials and fields created by a distribution of electrons 91
4. Equations for the mean values and Maxwell's macroscopic theory 95
5. Interpretation of the fields and the inductions of Maxwell's theory. Electromagnetic equations for the case of matter at rest 96
6. Lorentz' theory and electrodynamics of moving bodies 99

PART TWO / SPECIAL RELATIVITY

CHAPTER V / THE PRINCIPLE OF RELATIVITY 111

A. The Principle of Relativity before Einstein 111

1. The principle of relativity in classical mechanics 111
2. The principle of relativity in electrodynamics 113
3. Experimental possibilities of detecting absolute motion by optical means 114
4. First-order effects. The hypothesis of a partial dragging of light by transparent bodies 115
5. Lorentz's theory of electrons and first-order effects. The hypothesis of a motionless ether 119
6. Second-order effects 120
7. The Fitzgerald-Lorentz hypothesis 124
B. The Principle of Special Relativity 126
8. Einstein's basic postulate 126
9. Critique of the concept of simultaneity 128
10. The Lorentz transformation 129
11. Consequences of the transformation formulas 132
12. Proper time 135
13. Geometrical representation of the Lorentz formulas 136
14. Other expressions of the special Lorentz transformation 139
15. The general Lorentz transformation. C. Møller's method 140
16. Change of inertial system for a moving object. The clock paradox 146

CHAPTER VI / FOUR-DIMENSIONAL FORMALISM OF SPECIAL RELATIVITY 159

1. The pseudo-Euclidean universe of Special Relativity 159
2. Notational conventions 161
3. Reduced forms of the ds^2 in Special Relativity 162
4. Space-like four-vectors. Time-like four-vectors. Isotropic four-vectors 165
5. The invariance of the ds^2 under the displacement group in four-dimensional Euclidean space 166
6. The general Lorentz transformation and the special transformation 169
7. Expression of the coefficients in the general Lorentz transformation 171
8. Application to the special Lorentz transformation 174
9. Examples 176

10. The addition of velocities and the general Lorentz transformation 179
11. Application. Case where one of the systems is a proper system 181

CHAPTER VII / RELATIVISTIC KINEMATICS 184

A. Relativistic Law of Addition of Velocities 184
1. The velocity four-vector 184
2. The modification of velocities in a Lorentz transformation 185
3. The Lorentz transformation and the general formula for the addition of velocities 187
4. Length and direction of the velocity vector 190
5. The limiting velocity 193
6. Asymmetry of the parts played by the "relative" velocity and the "coordinate displacement velocity" 194
7. The special case of the addition of parallel velocities 195
B. Wave Propagation and Relativistic Kinematics 196
8. Propagation of a plane wave in refractive media moving uniformly with respect to each other 196
9. Huygens' principle and Special Relativity 198
10. Phase velocity and propagation velocity 203

CHAPTER VIII / RELATIVISTIC DYNAMICS 209

A. Relativistic Dynamics of a Point-Mass 209
1. Momentum, energy and proper mass of a particle 209
2. Minkowski force. The basic law of relativistic dynamics 211
3. Equivalence of mass and energy 213
4. Modification of velocities and basic quantities (momentum, energy, force) of dynamics in a Lorentz transformation 215
5. Systems of free particles 218
6. Systems of bound particles 223
B. The Relativistic Dynamics of Continuous Media 226
7. The non-relativistic equations of a fluid in a system of orthogonal coordinates 226
8. The relativistic equations of a continuous medium 227
9. The material energy-momentum tensor 230

10. The case of a perfect fluid 231
C. Use of Curvilinear Coordinates 232
11. Trajectory of a material point expressed in any arbitrarily chosen system of coordinates 233
12. The basic law of the dynamics of a point 235
13. Motion of a homogeneous fluid. The matter tensor 235
14. Equations of conservation and equations of motion 237
15. A special case: the equations of conservation and of motion for a perfect fluid 238

CHAPTER IX / RELATIVISTIC ELECTROMAGNETISM 244

A. The Covariant Form of Maxwell's Theory 244
1. The electromagnetic field, a tensor of second rank 244
2. The electromagnetic potential 247
3. Maxwell's equations and the general Lorentz transformation 248
4. The Lorentz electron theory. The energy momentum tensor 250
5. Lorentz equations and Maxwell's equations 254
6. The energy-momentum tensors 260
7. Use of arbitrary curvilinear coordinates 261
B. Extensions of Maxwell's Theory 265
8. The deduction of Maxwell's equations from a variational principle 265
9. Mie's Theory 267
10. The theory of M. Born and L. Infeld 272

CHAPTER X / THE EXPERIMENTAL VERIFICATIONS OF SPECIAL RELATIVITY 282

A. The Retardation of Moving Clocks 283
1. The theory of the Doppler effect and the slowing-down of clocks 283
2. Ives and Stillwell's experiments (1941) 286
3. The mean lifetime of mesons 289
B. The Variation of Mass with Velocity 292
4. The motion of a charged particle in an electromagnetic field 292
5. The deviation of charged particles subjected to the action of parallel electric and magnetic fields perpendicular to the initial velocity of the particles 294

6. The elastic collision of two particles 297
7. The Compton effect 301
C. The Equivalence of Mass and Energy 304
8. Mass defect and nuclear energy 304
9. The balance of energy and momentum in nuclear reactions 305

PART THREE / GENERAL RELATIVITY

CHAPTER XI / GENERAL RELATIVITY 313

A. The Newtonian Law of Gravitation 313
1. The Newtonian law of gravitation and observational data 313
2. The gravitational potential and its properties. The equivalence of gravitational mass and inertial mass 317
3. Poisson's law 320
4. Newton's law and the principle of Special Relativity 321
B. The Principle of Equivalence and the Introduction of a Non-Euclidean Universe 322
5. Accelerated reference systems and "fictitious" inertial forces. The limits of the principle of Special Relativity 313
6. The local equivalence of gravitational and inertial forces 325
7. Introduction of a non-Euclidean universe 327
8. Study of a special case: the problem of the rotating disc 330
C. Einstein's Law of Gravitation 341
9. The law of gravitation outside matter 342
10. The law of gravitation inside matter or in the presence of an electromagnetic field 343
11. The trajectories of a particle subjected to a gravitational field are the geodesics of a Riemannian space 344

CHAPTER XII / THE DEVELOPMENT OF GENERAL RELATIVITY AND SOME OF ITS CONSEQUENCES 351

A. The Equations in Various Approximations 351
1. The gravitational potential in the Newtonian approximation 351
2. The equations of the gravitational field in a system of De Donder and quasi-Galilean coordinates 353
3. Application to a continuous material medium treated as a perfect gas 357

4. Equations of the exterior case 363
5. Equations of the field and motion of the sources 365
B. Study of a Rigorous but Special Solution of the Field Equations: Schwarzschild's Solution 377
6. The gravitational field created in the neighbourhood of a static mass possesssing spherical symmetry 377
7. The field created in the neighbourhood of a spherically symmetric charged particle 382
8. The trajectory of a neutral particle in the neighbourhood of a static mass having spherical symmetry 384
9. The experimental verifications of Schwarzschild's solution 386

CHAPTER XIII / UNIFIED THEORIES OF ELECTROMAGNETISM AND GRAVITATION
Characteristics of a Pure Field Theory
Unified theories and Non-Dualistic Theories 400

A. Unified Theories 401
1. Unified theories until the advent of General Relativity 401
2. General Relativity and the construction of unified theories 402
3. Interpretation of the electromagnetic and gravitational fields proposed by unified theories 402
4. Classical unified theories and the possibilities of further predictions 405
5. Unified theories and quantum theories 406
B. Non-Dualistic Theories 409
6. The field and its sources 409
7. Non-linearity and the characteristics of a pure field theory 410
C. Unified and Non-Dualistic Theories 412

PART FOUR / MATHEMATICAL SUPPLEMENT

CHAPTER XIV / TENSOR CALCULUS IN AN EUCLIDEAN VECTOR SPACE 417

A. Rectilinear Axes 417
1. Covariance and contravariance 417
2. The norm of a vector. The scalar product of two vectors 419
3. Transformation of rectilinear axes 420

4. Invariants, four-vectors and tensors 422
5. Symmetry and antisymmetry 423
6. Transformation of the metric tensor. A special case: Utilization of orthogonal frames of reference 424
7. The rotations of axes in a four-dimensional Euclidean space 425

B. Use of Arbitrary Curvilinear Coordinates 427

8. Passage from one system of curvilinear coordinates to another in an Euclidean vector space 427
9. Differential relations between the components of the metric tensor 430
10. Covariant differentiation 432
11. Tensor densities 434

CHAPTER VX / TENSOR CALCULUS IN A NON-EUCLIDEAN METRIC MANIFOLD. APPLICATION TO A RIEMANNIAN SPACE 440

1. Metric space and tangent Euclidean space 440
2. Affine connection 441
3. Representation of the first order 442
4. Representation of the second order 444
5. Vectors and tensors associated with a metric manifold 446
6. Covariant derivation 448
7. The parallel transport of a vector 450
8. The conditions of integrability and the structure of space 452
9. The curvature of Riemannian space. The Riemann-Christoffel tensor 455
10. Properties of the Riemann-Christoffel tensor 461
11. The geodesics of Riemannian space as analogues of the straight lines of Euclidean space 463

BIBLIOGRAPHY 470

SUBJECT INDEX 473

PREFACE

The aim of this work is to study the principles upon which the classical and relativistic theories of the electromagnetic and gravitational fields are based. Thus, the primary object of the book is to present a simple exposition of Maxwell's theory, of General Relativity and of the link between those two concepts, namely, Special Relativity.

In the nineteenth century the notion of a continuous field gradually replaced the idea of action at a distance. The electromagnetic theory that was elaborated at that time covers a very large area of Physics, since it makes possible the description of permanent phenomena, electrostatics and magnetostatics, as well as of variable phenomena. It anticipates the existence of waves, and thereby the theory of light is annexed to this vast domain.

It was discovered that Maxwell's equations changed their form when they were related to reference systems associated with two observers in rectilinear uniform motion with respect to each other and each endowed with the absolute time required by classical mechanics. This was a most remarkable fact. Indeed, as soon as attempts were made to verify the results of classical kinematics by means of experiments with the propagation of light, there arose a whole series of contradictions.

This conflict between the old Newtonian mechanics and the newer electromagnetic theory was finally decided in favor of the latter. That choice is not surprising when we note that Mechanics – a physical science like all the others – is subject to experimental revision and is in no way based on immutable principles. The result was the reconstruction of mechanics on the basis of a much more critical and realistic conception of the notions of simultaneity at a distance, of space and of time. And thus, the way having been prepared by the critical works of Lorentz and Poincaré, the theory of Special Relativity was conceived in 1905 by Albert Einstein. The kinematics that it was possible to deduce from Einstein's theory very simply explained certain classical experiments, such as, for example, Fizeau's famous experiment. In addition, the new dynamics, assuming the equivalence of mass and energy, made it possible to predict the existence of great quantities of energy stored within the nuclei of atoms.

Special Relativity is, properly speaking, not a field theory, but it constitutes the basis upon which any classical or quantum field theory must rest. Maxwell's electromagnetic theory, well before 1905, was relativistic before that term became prevalent. The formalism of Maxwell's theory was to reveal itself in perfect harmony with the principles of Special Relativity.

Thus, at the beginning of the present century, a classical and relativistic theory of the electromagnetic field was, by and large, already in existence. On the other hand, we had to wait until 1916 or thereabouts, before a theory of the gravitational field at the same time classical and relativistic, was discovered. Up to that time gravitational phenomena were still explained in terms of Newtonian action at a distance. Notwithstanding rare and very slight disagreements with experimental data – the secular advance of the perihelion of Mercury being the most noteworthy of these – Newton's law of gravitation had made possible the successful elaboration of the whole structure of Celestial Mechanics.

Nevertheless, by virtue of its very guiding principle, Newton's law did not fit in with any field theory based on the existence of continuous actions transmitted contiguously. Whereas Coulomb's law, constructed along the lines of Newton's law but applicable to interactions between charged particles, had successfully been incorporated into the Maxwellian synthesis, Newton's law, in spite of numerous attempts, seemed to resist all efforts to deduce it simply from a hypothetical relativistic theory of the gravitational field.

It was at this point that Einstein, assuming the local equivalence of inertial forces and gravitational forces, succeeded in accomplishing two objectives:

First, he generalized the principle of Special Relativity to fit any given reference system; in other words, he extended to accelerated reference systems the equivalence which, by the special theory, was limited only to Galilean systems.

Second, as a result of that extension, he justified the identity between the gravitational and inertial masses – an identity that had been experimentally established but never satisfactorily explained.

Einstein's first theory concerning gravitation (1911) was a purely Euclidean one. In 1915 Einstein succeeded in demonstrating that the introduction of a non-Euclidean space would permit a very simple formulation of a relativistic law of the gravitational field and also confer a strictly local meaning to the inertial-gravitational identity. Newton's law then appears as an approximation of Einstein's relativistic theory. And finally, in an important special case – that of the gravitational field produced by a spherical body – Einstein's theory makes it possible to overcome the difficulties arising from the few points of disagreement with the Newtonian theory.

According to General Relativity, the relativistic laws of the gravitational field represent the structural conditions that a non-Euclidean space must satisfy. In this way the laws of gravitational phenomena are interpreted in the simplest and, at the same time, the most Cartesian fashion that one could imagine: a reduction to extension and motion. However, this geometrization deeply isolates these laws from the rest of physics, and especially from electromagnetism. As a result, attempts have been made – always within the purely classical framework – to construct so-called "unified" theories that would give analogous geometrical interpretations of both electromagnetic and gravitational phenomena. Together they would form a generalized field that would have to satisfy equations representing the structural conditions characteristic of a still more complex non-Euclidean universe.

If we wish to transpose General Relativity into the realm of quantum theories, we run into very serious difficulties. At the present time there exists no rigorous and wholly satisfactory quantum theory of the gravitational field. One may even be inclined to think that quantization of that field is simply not possible and to infer from this that geometrical theories should be restricted to General Relativity alone. Gravitational phenomena would in this way enjoy – or suffer from – a wholly special kind of treatment.

On the other hand, electromagnetic or nuclear phenomena, even though they took place in a non-Euclidean space, would remain entirely outside this framework. They would obey linear equations to which the usual methods of quantization would apply.

Indeed, this separation of gravitation from all other phenomena is far from satisfactory. A general field theory, could it be realized, would naturally involve elements that are largely unforeseeable but which would very likely tie in with one or the other of the two following lines of thought:

A suitable mathematical development might lead to the quantization of the gravitational equations, whether Euclidean or non-Euclidean, linear or non-linear. At the present time the mathematical development in question has not yet been worked out in a wholly satisfactory way, but whatever that may be, it would take on much greater interest if it led to experimentally verifiable predictions. It does indeed seem that the consequences of a possible quantization of gravitational waves still remain in the realm of pure speculation. The advantage of such a formalism thus remains chiefly methodological; it applies to the processes that might possibly be used in a non-linear quantum field theory.

On the other hand, one may believe – and this was Einstein's opinion – that recourse to a pure field theory would make possible a more rational explanation than the too-often purely formal and heuristic concepts suggested by further extensions of the quantum theories. It is obviously very tempting to suppose that the particles, which would be singularities of the field, do not constitute entities distinct from the field, and that their properties – especially their motion – are to be deduced entirely from the data of the field equations. The rigorous form of such equations would then be non-linear. Nevertheless, even though we assume that objective to have been achieved (which is far from being the case), we would still have to discover, if not a way of quantizing the generalized field, at least the developments that might be substituted for such a quantization with at least a reasonable chance of success.

My object here has been to point out the prospects in the development of gravitational field theories and to indicate the difficulties that these theories immediately run into. My book, however, has a much more modest object, for it is confined to presenting a general picture of the principles that have guided researchers in developing the two great classical field theories: the electromagnetic and the gravitational. As both the origin and final outcome of Special Relativity, the Maxwellian synthesis and General Relativity taken together form the most coherent and the most imposing unit in the ensemble of physical theories.

The work is thus quite naturally divided into three parts:

The presentation of the principles of electromagnetic theory (Chapters

I through IV) forms a brief review of the derivation of the basic equations and of the interpretation of these equations in the light of the Maxwell-Lorentz theory of the electron. It will be seen how the electrodynamics of moving bodies calls for the development of Special Relativity.

The second part is concerned with the bases and principal development of Special Relativity (Chapters V through X). Here we seek to demonstrate how the progressive development of that theory satisfied one of the imperative requirements in physics, when one considers that all other avenues of approach had in a very real sense been cut off.

Finally, the third part (Chapters XI through XIII) is concerned with General Relativity. It emphasizes certain developments in that theory: first, those that may be experimentally verifiable, and second, those that make General Relativity a field theory *par excellence.*

The last two chapters (XIV and XV) present certain supplementary mathematical considerations useful in reading the third part of the book. They do *not*, therefore, form a continuation of that third part but are intended as an aid – should the need for such aid be felt – in understanding it.

Numerous works on Special Relativity have been published during the last several years. The present one seeks to situate this theory in the over-all picture of what preceded it and what came after: namely, electromagnetism and General Relativity. It likewise seeks to formulate as clearly as possible the simplest basic ideas of these theories and to relate them to experimental evidence. The foundations of classical field theories reveal a remarkable chain of reasoning imposed by facts, guided by a rigorous formalism and verified by the results of that formalism.

We felt that an examination of the origin and value of the principles leading, on the one hand to the results of classical Electrodynamics of the present time, and on the other to the construction of Unified Theories of electromagnetism and gravitation, was indispensable. These extensions, which will involve further developments, are simply touched upon lightly in the present work.

PART ONE

ELECTROMAGNETIC THEORY

INTRODUCTION

The study of electrical and magnetic phenomena went on throughout the nineteenth century. Up until that time, only the laws of force governing gravitational phenomena were known. As Celestial Mechanics progressed, those laws found an immense domain in which they were applicable. But it remained for Coulomb, and then Faraday, to discover an exact expression for the forces of electrical and magnetic nature. It was then found that these forces were involved in far more phenomena than had at first been suspected. In fact, on the one hand, the domain in which electromagnetic theory was applicable was extended to the field of optics and was thus greatly enlarged, while on the other hand it was discovered that intramolecular forces as well as chemical binding forces were all electrical in nature. It was even possible to imagine that all forces were electromagnetic in nature, but the results of modern nuclear research reveal the existence of forces which, while obeying laws similar to those of electromagnetic phenomena, are very much larger and of a very special character. But even in spite of that fact, the area covered by electromagnetic theory remains extremely large.

The principles governing electromagnetic theory have undergone constant modification since the days of Newton's action at a distance up to the time when the concept of a field of force became prevalent. In any action-at-a-distance theory the force depends solely on the respective positions of the interacting particles. Such was Coulomb's law of two electrical charges. In a field theory, the force exerted on a test body depends on the field in its neighborhood, and this field can not be directly expressed in terms of the positions and velocities of the various bodies.

The action-at-a-distance theories began to undergo modification under the influence of Oersted's experiments. The laws of these actions were formulated by Ampère and concern the magnetic effects produced by a current. Every part of a circuit through which a current flows produces a magnetic force proportional, as in the case of the Coulomb forces, to $1/r^2$. (Biot and Savart's Law.)

It was Faraday who, in a decisive fashion, oriented the theory of electromagnetic action towards field-theory conceptions when he drew attention to the important properties of electrical and magnetic – or dielectric – media. These media are modified when interposed between charges and become polarized. They then contribute in a very active way – and one in which they alone are responsible – to the distribution of forces. So one has to assume that at any point in the medium there exist lines of force whose number is proportional to the intensity of that force. Any medium whatsoever, even empty space, when crossed by these lines of force then becomes the seat of a field. The discovery of the induction phenomena produced by variations of the magnetic flow through a closed circuit convinced Faraday that it was necessary to attach a very concrete meaning to lines of force in an electromagnetic field. Gauss began working on the mathematical formulation of the principles introduced by Faraday. Thirty years later, Maxwell was to give them a really satisfactory mathematical expression.

In order to state the laws of fields in as explicit a manner as possible, Maxwell had the choice of starting with Gauss' work or, with the formulation developed by Laplace and Poisson, of the Coulomb action-at-a-distance theory in terms of the field. But Maxwell's theory demonstrates that an electromagnetic action must have a finite velocity of propagation, and experimental work has proved that this velocity is equal to the velocity of light. So the theory of the electromagnetic field produces conclusions that are very different from those that one might have expected from an action-at-a-distance theory, since the forces to which an experimental body is subjected in a given field depend on the previous positions and state of motion of the other bodies.

The importance of the electromagnetic field theory also results from its extension to all of optics. Everyone knows that in Newton's time, light phenomena were explained by means of a corpuscular theory (Newton) or a wave theory (Huygens). In fact, these two interpretations were not so completely separate. Newton, who was familiar with interference and diffraction phenomena, explained them by introducing a periodic element in the light-particle's own behavior. The particle was supposed to pass through different states, being subject, for example, to sudden "fits" of easy reflection and sudden "fits" of easy transmission.[1] But Newton did believe that a corpuscular conception of light had to be maintained in order to explain rectilinear propagation and the formation of shadows.

Huygens' theory interprets correctly the phenomena of reflection and refraction by assuming the propagation of secondary spherical wavelets emanating from one principal spherical wave. Nevertheless, it was not until Young's discoveries and Fresnel's theory that rectilinear propagation was successfully and clearly explained in terms of a wave theory.

Fresnel assumed that light did not result from longitudinal vibrations, as had been thought up to that time, but from transverse vibrations perpendicular to the direction of propagation. This property explained the phenomena of double refraction, but the very existence of these waves required the presence of a vibrating medium, namely the ether. The properties of this medium were rather paradoxical; it was endowed with infinite rigidity and yet permitted bodies to move about freely in it. When Maxwell began to construct his electromagnetic theory, optics had reached just that point. Maxwell's synthesis transforms this mechanistic theory of a vibrating ether into a theory of perturbations in an electromagnetic field. The property of rectilinear propagation thus results from the extreme smallness of the wave-length of light.

The development of electrodynamics was to bring about a complete revision of the principles of classical kinematics. It is the discovery of Special Relativity which completely justifies Maxwell's electromagnetism and at the same time gives to electromagnetism its proper expression. Nevertheless, the propagation of light constitutes a limiting case in the new relativistic kinematics.

In a similar fashion, when the wave-particle dualism manifested by light phenomena led Louis de Broglie to the discovery of a Wave Mechanics of material particles, optics itself was integrated into this synthesis only with great difficulty, even though it was optics that had so strongly suggested the synthesis. A quantum and relativistic theory of photons was constructed quite late, and the adaptation of the formalism of the new quantum electrodynamics to the requirements of relativity was achieved only with great difficulty, even though the classical bases of the theory had done much to show how legitimate these requirements were.

Electromagnetic theory thus remains at once the pivot and the exceptional case in present field theories. Here we shall limit ourselves to the purely classical development of its principles.

ELECTROSTATICS

1. *The experimental laws. Coulomb's law*

The law for action at a distance between two charged particles was experimentally established by Coulomb (1780). Like the Newtonian force of gravity, the force exerted between two particles of charges q and q' is inversely proportional to the square of the distance separating the two particles. In vacuum, its magnitude is

$$F = \frac{1}{\varepsilon_0}\frac{qq'}{r^2}. \tag{I-1}$$

The value of the constant ε_0 depends solely on the system of units in which the particle's charge is expressed.

This type of a law of force can always be expressed as a function of the scalar:

$$V' = \frac{1}{\varepsilon_0}\frac{q'}{r} \tag{I-2}$$

called the *potential* associated with the charge q'. One thus has

$$\mathbf{F} = -\,q\,\mathbf{grad}\,V'. \tag{I-3}$$

If we consider a system made up of $n-1$ particles $q_1, q_2, \ldots q_{n-1}$, interacting with an nth test-particle of charge q, we still have

$$\mathbf{F} = -\,q\,\mathbf{grad}\,V' \tag{I-4}$$

with

$$V' = \frac{1}{\varepsilon_0}\sum_{i=1}^{n-1}\frac{q_i}{r_i} \tag{I-5}$$

$r_i = \sqrt{\sum_p (x_p - x_p^{(i)})(x_p - x_p^{(i)})}$ being the distance between every particle $(x_p^{(i)})$ of charge q_i and the test-particle x_p $(p = 1, 2, 3)$.

The quantities V' and $-\,\mathbf{grad}\,V'$ refer to the position of the test-particle but do not include the potential created by that particle itself – a potential which, moreover, would be infinite at that point.

At any given point of the medium one can, in a similar manner, define a potential:

(I-6) $$V = \frac{1}{\varepsilon_0} \sum_{1}^{n} \frac{q_i}{r_i}$$

and a field

(I-7) $$\mathbf{E} = -\,\mathbf{grad}\,V.$$

One will then have[2]

(I-8) $$\mathbf{F} = q\mathbf{E}$$

if there is a particle with charge q at that point.

If we consider a continuous distribution of charge in a volume $\mathscr{V}$, every charge q must be replaced by the charge $\rho d\mathscr{V}$ relative to the element of volume $d\mathscr{V} = d\xi\, d\eta\, d\zeta$. $\rho(\xi, \eta, \zeta)$ is the density of charge at the point $M(\xi, \eta, \zeta)$ of the volume $d\mathscr{V}$. If r is the distance between every point M and the point $P(x, y, z)$ where the potential is being calculated, (I-8) is replaced by the definition of a density of force

(I-9) $$\mathbf{F} = \rho\mathbf{E} = -\,\rho\,\mathbf{grad}\,V$$

with

(I-10) $$V = \frac{1}{\varepsilon_0} \int \frac{\rho}{r}\, d\mathscr{V}.$$

2. *The general laws of electrostatics*

The Coulomb law of action at a distance can be replaced by relations (I-9) and (I-10), which introduce a field and an electrostatic potential. From these expressions one could deduce the basic laws of electrostatics. But that procedure is contrary to the general idea of Maxwell's theory, which rejects the notion of action at a distance and retains only the notion of local actions. Maxwell's theory is based on the properties of the media: ether, and then, dielectrics. Now local actions depend on local relations that are partial differential equations. So we will try to substitute for the introduction of the potential (I-10), a definition that will involve only local data. In this way we will obtain partial differential equations that have a general validity and apply to physically meaningful magnitudes.

Integral equations may be deduced from the system thus obtained and, in certain special cases – in electrostatics, for example – they will coincide with the results that could be obtained from the action-at-a-distance hypothesis.

That is the way in which one can deduce Coulomb's law from the partial differential equations of the electrostatic field. But such a law is justified only insofar as a theory of action at a distance agrees with the principles of Maxwell's theory – that is, insofar as the relations provided by the action-at-a-distance theory coincide with the integral expressions deduced from the local equations of the field theory. So we will always take the local equations of the electromagnetic field as our starting point.

The basic principles of Maxwell's theory require that a distribution of charge modify the medium by creating a field of forces that is characterized at every point by the vector **E** tangent to the lines of force. The force exerted on a test-body is

$$\mathbf{F} = q\mathbf{E}, \tag{I-8}$$

q characterizing the test-body, and **E** being the force that would be exerted on a unit charge. **E** is the *electrical field* at a given point.

3. *The first law. Gauss' theorem*

Let there be a closed surface containing the charges q_i. At every point on the surface they create an electrical field **E**. The flux $d\Phi$ produced by **E** through an element dS of the surface is the product $E_n dS$ of that element of the surface multiplied by the component of **E** in the direction of the outward directed normal **n** at dS.

Gauss' theorem. The flux of force through a surface S which defines a volume $\mathscr{V}$ and which contains charges is proportional to these charges.

$$\boxed{\int_S E_n dS = \frac{4\pi}{\varepsilon_0} \sum_i q_i .} \tag{I-11}$$

The factor ε_0 alone determines the system of units selected for the measurement of q. If $\varepsilon_0 = 1$, it is the electrostatic system of units. This choice simplifies a great number of formulas.[3]

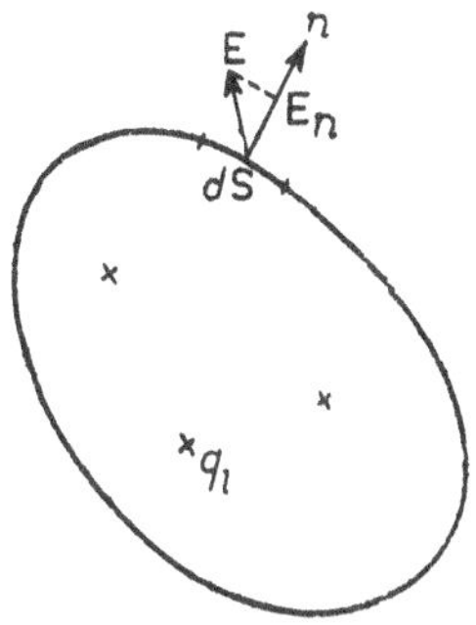

Fig. 1. The flux of force through a closed surface containing charges.

If, for the charges, we substitute a continuous charge distribution of density ρ, Gauss' theorem is written

(I-12) $$\boxed{\int_S E_n dS = \frac{4\pi}{\varepsilon_0} \int_{\mathscr{V}} \rho \, d\mathscr{V} .}$$

Now, according to Green's formula

(I-13) $$\int_S E_n dS = \int_{\mathscr{V}} \operatorname{div} \mathbf{E} \, d\mathscr{V} .$$

One will finally have

(I-14) $$\boxed{\operatorname{div} \mathbf{E} = \frac{4\pi}{\varepsilon_0} \rho .}$$

Gauss' theorem constitutes an *experimental law* that may be directly verified by measuring the charges (by means of a Faraday cylinder) and the fields (by means of a test-body). The validity of this theorem is experimentally proved, moreover, by the verification of all the consequences it entails, among which is Coulomb's law (Section 1).

4. *Applications. The electric field on the surface of a conductor. Electrostatic pressure*

If the charges are distributed in such a way that they make up a surface-distribution of density σ, one will have, on considering the tube of force Σ whose two ends are the two surface elements dS_1 and dS_2 situated on either side of dS and parallel to dS:

(I-15) $$\int_{\Sigma} E_n d\Sigma = \int_{S} (E_{n_1} + E_{n_2})\, dS = \frac{4\pi}{\varepsilon_0} \int_{S} \sigma dS$$

or

(I-16) $$E_{n_1} + E_{n_2} = \frac{4\pi}{\varepsilon_0} \sigma .$$

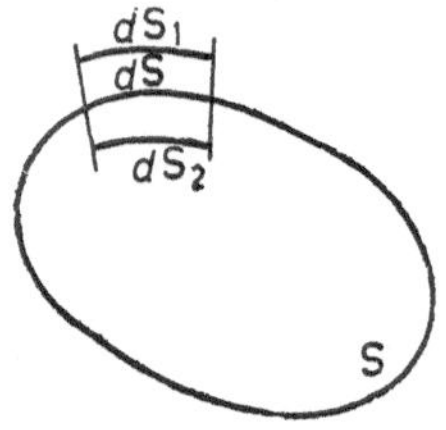

Fig. 2. The electrical field on the surface of a conductor

a) If dS is an element of the surface S of a conductor in equilibrium, $E_{n_2} = 0$. The electrical field in the vicinity of that conductor will then be:

(I-17) $$E = \frac{4\pi\sigma}{\varepsilon_0}.$$

b) If dS is a surface element which is not part of the closed conductor, $E_{n_1} = E_{n_2}$. The electrical field in the vicinity of the conductor is then:

(I-18) $$E' = \frac{2\pi\sigma}{\varepsilon_0}.$$

c) The field existing *on the electrified conductor S* is equal to the field E, produced at an infinitesimally separated point by the whole of that conductor and reduced by the amount of element dS considered by itself. In fact, the charge on that element produces no action on the element itself.

One will then have:

(I-19) $$E_s = E - E' = \frac{2\pi\sigma}{\varepsilon_0}.$$

On the charge σdS of that element, there will thus be exerted a force

$$dF = \frac{2\pi\sigma}{\varepsilon_0} . \sigma dS . \tag{I-20}$$

that is to say, a pressure[4]

$$p = \frac{dF}{dS} = \frac{2\pi\sigma^2}{\varepsilon_0} . \tag{I-21}$$

A mechanical measurement of p allows us to calculate σ and E, in electrostatic units ($\varepsilon_0 = 1$). This is the principle of Lord Kelvin's absolute electrometer.

5. *The second law. Definition of the potential*

If a test-body describes a closed circuit in an electric field, the work produced is:

$$W = q \int E_l dl \tag{I-22}$$

E_l being the projection of the electric field in the direction of the displacement.

Since we are dealing with a closed circuit, we must have

$$W = 0 \tag{I-23}$$

whatever the circuit that is covered. Now it is known that a necessary and sufficient condition for this to be the case – that is, for dW to be independent of the path followed between two fixed points, i.e., to be an exact total differential – is that

$$\boxed{\mathbf{curl\,E} = 0 .} \tag{I-24}$$

This condition is fulfilled if **E** is the gradient of a scalar $V(xyz)$ called the *electrostatic potential*

$$\boxed{\mathbf{E} = -\,\mathbf{grad}\, V .} \tag{I-25}$$

The comparison of (I-14) and (I-25) immediately leads us to *Poisson's equation:*

$$\boxed{\Delta V = -\frac{4\pi}{\varepsilon_0}\rho} \tag{I-26}$$

Δ being the Laplacian operator whose expression is

$$\Delta \equiv \frac{\partial^2}{\partial x^2} + \frac{\partial^2}{\partial y^2} + \frac{\partial^2}{\partial z^2} \tag{I-27}$$

in Cartesian coordinates.

Poisson's equation reduces to *Laplace's equation:*

$$\boxed{\Delta V = 0} \tag{I-28}$$

if the electrical density ρ is zero in the domain under consideration.

6. *The solutions of Laplace's and Poisson's equations*

Partial differential equations like those of Laplace and Poisson have a large number of possible solutions.

The general solution of the Poisson equation is the sum:

1) of a particular solution of that equation,

2) of a general solution of the equation without its right-hand member, that is, of the Laplace equation. If this general solution contains a sufficient number of arbitrary constants, it is possible to utilize them to satisfy certain limiting conditions.

Let us first try to find a particular solution of Poisson's equation.

In order to do this, let us consider a closed surface S limiting a volume $\mathscr{V}$. According to Green's theorem, one has for every vector $\mathbf{F}$:

$$\int_S F_n dS = \int_{\mathscr{V}} \operatorname{div} \mathbf{F} \, d\mathscr{V} . \tag{I-29}$$

If we put

$$\mathbf{F} = \varphi \, \mathbf{grad} \, \psi - \psi \, \mathbf{grad} \, \varphi .$$

ψ and φ being two scalars, we obtain:

$$\int_S (\varphi \, \mathrm{grad}_n \psi - \psi \, \mathrm{grad}_n \varphi) \, dS = \int_{\mathscr{V}} (\varphi \Delta \psi - \psi \Delta \varphi) \, d\mathscr{V} \tag{I-30}$$

and if

$$\psi = \frac{1}{r}$$

one has, in particular:

$$\int \left(\varphi \, \mathrm{grad}_n \frac{1}{r} - \frac{1}{r} \mathrm{grad}_n \, \varphi \right) dS = - \int_{\mathscr{V}} \frac{\Delta \varphi}{r} \, d\mathscr{V} \tag{I-31}$$

because

$$\Delta\psi = \Delta\left(\frac{1}{r}\right) \equiv 0 .$$

Let us suppose that the surface S surrounding the volume $\mathscr{V}$ is a small sphere with center P and radius R. Let us extend the integral (I-31) to the whole domain outside of the sphere. We have then:

$$\text{(I-32)} \qquad \int_R^\infty -\left(\frac{\varphi}{r^2} + \frac{1}{r}\frac{\partial\varphi}{\partial r}\right) dS = -\int_{\mathscr{V}} \frac{\Delta\varphi}{r} d\mathscr{V} .$$

If $\bar{\varphi}$ and $\overline{\partial\varphi/\partial r}$ represent the average values of φ and of $\partial\varphi/\partial r$ on the sphere of radius R, we then have:

$$\text{(I-33)} \qquad \left[\frac{\bar{\varphi}}{R^2} + \frac{1}{R}\left(\overline{\frac{\partial\varphi}{\partial r}}\right)\right] \int_s dS = -\int_{\mathscr{V}} \frac{\Delta\varphi}{r} d\mathscr{V}$$

or

$$\text{(I-34)} \qquad 4\pi\bar{\varphi} + 4\pi R\left(\overline{\frac{\partial\varphi}{\partial r}}\right) = -\int_{\mathscr{V}} \frac{\Delta\varphi}{r} d\mathscr{V} .$$

When the radius R approaches zero, $\bar{\varphi}$ becomes approximately equal to the potential V at the point P. One then has:

$$\text{(I-35)} \qquad 4\pi V = -\int \frac{\Delta V}{r} d\mathscr{V} .$$

If we now assume that the potential at the point P satisfies the Poisson equation (I-26) in the whole space outside the small sphere, we have:

$$\text{(I-36)} \qquad V = \frac{1}{\varepsilon_0}\int \frac{\rho}{r} d\mathscr{V} .$$

In order to find the general solution of the Laplace equation, it is convenient to write that equation in a system of spherical coordinates. One has:

$$\text{(I-37)} \qquad \Delta\psi \equiv \frac{1}{r^2}\frac{\partial}{\partial r}\left(r^2\frac{\partial\psi}{\partial r}\right) + \frac{1}{r^2\sin\theta}\frac{\partial}{\partial\theta}\left(\sin\theta\frac{\partial\psi}{\partial\theta}\right) + \frac{1}{r^2\sin^2\theta}\frac{\partial^2\psi}{\partial\varphi^2} = 0$$

by putting

(I-38) $$x = r \sin\theta \sin\varphi, \qquad y = r \sin\theta \cos\varphi, \qquad z = r\cos\theta.$$

The general solution of (I-37) may be put in the form of a product of three functions of r, θ, and φ. We find:

(I-39) $$\psi = \left(ar^l + \frac{b}{r^{l+1}}\right) P_l^m(\cos\theta)(C\sin m\varphi + D\cos m\varphi)$$

a, b, C and D being constants of integration, and the functions $P_l^m(\cos\theta)$ being Legendre polynomials.

The simplest solutions are thus obtained:

(I-40) $$\text{for} \quad l = 0 \quad m = 0, \qquad \psi = \left(a + \frac{b}{r}\right)\cdot D$$

(I-41) $$\text{for} \quad l = 1 \quad m = 0, \qquad \psi = \left(ar + \frac{b}{r^2}\right)\cdot D\cos\theta.$$

7. *Poisson's equation and boundary conditions*

Expression (I-39) determines the potential created at the point $P(\xi, \eta, \zeta)$ as a function of the charges with density ρ distributed in an unlimited region.

We can in a similar manner calculate the potential V created at a fixed point $P(\xi, \eta, \zeta)$ when the distribution of charge is limited by a surface Σ drawn within the thickness of a conductor. Let us suppose that within this field there is a series of conductors such as A, and let r be the distance from a fixed point P, where we wish to determine the potential, to a variable point $M(x,y,z)$ of the field.

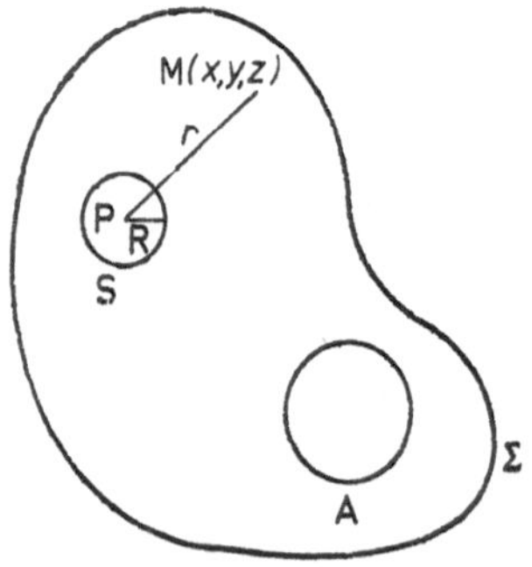

Fig. 3. The potential created at a point P in a region bounded by a surface Σ.

Let us surround the point P with a small sphere S of radius R. Green's formula (I-29) for $\psi = 1/r$ is still valid, and we still have the expression (I-31)

$$\int_S \left(\varphi \operatorname{grad}_n \frac{1}{r} - \frac{1}{r} \operatorname{grad}_n \varphi\right) dS = -\int_{\mathscr{V}} \frac{\Delta\varphi}{r}\, d\mathscr{V} \tag{I-42}$$

but now the integration must take place exclusively in the domain inside the small sphere S where we are seeking to determine the potential. We will still have, if the potential φ inside the conductor satisfies Poisson's equation:

$$-\int_{\mathscr{V}} \frac{\Delta\varphi}{r}\, d\mathscr{V} = \frac{4\pi}{\varepsilon_0} \int_{\mathscr{V}} \frac{\rho}{r}\, d\mathscr{V}. \tag{I-43}$$

In addition, the surface integral of the left-hand member must be calculated:

1) *on the small sphere S*. When its radius $R \to 0$, we still have, as in Section 6:

$$\int_S \left(\varphi \operatorname{grad}_n \frac{1}{r} - \frac{1}{r} \operatorname{grad}_n \varphi\right) dS \to 4\pi V. \tag{I-44}$$

2) *on the conductor A*. Let us put:

$$\frac{\partial\varphi}{\partial n} = \frac{4\pi\sigma}{\varepsilon_0} \tag{I-45}$$

n being the unit normal vector at the surface of the conductor, the positive direction along this normal being chosen outward. Let us designate by V_A the potential φ (constant) on the surface of the conductor A.

One has

$$\int_S \left(\varphi \operatorname{grad}_n \frac{1}{r} - \frac{1}{r} \operatorname{grad}_n \varphi\right) dS = -V_A \int_\Omega d\Omega - \frac{4\pi}{\varepsilon_0} \int \frac{\sigma}{r}\, dS \tag{I-46}$$

with:

$$\int_\Omega d\Omega = -\int_S \operatorname{grad}_n \frac{1}{r} \cdot dS = \int_S \frac{dS}{r^2} \cos\theta \quad (\theta = \widehat{\mathbf{r}, \mathbf{n}}) \tag{I-47}$$

Ω is the solid angle under which, from the point M, one sees the conductor A. When M is outside of this conductor, $\Omega = 0$.

3) *on the limiting surface* Σ. Expression (I-46) obtained for the conductor A is still valid. But this time the point M is inside Σ. So:

(I-48) $$\Omega = -\int_s \text{grad}_n \frac{1}{r} dS = 4\pi .$$

Moreover, calling the potential of the enclosure V_0 (a constant potential, since the field E_0 is zero), one has:

(I-49) $$\int \left(\varphi \,\text{grad}_n \frac{1}{r} - \frac{1}{r} \text{grad}_n \varphi \right) dS = 4\pi V_0 - \frac{4\pi}{\varepsilon_0} \int \frac{\sigma_0}{r} dS$$

with

$$\frac{4\pi\sigma_0}{\varepsilon_0} = \left(\frac{\partial V}{\partial n} \right)_0 = 0 .$$

Arbitrarily putting $V_0 = 0$ and substituting (I-43) for the right-hand member of (I-42), and then (I-44), (I-46) and (I-49) for the left-hand member of that same equation (I-42), one then obtains:

(I-50) $$\boxed{V = \frac{1}{\varepsilon_0} \int \frac{\rho}{r} d\mathscr{V} + \frac{1}{\varepsilon_0} \int \frac{\sigma}{r} dS .}$$

In particular, if there are no charges inside the conductor ($\rho = 0$):

(I-51) $$V = \frac{1}{\varepsilon_0} \int \frac{\sigma}{r} dS .$$

Finally, if each conductor is sufficiently small so that the distance between it and the point P, where one wishes to determine the potential, is practically the same for the whole conductor, one simply has:

(I-52) $$V = \frac{1}{\varepsilon_0} \Sigma_i \frac{q_i}{r_i}, \qquad \mathbf{E} = -\,\mathbf{grad}\, V = \frac{1}{\varepsilon_0} \Sigma_i \frac{q_i}{r_i^2} .$$

Under these conditions, if a test-body with a charge q' is located at a point P, it will be subjected to a mechanical force

(I-53) $$\boxed{\mathbf{F} = q'\mathbf{E} = \frac{1}{\varepsilon_0}\, q' \Sigma_i \frac{q_i}{r_i^2}.}$$

This is Coulomb's law. It has been deduced from definition (I-50) – that is, indirectly, from Gauss' theorem. So one can thus obtain Coulomb's law without assuming the presence of discontinuous point-masses generating action at a distance.

Historically, it is Coulomb's law that served to establish Gauss' theorem; and one can, as a matter of fact, beginning with the expression of Coulomb's law, deduce from it the basic laws of electrostatics. And most authors "prove" Gauss' theorem by taking for granted the existence of a field

$$E = \frac{q}{r^2}$$

deduced from Coulomb's law. Here, however, taking the opposite point of view, we state (I-12) as a general law *confirmed by experiment*, and with all its consequences likewise so confirmed, but *not itself susceptible of demonstration*. The local relation (I-14) is a rigorous relation from which Coulomb's law is deduced, first, in the particular case where integral expressions of the type (I-50) can be formed (as in electrostatics, for example); and second, in the case where it is applied to infinitely small charges.

8. *Applications*

1) *The potential created by a grounded spherical conductor situated in a constant field.* Let there be a grounded conducting sphere of radius R and likewise situated in a uniform electric field $\mathbf{E}$ with a direction that we will choose as our axis Oz. According to (I-41), the solution of Laplace's equation corresponding to $l = 1$, $m = 0$

(I-54) $$\psi = \left(ar + \frac{b}{r^2}\right)\cos\theta$$

permits us to express the potential at a point P by using the constants a and b, in order to take the boundary conditions into account.

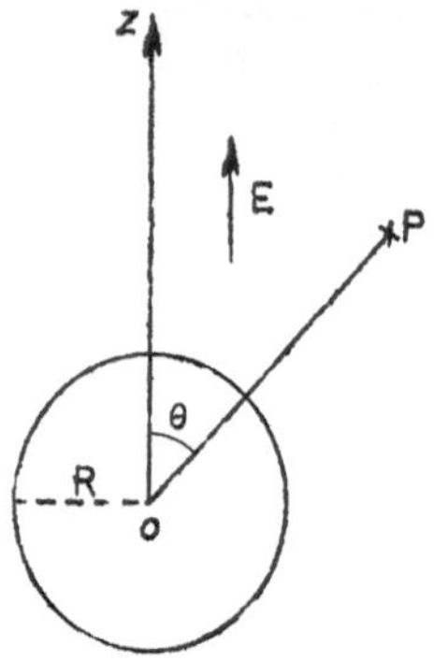

Fig. 4. The potential created by a homogeneous spherical conductor in a constant external field.

At a great distance from the origin:

(I-55) $$\mathbf{E} = -\,\mathbf{grad}\,V = -\frac{\partial V}{\partial z},$$

whence

(I-56) $$V = -\,Ez = -\,Er\cos\theta\,, \qquad a = -\,E\,.$$

On the sphere:

(I-57) $$V = 0$$

when, by comparison with (I-54),

$$-\,ER + \frac{b}{R^2} = 0\,, \qquad b = ER^3\,.$$

The potential at a point P

(I-58) $$V = -\left(r - \frac{R^3}{r^2}\right)E\cos\theta\,,$$

which is a solution of the Laplace equation, will satisfy the boundary conditions.

2) *The potential created by a dipole.* Let us now calculate the potential

created at P by a dipole $(+q, -q)$ oriented along Oz.

(I-59) $$V = \frac{q}{\varepsilon_0}\left(\frac{1}{r} - \frac{1}{r + d\cos\theta}\right)$$

or:

(I-60) $$V \simeq \frac{qd}{\varepsilon_0 r^2}\cos\theta \simeq -\frac{m_e}{\varepsilon_0}\operatorname{grad}_n\left(\frac{1}{r}\right) \qquad \text{with} \qquad m_e = qd$$

assuming that d, the distance between two charges, is small with respect to r,

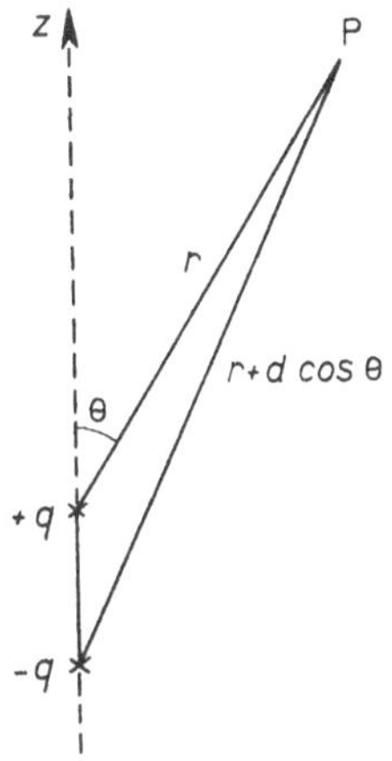

Fig. 5. Field created by a dipole.

and designating the direction of the dipole by **n** and the dipole moment by m_e.

The total potential created at P by a dipole placed in a constant field oriented in the direction of the dipole will then be:

(I-61) $$V = -\left(r - \frac{m_e}{\varepsilon_0 r^2 E}\right)E\cos\theta$$

for, to the potential created by the dipole in the absence of field, is added the potential $-rE\cos\theta$ of the external field. This potential corresponds to the solution (I-41) of Laplace's equation by selecting:

(I-62) $$a = -E \qquad b = \frac{m_e}{\varepsilon_0} = q\frac{d}{\varepsilon_0}.$$

By comparison with (I-58) we note that the grounded spherical conductor situated in the external field E behaves like a dipole whose moment $m_e = qd$ would be proportional to E:

(I-63) $$m_e = \varepsilon_0 R^3 E = \alpha E \qquad \text{with} \qquad \alpha = \varepsilon_0 R^3 .$$

A grounded spherical conductor in a uniform external field thus possesses a dipole moment. We say that it becomes polarized. It is then equivalent to a dipole of moment $qd = \varepsilon_0 R^3 E$ or to a double layer of electrical density σ such that $\sigma d = \varepsilon_0 R^3 E$.

9. *Dielectrics*

Faraday's experiments (1831) showed that if the air between the plates of a capacitor is replaced by an insulator, the difference in potential between the plates diminishes. This amounts to saying that the capacitance of the capacitor has increased. One can thus define by means of a factor $\kappa_e > 1$ the dielectric constant ε of the medium with respect to the vacuum

(I-64) $$\kappa_e = \frac{\text{dielectric capacitance}}{\text{vacuum capacitance}} = \frac{\varepsilon}{\varepsilon_0}.$$

If the electrostatic phenomena take place within a dielectric, and not in vacuum, the general laws are modified as follows:

According to Faraday's experiments, the first law, which expresses the conservation of the flux of force, is no longer valid. Nevertheless, one can postulate a law, similar to the one that would be satisfied by the vector **E** in vacuum, for a vector **D** which, by definition, will satisfy the relations:

(I-65) $$\int_S D_n dS = 4\pi \int_{\mathscr{V}} \rho d\mathscr{V}$$

that is

(I-66) $$\boxed{\operatorname{div} \mathbf{D} = 4\pi\rho}$$

D is connected with the electrical field by the relation:

(I-67) $$\boxed{\mathbf{D} = \varepsilon \mathbf{E}}$$

since, when the charge is constant, the field is ε-times smaller in the dielectric than in vacuum.

The second law of electrostatics is not modified, for it simply expresses the independence of the work accomplished by the electrostatic forces from the path that is followed. One still has

(I-68) $$\boxed{\mathbf{curl}\,\mathbf{E} = 0}$$

that is

(I-69) $$\boxed{\mathbf{E} = -\,\mathrm{grad}\,V.}$$

The vector **D** is generally called the *electric induction.* Maxwell called it *electrical displacement* because of the analogy suggested by the theory of elasticity: in an elastic medium, the force $\mathbf{E}'$ and the displacement $\mathbf{D}'$ are indeed linked by a relation of type (I-67), ε being the inverse of the coefficient of elasticity.

So we obtain from relation (I-66)

(I-70) $$\mathrm{div}\,\varepsilon\mathbf{E} = 4\pi\rho$$

that is

(I-71) $$\mathrm{div}\,\varepsilon\frac{\partial V}{\partial n} = -\,4\pi\rho\,.$$

We will thus have, by comparing (I-50), (I-53) and (I-21)

1) the expression of the potential as a function of the charges. If $\varepsilon = \text{constant}$

(I-72) $$\boxed{V = \frac{1}{\varepsilon}\int\limits_{\mathscr{V}}\frac{\rho\, d\mathscr{V}}{r} + \frac{1}{\varepsilon}\int\limits_{S}\frac{\sigma\, dS}{r}.}$$

2) Coulomb's law

(I-73) $$\boxed{f = \frac{1}{\varepsilon}\frac{qq'}{r^2}.}$$

3) the magnitude of the electrostatic pressure:

(I-74) $$p = E_S\sigma \qquad \text{with} \qquad E_S = \frac{2\pi\sigma}{\varepsilon}.$$

This pressure is then $\varepsilon/\varepsilon_0$-times smaller or greater than it is in vacuum, depending on whether one has a uniform density of charge ($p = 2\pi\sigma^2/\varepsilon$) or a uniform external field ($p = \varepsilon E_S^2/2\pi$).

The theory of dielectrics makes it very easy for us to arrive at the theory of electric currents. At first, there was a distinction made between "fictitious" charges developing within the dielectric and "real" charges appearing on the surface of the metals. In electronic theory this distinction is explained by the existence of bound and free electrons; but, so far as influence phenomena are concerned, a conductor is a dielectric whose dielectric constant ε becomes infinitely large. It thus seems that the concept of the dielectric, which constitutes so important an aspect of Maxwell's theory, has made considerably more general the dielectric theory of the vacuum.

It must, nevertheless, be pointed out that the notion of the dielectric as conceived by Maxwell, and as we have just now presented it, corresponds, in fact, only to appearances, for, on the electronic scale, there never exists anything but charges moving in vacuum. The notion of dielectric, then, is only the result of statistical observations – a purely macroscopic generalization of the theory of the vacuum. As soon as we try to interpret electromagnetic phenomena by means of a microscopic theory – as in Lorentz' theory, for example, or even in the theory of photons – inductions disappear and make room for the field, which alone retains a microscopic meaning.

10. *Dielectrics and dipoles*

A dielectric is made up of an assembly of dipoles whose moment is, in general, proportional to the external electric field that is applied to them.

1) The existence of these dipoles can, as a matter of fact, result from the external field itself. The electrons that surround the molecular nuclei are displaced by the external electric field and come to the surface of the molecule which is then approximately a charged sphere or, what amounts to the same thing, a dipole.

According to the results presented in Section 8, the dipole moment is

$$qd = \alpha \mathbf{E} \qquad \text{with} \qquad \alpha = \varepsilon_0 R^3 \tag{I-75}$$

R being of the order of the molecule's dimensions. If there exist N atoms in the volume $\mathscr{V}$, the dipole moment per unit-volume is:

$$\mathbf{P} = \frac{N}{\mathscr{V}} qd = \frac{N}{\mathscr{V}} \alpha \mathbf{E}. \tag{I-76}$$

P, the dipole moment per unit-volume, is also called the *polarization of the dielectric.*

2) Dipoles may exist in the absence of an external electrical field. Such is the case, for example, of molecules, in a gas or liquid, made up of one positive ion and one negative ion. In the absence of any external field, the orientation of each molecule is determined in a random way and the average moment is zero. The application of an electrical field has the effect of orienting these molecules parallel to the field, and the resulting dipole moment is likewise proportional to the external field.

The distinction between these two possible ways of formulating a resulting dipole moment may be stated as follows: The permanent dipoles undergo thermal agitation, and this agitation works against the establishing of resulting dipole moment. In a given field, this moment is inversely proportional to the absolute temperature. In contrast, the moment of the induced dipoles is totally independent of temperature.

11. *Polarization and displacement*

Let us consider a volume $\mathscr{V}$ enclosing a dielectric and limited by a surface S. The polarization of a molecule will be produced by the displacement d of a charge q. If there are $N/\mathscr{V}$ charges per unit-volume, the charge crossing a surface dS with the positive normal **n** is:

$$dq = \frac{N}{\mathscr{V}} q d_n dS . \tag{I-77}$$

But $Nqd/\mathscr{V}$ is the dipole moment per unit-volume, that is, the polarization P. The charge crossing the surface S is thus

$$\int_S P_n dS .$$

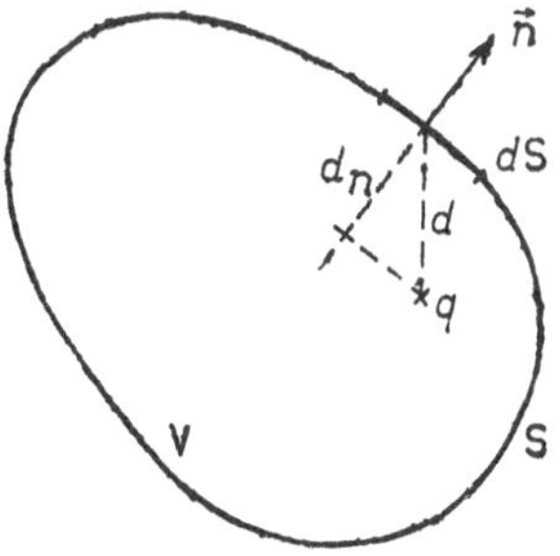

Fig. 6. Displacement of charges and polarization of a dielectric.

If ρ' represents the resulting density of charge inside the volume $\mathscr{V}$, one will have

$$\int_S P_n dS = -\int_{\mathscr{V}} \rho' d\mathscr{V}. \tag{I-78}$$

According to Green's theorem

$$\int_S P_n dS = \int_{\mathscr{V}} \operatorname{div} \mathbf{P} \, d\mathscr{V} \tag{I-79}$$

one has the local relation:

$$\operatorname{div} \mathbf{P} = -\rho'. \tag{I-80}$$

If the volume $\mathscr{V}$ contains, moreover, above and beyond any polarization, charges with the density ρ, the equation (I-14) resulting from the application of Gauss' theorem must understandably be written for the case of the vacuum, since we are adopting a microscopic description. And that equation will involve the constant ε_0 and the density of the total charge $\rho + \rho'$. One will then have:

$$\operatorname{div} \mathbf{E} = \frac{4\pi}{\varepsilon_0}(\rho + \rho'). \tag{I-81}$$

Whence

$$\operatorname{div} \mathbf{E} = \frac{4\pi\rho}{\varepsilon_0} - \frac{4\pi}{\varepsilon_0} \operatorname{div} \mathbf{P} \tag{I-82}$$

or

$$\operatorname{div} (\varepsilon_0 \mathbf{E} + 4\pi \mathbf{P}) = 4\pi\rho \tag{I-83}$$

and comparison with the definition (I-66) of **D** allows us to write:

$$\boxed{\mathbf{D} = \varepsilon_0 \mathbf{E} + 4\pi \mathbf{P}.} \tag{I-84}$$

In addition, **D** and **P** are proportional to **E**. One has, in fact, according to (I-67), (I-64) and (I-76):

$$\mathbf{D} = \varepsilon \mathbf{E} = \kappa_e \varepsilon_0 \mathbf{E}. \tag{I-85}$$

$$\mathbf{P} = \frac{N\alpha}{\mathscr{V}} \mathbf{E} = \chi_e \varepsilon_0 \mathbf{E} \tag{I-86}$$

putting

$$\chi_e = \frac{1}{\varepsilon_0} \frac{N\alpha}{\mathscr{V}}. \tag{I-87}$$

χ_e then defines *electrical susceptibility*.

Relation (I-84) has as a consequence:

$$\boxed{x_e = 1 + 4\pi\chi_e.} \tag{I-88}$$

Naturally, the values of α depend upon the presence of molecules in the neighbourhood. For a sufficiently dense dielectric, the values of electrical susceptibility and the dielectric constant are functions of experimental data. Relation (I-84) is still valid, but there is no longer any simple proportional relation between induction, polarization and the electrical field.

PROBLEMS

1. We consider a continuous distribution of charge having spherical symmetry. Calculate the density $\rho(r)$ that would produce the potential

$$V = -\frac{1}{\varepsilon_0} \frac{qe^{-ar}}{r}$$

(a and q being constants). Calculate the charge that a point-mass at the center would have in order to produce this same potential.

2. Calculate the electrical field created by a dipole.

3. a) Calculate the electrical field E_1 existing inside a sphere formed by a homogeneous dielectric and immersed in a constant electrical field E_0.

b) Limiting cases: α) the dielectric is eliminated; β) the dielectric is replaced by a conductor.

c) What is the moment of the dipole which would produce the same external field as the polarized sphere?

d) Calculate the field created under the same conditions in a spherical cavity inside a dielectric substance. What are the limiting values of this field?

Solution: a) We adopt, for the external potential, a solution V of the form (I-54), and we choose three constants a, b and ε_1 so that the following conditions are satisfied:

$$E_0 = -\left(\frac{\partial V}{\partial z}\right)_\infty ;$$

equality of the potentials

$$(\varphi_1)_{r=R} = (V)_{r=R} \qquad \left(E_1 = -\frac{\partial \varphi_1}{\partial z}\right);$$

equality of the normal components of the inductions

$$\varepsilon_0 \frac{\partial V}{\partial r} = \varepsilon_1 \frac{\partial V}{\partial r}.$$

b) The limiting cases correspond to

$$\varepsilon_1 = \varepsilon_0 . \qquad \varepsilon_1 = \infty .$$

c) One has recourse to $m_e = qd = \varepsilon_0 b$

d) ε_0 must be changed to ε_1, and vice-versa.

The field E_1 inside the cavity is included between E_0 and $(\varepsilon_1/\varepsilon_0)E_0$.

4. Determine the influence of an electrical charge $+e$ on an unspecified dielectric limited by a plane surface.

Solution: Calculate the potential resulting from the charge $+e$ and from the charge $-e$ situated at the image point (symmetrical to $+e$ with respect to the surface of the dielectric). Spell out the continuity conditions on the surface separating the vacuum and the dielectric.

NOTES

1. In a mechanistic theory such as Newton's, these sudden "fits" could result, for example, from a rotational motion associated with the propagation of ellipsoidal particles. The same configurations then recur periodically during the course of the motion.
2. As a matter of fact, in (I-6) and (I-5) $V' \neq V$ and $\mathbf{E}' \neq \mathbf{E}$ since V defines the potential created at a point by all the particles, whereas V' does not include the potential created by the particle located at the point under consideration. Because of this potential, which is infinite at the origin, V is itself infinite at the point-source, whereas V' is always finite. It thus seems that in the neighbourhood of the point-source, V and V' do not differ by merely a negligible quantity and cannot replace each other (since their difference becomes infinite at the point-source itself). But,

in practise, we know only the approximate position of the sources and the average value of the potential in their neighbourhood. In a region dV surrounding the point-source taken as the origin, the average value of the potential created by this source is

$$\frac{1}{\varepsilon_0 \mathscr{V}} \int_0^R \frac{q}{r} d\mathscr{V} = \frac{3}{4\varepsilon_0 \pi R^3} \int_0^R \frac{q}{r} 4\pi r^2 dr = \frac{3}{2\varepsilon_0} \frac{e}{R}.$$

The contribution of this potential becomes negligible if we agree to define both potential and field only outside a sphere of finite radius R surrounding the charged point.

3. It simplifies especially all formulas involving the hypothesis of spherical symmetry.
4. This pressure, directed towards the outside, is really a stress.

CHAPTER II

MAGNETOSTATICS

1. *Permanent states. The experimental law of Biot and Savart*

A magnetic field may be produced either by currents, that is to say, by charges in motion, or by magnetized bodies. We will first study magnetization produced by currents in order to obtain a simple model for representing the dipoles that occur in magnetized bodies.

The mutual independence of electric and magnetic fields is purely static. In variable states, the values of the electric and magnetic field are dependent upon each other. In magnetostatics we limit ourselves to the relatively simple case of a permanent state produced by an electric current of constant intensity i.

A constant current results from a uniform succession of charges. If $\mathbf{v}$ is the velocity of each charge in a wire with cross-section dS and v_n the projection of that velocity on the normal to dS, we have:

(II-1) $$\rho v_n \, dS = i$$

and for a length $\mathbf{dl}$ of the wire:

(II-2) $$i \, \mathbf{dl} = v_n \rho \, dS \, \mathbf{dl}$$

q being the charge contained in the volume element $dS \cdot dl$ of the wire.

If a charge q moves in the neighbourhood of an electric current or permanent magnet, one finds that this charge is acted on by a force

(II-3) $$\boxed{\mathbf{F} = q(\mathbf{v} \wedge \mathbf{B})}$$

$\mathbf{B}$ characterizes the field developed by the magnet or by the element of current.

Similarly, a circuit-element $\mathbf{dl}$, through which a current i passes and which is located in the neighbourhood of a current or magnet, is acted on by a force:

(II-4) $$\mathbf{F} = i(\mathbf{dl} \wedge \mathbf{B}).$$

The magnetic induction $\mathbf{B}$ is itself produced by the uniform motion of a charge q' of velocity $\mathbf{v}'$ or by a constant current i' in a circuit-element $\mathbf{dl}'$. Its value is given by Biot and Savart's experimental law:

$$\text{(II-5)} \qquad \mathbf{B} = \frac{q'(\mathbf{v}' \wedge \mathbf{r})}{|r|^3} = \frac{i'(\mathbf{dl}' \wedge \mathbf{r}')}{|r|^3}$$

whence

$$\text{(II-6)} \qquad \mathbf{F} = \frac{qq'}{|r|^3}[\mathbf{v} \wedge (\mathbf{v}' \wedge \mathbf{r})] = \frac{ii'}{|r|^3}[\mathbf{dl} \wedge (\mathbf{dl}' \wedge \mathbf{r})]$$

r being the distance between the test particle and the charge (or current-element) creating the induction **B**.

We have seen that in electrostatics we could obtain the value of the electric field from the expression of Coulomb's law, but we also saw that this method is of very limited applicability. It is preferable to introduce a potential, this being a solution of the Laplace or Poisson equations; i.e., to replace the integral equations (which in certain special cases are equivalent to the laws of action at a distance) by a system of partial differential equations. The same holds for magnetostatics. We must set up a system of local equations (partial differential equations) from which the basic laws of magnetostatics – in particular, Biot and Savart's law – may be deduced.

2. *The general laws of magnetism*

First Law. In the neighbourhood of a current or magnet there exists a flux of magnetic induction. The distribution of the lines of force shows that they issue from the magnet (or from the magnetic dipole layer equivalent to the circuit) and return to the magnet. A closed surface that does not pass through the magnet (or magnetic dipole layer) meets a tube of force $2n$ times. It is crossed by a zero total flux. We thus have:

$$\text{(II-7)} \qquad \int_S B_n \, dS = 0$$

that is

$$\text{(II-8)} \qquad \boxed{\operatorname{div} \mathbf{B} = 0 .}$$

The vector **B** defines the magnetic induction in the medium considered.

One can introduce, starting from **B**, a magnetic field **H**, by means of a relation of the type

(II-9) $$\boxed{\mathbf{H} = \frac{\mathbf{B}}{\mu}}$$

μ represents the *magnetic permeability* of the medium and **H** the limiting value assumed by the induction in the vacuum case. In spite of the

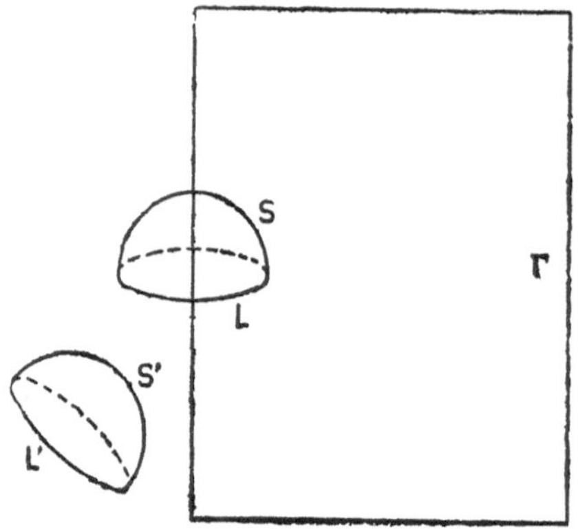

Fig. 7. Lines of force and closed paths.

similarity of relations (I-67) and (II-9), it will be seen that **H** (just like **D** in electrostatics) is not a primitive quantity of the theory. The basic quantities – which, in the microscopic theory, appear immediately – are the electric field **E** and the magnetic induction **B**.

Second Law. Let us compare the displacements of an electric charge and a hypothetical magnetic mass:

If we are concerned with a closed path L', the work done in both cases must be zero

(II-10) $$W_{L'} = 0 .$$

If we are concerned with displacement along a line of force L, the work done cannot be zero:

(II-11) $$W_L \neq 0 .$$

The comparison of (II-10) and (II-11) seems to lead to the conclusion that a line of force L cannot close in on itself, since the total work accumulated would have to be zero according to the very principle of the conservation of energy. And such is indeed the case for lines of force created by charges and by magnets. They do not close in on themselves and, in that case, (II-10) is valid for *all* surfaces bounded by closed contours (none of the surfaces being bounded by a line of force). One then has:

(II-12) $$W_{L'} = \int E_{L'}\, dL' = \int \mathrm{curl}_n\, \mathbf{E}\, dS' = 0$$

whence

(II-13) $$\boxed{\mathbf{curl}\,\mathbf{E} = 0\,, \qquad \mathbf{E} = \mathbf{grad}\, V.}$$

The field is then derived from a potential.

On the other hand, the lines of force created by a current close in on themselves. (II-11) is thus valid along certain closed trajectories that are lines of force. Nevertheless, the principle of conservation of energy is, to all intents and purposes, retained, since free magnetic masses cannot be displaced; but the field **H** is no longer derived from a potential as in (II-13). As a matter of fact, along the closed path that a line of force may in this case here describe, (II-12) is replaced by:

(II-14) $$W_L = \int H_L\, dL = \int \mathrm{curl}_n\, \mathbf{H}\, dS \neq 0\,.$$

Now, according to (II–10), **curl H** is everywhere zero for a surface S' bounded by L' and not cutting through the conductor. It is therefore only at the intersection of this surface with the conductor Γ that we can have **curl H** $\neq 0$. That being so, every surface S bounded by a line of force and, consequently, such that **curl H** $\neq 0$ must cut through the conductor at one point (or at an uneven number of points). If L is to be closed, then the conductor Γ must itself be closed; every current then passes through a closed circuit, and that justifies *a posteriori* this particular principle of Maxwell and also the introduction of displacement currents.

Let us now evaluate W_L at the intersection of S and L. We put:

(II-15) $$W_L = 4\pi i$$

thus defining a system of electromagnetic units for the measurement of i.[1] We will then have:

$$\int_L H_L \, dL = 4\pi \int_S I_n \, dS \qquad \text{with} \qquad i = \int I_n \, dS \tag{II-16}$$

I_n is the projection of a current-density vector $\mathbf{I}$ along the normal to the surface dS.

Thus, at every point of Γ:

$$\boxed{\mathbf{curl}\, \mathbf{H} = 4\pi \mathbf{I}\,.} \tag{II-17}$$

The comparison of equations (II-8), (II-9) and (II-17) shows that the magnetic equilibrium will be obtained for a solution $\mathbf{B}$ simultaneously satisfying

$$\boxed{\operatorname{div} \mathbf{B} = 0\,, \qquad \mathbf{curl}\, \mathbf{B} = 4\pi\mu \mathbf{I}\,.} \tag{II-18}$$

Such a solution is:

$$\boxed{\mathbf{B} = \mathbf{curl}\, \mathbf{A}} \tag{II-19}$$

if the vector $\mathbf{A}$ satisfies the relations[2]:

$$\Delta \mathbf{A} = -4\pi\mu \mathbf{I} \qquad \operatorname{div} \mathbf{A} = 0 \tag{II-20}$$

which is the case, in particular, for the harmonic solution:

$$\boxed{\mathbf{A} = \mu \int \frac{\mathbf{I}}{r} \, d\mathscr{V}\,.} \tag{II-21}$$

This solution determines the possibility of magnetic equilibrium[3] just as the solution

$$V = \frac{1}{\varepsilon} \int \rho \frac{d\mathscr{V}}{r} + \frac{1}{\varepsilon} \int \frac{\sigma dS}{r}$$

determines the possibility of electric equilibrium in a homogeneous dielectric.

Application. Biot and Savart's Law. Let us now consider the special case where the current-density $\mathbf{I}$ passes through a portion dl of a conductor of constant cross-section a. One has

(II-22) $\mathbf{I}\, d\mathscr{V} = \mathbf{I} a\, dl = i\, \mathbf{dl}$

and as a consequence of (II–21):

(II-23) $d\mathbf{A} = \dfrac{\mu i}{r} \mathbf{dl}.$

But according to (II-19)

(II-24) $\mathbf{dB} = \mathbf{curl}\, \mathbf{dA} = \dfrac{\mu i}{r} \mathbf{curl}\, \mathbf{dl} + \mu i \left[\mathbf{grad}\left(\dfrac{1}{r}\right) \wedge \mathbf{dl} \right] = \mu i \dfrac{\mathbf{dl} \wedge \mathbf{r}}{|r|^3}.$

This is the expression (II-5), as experimentally established by Biot and Savart. It is thus deduced from local relations involving partial derivatives resulting in part from the analogy postulated between the behavior of charges and that of hypothetical magnetic poles.

3. *Magnetic dipoles*

Let us consider an elementary closed circuit of surface dS. It subtends, as viewed from the point P, a solid angle Ω. By referring to the figure, one can see that a variation $\mathbf{dP}$ of the point P corresponds to a modification $d\Omega$ of the solid angle:

(II-25) $d\Omega = \int d\omega$ with $\omega = -\dfrac{(\mathbf{dP} \wedge \mathbf{dl})_n}{|r|} = \dfrac{(\mathbf{dP} \wedge \mathbf{dl}) \cdot \mathbf{r}}{|r|^2},$

$-(\mathbf{dP} \wedge \mathbf{dl})_n$ being the projection on the plane normal to $\mathbf{r}$ of the surface element constructed on $-\mathbf{dP}$ and $\mathbf{dl}$.

One will then have

$$d\Omega = -\int \frac{(\mathbf{dP} \wedge \mathbf{dl})\,\mathbf{r}}{|r|^3} = -\int \frac{\mathbf{dP}\,(\mathbf{dl} \wedge \mathbf{r})}{|r|^3}.$$

Now

(II-26) $d\Omega = \mathbf{grad}\, \Omega \cdot \mathbf{dP}.$

Consequently

(II-27) $\mathbf{grad}\, \Omega = -\displaystyle\int \frac{\mathbf{dl} \wedge \mathbf{r}}{|r|^3}$

and, according to (II-24):

(II-28) $$\mathbf{B} = -\mu i\, \mathbf{grad}\, \Omega .$$

In this particular case, the induction **B** is derived from a scalar potential:

(II-29) $$V = \mu i \Omega$$

and (II-28) brings about

(II-30) $$\mathbf{curl}\, \mathbf{B} = 0 .$$

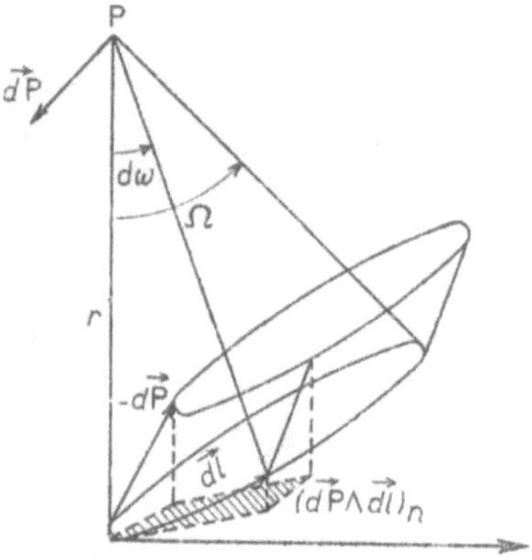

Fig. 8. Determination of **grad** Ω.

The expression of the potential V:

(II-31) $$V = \mu i \Omega = \frac{\mu i}{r^2} S \cos\theta$$

is in every way analogous to that of the potential created by an electric dipole

$$V = \frac{qd}{\varepsilon r^2} \cos\theta .$$

That is why one can say that a closed circuit is a magnetic dipole of moment iS: it is a *magnetic shell or layer*.

Once we have arrived at the concept of the layer, it is easy to come up with the consequences (II-18). One has, in fact, according to (II-28)

(II-32) $$\int B_L\, dL = -\mu i \int d\Omega\,.$$

For all paths L that cross the layer once and only once

(II-33) $$\int d\Omega = -4\pi$$

if Ω is positive.

One then has

(II-34) $$\int B_L\, dL = 4\pi\mu i\,.$$

For all the other paths L':

(II-35) $$\int B_{L'}\, dL' = 0\,.$$

But according to Stoke's theorem and (II-34), we obtain

(II-36) $$\int B_L\, dL = \int_S \mathbf{curl}_n \mathbf{B}\, dS = 4\pi \int_S \mu I_n\, dS$$

where $\mathbf{I}$ is the current density passing through the surface S, that is, $i = \int \mathrm{I}_n dS$. One will then have the partial differential equation

(II-37) $$\mathbf{curl}\,\mathbf{B} = 4\pi\mu\mathbf{I}\,.$$

Relation (II-34) is also called *Ampère's Law*. Expression (II-37) is its differential form.

In every region where $\mathbf{I} = 0$, that is, at every point in space outside of a conductor Γ, $\mathbf{curl}\ \mathbf{B} = 0$.

We have deduced Ampère's Law in (II-18) from the general laws of magnetostatics. Here we have applied the experimental law of Biot and Savart. This law thus permits us to treat a circuit as if it were a layer, and vice-versa, and to deduce from this equivalence the general laws (II-18). But here again, this procedure is hardly in keeping with the spirit of Maxwell's theory, which seeks to substitute local partial differential equations for actions at a distance. There are, thus, two alternatives:

1) We may assume Ampère's *hypothesis*, that is, arbitrarily treat a circuit element as if it were a layer or a magnetic dipole. In that case, as in the present section, we deduce the general laws from that hypothesis.

2) We may, following the electrostatic model, retain the hypothetical law (II-17) which assumes the existence and displacement of magnetic masses.

In any event, these postulates are justified by their consequences: namely, the deduction of Biot and Savart's law and the experimental confirmation of all its consequences.

4. *Magnetic media*

Just as electric media are formed by electric dipoles, so are magnetic media formed by magnetic dipoles. *A priori*, three sorts of magnetic media are known:

1) A *diamagnetic* medium does not contain permanent dipoles but only dipoles created by the external field. We have already interpreted the phenomena of the polarization of an atom in an external field assuming that the electrons (extra-nuclear) of the atom can move about more or less freely inside the atom and thus transform it into an electric dipole. In a similar manner, if this atom is placed in a region where the induction **B** varies with time, a current is induced within the atom and circulates in a direction such that the magnetic moment which it produces has a direction opposite to the induction **B**, which originally gave rise to the current. A medium possessing only this dipole moment opposite to **B** is a *diamagnetic* medium.

2) We have seen that certain media contain electric dipoles even in the absence of external field. Similarly, in the absence of any magnetic induction, one may still find permanent magnetic dipoles. In the absence of external field these dipoles are randomly oriented and their moment is zero. If an external field arises to give them orientation, a resultant dipole moment which is added to the induction, results. This effect is known as *paramagnetism*. When it exists, it is generally important enough to overshadow the diamagnetism which, nevertheless, always underlies it. The magnitude of the paramagnetic dipole moment is inversely proportional to the absolute temperature, just as is the corresponding electric dipole moment. (The diamagnetic moment is, on the contrary, independent of absolute temperature.) The existence of magnetic dipoles in the absence of any external field may be attributed to two causes and can be completely interpreted only by means of quantum theories.

a) The atomic electrons are propelled in a rotary motion around the nucleus. These motions may be in different directions and may compensate each other: that is the case of diamagnetic substances. If this is not the case, there is then a residual current whose existence was already postulated by Ampère and which, for that reason, was labelled the Ampère current. In actual fact, the predicting of the magnitude of the resultant magnetic moment and of the nature of these currents was accomplished by Sommerfeld's quantum theory.

b) The electrons, moreover, exhibit a characteristic rotational motion of their own. To this inherent angular momentum – or spin – there corresponds a magnetic moment. An atom can thus possess a residual inherent magnetic moment resulting from the values of the spin of its electrons. A diamagnetic substance is a substance in which the orbital magnetic moment and the spin magnetic moment both disappear for the ensemble of its electrons. The hypothesis of a magnetic and rotating electron had already been suggested in the earliest quantum theories, but that hypothesis was not given a completely correct interpretation until Dirac formulated his relativistic theory.

c) Ferromagnetism is a limiting case of paramagnetism. It finds valid explanation only in quantum theory. Specifically, it is shown in quantum theory that the permanent dipoles whose moment results, for example, from the spin of their electrons, are affected by exchange forces, which tend to orient the spins of two nearby molecules or atoms in the same direction. One will thus have, in certain substances, regions whose extent is of atomic proportions and which have a considerable magnetic momentum resulting from the addition of the spins of the various electrons in that region. In the absence of any field, these magnetic moments are distributed at random; but an external field, by orienting them all in its own direction, confers an appreciable magnetic moment on the substance. That moment reaches its limit when all the spins resulting from the different domains are parallel. This limit is the *saturation moment.* By reversing the direction of the external field, the resultant magnetic moment decreases, but with a certain delay resulting from a sort of intra-molecular friction. This is the phenomenon of *hysteresis.* So there exists a residual magnetic moment, even when the external field is zero. This is the source of *permanent magnetism.*

The properties of ferromagnetism are quite complicated, for in this case the magnetic moment is not proportional to the field, as in the case of dia- or para-magnetism. Ferromagnetism is a function of the absolute temperature and disappears when the absolute temperature exceeds a critical point θ known as the "Curie point". Above this critical point the magnetic moment is proportional to $1/(T-\theta)$; the thermal agitation that results at higher temperatures is opposed to the common orientation of the magnetic moments of the various domains.

5. *The magnetic moment of a layer. Magnetic permeability and susceptibility*

In a dia- or paramagnetic medium, an induction **B** produces a magnetic moment proportional to **B**. Let us call this magnetic moment, per unit volume, **M**. It is equal to the moment produced by a magnetic layer made up of a circuit element through which a permanent current of intensity j passes.

(II-38) $$\int_L M_L \, dL = \int_S \mathbf{curl}\,\mathbf{M}\, dS = j = \int_S J_n \, dS$$

with $$j = \int_S J_n \, dS .$$

One will then have

(II-39) $$\mathbf{curl}\,\mathbf{M} = \mathbf{J} .$$

Now the equation[4] that we have derived for the case of a steady state is

(II-40) $$\mathbf{curl}\,\mathbf{B} = 4\pi\mu_0 \mathbf{I}$$

and relates to the *total* current issuing from a closed surface bounded by a contour L. If one is not concerned with a magnetic medium, the circuit Γ forms a layer whose moment iS is characterized by the current density **I** such that $i = \int I_n dS$. This current arises from the usual motion of charges within a conductor. But, if the layer Γ is of a magnetic substance, there exists, in addition, a current density **J** produced by the magnetic polarization of that substance. Just as for a dielectric in which there exist both a conduction current and a current resulting from the polarization of this dielectric, we here obtain:

(II-41) $$\mathbf{curl}\,\mathbf{B} = 4\pi\mu_0 (\mathbf{I} + \mathbf{J}) .$$

Now, according to (II-39)

(II-39) $$\mathbf{curl}\,\mathbf{M} = \mathbf{J} .$$

We deduce from this

(II-42) $$\boxed{\operatorname{curl} \mathbf{H} = 4\pi \mathbf{I}}$$

putting

(II-43) $$\mathbf{H} = \frac{\mathbf{B}}{\mu_0} - 4\pi \mathbf{M}.$$

The magnetic field is here introduced just as the electric induction **D** is introduced in electrostatic theory. One thus obtains for all cases:

(II-44) $$\boxed{\mathbf{B} = \mu_0 (\mathbf{H} + 4\pi \mathbf{M}).}$$

But on the other hand, in dia- and paramagnetic media, we know that **M** is proportional to **B**.

One then has:

(II-45) $$\mathbf{M} = a\mathbf{B}$$

and, in addition, according to (II-9)

(II-46) $$\mathbf{B} = \mu \mathbf{H}.$$

And according to (II-44), (II-45) and (II-46):

(II-47) $$\mu = \mu_0 (1 + 4\pi a \mu).$$

It is convenient to define both the magnetic *permeability* of a medium:

(II-48) $$\boxed{x_m = \frac{\mu}{\mu_0}}$$

and its *magnetic susceptibility*:

(II-49) $$\boxed{\chi_m = \frac{M}{H}.}$$

The latter will be positive or negative, depending on whether the substance considered is para- or diamagnetic.

Furthermore, for dia- or paramagnetic substances, x_m is in the neighborhood of 1, while χ_m is very small with respect to unity (while χ_e can, on the other hand, be very large with respect to unity).

According to (II-45) and (II-46) one will have:

(II-50) $$\chi_m = a\mu .$$

Taking (II-47) and (II-50) into account, (II-48) is still written:

(II-51) $$\boxed{x_m = 1 + 4\pi\chi_m .}$$

PROBLEMS

1. Calculate the vector potential **A** created outside and inside a long straight wire of indefinite length and radius R through which a current of density I passes. Deduce from this the value of the magnetic field inside and outside the wire.

Solution: We choose the axis of the wire as our oz axis ($\mathbf{I} = J_z$, $\mathbf{A} = A_z$). We will write (II-20) as a function of the variable $r = \sqrt{x^2 + y^2}$. Integrating: $\mathbf{A} = C_1 Lr + C_2 - \pi\mu \mathrm{I} r^2$.

a) Outside the wire, $I = 0$; condition (II-15) gives us $C_1 = -2\mu i$.

b) Inside the wire, the three conditions $C = 0$ ($\mathbf{A}$ finite); $\chi R^2 I = i$; $\mathbf{A}_{\text{int}z} = A_{\text{ext}z}$ have as a consequence

$$\mathbf{A} = -2\mu i\left[LR + C_2 - \frac{1}{2}\left(1 - \frac{r^2}{R}\right)\right].$$

According to (II–19) one deduces from a) and b)

$$|\mathbf{H}|_{\text{ext}} = \frac{2i}{r} \qquad |\mathbf{H}|_{\text{int}} = 2i\,\frac{r}{R^2}.$$

2. Determine the magnitude and direction of the magnetic field **H** produced by the uniform rotation of a sphere that is charged around one of its diameters. (Let σ be the surface density of the charge and R the radius of the sphere.) Show that the field **H** does not vary inside the sphere.

3. An electron of charge e describes a circle of radius r with a velocity v. Calculate its magnetic moment. Show that this moment is proportional

to the angular momentum. Determine this magnetic moment when the angular momentum is $h/2\pi$.

NOTES

1. With that choice, Coulomb's law written for a medium with a dielectric constant ε,

$$F = \frac{1}{\varepsilon}\frac{qq'}{r^2},$$

leads us to choose for the vacuum a constant $\varepsilon_0 \neq 1$, q and q' being expressed in electromagnetic units.

2. In fact, according to (II-19), the first of the relations (II-18) is satisfied and the second is written

$$\mathbf{curl\ curl\ A} = \mathbf{grad}\ \mathrm{div}\ \mathbf{A} - \Delta\ \mathbf{A} = 4\pi\mu\mathbf{I}\,.$$

and results from relations (II-20).

3. The difference between **B** and another solution **B**′ satisfying (II-18) would be a harmonic function everywhere finite. So it would be identically zero. The magnetic equilibrium described by **B** is thus uniquely determined.
4. We must, of course, here put $\mu = \mu_0$. On the level that here interests us, magnetic dipoles exist only in the vacuum.

CHAPTER III

ELECTROMAGNETISM

Electrostatics and magnetostatics are concerned with two groups of phenomena which, up until the beginning of the nineteenth century, seemed completely different. The study of the magnetic field created by permanent currents demonstrated for the first time a relationship between the effects produced by magnets and those produced by the motion of charges (Oersted). Around 1830, Faraday undertook a series of experiments that were to reveal the reciprocal of the Oersted effect. It had seemed to Faraday, in fact, that a magnetic induction might be able to produce an electric current. If this hypothesis turned out to be correct, then the magnetic induction resulting from the existence of a current in a primary circuit would suffice to produce another current in a secondary circuit near to the first one. But what Faraday observed was that an induced current cannot result from magnetic induction alone, but arises from the variations of the magnetic induction. It is only at the moment of the setting up and breaking of the primary current that the variation of an induction flux can produce an electric current in a nearby secondary circuit. That was the result which Faraday established experimentally.

The discovery of an induced electromotive force had to be expressed in a mathematical formulation that could describe both electrical and magnetic phenomena and the new Faraday law. Maxwell achieved this formulation a few years later. He set up a system of partial differential equations which represents in a satisfactory manner the combined laws of electromagnetism. Now Maxwell's equations involved a new term that was indispensable in assuring their coherency: Maxwell interpreted this term as a displacement current, having assumed that every electric circuit is of necessity closed. The existence of this displacement current was proved experimentally. Its magnitude is exactly what was predicted by the theory. Maxwell's equations, while applying to all the electric and magnetic phenomena known at the time, further permitted the prediction of the existence of electromagnetic waves. So the theory applies to a vast domain, which was even further extended by the inclusion of optics.

The extent and success of Maxwell's electromagnetic theory were to be so great that they compelled assent even when the principles of classical kinematics contradicted those of electromagnetism. The Maxwellian synthesis, which implicitly presupposed a relativistic kinematics, was to permit the development of Special Relativity.

A. ELECTROMAGNETIC INDUCTION. DISPLACEMENT CURRENT

1. *Faraday's experimental law*

The magnetic field created by a constant current arises from permanent phenomena. If the magnitude of the current varies with time, or if the circuit is displaced, variable states are obtained. These latter are governed by Maxwell's theory of quasi-stationary states if the variation is slow, that is, if the duration of these states is long with respect to the time of the propagation of the electromagnetic perturbation in the system under consideration. This condition is satisfied in numerous cases which will now be examined.

If the magnetic induction varies in the neighbourhood of a circuit, the circuit becomes the seat of a current arising from the existence of an induced electromotive force. Experiments have shown that the magnitude of this induced electromotive force is proportional to the time rate of change of the flux through the circuit.

(III-1) $$e = -\frac{d}{dt}\int_S B_n \, dS .$$

e is the total electromotive force induced in the circuit. It is equal to the work:

(III-2) $$\int_L E_L \, dL = \int_S \mathrm{curl}_n E \, dS$$

of the electric field. So one will have the local equation:

(III-3) $$\frac{\partial \mathbf{B}}{\partial t} = -\,\mathbf{curl}\,\mathbf{E} .$$

This relation is the differential form of Faraday's experimental law.

Now we know that **B** is deduced from a vector potential **A** by the relation

(III-4) $$\mathbf{B} = \mathbf{curl}\,\mathbf{A}\,.$$

The comparison of (III-3) and (III-4) permits us to state that the vectors $-\partial\mathbf{A}/\partial t$ and **E**, having the same curl, differ only by a gradient. So we can put:

(III-5) $$\boxed{\mathbf{E} = -\frac{\partial \mathbf{A}}{\partial t} - \operatorname{grad}\psi}$$

ψ being an arbitrary scalar function.

2. *Conduction current. Displacement current*

a) *The conduction current.* The electric current produced by the motion of the charges in a conductor is governed by the presence of an electric field **E** which is derived from a potential V.

(III-6) $$\mathbf{E} = -\,\mathbf{grad}\,V\,.$$

If we are dealing with a linear conductor, we can define a current density:

(III-7) $$i = \frac{dq}{dt}$$

which is nothing more than the quantity of electricity going through a unit cross-section of wire per second.

Experimentally, Ohm's law makes it possible to characterize every conductor by a resistivity ρ_c and a conductivity

$$\sigma_c = \frac{1}{\rho_c}\,.$$

According to the definition of the field **E** and according to Ohm's law field and electric current are related by:

(III-8) $$\boxed{\mathbf{E} = \rho_c \mathbf{I} \qquad \text{or} \qquad \mathbf{I} = \sigma_c \mathbf{E}\,.}$$

I being the density of the current, that is, the current passing through a unit area.

At a constant rate, the flux entering through the surface S is equal to the flux going out at S'. In other words, electricity behaves like a non-compressible fluid, and this property is expressed by:

$$\text{div}\,\mathbf{I} = 0\,. \tag{III-9}$$

b) *The electric displacement current.* Definition (III-7) of the current, instead of being related to a closed circuit fed by an electromotive force, may also characterize the intensity of the discharge current of a capacitor:

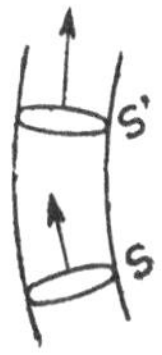

Fig. 9. Steady state. Flux conservation.

dq then represents the variation of the charge on the plates per unit time, but the current seems to be produced in an open circuit.

Maxwell introduced the hypothesis that open circuits do not exist, that is, that lines of force always close back on themselves. If we are concerned with the discharge of a condenser, we then have to assume that Gauss' theorem remains applicable, as it does in a state of equilibrium. We will then have:

$$\int_S D_n\, dS = 4\pi q \tag{III-10}$$

that is

$$\text{div}\,\mathbf{D} = 4\pi\rho \tag{III-11}$$

and, on the model of (III-7), one can define an electric displacement current

$$i' = \frac{dq}{dt} = \int I'_n\, dS \tag{III-12}$$

such that

$$\boxed{\mathbf{I}' = \frac{1}{4\pi}\frac{d\mathbf{D}}{dt}.} \tag{III-13}$$

In a medium that is both conductive (σ_c) and dielectric (ε), it will then be necessary to define a current:

$$\text{(III-14)} \qquad \boxed{\mathbf{I}_e = \mathbf{I} + \frac{1}{4\pi}\frac{d\mathbf{D}}{dt} = \sigma_c\mathbf{E} + \frac{\varepsilon}{4\pi}\frac{\partial\mathbf{E}}{dt}.}$$

Since the total electric current is incompressible, we will have the condition

$$\text{(III-15)} \qquad \boxed{\operatorname{div}\mathbf{I}_e = 0}$$

if there is a displacement current.

c) *Introduction of the displacement current into the field equations.* We have obtained the equation:

$$\text{(III-16)} \qquad \mathbf{curl}\,\mathbf{H} = 4\pi\mathbf{I}$$

by assuming, according to Ampère's law, that **I** is the current of *the conduction in the circuit under consideration.* If we apply this relation to the field existing in the circuit, which is apparently open when the discharge from a condenser takes place, we will have:

$$\text{(III-17)} \qquad \operatorname{div}\mathbf{curl}\,\mathbf{H} = 4\pi\operatorname{div}\mathbf{I} = 0$$

which means that, in reality, the circuit is closed.

Maxwell assumes that this is indeed the case, because equation (III-16) actually applies to a current $\mathbf{I}_e$ which is the sum of the current **I** and a supplementary current $\mathbf{I}'$, which we call the displacement current.

In the case of a steady state $\mathbf{I}' = 0$, and that is why we were able to apply Ampère's law. But for a varying rate of flow (for example, in the discharge from a condenser), one has to write:

$$\text{(III-18)} \qquad \boxed{\mathbf{curl}\,\mathbf{H} = 4\pi\mathbf{I}_e}$$

with

$$\text{(III-19)} \qquad \operatorname{div}\mathbf{I}_e = \operatorname{div}(\mathbf{I} + \mathbf{I}') = 0.$$

So we have, according to (III-13) and (III-11)

$$\text{(III-20)} \qquad \operatorname{div}\mathbf{I} = -\frac{1}{4\pi}\operatorname{div}\frac{d\mathbf{D}}{dt} = -\frac{\partial\rho}{\partial t}.$$

B. MAXWELL'S EQUATIONS

3. *Systems of units*

In all the preceding relations one can define a *system of electrostatic units* CGS by arbitrarily putting

(III-21) $$\varepsilon_0 = 1 .$$

In a similar manner, one can define a *system of electromagnetic units* CGS by choosing

(III-22) $$\mu_0 = 1 .$$

If we arbitrarily choose the dielectric constant of the vacuum as unity ($\varepsilon_0 = 1$), we define in this way a unit of charge which is incompatible with the definition that would result from the convention $\mu_0 = 1$.

Nevertheless, it is convenient to adopt a *mixed system*, provided we evaluate the quantities **E**, **D**, ε in the CGS electrostatic system ($\varepsilon_0 = 1$) and the quantities **H**, **B**, μ in the CGS electromagnetic system ($\mu_0 = 1$).

Let c be the relationship between the two systems of units. Let us designate by the indices (e) and (m) the system of electrostatic or electromagnetic units in which we express a certain quantity. We will haev

(III-23) $$\frac{q_{(e)}}{q_{(m)}} = \frac{j_{(e)}}{j_{(m)}} = \frac{\rho_{(e)}}{\rho_{(m)}} = c .$$

But, according to the expression of the laws of force

(III-24) $$|\mathbf{F}| = q\,|\mathbf{E}| = \frac{1}{\varepsilon}\frac{q^2}{r^2} \qquad \mathbf{F} = q\,(\mathbf{v} \wedge \mathbf{B}), \qquad d\mathbf{B} = \mu i\,\frac{d\mathbf{l} \wedge \mathbf{r}}{|r|^3}$$

which is valid whatever unit-system we choose, we will have further

(III-25) $$\frac{E_{(e)}}{E_{(m)}} = \frac{1}{c} \qquad \frac{\varepsilon_{(e)}}{\varepsilon_{(m)}} = c^2 \qquad \frac{D_{(e)}}{D_{(m)}} = c$$
$$\frac{B_{(e)}}{B_{(m)}} = \frac{1}{c} \qquad \frac{\mu_{(e)}}{\mu_{(m)}} = \frac{1}{c^2} \qquad \frac{H_{(e)}}{H_{(m)}} = c$$

by using the definitions $\mathbf{D} = \varepsilon\mathbf{E}$, $\mathbf{B} = \mu\mathbf{H}$ and the relation (III-18).

4. *Basic relations*

By analogy with the electric displacement current

(III-26) $$\mathbf{I}' = \frac{1}{4\pi}\frac{\partial \mathbf{D}}{\partial t},$$

it is convenient to introduce a magnetic displacement current:

(III-27) $$\mathbf{J}' = \frac{1}{4\pi}\frac{\partial \mathbf{B}}{\partial t} = \mathbf{J}_m .$$

As a matter of fact, this current $\mathbf{J}'$ represents the totality of the magnetic current $\mathbf{J}_m$, since there exist no free magnetic poles capable of producing a conduction current analogous to $\mathbf{I}$.

With these notations, and in the mixed system, equations (III-3) and (III-18) relative to the electromagnetic field are then written:

$$(\mathrm{I})\begin{cases} \text{a)}\ \mathbf{curl}\,\mathbf{H} = 4\pi \mathbf{J}_e \\ \text{b)}\ \mathbf{curl}\,\mathbf{E} = -4\pi \mathbf{J}_m \end{cases}$$

with

$$(\mathrm{II})\begin{cases} \text{a)}\ \mathbf{J}_e = \dfrac{\mathbf{I}}{c} + \dfrac{1}{4\pi c}\dfrac{\partial \mathbf{D}}{\partial t} \\ \\ \text{b)}\ \mathbf{J}_m = \dfrac{1}{4\pi c}\dfrac{\partial \mathbf{B}}{\partial t} \end{cases} \quad \text{and} \quad \begin{array}{l} \mathbf{D} = \varepsilon \mathbf{E} \\ \mathbf{I} = \sigma_c \mathbf{E} \\ \mathbf{B} = \mu \mathbf{H}. \end{array}$$

Equations (I) have as consequences:

$$(\mathrm{I}')\begin{cases} \text{a)}\ \operatorname{div} \mathbf{J}_e = 0 \\ \text{b)}\ \operatorname{div} \mathbf{J}_m = 0. \end{cases}$$

On the other hand, equations (III-11) and (III-4) relative to electric and magnetic inductions become:

$$(\mathrm{III})\begin{cases} \text{a)}\ \operatorname{div} \mathbf{D} = 4\pi\rho \\ \text{b)}\ \operatorname{div} \mathbf{B} = 0. \end{cases}$$

They result from the application of Gauss' theorem, and (III-b) indicates the absence of any free magnetism.

Finally, if we take into account definitions (II), equations (III) have as a consequence (I′) if:

$$\text{(III}'\text{)}\quad \operatorname{div}\mathbf{I} + \frac{\partial \rho}{\partial t} = 0$$

(III′) is still called the *continuity equation.* So the basic equations that all electromagnetic phenomena must obey are equations (I), (II), and (III).

5. *The potential*

In the case of static phenomena, we know that it is possible to deduce the electric field **E** and the magnetic induction **B** from a scalar potential V and a vector potential **A** such that

$$\text{(III-28)}\qquad \mathbf{E} = \mathbf{grad}\, V\,, \qquad \mathbf{B} = \mathbf{curl}\,\mathbf{A}\,.$$

In addition, the study of induction phenomena has shown us that the electric induction field can also result from a vector potential and be written:

$$\text{(III-29)}\qquad \mathbf{E} = -\frac{\partial \mathbf{A}}{\partial t} - \mathbf{grad}\,\Psi\,.$$

Combining these results and using the mixed system, let us try to find out what relations an electric field **E** and a magnetic induction **B** must satisfy such that:

$$\text{(IV)}\begin{cases} \text{a)}\ \mathbf{E} = -\dfrac{1}{c}\dfrac{\partial \mathbf{A}}{\partial t} - \mathbf{grad}\, V \\ \text{c)}\ \mathbf{B} = \mathbf{curl}\,\mathbf{A}\,. \end{cases}$$

Equations (I) and (III) then reduce to:

$$\text{(V)}\begin{cases} \text{a)}\ \mathbf{curl}\,\mathbf{H} = 4\pi \mathbf{J}_e \\ \text{b)}\ \operatorname{div}\mathbf{D} = 4\pi\rho\,. \end{cases}$$

6. *Equations of propagation. Retarded potentials*

Assuming that the dielectric constant ε and the permeability μ are constant, let us form, according to (IV), $-(\mu\varepsilon/c)(\partial\mathbf{E}/\partial t) + \mathbf{curl}\,\mathbf{B}$ and $\operatorname{div}\mathbf{E}$.

We find, taking (V) and definitions (II) into account:

$$\text{(VI)}\begin{cases}\dfrac{\mu\varepsilon}{c^2}\dfrac{\partial^2 \mathbf{A}}{\partial t^2} - \Delta\mathbf{A} = \dfrac{4\pi}{c}\mu\mathbf{I},\\[2ex] \dfrac{\mu\varepsilon}{c^2}\dfrac{\partial^2 V}{\partial t^2} - \Delta V = \dfrac{4\pi}{\varepsilon}\rho\end{cases} \qquad \Delta = \frac{\partial^2}{\partial x^2} + \frac{\partial^2}{\partial y^2} + \frac{\partial^2}{\partial z^2}$$

by introducing the condition

$$\text{(VII)}\quad \frac{\mu\varepsilon}{c}\frac{\partial V}{\partial t} + \operatorname{div}\mathbf{A} = 0$$

called the *Lorentz condition.* Equations (VI) are called *d'Alembert's equations,* and they permit us to calculate **A** and V. We know that Poisson's equation

$$\text{(III-30)}\qquad \Delta\varphi = -\frac{4\pi\rho}{\varepsilon}$$

has the particular solution:

$$\text{(III-31)}\qquad \varphi = \int_{\mathscr{V}} \frac{\rho}{\varepsilon r}\, d\mathscr{V}.$$

Every elementary potential

$$\text{(III-32)}\qquad d\varphi = \frac{1}{\varepsilon}\frac{\rho}{r}\, d\mathscr{V}$$

is the potential produced by a charge $\rho d\mathscr{V}$ at a distance r. This potential is a solution of Laplace's equation, except for $r = 0$. In order that a sum of potentials of type (III-32) should be a solution of d'Alembert's equation (VI), it is necessary that each one of them shall not be deduced from the value of the charge situated at $M(r)$ at the instant t, but rather from its value at the instant $t - (r/c)$.

This value of the charge:

$$\rho\left(\xi, \eta, \zeta, t - \frac{r}{c}\right) d\mathscr{V}$$

will produce at P (x, y, z) a potential

$$\frac{\rho}{r}\left(\xi, \eta, \zeta, t - \frac{r}{c}\right) d\mathscr{V}.$$

This potential is the one that exists at P at the instant t since r/c is precisely the time required by the disturbance to go from M to the point P where we are seeking to determine the potential.

The scalar potential, solution of (VI), will then be:

$$\text{(III-33)} \qquad V = \int_{\mathscr{V}} \frac{\rho}{\varepsilon r}\left(\xi, \eta, \zeta, t - \frac{r}{c}\right) d\mathscr{V} = \int_{\mathscr{V}} \frac{(\rho)_{t-\frac{r}{c}}}{\varepsilon r} d\mathscr{V}.$$

In a similar manner, one will have

$$\text{(III-34)} \qquad \mathbf{A} = \int_{\mathscr{V}} \frac{\mu(i)_{t-\frac{r}{c}}}{r} d\mathscr{V}.$$

These solutions satisfy the equations of propagation (VI) at every point except the origin. They are called *retarded potentials*, because the effects of the charges produce retarded phenomena at a point P due to the finite velocity of propagation of an electromagnetic disturbance.

Moreover, we know that the following solution is a solution of Poisson's equation in a limited region[1]:

$$\text{(III-35)} \qquad \varphi = -\frac{1}{4\pi}\int_{\mathscr{V}} \frac{\Delta\varphi}{r} d\mathscr{V} - \frac{1}{4\pi}\int_{S}\left(\varphi \cdot \operatorname{grad}_n\left(\frac{1}{r}\right) - \frac{1}{r}\operatorname{grad}_n \varphi\right) dS$$

taking boundary conditions into account.

It can be shown[2] that the following is a solution of d'Alembert's equations in a limited region:

(III-36)
$$\varphi = -\frac{1}{4\pi}\int_{\mathscr{V}} \frac{[\Delta\varphi - (1/c^2)(\partial^2\varphi/\partial t^2)]_{t-\frac{r}{c}}}{r} d\mathscr{V} + \frac{1}{4\pi}\int_S \frac{1}{r}\left\{\left[\frac{1}{c}\left(\frac{\partial\varphi}{\partial t}\right)_{t-\frac{r}{c}} + \frac{\varphi\left(t-\frac{r}{c}\right)}{r}\right]\cos(\mathbf{n},\mathbf{r}) + \left(\frac{\partial\varphi}{\partial n}\right)_{t-\frac{r}{c}}\right\} dS.$$

The first integral is connected with the distribution of charge in the volume. The second one results from a surface distribution of charge. It is this second one that occurs in optics and makes possible a mathematical representation of Huygens' theorem: every element of a spherical wave-surface behaves like a source and emits a secondary wavelet whose behavior is described by the formula (III-36).

C. ELECTROMAGNETIC ENERGY. ENERGY FLUX

7. *Electric and magnetic energy densities*

The energy of the system formed by two charges q_1 and q_2 separated by the distance r_{12} is

(III-37)
$$W_{12} = \frac{q_1 q_2}{\varepsilon_0 r_{12}}.$$

If there are n charges, the system will have an energy

(III-38)
$$W_{ij} = \frac{1}{2}\sum_{\substack{1\\ i\neq j}}^{n} \frac{q_i q_j}{\varepsilon_0 r_{ij}}.$$

One must, in fact, perform a summation over i and j only for $i \neq j$ (since each charge has no effect on itself) and take the half-sum, since the terms $W_{ij} = W_{ji}$ supply only a single contribution. One will then have:

(III-39)
$$W = \tfrac{1}{2}\,\Sigma_i q_i V_i$$

with

(III-40) $$V_i = \sum_{\substack{1 \\ i \neq j}}^{n} \frac{q_j}{\varepsilon_0 r_{ij}}.$$

If we are concerned with a continuous distribution:

(III-41) $$W_e = \tfrac{1}{2} \int_{\mathscr{V}} \rho V \, d\mathscr{V}.$$

Let us calculate that expression as a function of the inductions and fields:

(III-42) $$\mathbf{E} = -\,\mathbf{grad}\, V, \qquad \operatorname{div} \mathbf{D} = 4\pi\rho.$$

One has, according to (III-42):

(III-43)
$$\begin{aligned} W_e &= \frac{1}{8\pi} \int_{\mathscr{V}} V \operatorname{div} \mathbf{D} \, d\mathscr{V} = \\ &= \frac{1}{8\pi} \int_{\mathscr{V}} \left[\operatorname{div}(V\mathbf{D}) - \mathbf{D}\cdot\mathbf{grad}\, V\right] d\mathscr{V} = \\ &= \frac{1}{8\pi} \int_{\mathscr{V}} \mathbf{E}\cdot\mathbf{D} \, d\mathscr{V} + \frac{1}{8\pi} \int_{S} V D_n \, dS. \end{aligned}$$

The second integral reduces to:

(III-44) $$\frac{1}{8\pi} \int_S V D_n \, dS = -\int_S \tfrac{1}{2} V \sigma \, dS \qquad \text{for} \qquad D_n = \frac{\partial \mathbf{D}}{\partial n} = -4\pi\sigma.$$

It is negligible for a surface S limiting a sufficiently large volume $\mathscr{V}$. One will then have[3]

(III-45) $$W_e = \frac{1}{8\pi} \int_{\mathscr{V}} \mathbf{E}\cdot\mathbf{D} \, d\mathscr{V}.$$

The electrostatic energy density is then:

(III-46) $$u_e = \frac{1}{8\pi} \mathbf{E}\cdot\mathbf{D} = \frac{1}{8\pi} \varepsilon \mathbf{E}^2$$

if the relation $\mathbf{D} = \varepsilon \mathbf{E}$ is valid.

The rigorous evaluation of the energy density of the magnetic field is much more complicated.[4] We will simply assume here that the analogy with the electrostatic case is still valid. The energy density of the magnetic field is then:

(III-47) $$u_m = \frac{1}{8\pi} \mathbf{B} \cdot \mathbf{H} = \frac{1}{8\pi} \mu \mathbf{H}^2$$

if relation $\mathbf{B} = \mu\mathbf{H}$ is obeyed.

Naturally, one cannot represent by this expression the internal energy of the magnets (an energy we are incapable of determining), but simply the energy of the external media that may be the site of a field and of a magnetic induction. Specifically, in the case of a magnetic field created by currents, we can define an energy $W_m = \int u_m d\mathscr{V}$, with the energy distribution being this time known throughout the whole space and well determined as a function of the intensity of the current. That the experimentally measured values agree very well with definition (III-47) is easily verifiable.[5]

In the general case of an electromagnetic field, we assume that the energies of the electric field and the magnetic field may be added without reciprocal modification of their values. The electromagnetic energy density is then

(III-48) $$\boxed{u = \frac{1}{8\pi} (\mathbf{E} \cdot \mathbf{D} + \mathbf{H} \cdot \mathbf{B}) = \frac{1}{8\pi} (\varepsilon \mathbf{E}^2 + \mu \mathbf{H}^2).}$$

It is easily shown that this expression satisfies the principles of the conservation of energy.

8. *Poynting vector and Poynting's theorem*

The preceding calculation of the energy of the electromagnetic field was made only for a purely static case. But in the case of variable states, the expression is still valid. In order to demonstrate this, let us calculate the vector expression:

(III-49) $$\boxed{\mathbf{S} = \frac{1}{4\pi} (\mathbf{E} \wedge \mathbf{H})}$$

known as the *Poynting vector*. We still have:

$$4\pi \operatorname{div} \mathbf{S} = \mathbf{H} \operatorname{\mathbf{curl}} \mathbf{E} - \mathbf{E} \operatorname{\mathbf{curl}} \mathbf{H} =$$

(III-50)
$$= -\mathbf{H}\left(\frac{\partial \mathbf{B}}{c\partial t}\right) - \mathbf{E}\left(\frac{4\pi \mathbf{I}}{c} + \frac{1}{c}\frac{\partial \mathbf{D}}{\partial t}\right) =$$

$$= -\mathbf{H}\left(\frac{\partial \mu \mathbf{H}}{c\partial t}\right) - \mathbf{E}\left(\frac{\partial \varepsilon \mathbf{E}}{c\partial t}\right) - \frac{4\pi \mathbf{I}}{c}\cdot\mathbf{E}$$

or

(III-51)
$$\operatorname{div} \mathbf{S} = -\frac{1}{8\pi}\frac{\partial}{c\partial t}(\mu \mathbf{H}^2 + \varepsilon \mathbf{E}^2) - \frac{\mathbf{I}\cdot\mathbf{E}}{c}.$$

We will then have, according to (III-48), the following relation known as *Poynting's theorem*

(III-52)
$$\boxed{\frac{1}{c}\frac{\partial u}{\partial t} + \operatorname{div} \mathbf{S} = -\frac{\mathbf{I}\cdot\mathbf{E}}{c}}$$

that is a continuity equation for the energy. The increase $\partial u / c\partial t$ of that energy per unit volume is equal to the production of an energy $-\mathbf{IE}/c$ per unit volume minus the flux of the vector **S** through the limiting surface of the volume considered; the Poynting vector **S** thus represents the *energy flux*, that is, the quantity of energy that passes perpendicularly per unit cross-sectional area per second.

D. ELECTROMAGNETIC WAVES

9. *Equations for the propagation of fields*

One of the greatest successes of Maxwell's theory was the prediction of the existence of electromagnetic waves with a propagation velocity that could be known theoretically and measured.

According to relations (I) and (II) the equations of propagation of the fields **E** and **H** are as follows:

(III-53)
$$\operatorname{\mathbf{curl}} \operatorname{\mathbf{curl}} \mathbf{E} = -\frac{\mu}{c}\frac{\partial}{\partial t}(\operatorname{\mathbf{curl}} \mathbf{H}) = -\frac{\mu}{c^2}\frac{\partial}{\partial t}\left(4\pi \mathbf{I} + \frac{\partial \mathbf{D}}{\partial t}\right) =$$

$$= -\frac{4\pi\mu}{c^2}\frac{\partial \mathbf{I}}{\partial t} - \frac{\mu\varepsilon}{c^2}\frac{\partial^2 \mathbf{E}}{\partial t^2}$$

(III-54) $$\mathbf{curl}\,\mathbf{curl}\,\mathbf{H} = \frac{4\pi}{c}\,\mathbf{curl}\,\mathbf{I} + \frac{\varepsilon}{c}\frac{\partial}{\partial t}(\mathbf{curl}\,\mathbf{E}) = \frac{4\pi}{c}\,\mathbf{curl}\,\mathbf{I} - \frac{\varepsilon\mu}{c^2}\frac{\partial^2\mathbf{H}}{\partial t^2}.$$

But we continue to have for an arbitrary vector **A**:

(III-55) $$\mathbf{curl}\,\mathbf{curl}\,\mathbf{A} = \mathbf{grad}\,\mathrm{div}\,\mathbf{A} - \Delta\mathbf{A}.$$

One then obtains, by taking (III-55) into account:

(III-56) $$-\Delta\mathbf{E} + \frac{4\pi}{\varepsilon}\,\mathbf{grad}\,\rho = -\frac{4\pi}{c^2}\mu\frac{\partial\mathbf{I}}{\partial t} - \frac{\mu\varepsilon}{c^2}\frac{\partial^2\mathbf{E}}{\partial t^2}$$

(III-57) $$-\Delta\mathbf{H} = \frac{4\pi}{c}\,\mathbf{curl}\,\mathbf{I} - \frac{\varepsilon\mu}{c^2}\frac{\partial^2\mathbf{H}}{\partial t^2}.$$

Let us assume that there are no charges ($\rho = 0$). In that case, according to

(III-8) $$\mathbf{I} = \sigma_c\mathbf{E}$$

the equations for the propagation of the fields finally take on the following form:

(III-58) $$\frac{\mu\varepsilon}{c^2}\frac{\partial^2\mathbf{E}}{\partial t^2} - \Delta\mathbf{E} + \frac{4\pi\sigma_c}{c^2}\mu\frac{\partial\mathbf{E}}{\partial t} = 0.$$

(III-59) $$\frac{\mu\varepsilon}{c^2}\frac{\partial^2\mathbf{H}}{\partial t^2} - \Delta\mathbf{H} + \frac{4\pi\sigma_c}{c^2}\mu\frac{\partial\mathbf{H}}{\partial t} = 0$$

taking (III-8), $(\mathrm{I})_b$ and $(\mathrm{II})_b$ (p. 54) into account.

Each of the quantities **E** and **H** thus satisfies an equation of propagation of the following form:

(III-60) $$\boxed{\frac{\varepsilon\mu}{c^2}\frac{\partial^2\mathbf{a}}{dt^2} - \Delta\mathbf{a} + \frac{4\pi\sigma_c}{c^2}\mu\frac{\partial\mathbf{a}}{\partial t} = 0}$$

which is reducible to

(III-61) $$\Box\,\mathbf{a} = 0, \qquad \Box = \frac{1}{c^2}\frac{\partial^2\mathbf{a}}{\partial t^2} - \Delta\mathbf{a} = 0$$

in a non-conducting medium ($\sigma_c = 0$).

10. *Plane waves*

Equations (III-60) allow the following solution in particular:

(III-62) $$a = a_0 e^{i\omega t - \gamma z}$$

which represents a plane wave propagated along an axis Oz. By substitution in (III-60) one obtains the value of the complex constant γ:

(III-63) $$\gamma = \pm i \frac{\omega}{c} \sqrt{\left(\varepsilon - \frac{4\pi i \sigma_c}{\omega}\right)\mu}$$

In a non-conducting medium ($\sigma_c = 0$), one has in particular:

(III-64) $$\gamma = \pm \frac{i\omega}{c} \sqrt{\varepsilon\mu}.$$

Then, one gets

(III-65) $$a = a_0 e^{i\omega\left(t \pm \frac{z}{u}\right)}$$

putting

(III-66) $$\boxed{u = \frac{c}{\sqrt{\varepsilon\mu}} = \frac{c}{n}}$$

with

(III-67) $$n = \sqrt{\varepsilon\mu} = \sqrt{x_e x_m}$$

for $\varepsilon_0 = \mu_0 = 1$ according to the unit-system selected (the mixed system).

(III-68) $$\varepsilon = \varepsilon_0 x_e, \qquad \mu = \mu_0 x_m.$$

Moreover, because of the very form of the equation of propagation (III-60)

(III-69) $$\frac{\varepsilon\mu}{c^2} = \frac{1}{V^2},$$

with V designating, in this section, the velocity of propagation of a plane wave.

Now, according to expression (III-65) relative to the plane wave, it is also the phase-velocity u which determines the times

$$t = \frac{z + k\lambda}{u} \qquad [k = \text{integer}],$$

at the end of which the phase

$$\varphi = \omega\left(t - \frac{z}{u}\right)$$

again takes on the same value, that is, the period

$$T = \frac{\lambda}{u}$$

of the vibration.

In expression (III-66), c represents (as we pointed out in Section 3) the relation between the electromagnetic units and the electrostatic units CGS

$$c = \frac{q_e}{q_m} = \frac{[Q]_m}{[Q]_e}$$

But, according to (III-69), it is also the velocity of propagation of an electromagnetic perturbation in vacuum.

In that case, one has, as a matter of fact:

$$\text{(III-70)} \qquad V_0 = \frac{c}{\sqrt{\varepsilon_0 \mu_0}} = c$$

since, in the mixed system, $\varepsilon_0 = \mu_0 = 1$.

According to Maxwell's theory, the velocity V_0 of the plane electromagnetic waves in the vacuum is thus equal to their phase velocity u_0 and to the ratio between electrostatic and electromagnetic units.

Electromagnetic waves and light waves. At the time when Maxwell's theory was being developed, the velocity of light in air was known very accurately, which meant that its velocity in the vacuum was approximately known. To determine the velocity of light, both extra-terrestial (Römer, Bradley) and terrestial sources (Fizeau, Foucault, Michelson) were used. In recent years, this velocity has been determined with increasing accuracy.[6,7] These measurements were made either on a signal velocity V – direct measurements carried out on light waves[8], Hertzian waves[9], or γ-rays[10]; or on the phase-velocity u – indirect measurements involving the frequencies or lengths of Hertzian waves[11,12,13,14] or light waves[15,16,17,18,19]. We know that the values of u and V coincide in non-dispersing media. The best evaluations have been achieved either by the resonant cavity method (ESSEN[11], HANSEN-BOL[20]) or by use of wave-guides (FROOME[13], FLORMAN[14]) or even by using infrared and Hertzian spectroscopic methods (RANK[17,18], PYLER[21]). And finally let us not forget to mention the measurements achieved by radar methods (ASLAKSON[9]) and by means of BERGSTRAND's[8] "geometric" optics. This latter method, which renews the toothed-wheel method of Fizeau, uses a beam of polarized light which a Kerr cell can break down into very high frequency pulses. One obtains in this way:

$$V_0 \neq 299\,790 \pm 1 \text{ km/sec}.$$

It should be noted that most of these methods amount to taking relative measurements of the wave-length and the period of the same electromagnetic oscillation. Now such measurements are more accurate than the optical determination of a wave-length relative to a standard meter could possibly be. So it seems possible and even desirable[6] to postulate that the value of V_0 is rigorously the value supplied to us by the best measurements and to link together in this way the length- and time-standards. The measurement of either one of the two (preferably of the time-standard, since the Hertzian chronometer is more exact than optical metrology) then permits us to determine the other.

On the other hand, the first determination of the c relation between electrostatic and electromagnetic units CGS is due to WEBER and KOHLRAUSCH (1864)[22]. It consisted of a comparison in the two unit-systems of the values of an identical difference in potential. Its expression in electrostatic units is calculated from a measurement obtained by an electrometer, and the direct determining of it in electromagnetic units is effected by means of an electrodynamometer. One can also have recourse to measurements of one and the same capacitance. The most recent experiments performed in this way give us, within one ten-thousandth, the following value:

$$c = 299\,790 \text{ km/sec}.$$

The experimentally established identity between V_0 and c thus leads us to believe that light phenomena obey Maxwellian equations which, theoretically, impose that result. Light-waves would then constitute a part of the domain of electromagnetic waves.

Indeed, a connection between luminous and electromagnetic phenomena had long been foreseen; Faraday had pointed out the rotation of the polarization plane of a light-beam under the influence of a magnetic field. That phenomenon, which at that time had received only a qualitative explanation, is really a special case of the Zeeman effect, discovered some fifty years later.[23]

It was Hertz who supplied the proof of the identity of luminous and electromagnetic phenomena. He succeeded in producing electromagnetic waves by using the oscillating discharge that occurs between two electrodes connected with two spheres of large capacity and brought to a high potential. The system is really an oscillator that emits high-frequency waves. And indeed, experimental measurement of that frequency coincides with the frequency that may be theoretically deduced from Thomson's formula:

$$T = \frac{1}{\nu} = 2\pi\sqrt{LC}\,.$$

It has been proved that Hertzian waves produce the phenomena of interference, diffraction and polarization characterizing light-waves. The production of ultra-short electromagnetic waves has made it possible to reach the ultraviolet spectrum and to unify the domain of luminous vibrations with that of the electromagnetic field.

Relation (III-66) is, of course:

(III-71) $$u = \frac{c}{n}$$

valid only for a perfect dielectric ($\sigma_c = 0$). If this dielectric constitutes a dispersive medium, (III-71) is valid only for a rigorously monochromatic wave. It is, thus, a limiting case. We will see in the next section what becomes of this relationship in the case of an only slightly dispersive medium if the waves have only slightly different frequencies.

Most transparent media have a magnetic permeability close to unity. In that case (III-67) is still written:

(III-72) $$n^2 = \varepsilon$$

Maxwell verified the accuracy of this relation in the case of gases and of certain dielectrics (sulfur, paraffin). In several instances (Boltzmann's experiment on carbon dioxide) one can deduce n by determining ε measured electrically. Most frequently ε varies rapidly with the wavelength and reaches its limiting value for T_∞. One should thus be able to obtain good agreement with experiment by using large wave-lengths.[24] And as a matter of fact, Maxwell's formula has indeed been verified by using Hertzian waves. Even in this limiting case, certain dielectrics still present marked dispersion. One must then have recourse to the conclusions obtained by applying Maxwell's theory, not to perfect dielectrics (characterized solely by ε, μ and the displacement currents), but to semi-conducting media wherein, along with the constant σ_c, conduction currents occur. In semi-conducting media there invariably arises selective absorption and dispersion resulting from a resonance phenomenon. They are functions of the difference between the frequency of the incident wave and the natural frequency of the electrons of the medium.

11. *Wave-trains*

Let us consider two plane monochromatic waves having the same amplitude a_0 and frequencies $\nu + d\nu$ and $\nu - d\nu$, differing only very slightly from each other. These frequencies correspond to the angular velocities $\omega + d\omega$ and $\omega - d\omega$. And the phase-velocities are respectively $u + du$ and $u - du$ in a dispersive medium. The resulting vibration will then be:

$$\text{(III-73)} \quad \begin{aligned} A &= a_0 e^{i(\omega + d\omega)\left(t - \frac{z}{u + du}\right)} + a_0 e^{i(\omega - d\omega)\left(t - \frac{z}{u - du}\right)} \simeq \\ &\simeq a_0 e^{i\left[\omega t + t d\omega - \frac{z}{u} d\omega - \frac{\omega z}{u}\left(1 - \frac{du}{u}\right)\right]} + \\ &+ a_0 e^{i\left[\omega t - t d\omega + \frac{z}{u} d\omega - \frac{\omega z}{u}\left(1 + \frac{du}{u}\right)\right]} \end{aligned}$$

the infinitesimal products of du and $d\omega$ being neglected. We then have:

$$\text{(III-74)} \quad \begin{aligned} A &\simeq 2a_0 \cos\left(t d\omega - \frac{z}{u} d\omega + \frac{\omega z}{u^2} du\right) e^{i\omega\left(t - \frac{z}{u}\right)} \\ &\simeq 2a_0 \cos 2\pi\left(t d\nu - \frac{z}{u} d\nu + \frac{\nu z}{u^2} du\right) e^{i\omega\left(t - \frac{z}{u}\right)} \end{aligned}$$

and one can also put:

(III-75) $$A \simeq 2a_0 \cos 2\pi d\nu\left(t - \frac{z}{U}\right)e^{i\omega\left(t-\frac{z}{u}\right)}$$

with

(III-76) $$\frac{1}{U} = \frac{1}{u} - \frac{\nu}{u^2}\frac{du}{d\nu} = \frac{d}{d\nu}\left(\frac{\nu}{u}\right)$$

that is

(III-77) $$\frac{1}{U} = \frac{d}{d\nu}\frac{1}{\lambda}$$

or

(III-78) $$U = \frac{d\nu}{d\left(\frac{1}{\lambda}\right)}.$$

The wave resulting from the superposition of two vibrations having the same amplitude and almost the same frequency appears, then, like a vibration of frequency ν and with an amplitude that is a periodic function of frequency $d\nu$ and velocity U. The original wave-train is thus resolved into a so-called *modulated* vibration. Its frequency ν is unchanged, but *its amplitude vibrates with a new frequency $d\nu$.* U is called the *group-velocity*. It is the propagation velocity of the amplitude and, consequently it is likewise the propagation velocity of the energy of the resultant vibration. On the other hand, the phase velocity retains the value u.

One must have, according to (III-75)

(III-79) $$-\frac{\pi}{2} \leqslant 2\pi d\nu\left(t - \frac{z}{U}\right) \leqslant \frac{\pi}{2}$$

that is:

(III-80) $$Ut - l \leqslant z \leqslant l + Ut$$

putting

(III-81) $$l = \frac{U}{4d\nu}$$

The wave-train is thus included between the points $Ut-l$ and $Ut+l$; $2l$ is the length of the wave-train.

In a non-dispersive medium, the velocity u is the same no matter what λ is. One obtains, then, from (III-76):

(III-82) $$\frac{du}{d\nu} = \frac{u^2}{\nu}\left(\frac{1}{u} - \frac{1}{U}\right) = 0$$

that is:

(III-83) $$u = U$$

The group-velocity is equal to the phase velocity. This condition obviously holds for the propagation of an electromagnetic wave in the vacuum.

The preceding calculations presuppose

$$\frac{d\nu}{\nu} \leqslant 1.$$

Now, according to (III-81), one has:

(III-84) $$l = \frac{U}{4d\nu} = \frac{1}{4d\left(\frac{1}{\lambda}\right)} = -\frac{\lambda^2}{4d\lambda}.$$

Thus:

(III-85) $$\frac{l}{\lambda} = -\frac{\lambda}{4d\lambda} = \frac{\nu}{4d\nu} \geqslant 1.$$

In order for the preceding hypotheses to be valid, the length l of the wave-train must be large in relation to λ. In that case the wave-train moves along without deformation, since its amplitude is a periodic function that is propagated with the velocity U.

Let us call n the index of the dispersive medium corresponding to the wave-length λ_0 in the vacuum:

(III-86) $$\lambda_0 = \frac{u}{\nu}, \qquad u = \frac{c}{n}$$

Now, a result of (III-78) is

(III-87) $$U = \frac{d\nu}{\left(\frac{1}{d\lambda}\right)} = \frac{d\left(\frac{u}{\lambda_0}\right)}{d\left(\frac{1}{\lambda_0}\right)} = u - \lambda_0 \frac{du}{d\lambda_0} = $$
$$= u\left(1 - \frac{\lambda_0}{u}\frac{du}{d\lambda_0}\right) = u\left(1 + \frac{\lambda_0}{n}\frac{dn}{d\lambda_0}\right)$$

for, taking (III-86) into account

$$\frac{du}{u} = -\frac{dn}{n}. \tag{III-88}$$

and (III-87) is still written

$$U = \frac{c}{n}\left(1 + \frac{\lambda_0}{n}\frac{dn}{d\lambda_0}\right). \tag{III-89}$$

If light is made up of a train of monochromatic waves, the measurements taken are always those of the group-velocity U, which is also the velocity of the energy.

In a non-dispersive or only slightly dispersive medium (such as air), this amounts to measuring the phase-velocity as well, for we still have

$$U = u = \frac{c}{n}. \tag{III-90}$$

In a dispersive medium (carbon disulphide, for example), one can, in contrary fashion, verify (III-89).

12. *Spherical waves*

The Maxwell equations may be written in spherical coordinates and stated accordingly. Here we will limit ourselves to one particular case.

Let us consider the equation of propagation $(VI)_b$, which is obeyed by the scalar potential. In the vacuum ($\mu_0 = \varepsilon_0 = 1$), and if there is no charge ($\rho = 0$), the scalar potential V will satisfy the equation of propagation:

$$\Box V \equiv \frac{1}{c^2}\frac{\partial^2 V}{\partial t^2} - \Delta V = 0. \tag{III-91}$$

In spherical polar coordinates

$$x = r\sin\theta\cos\varphi, \qquad y = r\sin\theta\sin\varphi, \qquad z = r\cos\theta, \tag{III-92}$$

and one has for any arbitrary scalar V

$$\Delta V = \frac{1}{r^2}\frac{\partial}{\partial r}\left(r^2\frac{\partial V}{\partial r}\right) + \frac{1}{r^2\sin\theta}\frac{\partial}{\partial\theta}\left(\sin\theta\frac{\partial V}{\partial\theta}\right) + \frac{1}{r^2\sin^2\theta}\frac{\partial^2 V}{\partial\varphi^2}. \tag{III-93}$$

Equation (III-91) has solutions of the form:

$$V = R(r)\,\Theta(\theta)\,\Phi(\varphi)\,e^{i\omega t}. \tag{III-94}$$

In the special case where V depends only on r and t, (III-91) is written:

$$\frac{1}{c^2}\frac{\partial^2}{\partial t^2}(rV) - \frac{\partial^2 (rV)}{\partial r^2} = 0 \tag{III-95}$$

and may have solutions of the form

$$V = \frac{l}{r} f\left(t \pm \frac{r}{c}\right) \tag{III-96}$$

where f is an arbitrary function. This solution represents a particularly simple kind of spherical wave. In expressing it, we may choose the minus sign so that the wave is propagated in the sense of increasing r's. And finally, for f, one can put the simple form of a plane wave:

$$f\left(t - \frac{r}{c}\right) = a_0 e^{i\omega\left(t - \frac{r}{c}\right)} \tag{III-97}$$

so that the potential at a point P (r, t) created by S at the origin will be:

$$V(r,t) = a_0 \frac{e^{i\omega\left(t - \frac{r}{c}\right)}}{r}. \tag{III-98}$$

Let us now assume that, according to Huygens' principle, a wave surface with the center S sends out wavelets at an instant that we choose as our origin. Each of these wavelets has, according to Huygens, a radius ct at the instant t, and the envelope of all the wavelets forms a new surface, which constitutes the wave-front at the instant t. The disturbance trans-

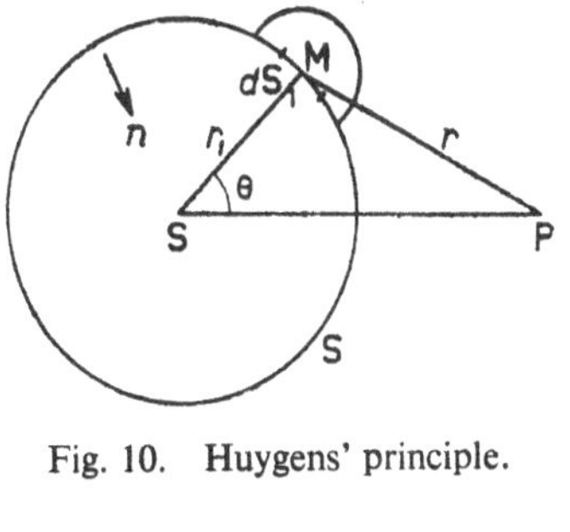

Fig. 10. Huygens' principle.

mitted to P by the surface element dS_1 will be proportional to S_1. Now according to (III-98), this amplitude at M (r_1, t) is given by:

$$V(r_1, t) = a_0 \frac{e^{i\omega\left(t - \frac{r_1}{c}\right)}}{r_1} \tag{III-99}$$

r_1 being the distance from the source to the point M.

The potential at the point P of distance r from the point M is then

$$V\left(t - \frac{r}{c}\right) = a_0 \frac{e^{i\omega\left(t - \frac{r+r_1}{c}\right)}}{r_1}. \tag{III-100}$$

This potential satisfies d'Alembert's equation; so it is of the form (III-36). Taking boundary conditions into consideration:

$$\begin{aligned} V = &- \frac{1}{4\pi} \int \left[\frac{\Delta V - \frac{1}{c^2} \frac{\partial^2 V}{\partial t^2}}{r} \right]_{t - \frac{r}{c}} d\mathcal{V} + \\ &+ \frac{1}{4\pi} \int \frac{1}{r} \left\{ \left[\frac{1}{c} \left(\frac{\partial V}{\partial t} \right)_{t - \frac{r}{c}} + \right. \right. \\ &\left. \left. + \frac{V\left(t - \frac{r}{c}\right)}{r} \right] \cos(\mathbf{n}, \mathbf{r}) + \left(\frac{\partial V}{\partial n} \right)_{t - \frac{r}{c}} \right\} dS \end{aligned} \tag{III-101}$$

Now let us calculate, according to (III-100):

$$\frac{\partial V\left(t - \frac{r}{c}\right)}{\partial t} = i\omega \frac{a_0 e^{i\omega\left(t - \frac{(r+r_1)}{c}\right)}}{r_1} \tag{III-102}$$

$$\frac{\partial V\left(t - \frac{r}{c}\right)}{\partial n} = - a_0 \cos(\mathbf{n}, \mathbf{r}_1) \left(\frac{1}{r_1} + \frac{i\omega}{c} \right) \frac{e^{i\omega\left[t - \frac{(r+r_1)}{c}\right]}}{r_1}. \tag{III-103}$$

By substitution in (III-101) we thus find:

$$\begin{aligned} V = \frac{1}{4\pi} \int \frac{a_0}{r r_1} e^{i\omega\left(t - \frac{(r+r_1)}{c}\right)} &\left[\left(\frac{1}{r} + \frac{i\omega}{c} \right) \cos(\mathbf{n}, \mathbf{r}) \right. \\ &\left. - \left(\frac{1}{r_1} + \frac{i\omega}{c} \right) \cos(\mathbf{n}, \mathbf{r}_1) \right] dS \end{aligned} \tag{III-104}$$

Let us suppose that r and r_1 are large with respect to the wave length – which is always the case in optics. (III-104) is then written:

$$V = \int \frac{ia_0}{2\lambda r_1 r} e^{i\omega\left[t - \frac{r+r_1}{c}\right]} \left[\cos(\mathbf{n}, \mathbf{r}) - \cos(\mathbf{n}, \mathbf{r}_1)\right] dS \tag{III-105}$$

with

$$\omega = \frac{2\pi c}{\lambda}.$$

Now

$$a_0 \frac{e^{i\omega\left[t - \frac{r_1}{c}\right]}}{r_1}$$

is the incident amplitude a_1 at the point M. If a wave of that amplitude starts up at M, it will produce at P a disturbance:

$$\int \left(\frac{a_1 dS_1}{r}\right)_{t-\frac{r}{c}} = \int \frac{a_0}{rr_1} e^{i\omega\left[t - \frac{r+r_1}{c}\right]} dS, \qquad r = MP. \tag{III-106}$$

This is approximately the value found in (III-105). Nevertheless, the corrective factor

$$\frac{t}{2\lambda}\left[\cos(\mathbf{n}, \mathbf{r}) - \cos(\mathbf{n}, \mathbf{r}_1)\right]$$

was not provided for in Huygens' principle. It shows that the amplitude of the wavelets arriving at the point P depends upon the angles which $\mathbf{r}$ and $\mathbf{r}_1$ make with the normal $\mathbf{n}$.[25] It is also inversely proportional to the wave-length.

Formula (III-106) makes it possible for us to solve ordinary diffraction problems. These problems present themselves as applications of Huygens' principle. The solution of d'Alembert's equation thus gives us an exact mathematical representation of that principle and at the same time provides it with the necessary corrections.

E. ELECTROMAGNETIC EQUATIONS VALID FOR NON-MAGNETIC BODIES IN SLOW MOTION

13. *Application of Maxwell's theory to moving media*

In applying Maxwell's conceptions to permanent or quasi-stationary phenomena, we have constantly taken for granted that the material media were at rest. So we have systematically put off the study of fields produced by the over-all motion of a charged conductor or a polarized dielectric.

Nevertheless, the study of variable states has shown us that an electromagnetic induction is produced by a variation in flux that may result from the change in intensity of a current or even from the motion of a circuit (or a magnet) through which a constant current passes. The equivalence between these two possible origins of induced electromotive forces is an experimental fact. It is implied by Faraday's law

(III-1) $$e = -\frac{d}{dt}\int_S B_n \, dS,$$

since the variations in time of $B_n dS$ have as much bearing on the intensity of the current producing induction **B** as they do on the modification of dS. Nevertheless, we do not know if the circuit or dielectric in motion is subject to the same action from the magnetic field as when it is at rest. In fact, the explanations proposed by Hertz, Lorentz and Einstein are radically different. But first of all we will examine the experimental effects that we are seeking to interpret.

14. *Motion of a conductor or insulator in an electric field*

1) The first experiments on the displacement of a conductor in an electric field were performed by H. A. Rowland (1875) and confirmed by Eichenwald. They consisted of comparing magnetic effects resulting from the uniform rotation of a metallic plate, bearing a charge of surface density σ, to the effects produced by a conduction current of density **I** circulating in the same plate when at rest. An equivalence of effects is achieved when

(III-107) $$\mathbf{I} = \sigma \mathbf{v}$$

a relation which states precisely, as in Chapter II, the equivalence of the conduction currents and convection currents produced by the uniform motion of any conductor.

2) The motion of an insulator in an electric field was first studied by W. C. Röntgen (1885).[26] His experiments were continued and perfected by A. Eichenwald (1903).[27]

Eichenwald used a condensor made up of two flat metallic rings, each cut by a thin break and separated from each other by a rubber insulator (Figure 11). The insulator revolves around the axis zz', and the metallic plates may, when necessary, be made to share its motion.

Now the density of the charge of a metallic plate is:

$$\sigma_p = \frac{\varepsilon V}{4\pi d} = \frac{\varepsilon E}{4\pi}. \tag{III-108}$$

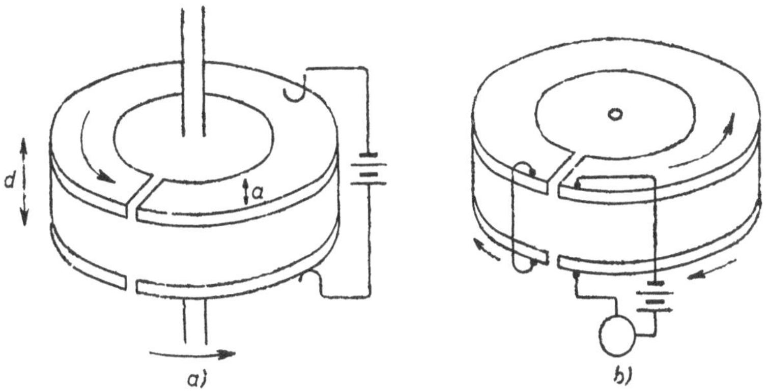

Fig. 11. Röntgen and Eichenwald's experiment (after R. BECKER [1]).

And that is also the density of charge (with the sign reversed) which the insulating layer possesses when at rest. When this insulator alone moves with the velocity $\mathbf{v}$, one will expect to observe a convection current of density

$$i = \frac{\varepsilon E}{4\pi} a\mathbf{v} = \frac{\varepsilon V}{4\pi d} a\mathbf{v} \tag{III-109}$$

according to the conclusions of the preceding section. Now Röngten's experiments showed that the value of the current obtained is very low. The density of the conduction current measured by Röntgen is:

$$i = \frac{(\varepsilon - 1)}{4\pi d} Va\mathbf{v} = \frac{(\varepsilon - 1)}{4\pi} Ea\mathbf{v}. \tag{III-110}$$

Everything takes place as if the density of charge of the insulator in motion were:

(III-111) $$\sigma_i = (\varepsilon - 1)\frac{E}{4\pi}$$

that is to say, as if one had to substitute for the field E, the field

(III-112) $$E' = \left(1 - \frac{1}{\varepsilon}\right)E.$$

If the insulator and the metallic plates are made to turn at the same time, as in Eichenwald's experiment, one should observe a total current resulting from the contributory quantities σ_p and of $-\sigma_i$. According to Röntgen's results this current will be determined by (III-108) and (III-111). Its intensity will then be:

(III-113) $$(\sigma_p - \sigma_i)\,a\mathbf{v} = \frac{\varepsilon E}{4\pi}a\mathbf{v} - \frac{(\varepsilon - 1)}{4\pi}Ea\mathbf{v} = \frac{E}{4\pi}a\mathbf{v}.$$

If a is the breadth of the metallic rings, one will observe a current:

(III-114) $$\mathbf{i} = (\sigma_p - \sigma_i)\,\mathbf{v}a = \frac{E}{4\pi}\mathbf{v}a = \frac{V}{4\pi d}\mathbf{v}a.$$

Its value is independent of the dielectric constant of the insulator.

These predictions, made on the basis of the results of Rowland's experiment, have been fully confirmed by experiment. Eichenwald observed the deviation produced by this current on a magnetized needle. In a second experiment he measured the intensity of the conduction current which produced the same deviation in the plates when they were held motionless (Figure 11). The value obtained, independent of the nature of the insulator, is exactly what expression (III-114) allows us to predict.

15. *Displacement of a conductor or an insulator in a magnetic field*

a) The displacement of a conductor in a magnetic field was the object of Faraday's well-known experiments (motion of a circuit in the air-gap of a magnet). This is the principle of dynamos and current-generators of every sort.

b) The displacement of a dielectric in a magnetic field was the object of the experiments of H. A. Wilson.[28] He uses a hollow cylindrical dielectric of thickness a placed in the air-gap of a magnet. The internal and external surfaces of the cylinder are lined with a metal coating that may be connected to the terminals of an electrometer.

Now a charge q placed in a magnetic field experiences the Lorentz force (cf. II-3)

$$\text{(II-3)} \qquad \mathbf{F} = q\left[\frac{\mathbf{v}}{c} \wedge \mathbf{B}\right]$$

(mixed-system of units). Everything then takes place as if the motion of the particle were in an electric field:

$$\text{(III-115)} \qquad \mathbf{E} = \left[\frac{\mathbf{v}}{c} \wedge \mathbf{B}\right].$$

The rotation of the cylinder around its axis oriented parallel to the magnetic field of the magnet must then produce a field of value E in a radial direction.

In a conductor one would thus have the creation of a difference in potential $V = aE$ between the two surfaces of the cylinder.

In an insulator one must observe a polarization such that

$$\text{(III-116)} \qquad \mathbf{D} = \varepsilon\mathbf{E} = \mathbf{E} + 4\pi\mathbf{P}$$

that is

$$\text{(III-117)} \qquad \mathbf{P} = \left(\frac{\varepsilon - 1}{4\pi}\right)\mathbf{E} = \left(\frac{\varepsilon - 1}{4\pi}\right)\left[\frac{\mathbf{v}}{c} \wedge \mathbf{B}\right].$$

Now the charge on the cylinder, measured experimentally by means of an electrometer, has exactly the density

$$\text{(III-118)} \qquad \sigma = \left(\frac{\varepsilon - 1}{4\pi}\right)\left[\frac{\mathbf{v}}{c} \wedge \mathbf{B}\right].$$

This, then, coincides with the density of charge that would be produced, not by the field $\mathbf{E}$, but by a field

$$\mathbf{E}' = \left(1 - \frac{1}{\varepsilon}\right)\mathbf{E}.$$

So it is this $\mathbf{E}' < \mathbf{E}$ field – or, if one prefers, the polarization $\mathbf{P}$ – which intervenes to produce the charge of a dielectric in motion in a magnetic field.

16. *Hertz' and Lorentz' hypotheses*

In Chapter V we will discuss optical experiments involving media in relative motion (Doppler's, Fizeau's and Zeeman's experiments). We will see that, long before the discovery of Special Relativity, these experiments led to the assumption of a kind of equivalence in the descriptions of motions with respect to frames of reference whose relative uniform velocities were very small in comparison to the velocity c of light (so that terms involving v^2/c^2 could be neglected). This principle found expression either in the hypothesis of a local drag on light waves by matter in motion (Stokes) or in the hypothesis of partial drag (Fresnel).

Hertz was eager to extend Stokes' hypothesis to the ensemble of electromagnetic phenomena. So he assumed that Maxwell's ether, characterized by the fields and likewise by the inductions **E**, **D**, **H** and **B**, participated fully in the motion of matter. However, the experiments we have been describing invalidate Hertz' hypothesis.[29] It so happens that the induction **D**, for example, always has the expression $\mathbf{D} = \mathbf{E} + 4\pi\mathbf{P}$.

The polarization **P**, the definition of which is bound up with the presence of matter, completely shares the motion of matter. But the electrical field **E**, on the contrary, in no way shares that motion.

These conclusions follow naturally from Lorentz' microscopic theory based on the hypothesis of a strictly motionless ether. The microscopic fields (**e**, **h**), which then permit definition of the macroscopic magnitudes **E**, **B**, are bound up with the motionless ether. The electric polarization vector **P** and the magnetic moment vector **M** are, on the contrary, completely involved in the motion of matter. One is thus led to define macroscopic quantities **D** (**E**, **P**), **H** (**B**, **M**) which will have to be partially involved in this motion. Everything takes place as if the field **E**, for example, were replaced by a field

$$\mathbf{E}' = \left(1 - \frac{1}{\varepsilon}\right)\mathbf{E},$$

following the experiments cited in this chapter. Thus, Lorentz' postulate permits us to arrive, at least for low velocities ($v \ll c$) of conductors or dielectrics, at the results which Fresnel obtained, in optics, from an hypothesis of partial drag (cf. p. 120).

We will here limit ourselves to qualitative indications concerning theories whose bases have been superseded by the conclusions of Special Relativity. The principles of these theories, nevertheless, inevitably arise when one wishes to verify Einstein's theory physically and to demonstrate that it is imposed completely by necessity – that is to say, both logically and experimentally. We will see, in fact, that Maxwell's theory and Lorentz' developments find their natural outcome in the very principles of Special Relativity.

PROBLEMS

1. Beginning with a plane-wave solution of the form (III-62)

$$\mathbf{E} = \mathbf{E}_0 e^{i\omega t - \gamma z}, \qquad \mathbf{H} = \mathbf{H}_0 e^{i\omega t - \gamma z} \qquad (i = \sqrt{-1})$$

set up, by substitution in (I), (II), and (III), the relations between the components of **E** and **H**. Then deduce the following consequences:

a) The field constituted by **E** and by **H** is necessarily transverse ($E_z = H_z = 0$).

b) The fields **E** and **H** are orthogonal and such that

$$\frac{E_x}{H_y} = -\frac{E_y}{H_x} = \pm \sqrt{\frac{\mu}{\varepsilon - \dfrac{4\pi\sigma_c}{\omega}}}.$$

2. We consider Maxwell's equations, modified by the introduction of a small term ($\mathbf{k}_0^2 \ll 1$) which brings in the electromagnetic potential **A**, V (mesonic or photonic field with mass different from zero)

$$\mathbf{curl}\,\mathbf{H} = \frac{\varepsilon}{c}\frac{\partial \mathbf{E}}{\partial t} + k_0^2 \mathbf{A}, \qquad \operatorname{div} \mathbf{E} = k_0^2 V$$

$$\mathbf{curl}\,\mathbf{E} = -\frac{\mu}{c}\frac{\partial \mathbf{H}}{\partial t}, \qquad \operatorname{div} \mathbf{H} = 0.$$

Show that these equations have plane-wave solutions permitting the existence of a longitudinal electric field ($E_z \neq 0$).

Deduce from the Maxwell's equations so modified the Lorentz condition between the potentials (a condition imposed *a priori* in a strictly Maxwellian theory). Set up the equations of propagation.

3. Show that the plane-wave solution

$$\mathbf{E} = \mathbf{E}_0 e^{i\omega t - \gamma z}$$

of equation (III-58) is also a solution of

$$\frac{\mu\varepsilon'}{c^2}\frac{\partial^2 \mathbf{E}}{\partial t^2} - \Delta\mathbf{E} = 0 \quad \text{with} \quad \varepsilon' = \varepsilon - 4\pi i \sigma_c \tau \quad \left(\tau = \frac{1}{\omega}\right).$$

Show that $\mathbf{E}$ may be put in a damped sinusoidal form:

$$\mathbf{E} = \mathbf{E}_0 e^{\frac{-p'z}{\tau}} e^{\frac{1}{\tau}(t - pz)}$$

with

$$p = \frac{1}{u} = \frac{n}{c}, \qquad p' = \frac{k}{c}$$

$$k = \frac{2\pi\mu\sigma_c\tau}{n}$$

being the absorption coefficient of the medium.

NOTES

1. This is equivalent to solution (I-72), if one puts:

$$\Delta\varphi = 4\pi p, \frac{\partial\varphi}{\partial n} = -4\pi\sigma \int, \varphi\cdot\text{grad}_n\left(\frac{1}{r}\right) = 4\pi V_0$$

and if one arbitrarily cancels out the potential V_0 of the enclosure.
2. Cf. for example, C. Slater [6], p. 171.
3. The same expression would have been obtained by evaluating the work supplied by the reversible transfer of a charge q. If this charge is brought from a region that is infinitely far away and of zero potential, to a neutral conductor, there are communicated to this conductor a charge and a potential varying from zero to q and zero to V. So both charge and potential may at every instant be represented by qa and Va, a varying from zero to 1 during the course of the transfer. The elementary work supplied is then:

$$d\mathcal{T} = qda\cdot V$$

and the energy of the conductor is finally:

$$W_e = qV\int_0^1 ada = \tfrac{1}{2}qV.$$

Whence we deduce, for a dielectric:

$$W_e = \frac{1}{8\pi}\int_{\mathcal{V}} E\cdot D\, d\mathcal{V}$$

according to (III-42) and (III-44).

4. Cf. for example, J. A. STRATTON [7], Lec. 2.14 to 2.18.
5. For example, by applying (III-47) to the definition of the energy density resulting from the self-induction or mutual induction of permanent currents.
6. BIRGE: *Rep. Prog. Phys.* **8** (1941) 90;
   BERGSTRAND, E.: *Handbuch der Physik* **24** (1956) 1;
   O. COSTA DE BEAUREGARD: *Rev. des questions scientifique,* 1957, 5.
7. E. BERGSTRAND: *N. P. L., Rec. Dev. Stand. London,* 1952, 75.
8. E. BERGSTRAND: *Arkiv för Physik* **2** (1950) 119.
9. C. I. ASLAKSON: *Proc. Am. Soc. Civ. Eng.* **77** (1951) 1; *Trans. Amer. Geophys. Un.* **32** (1951) 813.
10. CL. JASTRAM: *Phys. Rev.* **84** (1951) 271.
11. L. ESSEN, A. C. GORDON-SMITH: *Proc. Roy. Soc.* **194** (1943) 348; **204** (1950) 260; *Nature* **175** (1955) 793.
12. CULSHAW: *Proc. Phys. Soc.* B **66** (1953) 597.
13. K. D. FROOME: *Proc. Roy. Soc.* **213** (1952) 123, 223; **215** (1954) 195.
14. E. F. FLORMAN: *Journ. of Res. N. B. S.* **54** (1955) 235.
15. D. H. RANK, R. VAN DER SLUIS: *Phys. Rev.* **86** (1952) 799; *J. Opt. Soc. Amer.* **42** (1952) 693.
16. N. KLEIN TOWNES: *Phys. Rev.* **86** (1952) 798.
17. D. H. RANK, SHEARER, WIGGINS: *Phys. Rev.* **94** (1954) 575.
18. D. H. RANK, BENNETT, BENNETT: *Phys. Rev.* **100** (1955) 993.
19. D. H. RANK, GUENTHER, SHEARER: *J. Opt. Soc. Amer.* **47** (1957) 148.
20. HANSEN-BOL: *Phys. Rev.* **80** (1950) 298.
21. E. K. PLYLER: *J. Opt. Soc. Amer.* **44** (1954) 507.
22. W. WEBER and R. KOHLRAUSCH: *Ostwalds Klassiker der exakten Wissenschaften,* No. 2, 1856.
23. The Zeeman effect, like diamagnetism, results from a modification produced by a magnetic field on the rotation of the electrons constituting certain substances. It superimposes upon the original rotation of the electron a uniform rotation of constant direction and having an angular velocity

$$\omega_2 = \frac{-e}{2mc} H.$$

This rotation is superimposed upon the rotations of angular velocity $\pm\omega$ which the two circular components of the polarized incident light possess.

24. Let us recall the basic definitions of wave theory. If ω, ν, T, λ and u represent respectively the angular velocity, frequency, period, wavelength and phase-velocity of the wave, one has

$$\omega = 2\pi\nu\,; \qquad T = \frac{1}{\nu} = \frac{\lambda}{u}.$$

25. If S_1 is a sphere, $\cos(\mathbf{n}, \mathbf{r}_1) = -1$. The approximate calculation of (III-105) by Fresnel's zone method gives us

$$V = A \exp\left(i\omega\left(t - \frac{R}{c}\right)\right)$$

(putting $R = SP$), which is an expression representing the perturbation produced directly at P by the source S — thus justifying Huygens' principle. The calculation of (III-105) for the case in which slitted screens are interposed between *S* and *P* leads to the correct prediction of the diffraction figures.

26. W. C. Röntgen: *Ann. d. Phys.* **35** (1888) 268.
27. A. Eichenwald: *Ann. d. Phys.* **11** (1903) 1 and 421.
28. H. A. Wilson: *Phil. Trans.* **204** (1904) 121.
29. Hertz's theory predicted, for example, exact compensation between the Rowland and Röntgen currents which arise in the Eichenwald experiment, and that conclusion is nullified by experiment. (On this subject the developments set forth in the work of L. Bloch [2], pp. 387 and 397, should be consulted.)

SOURCES OF THE ELECTROMAGNETIC FIELD. LORENTZ' THEORY

LORENTZ' theory (1892)[1] was motivated by numerous experiments which, at the end of the nineteenth century, revealed the corpuscular nature of matter and electricity. The hypothesis of an atomic structure of matter entails, as a matter of fact, the prevision of a discontinuous structure for electricity, when the hypothesis is applied to such well-known phenomena as, for example, electrolysis. Thus, every charge must be considered as a multiple of an elementary charge e which may be measured either directly or indirectly.

Direct measurements consist of determining the charge acquired by extremely small droplets suspended between the plates of a condenser (MILLIKAN[2], Ehrenhaft, REGENER[3]), or else of measuring the charge carried by the alpha rays of radium C (REGENER[4]).[5]

Indirect measurements made in Lorentz' time were based on the previous knowledge of Avogadro's number N. This number was itself based on the determination of Boltzmann's constant, that is, on certain consequences of the kinematic theory of gases. Such measurements have been the object of numerous methods applied mostly to the Brownian movement and developed chiefly by Jean Perrin.

If $F = 96\,600$ coulombs is the equivalent involved in Faraday's law, one will have:

$$e = \frac{96\,600}{N} = 4.77 \cdot 10^{-10} \text{electrostatic CGS units}$$

$$(N = 6.10^{23}).$$

Moreover, experiments on the bending of beams of cathode rays in associated electric and magnetic fields make it possible for us to measure the ratio e/m. Particles of the same nature but of different parallel velocities, when subjected to an electric field and a magnetic field that are parallel to each other (perpendicular to the original direction of the beam), are distributed, after deviation, along a parabola

$$\frac{y^2}{x} = \frac{1}{2}\frac{e}{m}\frac{H^2}{c^2}\frac{l^2}{E}.$$

When this experiment is performed on cathode rays one still finds the same ratio:

$$\frac{e}{m} = 1.76{\cdot}10^7 \text{ electromagnetic CGS units}.$$

1. *"Microscopic" fields and potentials connected with an electron*

The existence of electrons imposes the reconstruction of the theory on the basis of the microscopic fields that each of the charges individually gives rise to. The description of the electromagnetic field must, in fact, reduce to a description of the properties exhibited by the propagation of elementary charges in vacuum. It is from these admitted facts that one should explain the origin and the properties of the fields, as well as the inductions, which enter into Maxwell's theory.

Lorentz assumes that the existence and motion of a charge permit us to define the microscopic fields **e** and **h** bound to that charge. As we have already pointed out, the magnetic induction corresponds to a global phenomenon and need not be considered in the description of elementary phenomena. Let us put ρ and **v** as the density and velocity of the charges. The only possible form of current is the convection current produced by the motion of the electrons.

Lorentz' hypothesis consists of supposing that the electrons move in a motionless ether and that relations of a Maxwellian type are still valid with respect to the quantities **e**, **h**, ρ and ρ**v**. We will then have, by transposing Maxwell's equations (I), (II) and (III):

(IV-1) $$\mathbf{curl}\,\mathbf{h} = \frac{4\pi}{c}\rho\mathbf{v} + \frac{1}{c}\frac{\partial \mathbf{e}}{\partial t}$$

(IV-2) $$\mathbf{curl}\,\mathbf{e} = -\frac{1}{c}\frac{\partial \mathbf{h}}{\partial t}$$

and

(IV-3) $$\operatorname{div}\mathbf{e} = 4\pi\rho$$

(IV-4) $$\operatorname{div}\mathbf{h} = 0.$$

Likewise, the density of the force acting on a charge is

(IV-5) $$\boxed{\mathbf{f} = \rho\left[\mathbf{e} + \frac{1}{c}[\mathbf{v} \wedge \mathbf{h}]\right].}$$

This force acting on the electron itself must be compensated by actions originating from some other source if the electron has finite dimensions. If that were not so, the electron could not be stable and would explode. Let us note [6] that relations (IV-1), (IV-2), (IV-3), (IV-4) and (IV-5) – equations which are absolutely basic in any electronic theory – are equations between fields and involve only quantities of a continuous character. The intrinsic properties of the particle are not directly involved, and the very existence of the particle seems to be compromised if the force described in (IV-5) is the only force involved.

The expressions of the energy densities and the energy flux associated with an electron are generalized in a similar manner:

(IV-6) $$u = \frac{1}{8\pi}(\mathbf{e}^2 + \mathbf{h}^2)$$

(IV-7) $$\mathbf{s} = \frac{1}{4\pi}[\mathbf{e} \wedge \mathbf{h}].$$

Starting with (IV-5), one can calculate $\mathbf{f}$ as a function of the fields $\mathbf{e}$ and $\mathbf{h}$. By introducing expressions (IV-3) and (IV-1) of ρ and $\rho\mathbf{v}$ into (IV-5). and taking into account (IV-2), one thus finds:

(IV-8)$_1$ $$f_x = -\frac{1}{c}\frac{\partial s_x}{\partial t} + \frac{\partial}{\partial x}T_{xx} + \frac{\partial}{\partial y}T_{xy} + \frac{\partial}{\partial z}T_{xz}$$

(IV-8)$_2$ $$f_y = -\frac{1}{c}\frac{\partial s_y}{\partial t} + \frac{\partial}{\partial x}T_{yx} + \frac{\partial}{\partial y}T_{yy} + \frac{\partial}{\partial z}T_{yz}$$

(IV-8)$_3$ $$f_z = -\frac{1}{c}\frac{\partial s_z}{\partial t} + \frac{\partial}{\partial x}T_{zx} + \frac{\partial}{\partial y}T_{zy} + \frac{\partial}{\partial z}T_{zz}$$

putting

(IV-9)
$$T_{pq} = \frac{1}{4\pi}\begin{vmatrix} e_x^2 + h_x^2 - \frac{1}{2}(\mathbf{e}^2 + \mathbf{h}^2), & e_xe_y + h_xh_y & e_xe_z + h_xh_z \\ e_xe_y + h_xh_y & e_y^2 + h_y^2 - \frac{1}{2}(\mathbf{e}^2 + \mathbf{h}^2), & e_ye_z + h_yh_z \\ e_xe_z + h_xh_z & e_ye_z + h_yh_z & e_z^2 + h_z^2 - \frac{1}{2}(\mathbf{e}^2 + \mathbf{h}^2) \end{vmatrix}.$$

By integrating (IV-8) over the volume $\mathscr{V}$, and by putting

$$\text{(IV-10)} \qquad \mathbf{F} = \int \mathbf{f}\, d\mathscr{V}, \qquad \mathbf{S} = \int \mathbf{s}\, d\mathscr{V},$$

one will then have:

$$\text{(IV-11)} \qquad F_x = -\frac{1}{c}\frac{\partial S_x}{\partial t} + \\ + \int \left[T_{xx}\cos(\mathbf{n},\mathbf{x}) + T_{xy}\cos(\mathbf{n},\mathbf{y}) + T_{xz}\cos(\mathbf{n},\mathbf{z}) \right] dS,$$

and two similar relations.

Assuming that only forces of electromagnetic origin enter into the picture, $\mathbf{F}$ represents the total force exerted on the volume $\mathscr{V}$. So it must be equal to the rate of change in the momentum $\mathbf{P}^{(m)}$ resulting from the matter contained in the volume $\mathscr{V}$. One will then have:

$$\text{(IV-12)} \qquad \mathbf{F} = \frac{dP^{(m)}}{dt}, \qquad \mathbf{P}^{(m)} = (P_x^{(m)}, P_y^{(m)}, P_z^{(m)}).$$

By then putting:

$$\text{(IV-13)} \qquad \mathbf{p}^{(r)} = \frac{\mathbf{s}}{c} \quad \text{and} \quad \mathbf{P}^{(r)} = \int \mathbf{p}^{(r)}\, d\mathscr{V}, \qquad \mathbf{P}^{(r)} = P_x^{(r)}, P_y^{(r)}, P_z^{(r)},$$

one will obtain, according to (IV-11):

$$\text{(IV-14)} \qquad \frac{d}{dt}(P_x^{(m)} + P_x^{(r)}) = \\ \int_S \left[T_{xx}\cos(\mathbf{n},\mathbf{x}) + T_{xy}\cos(\mathbf{n},\mathbf{y}) + T_{xz}\cos(\mathbf{n},\mathbf{z}) \right] dS$$

and two other analogous equations.

T_{pq} represents the "electromagnetic stresses"; (IV-14) expresses the conservation of momentum if to the momentum $\mathbf{P}^{(m)}$, resulting from the presence of matter, is added the momentum $\mathbf{P}^{(r)}$ produced by the radiation. The quotient, after division by c of the Poynting vector:

$$\text{(IV-15)} \qquad \mathbf{p}^{(r)} = \frac{\mathbf{s}}{c} = \frac{1}{4\pi c}[\mathbf{e} \wedge \mathbf{h}],$$

then represents the momentum connected with the presence of the microscopic electromagnetic field $\mathbf{e}$, $\mathbf{h}$.

2. *"Structure" of the Lorentz electron*

The simplest hypothesis was to assume that the charged particle resembled a small sphere of finite dimensions. The earlier electronic theories (Heaviside, Searle, J. J. Thomson, Abraham) therefore assumed the existence of a rigid, spherical electron.[7] Now, according to (IV-13), the momentum density produced by the electromagnetic field of this electron is[8]:

$$\text{(IV-15)} \qquad \mathbf{p}^{(r)} = \frac{\mathbf{s}}{c} = \frac{1}{4\pi c}[\mathbf{e} \wedge \mathbf{h}] \qquad \text{with} \qquad \mathbf{h} = \frac{1}{c}[\mathbf{v} \wedge \mathbf{e}].$$

One will then have:

$$\text{(IV-16)} \qquad \mathbf{p}^{(r)} = \frac{1}{4\pi c^2}[\mathbf{e} \wedge [\mathbf{v} \wedge \mathbf{e}]] = \frac{1}{4\pi c^2}(\mathbf{v}\cdot\mathbf{e}^2 - \mathbf{e}(\mathbf{v}\cdot\mathbf{e})).$$

If the motion takes place along oz ($|\mathbf{v}| = v_z = v$).

$$\text{(IV-17)} \qquad p_z^{(r)} = \frac{v}{4\pi c^2}(e_x^2 + e_y^2)$$

and the total momentum is then:

$$\text{(IV-18)} \qquad P_z^{(r)} = \frac{v}{4\pi c^2}\int_{\mathscr{V}} (e_x^2 + e_y^2)\, d\mathscr{V}.$$

Now, for a spherically symmetric distribution of charge,

$$\text{(IV-19)} \qquad e_x^2 + e_y^2 = \frac{2}{3}\mathbf{e}^2$$

and, if the field is produced by the charge q distributed over the sphere of radius r:

$$\text{(IV-20)} \qquad \mathbf{e} = \frac{q}{r^2}, \qquad d\mathscr{V} = 4\pi r^2\, dr.$$

We must integrate over the whole space outside a sphere of radius r_0. This quantity r_0 represents the "radius of the electron". We will then have:

$$\text{(IV-21)} \qquad \mathbf{P}^{(r)} = \frac{\mathbf{v}}{4\pi c^2}\int_{\infty}^{r_0} \frac{2}{3}\frac{q^2}{r^4}\cdot 4\pi r^2\, dr = \frac{2\mathbf{v}q^2}{3c^2 r_0}.$$

Since $\mathbf{P}^{(r)}$ represents a momentum, everything takes place as if the particle had an inert mass:

$$m_0 = \frac{2q^2}{3c^2 r_0}. \tag{IV-22}$$

The mass m_0 is called the *electromagnetic mass of the particle.* Its determination would have meaning only if one could directly measure r_0, the radius of the electron.

On the other hand, one can still define a length r_0 as a function of the experimental mass m_0 and of the charge q of the particle, separately determined[9]:

$$r_0 = \frac{2q^2}{3c^2 m_0} \simeq 1.9 \cdot 10^{-13}. \tag{IV-23}$$

By definition, r_0 would be the radius of the particle if the whole mass were of electromagnetic origin. We know that this length determines the dimensions of a region where the usual laws of electromagnetism are applicable only with certain restrictions.

On the other hand, according to (IV-6), the energy of a particle is still

$$U_0 = \frac{1}{8\pi} \int_{\mathscr{V}} \mathbf{e}^2 \, d\mathscr{V} = \frac{1}{8\pi} \int_{r_0}^{\infty} \frac{q^2}{r^4} \cdot 4\pi r^2 \, dr = \frac{q^2}{2r_0}. \tag{IV-24}$$

By comparing (IV-22) and (IV-24) one then obtains:

$$m_0 = \frac{4}{3} \frac{U_0}{c_l^2}. \tag{IV-25}$$

The hypothesis of an electromagnetic origin of mass is not surprising. We know, as a matter of fact, that the uniform motion of a charge creates a magnetic field. If we decelerate the charge, the variation in this magnetic field sets up an electric field. This electric field, according to the laws of induction, causes an acceleration of the particle. The deceleration of an electron thus causes inertial forces to appear which are entirely the result of electromagnetic induction.

However, even though the electromagnetic origin of mass is a plausible hypothesis, one cannot say that it is confirmed (or, furthermore, disproved) by facts. For that to be so, it would have to be possible to measure r_0

directly. Moreover, the factor $\frac{4}{3}$ is arbitrary, because it depends upon the hypotheses made concerning the distribution of charges.

The theory of Abraham's rigid spherical electron was quite rapidly replaced by Bücherer's and Lorentz' deformable electron. Bücherer and Lorentz assume that the electron in motion is capable of contraction in ratio

$$\frac{1}{\sqrt{1-\beta^2}} \qquad \left(\beta = \frac{v}{c}, \qquad v \text{ being the electron's velocity}\right).$$

This hypothesis, which was hit upon as a result of Michelson's experiment, is intimately bound up with the concepts of Special Relativity. We will, therefore, not undertake justifying it here; that will be done later, proceeding from more satisfactory considerations. (Cf. Chapter V, Section 11). Let us merely indicate to what degree Lorentz' conclusions differed from Abraham's.

The introduction of a density of charge:

$$\text{(IV-26)} \qquad \rho(xyz) = \frac{1}{\sqrt{1-\beta^2}}\rho_0\left(x, y, \frac{z}{\sqrt{1-\beta^2}}\right)$$

in the case of an electron moving along oz with a velocity v, leads us to define a momentum density:

$$\text{(IV-27)} \qquad p_z^{(r)} = \frac{v}{4\pi c^2\sqrt{1-\beta^2}}(e_x^2 + e_y^2)$$

along oz. By a calculation analogous to the preceding one, we find that this density corresponds to a total momentum:

$$\text{(IV-28)} \qquad \mathbf{P}^{(r)} = \frac{\mathbf{v}}{4\pi c^2\sqrt{1-\beta^2}}\int_{r_0}^{\infty}\frac{2}{3}\frac{q^2}{r^4}\cdot 4\pi r^2\,dr = \frac{2}{3}\frac{\mathbf{v}}{c^2}\frac{q^2}{r_0\sqrt{1-\beta^2}}$$

that is, instead of $\mathbf{P} = m_0\mathbf{v}$,

$$\text{(IV-29)} \qquad \mathbf{P}^{(r)} = \frac{m_0\mathbf{v}}{\sqrt{1-\beta^2}}$$

provided that we put, as in (IV-25):

$$\text{(IV-25)} \qquad m_0 = \frac{2}{3}\frac{q^2}{c^2 r_0} = \frac{4}{3}\frac{U_0}{c^2}.$$

Now it is Lorentz' relation (IV-29), and not the relation $\mathbf{P} = m_0 \mathbf{v}$ deduced from Abraham's hypotheses, that is confirmed by experiment.

Indeed, if we once more consider the experiments on the deviation of charged particles by two associated electric and magnetic fields, and if we assume in addition the validity of formula (IV-29), one notes that for sufficiently high velocities, the deviating particles should be distributed along a curve

$$\text{(IV-30)} \qquad \frac{y^2}{x} = \frac{1}{2}\frac{e}{m}\frac{H^2}{c^2}\frac{l^2}{E}\sqrt{1-\left(\frac{v^2}{c^2}\right)}$$

which, for the given e/m, departs increasingly from the parabola envisaged by the earlier theories as the velocity $\mathbf{v}$ draws nearer to the velocity of light. The results obtained are in complete harmony with formula (IV-30). (Cf. Chapter X, Section 2.)

Moreover, if $\mathbf{F}$ is the force acting on an electron, one should have, according to (IV-29):

$$\text{(IV-31)} \qquad \mathbf{F} = \frac{d\mathbf{P}}{dt} \qquad \text{with} \qquad \mathbf{P} = \mathbf{v}f(v^2).$$

By decomposing $\mathbf{F}$ into a component $\mathbf{F}_l$ parallel to $\mathbf{v}$ and another component $\mathbf{F}_t$ perpendicular to $\mathbf{v}$, one then obtains:

$$\text{(IV-32)} \qquad \mathbf{F} = \mathbf{F}_l + \mathbf{F}_t = \frac{d\mathbf{v}}{dt}f(v^2) + 2\mathbf{v}f'(v^2)\cdot\mathbf{v}\cdot\frac{d\mathbf{v}}{dt}$$

that is

$$\text{(IV-33)} \qquad \mathbf{F}_l = \left(\frac{d\mathbf{v}}{dt}\right)_l [f(v^2) + 2v^2 f'(v^2)] = \left(\frac{d\mathbf{v}}{dt}\right)_l \frac{d\mathbf{P}}{d\mathbf{v}}$$

$$\text{(IV-34)} \qquad \mathbf{F}_t = \left(\frac{d\mathbf{v}}{dt}\right)_t f(v^2) = \left(\frac{d\mathbf{v}}{dt}\right)_t \frac{\mathbf{P}}{\mathbf{v}}.$$

The longitudinal and transverse components $\mathbf{F}_l$ and $\mathbf{F}_t$ of the force are proportional to the corresponding accelerations $\gamma_l = (d\mathbf{v}/dt)_l$, $\gamma_t = (d\mathbf{v}/dt)_t$, provided we introduce inertial masses differing from each other. We will then define a longitudinal mass:

$$\text{(IV-35)} \qquad m_l = \frac{d\mathbf{P}}{d\mathbf{v}} = \frac{m_0}{(1-\beta^2)^{3/2}}$$

and a transverse mass:

$$m_t = \frac{\mathbf{P}}{\mathbf{v}} = \frac{m_0}{\sqrt{1 - \beta^2}}. \tag{IV-36}$$

In fact, these results are profoundly modified by Special Relativity which, as we will see, no longer rests upon the basic law $\mathbf{F} = m\gamma$. Only relation (IV-29) obtained by Lorentz and confirmed by experiment remains valid. But even then, it does not retain the significance that was originally attributed to it. It was thought that, as a matter of fact, only the electronic mass defined by (IV-25) varied with velocity and that the mechanical mass remained unchanged. The importance of spectrographic experiments on mass which were undertaken in this regard was due to the hope of revealing the electromagnetic or the purely mechanical contributions which intervened in forming the mass.

Experimental verification of Lorentz' formula would then lead to the conclusion that the whole mass of the particle was of electromagnetic origin. Under these circumstances, it was necessary to deny the existence of a purely mechanical mass and also of a basic dualism in the concepts of particle and field. But, according to the concepts of Special Relativity, a mass varies with velocity, no matter what its origin. Mass spectrography makes it possible to verify this conclusion but cannot furnish a criterion for detection of the "real" nature of mass.

3. *Potentials and fields created by a distribution of electrons*

Equations (IV-2) and (IV-4) are always satisfied by defining the fields $\mathbf{e}$ and $\mathbf{h}$ as a function of the potentials $\mathbf{A}$ and φ:

$$\boxed{\begin{aligned} \mathbf{h} &= \mathbf{curl}\,\mathbf{A} && \text{(IV-37)} \\ \mathbf{e} &= -\frac{1}{c}\frac{\partial \mathbf{A}}{dt} - \mathbf{grad}\,\varphi && \text{(IV-38)} \end{aligned}}$$

provided we establish the following condition between potentials:

$$\boxed{\frac{1}{c}\frac{\partial \varphi}{\partial t} + \operatorname{div} \mathbf{A} = 0.} \tag{IV-39}$$

This condition combined with relations (IV-1) and (IV-3) allows us to determine $\mathbf{A}$ and φ completely.

Taking into account (IV-1), (IV-3), and the condition (IV-39), one obtains the equations of propagation:

$$\text{(IV-40)} \qquad \frac{1}{c^2}\frac{\partial^2 \mathbf{A}}{\partial t^2} - \Delta \mathbf{A} = \frac{4\pi}{c}\rho\mathbf{v}$$

$$\text{(IV-41)} \qquad \frac{1}{c^2}\frac{\partial^2 \varphi}{\partial t^2} - \Delta \varphi = 4\pi\rho\,.$$

Their solution is the sum of the solution of the homogenous equation and a special solution of the inhomogenous equation. In this way one obtains:

$$\text{(IV-42)} \qquad \boxed{\mathbf{A}(x, y, z, t) = \frac{1}{c}\int_{\mathscr{V}} \frac{[\rho\mathbf{v}]_{t-\frac{r}{c}}}{r}\, d\mathscr{V}}$$

$$\text{(IV-43)} \qquad \boxed{\varphi(x, y, z, t) = \int_{\mathscr{V}} \frac{[\rho]_{t-\frac{r}{c}}}{r}\, d\mathscr{V}}$$

A and φ are the Maxwell-Hertz potentials created at the instant t and at point $P(x, y, z)$ by the element $d\mathscr{V} = d\xi\, d\eta\, d\zeta$ having a density of charge

$$\rho\left(\xi, \eta, \zeta, t - \frac{r}{c}\right)$$

and a current density

$$\rho\mathbf{v}\left(\xi, \eta, \zeta, t - \frac{r}{c}\right).$$

In order to calculate the integrals (IV-42) and (IV-43), the elements of which have to be evaluated at differing times $t - r/c$, let us consider the following representation, which was imagined by Planck. Let us suppose that a sphere with the center $P(x, y, z)$ encloses a volume that is constantly decreasing. The surface of the sphere passes successively through the whole region $d\mathscr{V}$. At the instant $t - r/c$, this surface passes through the point $M(\xi, \eta, \zeta)$ corresponding to the density ρ and the momentum $\rho\mathbf{v}$. The sphere thus "gathers up" in the course of its contraction the contributions made successively by the various regions of space to the potential created at P.

Between two instants τ and $\tau + d\tau$ corresponding to the radii $r = c(t - \tau)$ and $r' = r - cd\tau$, the quantity of charge contained in that portion of the spherical ring defined by the arc ds and the thickness dr will be[10]

$$\text{(IV-44)}_1 \qquad \int \rho(x, y, z, \tau)\, ds\, dr\,.$$

On the other hand, the charges of velocity $\mathbf{v}$ can pass through the surface of the sphere from the inside out – or vice-versa – during the time $d\tau$. The flux passing through the surface during the time $d\tau$ and corresponding to this velocity $\mathbf{v}$ is then

$$\text{(IV-44)}_2 \qquad \int \rho ds \cdot v \cos\theta \cdot d\tau = \int \rho \frac{(\mathbf{v}\cdot\mathbf{r})}{r}\, ds\, d\tau = \int \rho \frac{(\mathbf{v}\cdot\mathbf{r})}{cr}\, ds\, dr\,.$$

The quantity of electricity gathered up by the contracting sphere during the time $d\tau$ is, thus, the difference between (IV-44)_1 and (IV-44)_2:

$$\text{(IV-45)} \qquad q = \int \rho \left[1 - \frac{(\mathbf{v}\cdot\mathbf{r})}{cr}\right] ds\, dr\,.$$

Now the expression

$$\text{(IV-46)} \qquad \rho \left[1 - \frac{(\mathbf{v}\cdot\mathbf{r})}{cr}\right]$$

also represents the charge dq contained in the volume $d\mathscr{V} = ds\, dr$, provided that ds is large with respect to dr. Under those circumstances, the flux of charges through dr is negligible with respect to the charges through ds. One thus has:

$$\text{(IV-47)} \qquad \rho\, d\mathscr{V} = \frac{dq}{1 - \dfrac{(\mathbf{v}\cdot\mathbf{r})}{cr}}$$

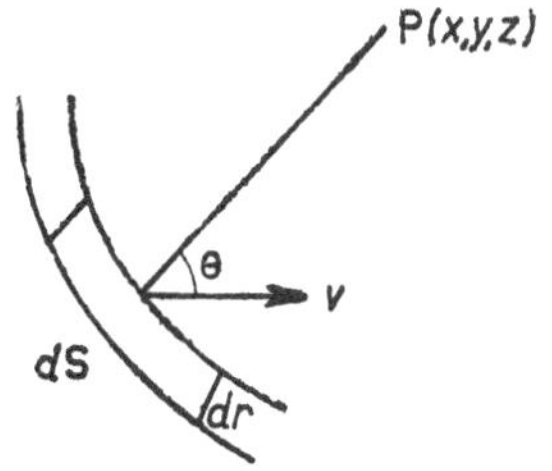

Fig. 12. Method for the calculation of Maxwell-Hertz potentials.

By comparison with expressions (IV-42) and (IV-43), we may write:

(IV-48) $$\mathbf{A} = \frac{1}{c}\int \frac{\mathbf{v}\,dq}{\left[r - \frac{(\mathbf{v}\cdot\mathbf{r})}{c}\right]_{t-\frac{r}{c}}}$$

(IV-49) $$\varphi = \int \frac{dq}{\left[r - \frac{(\mathbf{v}\cdot\mathbf{r})}{c}\right]_{t-\frac{r}{c}}}$$

If the particle is extremely small, the quantities

$$r - \frac{(\mathbf{v}\cdot\mathbf{r})}{c}$$

are approximately constant within a domain of the order r_0^3. Hence:

(IV-50) $$\mathbf{A} = \frac{q\mathbf{v}}{c\left[r - \frac{\mathbf{v}\cdot\mathbf{r}}{c}\right]_{t-\frac{r}{c}}}$$

(IV-51) $$\varphi = \frac{q}{\left[r - \frac{\mathbf{v}\cdot\mathbf{r}}{c}\right]_{t-\frac{r}{c}}}$$

These potentials connected with the electron are also called *Liénard-Wiechert potentials*, after those who introduced these particular expressions. They allow us to calculate the fields created by a distribution of charge characterized by **A** and φ. Indeed, the Maxwell equations:

(IV-52) $$\mathbf{E} = -\,\mathbf{grad}\,\varphi - \frac{1}{c}\frac{\partial \mathbf{A}}{\partial t}$$

(IV-53) $$\mathbf{H} = \mathbf{curl}\,\mathbf{A}$$

applied to (IV-50) and (IV-51) lead, after a few calculations[11], to the following expressions:

(IV-54) $$\mathbf{E} = \mathbf{E}_1 + \mathbf{E}_2 \qquad \mathbf{H} = \mathbf{H}_1 + \mathbf{H}_2$$

with

$$\text{(IV-55)} \qquad \mathbf{E}_1 = \frac{q(1-\beta^2)}{\left(r - \frac{\mathbf{r}\cdot\mathbf{v}}{c}\right)^3}\left(\mathbf{r} - \mathbf{v}\frac{r}{c}\right), \qquad \mathbf{H}_1 = \frac{\mathbf{v}}{c} \wedge \mathbf{E}_1$$

$$\text{(IV-56)} \qquad \mathbf{E}_2 = \frac{q}{c^2\left(r - \frac{\mathbf{r}\cdot\mathbf{v}}{c}\right)^3}\left[\mathbf{r} \wedge \left[\mathbf{r} - \frac{\mathbf{v}}{c} r\right] \wedge \frac{\partial \mathbf{v}}{\partial t}\right], \qquad H_2 = \frac{\mathbf{r}}{r} \wedge \mathbf{E}_2$$

The fields $\mathbf{E}_1$ and $\mathbf{H}_1$ refer to the electrostatic field resulting from an electron of constant velocity. Such an electron produces an electrostatic field that is carried along by it through convection.

The fields $\mathbf{E}_2$ and $\mathbf{H}_2$ concern solely states of variable velocity. They are thus connected with acceleration and deceleration phenomena. They vary as $1/r$ and become preponderent at great distances. Finally, $\mathbf{E}_2$ and $\mathbf{H}_2$ are perpendicular to the direction of propagation. They characterize a spherical electromagnetic wave involved in the accelerated states of the motion.

In developing Maxwell's theory we have taken into account the electronic properties that make it possible for us to explain the polarization of dielectric and the magnetization of magnetic media. Maxwell's theory does indeed become much clearer when the constitution of the magnets and dielectrics can be deduced from the properties of charges and currents. Nevertheless, the complete theories of polarization and magnetism involve the problems of the motion of electrons inside atoms. As a matter of fact, these questions, which were first considered on the basis of the classical hypothesis of the electron having elastic bondings inevitably find their developments in the quantum theories. Here we shall limit ourselves to the deduction of the formalism of Maxwell's theory from the principles of Lorentz' theory.

4. *Equations for the mean values and Maxwell's macroscopic theory*

According to Lorentz, relations (IV-1), (IV-2), (IV-3) and (IV-4), which apply to the microscopic fields **e** and **h** are valid in all points, outside and inside the electron.

We here apply these relations to a domain that is at once large enough to contain numerous molecules and small enough that the total **E**, **H**, **D** and **B** fields of Maxwell's theory remain constant within that domain, to all intents and purposes. (The fields and inductions are indeed slowly variable over distances that are of the same order as molecular dimensions.) Under these circumstances we can replace the heterogeneity of the microscopic domain by an apparent continuity.

To calculate the mean value of a quantity A at a point $P(x, y, z)$ and at the instant t_1, we surround the point P with an infinitesimal sphere of radius a and volume $\mathscr{V}$. The mean value $\bar{A}$ will be given by the integral:

$$\bar{A} = \frac{1}{2\tau}\frac{1}{\mathscr{V}}\int_{-\tau}^{+\tau}\int_{\mathscr{V}} A(x+\xi, y+\eta, z+\zeta, t+\theta)\, d\xi\, d\eta\, d\zeta\, d\theta \tag{IV-57}$$

over the volume, contained within the sphere, and the time interval $t-\tau, t+\tau$. One then has:

$$\frac{\partial \bar{A}}{\partial t} = \overline{\frac{\partial A}{\partial t}}, \qquad \frac{\partial \bar{A}}{\partial x^p} = \overline{\frac{\partial A}{\partial x^p}}. \tag{IV-58}$$

In addition, within the domain previously delimited – that is, except for very short wave-lengths and very rarefied media – one can always obtain, starting with the microscopic equations (IV-1), (IV-2), (IV-3), (IV-4), the mean-value equations:

$$\mathbf{curl}\,\bar{\mathbf{H}} - \frac{1}{c}\frac{\partial \bar{\mathbf{E}}}{\partial t} = \frac{4\pi}{c}\overline{\rho \mathbf{v}} \tag{IV-59}$$

$$\mathbf{curl}\,\mathbf{E} + \frac{1}{c}\frac{\partial \bar{\mathbf{H}}}{\partial t} = 0 \tag{IV-60}$$

$$\operatorname{div} \bar{\mathbf{E}} = 4\pi\bar{\rho} \tag{IV-61}$$

$$\operatorname{div} \bar{\mathbf{H}} = 0. \tag{IV-62}$$

5. *Interpretation of the fields and inductions of Maxwell's theory. Electromagnetic equations for the case of matter at rest*

If matter is at rest with respect to the ether, **v** represents the velocity of the electrons with respect to the immobile material system we are considering.

a) *In a conductor*, the density ρ and the conduction current $\mathbf{I}$ are written thus:

(IV-63) $\rho = (\bar{\rho})_1 = nq$

(IV-64) $\mathbf{I} = (\overline{\rho \mathbf{v}})_1 = nq\mathbf{v}$

with $n = \dfrac{N}{\mathscr{V}}$.

This conduction current is produced by the "free electrons".

b) *In a dielectric*, the displacement of electrons will produce no additional charge on the molecule but, as we have postulated (Chapter I, Section 10), it will produce a polarization

(IV-65) $\mathbf{P} = nq\mathbf{d}$.

This polarization sets up a charge $\int P_n dS$ over the surface of the molecule. As we have noted in Chapter I, this surface charge balances a charge $-\int \rho' d\mathscr{V}$ of density ρ' inside the molecule.

One will then have:

(IV-66) $\int P_n\, dS = \int \operatorname{div} \mathbf{P}\, d\mathscr{V} = -\int \rho'\, d\mathscr{V}$.

Whence the setting up of an additional density of charge:

(IV-67) $\rho' = (\bar{\rho})_2 = -\operatorname{div} \mathbf{P}$.

On the other hand, the variation of $\mathbf{P}$ with respect to time creates the polarization current:

(IV-68) $(\overline{\rho \mathbf{v}})_2 = \dfrac{\partial \mathbf{P}}{\partial t}$.

The quantities $(\bar{\rho})_2$ and $(\overline{\rho \mathbf{v}})_2$ represent the contribution of the charges that exist in the dielectric – charges that were called "fictitious" and which represented the contribution of the bound electrons.

c) Finally, a certain number of molecules possess a *magnetic moment* whose origin may be explained by quantum theory. The magnetic moment per unit volume is:

(IV-69) $\mathbf{M} = n\mathbf{m}$

$\mathbf{m}$ being the magnetic moment of each molecule. Each of these molecules is equivalent to the magnetic layer produced by an intra-molecular convection current j'. One will then have

(IV-70) $\int m_l \, dl = j'$

that is:

(IV-71) $\int \mathbf{curl\,m} \, dS = \int \mathbf{J}' \, dS$

or:

(IV-72) $\mathbf{curl\,m} = \mathbf{J}'$.

Within a unit volume of a magnetic substance, one will then be able to define the convection current:

(IV-73) $(\overline{\rho \mathbf{v}})_3 = n\mathbf{J}' = c\,\mathbf{curl\,M}$

within the system of mixed units we have adopted. In terms of mean value, we will then have the following relations:

(IV-74) $\bar{\rho} = (\bar{\rho})_1 + (\bar{\rho})_2 = \rho - \operatorname{div} \mathbf{P}$

(IV-75) $\overline{\rho \mathbf{v}} = (\overline{\rho \mathbf{v}})_1 + (\overline{\rho \mathbf{v}})_2 + (\overline{\rho \mathbf{v}})_3 = \mathbf{I} + \dfrac{\partial \mathbf{P}}{\partial t} + c\,\mathrm{curl}\,\mathbf{M}$.

If we now put:

(IV-76) $\bar{\mathbf{E}} = \mathbf{E}, \qquad \bar{\mathbf{H}} = \mathbf{B},$

equations (IV-59) through (IV-62), expressed as functions of mean values, are still written:

(IV-77) $\mathbf{curl\,B} - \dfrac{1}{c}\dfrac{\partial \mathbf{E}}{\partial t} = 4\pi\left(\dfrac{\mathbf{I}}{c} + \dfrac{\partial \mathbf{P}}{c\partial t} + \mathbf{curl\,M}\right)$

(IV-78) $\mathbf{curl\,E} = -\dfrac{1}{c}\dfrac{\partial \mathbf{B}}{\partial t}$

(IV-79) $\operatorname{div} \mathbf{E} = 4\pi\rho - 4\pi \operatorname{div} \mathbf{P}$

(IV-80) $\operatorname{div} \mathbf{B} = 0$.

Let us put

(IV-81) $\mathbf{D} = \mathbf{E} + 4\pi\mathbf{P}$

(IV-82) $\mathbf{H} = \mathbf{B} - 4\pi\mathbf{M}$.

Equations (IV-77) and (IV-79) are then written:

(IV-83) $$\mathbf{curl}\,\mathbf{H} = 4\pi\frac{\mathbf{I}}{c} + \frac{1}{c}\frac{\partial \mathbf{D}}{\partial t}$$

(IV-84) $$\operatorname{div}\mathbf{D} = 4\pi\rho$$

and the system formed by (IV-78), (IV-80), (IV-83) and (IV-84) coincides well with Maxwell's macroscopic equations (Systems [I], [II] of the relations developed in Chapter III). It is worth noting that the relations directly define the electric field and the magnetic induction (and not the magnetic field) as a function of the mean values arising from the microscopic theory. So it is clearly the electromagnetic induction that constitutes the true counterpart of the electric field.

6. *Lorentz' theory and the electrodynamics of moving bodies*

Let us now suppose that matter moves with a velocity u, which we will assume to be uniform within a domain that is of the order of molecular dimensions. If[12]

(IV-85) $$\mathbf{u} \ll c$$

we can apply the non-relativistic principle of the composition of velocities. In that way we obtain:

(IV-86) $$\mathbf{v} = \mathbf{u} + \mathbf{v}'$$

(IV-87) $$\overline{\rho\mathbf{v}} = \overline{\rho\mathbf{u}} + \overline{\rho\mathbf{v}}'\,.$$

The current $\rho\mathbf{v}$ is written, as we did previously:

(IV-88) $$\overline{\rho\mathbf{v}}' = (\overline{\rho\mathbf{v}}')_1 + (\overline{\rho\mathbf{v}}')_2 + (\overline{\rho\mathbf{v}}')_3 = \mathbf{I}' + \frac{d\mathbf{P}'}{dt} + c\,\mathbf{curl}\,\mathbf{M}'$$

provided that we measure $\mathbf{I}'$, $d\mathbf{P}'/dt$, $\mathbf{curl}\,\mathbf{M}'$ with instruments that are moving along with the matter.

Now the polarization current

$$(\overline{\rho\mathbf{v}}')_2 = \frac{d\mathbf{P}'}{dt}$$

is defined by the surface integral:

$$\text{(IV-89)} \qquad \int_S (\overline{\rho v_2'})_n \, dS = \frac{d}{dt} \int_S P_n \, dS$$

and, following the usual procedures of vector calculus[13]

$$\text{(IV-90)} \qquad \frac{d}{dt} \int_S \mathbf{P}_n' \, dS = \int_S \left(\frac{\partial \mathbf{P}'}{\partial t} + \mathbf{curl}\,[\mathbf{P}' \wedge \mathbf{u}] + \mathbf{u} \operatorname{div} \mathbf{P}' \right)_n dS.$$

One then has:

$$\text{(IV-91)} \qquad (\overline{\rho \mathbf{v}'})_2 = \frac{d\mathbf{P}'}{dt} = \frac{\partial \mathbf{P}'}{\partial t} + \mathbf{curl}\,[\mathbf{P}' \wedge \mathbf{u}] + \mathbf{u} \operatorname{div} \mathbf{P}'.$$

The equations with mean values (IV-59) and (IV-61) are then written:

$$\text{(IV-92)} \qquad \begin{aligned} &\mathbf{curl}\,\mathbf{B} - \frac{1}{c}\frac{\partial \mathbf{E}}{\partial t} = \\ &= \frac{4\pi}{c}\left(\mathbf{I} + \frac{\partial \mathbf{P}'}{\partial t} + \mathbf{curl}\,[\mathbf{P}' \wedge \mathbf{u}] + \mathbf{u} \operatorname{div} \mathbf{P}' + c\,\mathbf{curl}\,\mathbf{M}' + \overline{\rho \mathbf{u}} \right) \end{aligned}$$

$$\text{(IV-93)} \qquad \operatorname{div} \mathbf{E} = 4\pi\rho - 4\pi \operatorname{div} \mathbf{P}'$$

$\mathbf{P}'$ and $\mathbf{M}'$ represent the polarization and the magnetic moment, respectively, bound to the moving system. By now putting:

$$\text{(IV-94)} \qquad \mathbf{D} = \mathbf{E} + 4\pi \mathbf{P}'$$

$$\text{(IV-95)} \qquad \mathbf{H} = \mathbf{B} - 4\pi \mathbf{M}'$$

one will still have:

$$\text{(IV-96)} \qquad \mathbf{curl}\,\mathbf{H} - \frac{1}{c}\frac{\partial \mathbf{D}}{\partial t} = \frac{4\pi}{c}(\mathbf{I} + \rho \mathbf{u} + \mathbf{curl}\,[\mathbf{P}' \wedge \mathbf{u}])$$

$$\text{(IV-97)} \qquad \operatorname{div} \mathbf{D} = 4\pi\rho.$$

In fact, according to (IV-74):

$$\text{(IV-98)} \qquad \overline{\rho \mathbf{u}} = \rho \mathbf{u} - \mathbf{u} \operatorname{div} \mathbf{P}'.$$

If we then define vectors **P** and **M** in such a way that:

$$\text{(IV-99)} \qquad \mathbf{M}' - \mathbf{M} = -\left[\frac{\mathbf{P}' \wedge \mathbf{u}}{c}\right]$$

$$\text{(IV-100)} \qquad \mathbf{P}' - \mathbf{P} = \left[\frac{\mathbf{M}' \wedge \mathbf{u}}{c}\right]$$

we can introduce the auxiliary quantities:

$$\text{(IV-101)} \qquad \mathbf{B} = \mathbf{H}_1 + 4\pi\mathbf{M}$$

$$\text{(IV-102)} \qquad \mathbf{D}_1 = \mathbf{E} + 4\pi\mathbf{P}$$

that is, the field $\mathbf{H}_1$, and the induction $\mathbf{D}_1$ which are deduced from **B** and **M** and from **E** and **P** by the usual expressions. It will be noted that (IV-83) and (IV-84) take on the following form:

$$\text{(IV-103)} \qquad \operatorname{curl} \mathbf{H}_1 - \frac{1}{c}\frac{\partial \mathbf{D}}{\partial t} = \frac{4\pi}{c}(\mathbf{I} + \rho\mathbf{u})$$

$$\text{(IV-104)} \qquad \operatorname{div} \mathbf{D} = 4\pi\rho\,.$$

Minkowski arrived at these equations following a different approach. But, whereas $\mathbf{H}_1$ and $\mathbf{P}'$ represent, according to Minkowski, the magnetic field and the polarization for a stationary observer, it is the quantities **H** and **P** that play these roles in Lorentz' theory. They differ, moreover, from $\mathbf{H}_1$ and $\mathbf{P}'$ only by the quantities

$$\text{(IV-105)} \qquad \mathbf{H}_1 - \mathbf{H} = 4\pi(\mathbf{M}' - \mathbf{M}) = -\frac{4\pi}{c}[\mathbf{P}' \wedge \mathbf{u}]$$

which are small with respect to the field itself if $u \ll c$.

The Lorentz force. The expression of the force as a function of the microscopic fields **e** and **h**:

$$\text{(IV-106)} \qquad \boxed{\mathbf{f} = q\left(\mathbf{e} + \frac{1}{c}[\mathbf{v} \wedge \mathbf{h}]\right)}$$

is written in the following manner when we proceed to mean values and then, following (IV-76), to the fields **E** and **B**:

$$\text{(IV-107)} \qquad \mathbf{f} = q\left[\mathbf{E} + \frac{1}{c}[\mathbf{v} \wedge \mathbf{B}]\right].$$

One will then have, according to (IV-86):

$$\mathbf{f} = q\left(\mathbf{E} + \frac{1}{c}[\mathbf{u} \wedge \mathbf{B}] + \frac{1}{c}[\mathbf{v}' \wedge \mathbf{B}]\right). \tag{IV-108}$$

If the charge is carried along by the moving matter, ($\mathbf{v}' = 0$), what happens is equivalent with the case in which the moving charge is submitted to a field $\mathbf{E}'$, measured in the moving system. This field is given by

$$\mathbf{f} = q\mathbf{E}' = q\left(\mathbf{E} + \frac{1}{c}[\mathbf{u} \wedge \mathbf{B}]\right). \tag{IV-109}$$

One will then have:

$$\mathbf{E}' = \mathbf{E} + \frac{1}{c}[\mathbf{u} \wedge \mathbf{B}]. \tag{IV-110}$$

In the moving system, one continues to have the relations:

$$\mathbf{D}' = \varepsilon\mathbf{E}' = \mathbf{E}' + 4\pi\mathbf{P}' \tag{IV-111}$$

that is:

$$\mathbf{P}' = \frac{\varepsilon - 1}{4\pi}\mathbf{E}'. \tag{IV-112}$$

One defines a polarization thus:

$$\mathbf{P}' = \frac{\varepsilon - 1}{4\pi}\left[\mathbf{E} + \frac{1}{c}[\mathbf{u} \wedge \mathbf{B}]\right]. \tag{IV-113}$$

Let us consider in particular a non-magnetic medium. According to (IV-100), $\mathbf{M}' = 0$, and $\mathbf{P}' = \mathbf{P}$. From the Lorentz equations (IV-94) and (IV-96) we then have:

$$\mathbf{curl}\,\mathbf{B} = \frac{1}{c}\frac{\partial \mathbf{E}}{\partial t} + \frac{4\pi}{c}\left(\mathbf{I} + \rho\mathbf{u} + \frac{\partial \mathbf{P}}{\partial t} + \mathbf{curl}\,[\mathbf{P} \wedge \mathbf{u}]\right). \tag{IV-114}$$

The total current produced by a charged dielectric in motion is, according to (IV-114) and (IV-91):

$$\mathbf{I} + \rho\mathbf{u} + \frac{\partial \mathbf{P}}{\partial t} + \mathbf{curl}\,[\mathbf{P} \wedge \mathbf{u}] = \mathbf{u}(\rho - \operatorname{div}\mathbf{P}) + \frac{d\mathbf{P}}{dt} + \mathbf{I}. \tag{IV-115}$$

If we limit ourselves to stationary states – and such is the case in Röntgen and Eichenwald's experiment – the convection current observed has a density

$$\mathbf{u}(\rho - \operatorname{div}\mathbf{P}) \tag{IV-116}$$

in the first experiment, where the insulator and its metallic plates turn simultaneously. On the other hand, its value is **I** in the second experiment where the system is at rest but at the same time is the seat of a conduction current. If we adjust things in such a way (by modifying the resistance of the outside circuit) that these two currents are equal, one then has:

$$i = \int_S \mathbf{u}(\rho - \operatorname{div}\mathbf{P})\, dS\,. \tag{IV-117}$$

By taking (III-108) into account, as well as the notations of Figure 11, the total current $i = \int \mathbf{I} dS$ must then be the sum of a convection current:

$$i_1 = \int \mathbf{u}\rho\, dS_1 = \mathbf{u}\rho \cdot ad = a\mathbf{u}\sigma_p = a\mathbf{u}\frac{\varepsilon E}{4\pi} \tag{IV-118}$$

and a Röntgen current:

$$i_2 = -\int \mathbf{u}\operatorname{div}\mathbf{P}\cdot dS_2 = -\int \mathbf{u}\operatorname{div}\mathbf{P}adx = -a\mathbf{u}\int \frac{\partial \mathbf{P}}{\partial x}dx = -a\mathbf{u}|P| \tag{IV-119}$$

P being parallel to the axis of rotation in Eichenwald's experiment. One will then have according to (IV-112)

$$i_2 = -a\mathbf{u}\frac{(\varepsilon - 1)}{4\pi}E\,. \tag{IV-120}$$

So we should observe a total current:

$$i = i_1 + i_2 = a\mathbf{u}\frac{E}{4\pi} = a\mathbf{u}\frac{V}{4\pi d}\,. \tag{IV-121}$$

This is, in fact, the value found experimentally by Eichenwald. The current produced by the motion of a polarized dielectric is characterized by a density:

$$\mathbf{I} = -\mathbf{u}\operatorname{div}\mathbf{P} = -\mathbf{u}\frac{(\varepsilon - 1)}{4\pi}E\,. \tag{IV-122}$$

This current likewise is referred to as Röntgen's current. In all the experiments in which matter in slow motion ($u/c \ll 1$) is involved, the convection current

$$i = -\mathbf{u}\frac{a\varepsilon}{4\pi}E$$

must, consequently, be replaced by the Röntgen current

$$i' = -\mathbf{u}a\frac{\varepsilon - 1}{4\pi}E.$$

In practical terms, this amounts to replacing the field E with the field

$$E' = \left(1 - \frac{1}{\varepsilon}\right)E.$$

And that is, in fact, the conclusion to which Röntgen's, Eichenwald's and Wilson's experiments lead.

The substitution of E' for E might lead one to conclude that a partial carrying-along or drag of the fields took place – that is, a carrying-along of the electromagnetic ether by matter in motion. And indeed, Lorentz' conclusions are much the same as those that may be deduced from the hypothesis – suggested by Fresnel – of the partial carrying-along of the ether. But in reality the explanation proposed by Lorentz' theory is entirely different: the introduction of the field E' is only a convenient way of interpreting formula (IV-122), and the Röntgen current is linked, in Lorentz' theory, with the occurrence of the polarization of the dielectric. The quantities **P** and **M** characterize the dielectrics and the magnetic media, which are carried along, whereas the ether remains strictly motionless. We will have occasion to take up this conclusion again in connection with Special Relativity (Cf. Chapter V, Section 5).

In its final form, the Maxwell-Lorentz theory thus succeeds in interpreting correctly the experiments on the electrodynamics of bodies in slow motion, that is, experiments in which the terms in u^2/c^2 are negligible. By taking into account only terms in u/c (terms of the first order), one can retain the form of the Maxwellian equations and make them harmonize with Lorentz' hypotheses concerning the properties of their sources.

Moreover, the very structure of the sources introduces a mass that varies with velocity. This property, confirmed by experiment but interpreted with the aid of pre-relativistic conceptions, seems to suggest that the whole of mass is of electromagnetic origin. So the characteristics of the sources seem, provisionally, to resolve themselves into purely electromagnetic data.

In addition, in its final form the Maxwell-Lorentz theory seems to have succeeded in preserving the ideas of contiguous action and of finite propagation, and also in establishing a link between the classical theory of fields and the existence of sources. Furthermore, the characteristics of these sources seem finally to resolve into data that are not radically foreign to the field. All the phenomena – except gravitation – should, it seems, be reducible to electromagnetic actions of a Maxwellian type. The synthesis that theories of action at a distance had so vainly striven to achieve seems now to revolve around the concept of the field. The Maxwell-Lorentz equations, which were relativistic *avant la lettre* but still bound up with a pre-relativistic interpretation, constitute the bases of a classical theory of fields.

PROBLEMS

1. An electron moves in a constant magnetic field **H** oriented along *oz*.

a) Show that the trajectory is a spiral with axis parallel to *oz*.

b) The projection of the trajectory on the plane *xoy* is a circle. Calculate its radius, as a function of e/m from the initial velocity v of the field H.

c) Assuming that the initial velocity is directed along *ox*, calculate the deviation produced by a screen perpendicular to *ox* and situated at a distance l from the last diaphragm.

Solution: a) We use the definition of the Lorentz force:

$$\mathbf{f} = e\left[\frac{\mathbf{v}}{c} \wedge \mathbf{H}\right]$$

$$\left(m\ddot{x} = \frac{e}{c}\dot{y}H\,, \qquad m\ddot{y} = -\frac{e}{c}\dot{x}H\,, \qquad \ddot{z} = 0\right).$$

Putting

$$\xi = x + iy, \qquad \ddot{\xi} = -i\omega\dot{\xi} \qquad \text{with} \qquad \omega = \frac{eH}{mc},$$

one obtains by integration

$$\dot{\xi} = \dot{\xi}_0 e^{-i\omega t}, \qquad \xi = \xi_0 + \frac{\dot{\xi}_0}{i\omega}(1 - e^{-i\omega t}).$$

b) In the plane xoy the trajectory is a circle:

$$x = x_0 + \frac{1}{\omega}(\dot{y}_0(1 - \cos\omega t) + \dot{x}_0 \sin\omega t),$$

$$y = y_0 - \frac{\dot{x}_0}{\omega}(1 - \cos\omega t) + \frac{\dot{y}_0}{\omega}\sin\omega t.$$

The radius of the circle is:

$$R^2 = \frac{\dot{x}_0^2 + \dot{y}_0^2}{\omega^2} \qquad \text{or} \qquad R = \frac{v}{\omega} = \frac{vmc}{eH}.$$

c) From $(\dot{x} + i\dot{y}) = (\dot{x}_0 + i\dot{y}_0)e^{-i\omega t}$ results $\begin{cases} \dot{x} = \dot{x}_0 \cos\omega t + \dot{y}_0 \sin\omega t \\ \dot{y} = \dot{y}_0 \cos\omega t + \dot{x}_0 \sin\omega t \end{cases}$

If the initial velocity is directed along ox ($\dot{x}_0 = v, \dot{y}_0 = 0$), the equations of motion

$$\left(m\frac{d^2x}{dt^2} = 0, \qquad m\frac{d^2y}{dt^2} = 0\right)$$

applied to the displacement of the electron after passing beyond the diaphragm, lead to

$$\frac{dx}{dt} = v\cos\omega t \simeq v, \qquad \frac{dy}{dt} = v\sin\omega t \simeq v\omega t$$

that is

$$x \simeq vt \qquad y = -\frac{v\omega t^2}{2} = \frac{v\omega}{2}\left(\frac{l}{v}\right)^2 = \frac{1}{2}\frac{e}{mc}\frac{Hl^2}{v}$$

2. Determine the trajectory of an electron subjected to both an electric

and a magnetic field parallel to each other and perpendicular to the electron's initial trajectory.

$$(\mathbf{H} = H_z, \qquad \mathbf{E} = E_z, \qquad \mathbf{v} = v_x).$$

3. Determine the trajectory of an electron in an electric field and in a magnetic field perpendicular to each other ($\mathbf{E} = E_x, \mathbf{H} = H_y$). If the electron is at the origin of the coordinates at the instant $t = 0$ and its initial velocity v_0 is zero, show that its trajectory is a cycloid.

4. A cyclotron produces a magnetic field of 20 000 gauss. What is the angular velocity of a proton accelerated by this field?

NOTES

1. Cf. H. A. LORENTZ [5]: W. GERLACH: *Handbuch der Physik* **22** (1933) II, 2; L. ROSENFELD: *Theory of electrons*, Amsterdam, 1951; R. BECKER: [1].
2. A. MILLIKAN: *Phys. Rev.* **14** (1913) 136, 796.
3. E. REGENER: *Z. Physik* **39** (1926) 247.
4. E. REGENER: *Berl. Ber.* 1909, 948.
5. The determination of e by means of the Schrott effect may also be mentioned here. W. SCHOTTKY: *Ann. d. Phys.* **65** (1918) 541; **68** (1922) 157. Cf. also R. BECKER [1], p. 6.
6. Cf. R. BECKER [1], p. 39.
7. We will see that this model, developed especially by Abraham, is incompatible with the conclusions of special relativity.
8. As a matter of fact, the relative field **e** corresponding to the motion of the charges would have to be substituted for **e**. But these two fields (Cf. Section 3) differ by terms containing $\beta = v/c$ depending on velocity. This difference is negligible in most cases, and we here assume that this condition is fulfilled.
9. The charge q is here expressed in CGS electrostatic units.
   We must therefore put
   $$q = 4.77 \cdot 10^{-10} \text{ e.s.u.}$$
   $$\left(\frac{q}{m_0}\right) \text{e.s.u.} = c\left(\frac{q}{m_0}\right) \text{e.m.u.} = 1.76 \cdot 10^{7} \cdot 3 \cdot 10^{10} \text{ e.s.u.}$$
   in keeping with (III-23) and the numerical values indicated on p. 83.
10. This integral calculated at the instants τ and $\tau + d\tau$ has approximately the same value. The difference that one would obtain disappears at the limit when $dr \to 0$.
11. Cf. e.g. R. BECKER [1], p. 72.
12. Such is the case of the orbital velocity of the earth (30 km/sec).
13. As a matter of fact

$$(1) \qquad \frac{d}{dt}\int P_n dS = \int \frac{\partial P_n}{\partial t} dS + \int \frac{P_n}{dt}\left[(dS)_{t+dt} - (dS)_t\right].$$

Let us apply Green's theorem to the volume delimited by the closed surface $d\Sigma$ generated by the motion of dS. $d\Sigma$ is made up of $(dS)_{t+dt}$, $(dS)_t$ and of the sum of the areas $d\sigma = \mathbf{dl} \wedge \mathbf{u}\, dt$ described by the contour dl limiting dS. One has

$$(2) \qquad \int P_n(dS)_{t+dt} - \int P_n(dS)_t + \int \mathbf{P} \cdot [\mathbf{dl} \wedge \mathbf{u}\, dt] = \int \operatorname{div} \mathbf{P}\, d\mathscr{V}$$

that is

$$(3) \qquad \int P_n(dS)_{t+dt} - \int P_n(dS) - \int \operatorname{div} \mathbf{P}\, d\mathscr{V} + \int [\mathbf{P} \wedge \mathbf{u}]\, \mathbf{dl}\, dt$$
$$= dt \int \mathbf{u} \operatorname{div} \mathbf{P}\, dS + dt \int \mathbf{curl}\, [\mathbf{P} \wedge \mathbf{u}]\, dS$$

when we apply Stokes' theorem to the last term. By substituting (3) in (1) we are lead immediately to (IV-90).

PART TWO

SPECIAL RELATIVITY

THE PRINCIPLE OF RELATIVITY

A. THE PRINCIPLE OF RELATIVITY BEFORE EINSTEIN

1. *The principle of Relativity in classical mechanics*

Newtonian dynamics postulates the existence of an absolute space "independent of the bodies contained in it", and of a universal time that "flows uniformly".

The hypothesis of a universal time means that the motion of a spatial reference system does not affect the course of events taking place within that system. Analytically, this hypothesis is expressed by the transformations of coordinates necessarily having the form:

$$x_p = x'_p(x_q, t) \qquad t' = t \qquad (p, q = 1, 2, 3).$$

The notion of absolute space is more confused. If we confine ourselves to the principles of classical kinematics, it is possible to relate the motion of a solid body to the reference system constituted by another solid body, and that operation is completely reciprocal. The equations of the relative motion are the same no matter which of the two solid bodies is arbitrarily chosen as a reference system. Classical kinematics thus introduces a wholly reciprocal relationship into the description of the motions and allows the existence of a very broad sort of Relativity principle. In this case the notion of absolute space merely expresses a mental need. This absolute space is the motionless framework in which the displacement of all the solid bodies takes place, but it cannot in fact be defined by any special reference system. Its importance is essentially metaphysical or psychological, but in kinematics it is of no real importance.

It is dynamics which, by limiting the principle of Relativity, begins to bestow a physical meaning on the notion of absolute space. In this case the concept can be connected with the possibility of defining at least one category of special reference systems: those in which a free point-mass moves in a straight line with uniform motion. Such a system is said to be an *inertial system*, and the possibility of determining such a system constitutes the principle of inertia. On that basis, one can, along with Euler, make the notion of absolute space a guarantee of the validity of the principle of inertia.

Physically, there exist only approximate inertial systems. The walls of a laboratory form an inertial system for the phenomena taking place in that laboratory. The directions that connect the center of the earth to three fixed stars constitute an inertial system, referred to as "the Galilean system", for the ensemble of terrestrial phenomena. It was through consideration of the special case of inertial systems that the idea was advanced, long before Newton, of an ideal reference system adapted to a universal and rigorous verification of the principle of inertia.

Starting with one inertial system, one can define an infinite number of other inertial systems by means of a uniform translation expressed by the *Galilean transformation:*

$$\text{(V-1)} \qquad \boxed{x' = x - vt, \qquad t' = t.}$$

This transformation makes it possible to narrow down the relativity allowed in classical kinematics by limiting it to Galilean systems. By means of a Galilean transformation one can always choose a reference system such that a body moving with a uniform rectilinear motion behaves as if it were at rest. "It can be said that one and the same thing moves or does not move, depending on the different ways one determines its position", Descartes had already declared.

The principle of inertia thus makes it possible in Mechanics to establish an equivalence among the special frames of reference or, what amounts to the same thing, to define a relativity of velocity. But the very notions of a special frame of reference and of uniform motion are related to a particular case that has meaning only within the general framework of dynamics. The statement of the principle of inertia does not take on its full meaning until it is assimilated into Newtonian dynamics, which the principle implicitly postulates.

Newtonian dynamics is based on the definition of force

$$\text{(V-2)} \qquad \mathbf{f} = m\frac{d\mathbf{v}}{dt} = m\gamma$$

which is a reduction of the more general law

$$\text{(V-3)} \qquad \mathbf{f} = \frac{d\mathbf{p}}{dt} = \frac{d}{dt}(m\mathbf{v})$$

when the mass m is an intrinsic property of the material point.

For Newton, the appearance of accelerations always indicates the existence of absolute motion – absolute motion of matter, where *real* forces are concerned; absolute motion of the reference system, where so-called *fictitious* forces, such as inertial forces or those of Coriolis, are concerned.

The qualifier "fictitious" simply indicates that these forces could be eliminated through a judicious choice of the reference system and that a change of systems would continue to assure the validity of the principle of inertia. In reality, however, the application of the principle of inertia in dynamics is not quite so simple. If it is not verified, that may result from a defective choice of the reference system, but it may also arise from the existence of real forces that we are unable to discern. Newton's mechanics assumes that it is always possible to define a free point mass or, what amounts to the same thing, that it is always possible to distinguish real forces from imaginary forces. The criticism of the General Relativity will show that it is not at all the case.

If an inertial system is determined, it is possible to use that system as the starting point for establishing an infinite number of other systems, according to the principle of Relativity. The basic law of dynamics remains invariant within a Galilean transformation and thus preserves the same form in all inertial systems.

2. *The principle of Relativity in electrodynamics*

The principle of Special Relativity pertains to the realm of Mechanics. One may legitimately ask whether it is also valid in the other domains of physics.

In optics the question is put in the following manner. Up until the advent of Special Relativity, it seemed obvious that an electromagnetic wave, which is isotropic in a given reference system, cannot preserve this isotropy in a new reference system moving in a rectilinear and uniform manner with a velocity v with respect to the first system. Indeed, according to the law of the addition of velocities assumed in Classical Mechanics, if the velocity of light is c in the first reference system, it will be $c + v$ in the second. This property thus seems to contravene the principle of Relativity resulting from classical dynamics. By means of an optical experiment, it should be possible to determine the motion of the combined source-measuring apparatus with respect to a supposedly motionless ether. In the electromagnetic theory this possibility arises immediately from the fact that the Maxwell equations are not invariant in a Galilean transformation.

3. *Experimental possibilities of detecting absolute motion by optical means*

If v is the absolute velocity of a motion that is uniform with respect to the ether, we will say that it produces a first-order effect if it depends on the relation $\beta = v/c$, a second-order effect if it involves only terms in β^2. Now, by optical methods, it is quite easy to reveal an effect of the first order, and with much more difficulty – though only in certain areas – to detect effects of the second order. The earth covers about thirty kilometers per second as it turns around the sun, and during that time (which is very short in comparison to a complete revolution) its motion is approximately rectilinear and uniform. Here $\beta = 1/1000$ and one is justified in hoping that, by optical means, an "ether-drift" may be detected, arising in the ecliptic plane and, by this very fact, may reveal an absolute reference system.

One is obliged to point out, however, that the best known experiments on the optics of moving bodies give no evidence of an ether-drift arising from first-order effects.

a) *The measurement of the "round-trip" time of light rays.* It would seem that the "ether-drift" could be easily detected by measuring the velocity of light $c \pm v$ along a rigid scale. But no experimental procedure proves usable. All of them assume a synchronization of clocks along the whole "route" and amount simply to measuring the "round-trip" time of light.[1] But such a measurement automatically eliminates all first-order effects.[2]

b) *The Doppler effect and aberration.* The best-known effects of the motion of sources are *aberration phenomena* and the *Doppler effect.*

The latter was discovered by DOPPLER in 1842 and consists of a variation in the frequency of light-waves emitted from a source in motion.[3]

Aberration phenomena, pointed out by BRADLEY as early as 1728, are explained by a modification in the direction of light-rays as a result of the relative source-observer motion.[4] Experimentally, for example, one notes a variation in the position of the image of a fixed star observed through a telescope. During the course of a year the motion of the earth does indeed modify the position of this image, which describes a small ellipse.

The first-order consequences produced by the Doppler effect and by aberration concern the relative motion of source and measuring apparatus. The absolute motion with respect to the ether can produce only second-order effects, and up until very recent times, this method left such effects beyond the realm of experience.[5] In the celebrated experiments based on a prefiguring of the Doppler effect and of aberration, Römer (1676) and then Bradley (1728)[6] both determined the velocity of light by means of measurements made exclusively on a simple unidirectional trajectory. It is precisely the disappearance of ether-drift effects of the first order that allows us to consider these experiments as effective measurements of c, without there being any chance of influence from a possible "ether-drift".

4. *First-order effects. The hypothesis of a partial dragging of light by transparent bodies*

In order to reveal an effect of the first order produced by absolute motion, one must have recourse to experiments that involve a possible drag of the ether and on the light-waves propagated therein, in a transparent medium.[7] Whether that effect be zero or total, it gives rise to first-order effects attributable to "ether-drift". So one should be able to determine the existence of an absolute motion with the help of experiments devised to study the propagation of light inside a transparent medium. Such experiments have been numerous: Arago, then later on Fizeau, Hoek, Mascart, Michelson, and finally Zeeman (1914) all consistently came up with negative results.[8]

The first experiment of this type (Arago's in 1818) used the refraction of light-rays through a system of lenses. In that same year, FRESNEL[8a] explained the negative result of Arago's experiment[9] by postulating a partial drag on the ether by transparent media. The drag coefficient must have the value:

(V-4) $$\boxed{\alpha = 1 - \frac{1}{n^2}}$$ n = index of medium

Fresnel's formula was deduced from theoretical considerations arising from his hypothesis about the mechanistic structure of the ether. These hypotheses were not very convincing; nevertheless, they explained the negative result of Arago's experiment and they seemed to be confirmed in a very satisfactory way by the experiments, imagined by Fizeau, of the drag on light-rays by a stream of water (1851).

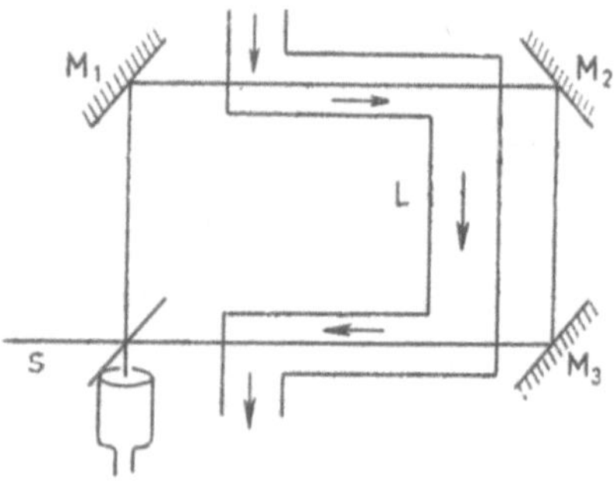

Fig. 15. Fizeau's experiment.

While observing the displacement of the fringes, FIZEAU[10] actually noted a partial drag on the light-waves propagated in a rather rapid watercurrent inside a tube L (Figure 15). The displacement had the value predicted by Fresnel's formula.

In a still more rigorous way, ZEEMAN confirmed Fresnel's predictions by measuring the velocity of light in a quartz measuring-rod in rapid motion.[11] Zeeman's experiments, moreover, were sufficiently accurate to detect any dispersion effect of the medium. (Cf. p. 206.)

In our discussion of HOEK's experiment[12], which is equivalent to Fizeau's, we are going to show that *the partial drag on light-waves resulting from Fresnel's formula must automatically compensate for any first-order appearance of an ether-drift effect.*

Let us imagine, in fact, the following experimental arrangement (Figure 16): the light emitted from S strikes a semi-reflecting mirror M at an angle of 45°. The incident beam is split in two and each of the resulting beams travels in an opposite direction along the circuit M_1 M_2 M_3. Since they are reflected by mirrors at an angle of 45°, they interfere with each other at F.

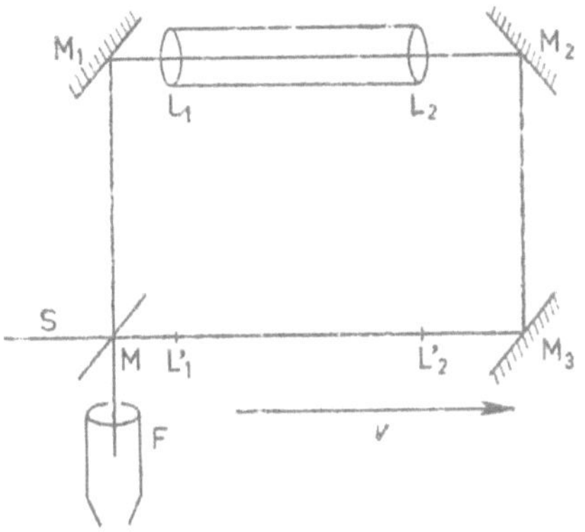

Fig. 16. Hoek's experiment.

The whole system is carried along by the earth's motion at a velocity v. If at L_1 L_2 we place a tube full of water in which the light-rays are propagated, we thus interpose between the two separated parts of the beam an optical path difference, the value of which depends upon the possible drag on the ether by the motion of the L_1 L_2 medium, which is attached to the earth.

As a matter of fact, if the ether in L_1 L_2 is not "dragged" at all by the motion of the water, the light-waves that are propagated in that ether will still have a constant velocity c_1 with respect to interstellar ether and a velocity $c_1 \pm v$ with respect to the earth. On the other hand, if the ether contained in L_1 L_2 is entirely carried along by the earth's motion, the velocity of the light becomes $c_1 \pm v$ with respect to the motionless interstellar ether but will have the value c_1 with respect to the earth.

In the intermediate case of a partial drag on the ether, the velocity of light with respect to motionless interstellar ether lies somewhere between c_1 and $c_1 \pm v$. With respect to the earth, the velocity will lie somewhere between $c_1 + v$ and c_1. In the case of partial dragging, the velocity of light can thus be represented by

$c_1 + \varphi$ with respect to motionless ether

$c_1 + \varphi - v$ with respect to the earth, putting

$$\varphi = \alpha v \quad \text{with} \quad 0 < \alpha < 1.$$

α is the drag coefficient of the ether; if it equals 1, the drag is total.

With the apparatus we are here considering, part (1) of the beam follows the optical path $MM_1 M_2 M_3 M$. The time taken by (1) to travel the length $L_1 L_2 = l$ through the water, and then the equal length $L_1' L_2'$ in the air, is:

$$t_1 = \frac{l}{c_1 + \varphi - v} + \frac{l}{c + v}. \tag{V-5}$$

The time taken by part (2) of the beam, travelling along the optical path $MM_3 M_2 M_1 M$, to cover the trajectory $L_1' L_2'$ in the air and then $L_2 L_1$ through the water, is:

$$t_2 = \frac{l}{c - v} + \frac{l}{c_1 - \varphi + v}. \tag{V-6}$$

The difference between these travel-times is thus

$$\begin{aligned} \Delta t &= t_1 - t_2 = \\ &= l\left\{\frac{1}{c_1 + \varphi - v} + \frac{1}{c + v} - \frac{1}{c - v} - \frac{1}{c_1 - \varphi + v}\right\} \\ &= 2l\left\{\frac{-\varphi + v}{c_1^2 - (\varphi - v)^2} - \frac{v}{c^2 - v^2}\right\} = \\ &= \frac{2l(v\varphi^2 - v^2\varphi - c^2\varphi + c^2 v - vc_1^2)}{(c^2 - v^2)[c_1^2 - (\varphi - v)^2]} \\ &= \frac{2l\left(\frac{\varphi^2}{c^2} - \beta\frac{\varphi}{c} - \frac{\varphi}{v} + 1 - \frac{1}{n^2}\right)}{v(1 - \beta^2)\left[\left(\frac{c_1}{v}\right)^2 - \left(1 - \frac{\varphi}{v}\right)^2\right]} \end{aligned} \tag{V-7}$$

where we define the index n of the medium by the usual formula

$$n = \frac{c}{c_1}. \tag{V-8}$$

We will then have approximately

$$\Delta t \simeq \frac{2l}{v}\left(-\frac{\varphi}{v} - \frac{1}{n^2} + 1\right)n^2\beta^2. \tag{V-9}$$

that is, a phase difference[13]

$$\Delta\psi = \nu\Delta t = \frac{c\,2l}{\lambda\, v}\, n^2\beta^2\left(1 - \frac{1}{n^2} - \frac{\varphi}{v}\right) = $$
$$= \frac{2ln^2}{\lambda}\left(1 - \frac{1}{n^2} - \frac{\varphi}{v}\right)\beta.$$

If we assume with FRESNEL[14] that the ether is partially carried along by the motion of transparent bodies and in such a way that

(V-10) $$\varphi = \mathrm{v}\left(1 - \frac{1}{n^2}\right)$$

that is, in such a way that the drag coefficient has the value

(V-4) $$\alpha = 1 - \frac{1}{n^2}$$

we have, according to (V-9), $\Delta t = 0$. One is thus completely assured of the impossibility, at least by means of any experiment of the preceding kind, of detecting any first-order effects (but *only* first-order effects) of ether drag.

FRESNEL's formula[14] and the countless experiments that have verified it destroy in advance all possible hope of observing any first-order effect that may be attributed to ether drift; the drag is of such a nature that it automatically corrects any such effect.

It was not until 1874 that MASCART[15], VELTMANN[16], and then POTIER[17] clearly established the general validity of this conclusion – that is the certain and preordained failure of any attempt to detect an absolute motion with respect to the ether by optical methods. Of course, these conclusions rest solely on the possible detection of first-order effects, but, even that early, Mascart was suggesting that in optics, as in dynamics, it was impossible to distinguish any special Galilean system by any experimental means whatsoever.

5. *Lorentz' theory of electrons and first-order effects. The hypothesis of a motionless ether*

The negative results obtained from the study of light within transparent media may be explained by the hypothesis of a partial drag on the ether, with the drag coefficient of the ether being determined by Fresnel's formula.

The development of Maxwell's theory leaves these results intact. In spite of Hertz' attempt to extend Stokes' hypothesis (of a total carrying along of the ether by matter in motion) to electromagnetic theory, experiment[18] makes it clear that we must go back to the concept of a partial carrying-along of the ether by matter in motion. The value of the drag coefficient is, naturally, retained.

However, H. A. LORENTZ' theory of electrons makes possible a microscopic interpretation of Maxwell's theory and succeeds in predicting the disappearance of any ether-drift effect of the first order by assuming that this ether is strictly motionless.[19] And, as a matter of fact, the deduction of Maxwell's equations from Lorentz' theory is valid, not only for the electrodynamics of bodies at rest, but also for the electrodynamics of bodies in motion, provided we confine ourselves to velocities sufficiently low to permit the neglect of the second-order terms involving $\beta^2 = v^2/c^2$. Now Maxwell's equations were set up for media at rest, in other words, for an observer who is motionless with respect to matter and consequently in uniform rectilinear motion with respect to the ether. On the other hand, Lorentz' microscopic equations presuppose a strictly motionless ether, with the observer fixed with respect to that ether. Matter (electrons) is in motion with respect to both ether and observer. The equivalence with Maxwell's theory means that it is impossible, within the first order, to detect by means of any electromagnetic experiment whatever, any uniform rectilinear motion with respect to the ether. For example, the aberration experiments using the interposing of a refractive medium (water-filled tube) must inevitably be negative – and that without any need for assuming a partial carrying along of the ether in the vicinity of matter in motion.[20] Fizeau's experiment thus brings out the following phenomenon: although the ether is strictly motionless, there does exist a partial drag on the electromagnetic waves propagated within matter in motion[21], and the value of the coefficient α is retained.[22]

6. *Second-order effects*

In the wake of Lorentz' theory, the hope of detecting any "ether-drift" was reduced to the possibility of detecting a second-order effect.[23] Such was the purpose of the experiments performed first by MICHELSON (1881)[24] and then by MICHELSON and MORLEY[25].

The Michelson experiment. An interferometer is constructed in the following manner: the light emitted from a source S is divided into two beams by means of a semi-reflecting mirror M. Beam (1), which is transmitted through M, is reflected by the mirror M_1 and then by M. So it describes the trajectory SMM_1MF. Beam (2), reflected by M, is reflected a second time by M_2 and, on its return-trip, passes through M;

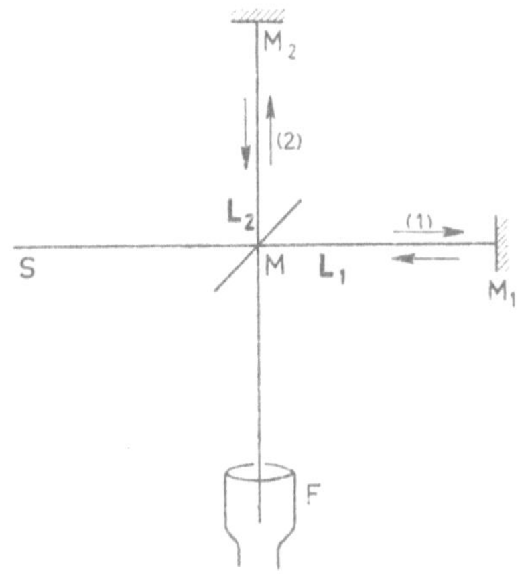

Fig. 18. Michelson's experiment.

its trajectory is SMM_2MF. The two beams interfere at F, and the fringes are observed by means of a telescope. The displacement of the fringes will indicate the difference in travel-time of the two beams.[26]

In this experiment the apparatus is fastened to a fixed base that floats on mercury, so that it can be easily oriented. We will make the calculations by relating them to a reference system bound to the ether – a system in which light always travels at the velocity c.

1) Let us first turn the apparatus in such a way that the arm L_1, of length l_1, is oriented in the direction of the earth's movement with respect to the ether.

a) Let t_A be the time it takes light to cover the forward trajectory, and t_B the time required for the backward trajectory. In the system bound to the ether we then have

$$t_A = \frac{l_1 + vt_A}{c}, \qquad t_B = \frac{l_1 - vt_B}{c}$$

that is

$$t_A(1-\beta)=\frac{l_1}{c}, \qquad t_B(1+\beta)=\frac{l_1}{c}$$

or

$$t_A=\frac{l_1}{c(1-\beta)}=\frac{l_1}{c-v}, \qquad t_B=\frac{l_1}{c(1+\beta)}=\frac{l_1}{c+v}$$

The time t_1 required for light to cover the trajectory MM_1 is then [26a]

(V-11) $$t_1=t_A+t_B=\frac{l_1}{c+v}+\frac{l_1}{c-v}=\frac{2l_1}{c}\frac{1}{1-\beta^2}.$$

b) Since the apparatus is carried along by the earth's motion, beam (2) travels along a path MM_2M', M' being the position of the mirror M at the end of the time t_2, which it takes light to make the roundtrip to and from M_2. One then has

$$MM'=vt_2$$

and

$$MM_2+M_2M'=2\sqrt{l_2^2+\left(\frac{vt_2}{2}\right)^2}=\sqrt{4l_2^2+v^2t_2^2}.$$

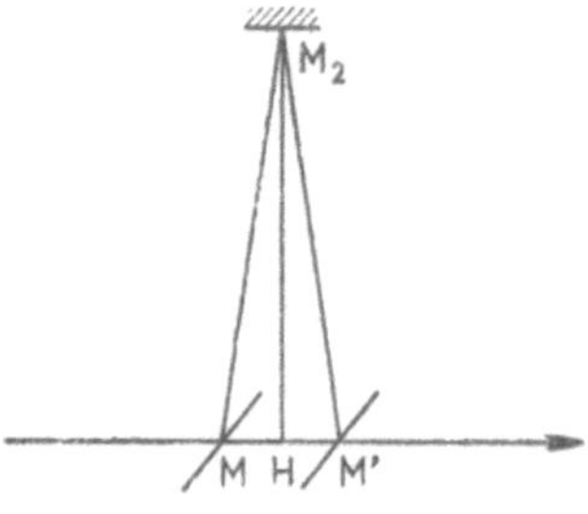

Fig. 19.

One will thus have

(V-12) $$t_2=\frac{MM_2+M_2M'}{c}=\sqrt{\frac{4l_2^2}{c^2}+\beta^2t_2^2}$$

since the velocity of light in the perpendicular direction is, to all intents and purposes, the same as its velocity along M_2M or M_2M'. From

(V–12) we obtain[26b]:

$$t_2^2(1-\beta^2)=\frac{4l_2^2}{c^2} \tag{V-13}$$

or

$$t_2=\frac{2l_2}{c}\frac{1}{\sqrt{1-\beta^2}}\,. \tag{V-14}$$

The difference in travel-time between the two beams (1) and (2) is thus:

$$\Delta_1 t=t_2-t_1=\frac{2}{c}\left(\frac{l_1}{\sqrt{1-\beta^2}}-\frac{l_1}{1-\beta^2}\right). \tag{V-15}$$

2) Now let us turn the apparatus by $\pi/2$ so that the respective roles of the interferometer-arms L_1 and L_2 are reversed.[27] L_2 then lies along the direction of the motion, and the traveltimes are, respectively:

$$t_2'=\frac{2l_2}{c}\frac{1}{1-\beta^2}, \qquad t_1'=\frac{2l_1}{c}\frac{1}{\sqrt{1-\beta^2}} \tag{V-16}$$

and give rise to a time-lag:

$$\Delta_2 t=t_2'-t_1'=\frac{2}{c}\left(\frac{l_2}{1-\beta^2}-\frac{l_1}{\sqrt{1-\beta^2}}\right). \tag{V-17}$$

And the displacement of the fringes produced by this rotation corresponds to the time-lag:

$$\Delta t=\Delta_2 t-\Delta_1 t=\frac{2}{c}(l_1+l_2)\left(\frac{1}{1-\beta^2}-\frac{1}{\sqrt{1-\beta^2}}\right). \tag{V-18}$$

It is thus approximately equal to:

$$\Delta t\simeq\frac{2}{c}(l_1+l_2)\left[(1+\beta^2)-\left(1+\frac{\beta^2}{2}\right)\right]=\frac{(l_1+l_2)}{c}\beta^2\,. \tag{V-19}$$

Now a displacement equal to the width of one fringe separation corresponds to the period $\tau=\lambda/c$ of light. Such a displacement will then be observed for

$$\Delta t\simeq\tau\,, \qquad l_1+l_2\simeq\frac{\lambda}{\beta^2}\,. \tag{V-20}$$

If $\lambda=5\cdot10^{-5}$ cm, $v=30$ km/sec, that is $\beta\simeq10^{-4}$ one should be able to achieve perceptible results by setting up a trajectory

(V-21) $$l_1 + l_2 = 5 \cdot 10^3 \text{ cm} = 50 \text{ m}.$$

By means of a system of multiple reflections, it is possible to achieve a trajectory of that order of magnitude. One can even improve upon the precision of these measurements and contrive experiments that would make it possible to detect an ether-drift having a velocity of 1.5 km/sec [KENNEDY (1926)[28], ILLINGSWORTH (1927)[29], PICCARD and STAHEL (1928)[30], JOOS (1930)[31]].

The results of the Michelson-Morley experiments were entirely negative, and the same is true of all subsequent experiments duplicating Michelson's, but with even greater precision.[32] A series of experiments of a different sort performed by TROUTON and NOBLE (1903)[33], TROUTON and RANKINE (1908)[34] and later taken up by CHASE (1927)[35] and by TOMASHEK[36] have all confirmed these negative results with a precision of four to five km/sec.

So the hypothesis of a motionless ether, which is the basis of the Lorentz theory, seems justified for first-order phenomena but seems inadequate for second-order phenomena.

By assuming the total carrying along of ether by matter in motion (Hertz), or even by assuming a modification of the velocity of light by sources in motion (RITZ)[37], one can explain the negative result of the Michelson experiment. Nevertheless, either one or the other of these hypotheses is unacceptable. The first is already difficult to admit simply on theoretical grounds and is, moreover, in contradiction with the existence of aberration phenomena and the Fizeau effect. But the second hypothesis contradicts the results deduced from the study of double stars and from Tomashek's experiment. The velocity of light seems, on the contrary, quite independent of the velocity of the bodies emitting the light (DE SITTER, 1912)[38] and also quite independent of the bodies found in its vicinity (LODGE, 1892)[39].

7. *The Fitzgerald-Lorentz hypothesis*

FITZGERALD[40] and LORENTZ[41] succeeded in saving the hypothesis of a strictly motionless ether by admitting a supplementary effect:

All bodies in uniform rectilinear motion undergo a contraction in the ratio of $\sqrt{1-\beta^2}$ along the direction of their motion.

This postulate completely explains the negative result of the Michelson experiment. In fact, one must then replace l_1 by $l_1 \sqrt{1-\beta^2}$ in calculating t. One then has:

$$t_1 = \frac{2l_1\sqrt{1-\beta^2}}{c\ 1-\beta^2} = \frac{l_1}{l_2} t_2 . \tag{V-22}$$

Whence

$$\Delta_1 t = \frac{2}{c}(l_2 - l_1) \frac{1}{\sqrt{1-\beta^2}} = \Delta_2 t . \tag{V-23}$$

The hypothesis of a contraction in the ratio of $\sqrt{1-\beta^2}$, affecting all measuring apparatus to the same degree, automatically assures, not only in the first order, but in *any* order of approximation, the negative result of any experiment designed to reveal ether drift.

One might think that this contraction, in its turn, is a measurable phenomenon and that, thus, experiments designed to detect this contraction might be imagined. For example, the refraction index of a refracting solid ought to be modified by motion. But experiments that were attempted towards that end by RAYLEIGH[42] and by BRACE[43] were, in their turn, negative. And the same fate was in store for the experiments of TROUTON and RANKINE[44] (electrical resistance of a wire conductor) and of WOOD, TOMLINSON and ESSEX[45] (measurement of the vibration frequency of a quartz measuring rod).

So one was forced to think that the effects of the contraction, just like those of the ether drift, were likewise hidden by another effect of motion – namely, an increase in rest-mass.

And indeed, a simultaneous variation in lengths and masses such that:

$$l = l_0 \sqrt{1-\beta^2} \tag{V-24}$$

$$m = \frac{m_0}{\sqrt{1-\beta^2}} \tag{V-25}$$

gives rise to the impossibility of detecting the influence of any uniform rectilinear motion on any optical phenomenon.

Now formula (V-25), which was to result quite naturally from the future Einsteinian dynamics, also arises from the hypothesis of the Lorentz contraction applied to the electron itself. And one may, moreover, wonder if condition (V-24), which is essential to explain the negative result of Michelson's experiment, is also adequate to establish the independence of all electromagnetic phenomena from the uniform rectilinear motion of the reference system with which they are connected.

Lorentz and, independently, Poincaré showed that it was further necessary to define a time-measurement appropriate to each Galilean system. If a given reference system is moving in a uniform and rectilinear manner along Ox with respect to a second system, one must pass from one system to the other by means of the transformation formulas:

(V-26) $$x' = \frac{x - vt}{\sqrt{1 - \beta^2}} \qquad t' = \frac{t - \frac{\beta}{c}x}{\sqrt{1 - \beta^2}}$$

if the Maxwell equations are to be entirely independent of the Galilean system used. Equations (V-26) define a transformation of coordinates referred to as the *Lorentz transformation*, which will also arise from the hypotheses of Einstein that we are about to examine.

If we accept Lorentz' theory of electrons, along with equation (V-25), which is one of its consequences, formulas (V-26) insure the invariance of Maxwell's equations in a change of Galilean systems[46], and consequently, these formulas impose the impossibility of detecting absolute motion with respect to the ether by any electromagnetic experiment whatsoever. So, the ultimate result of Lorentz' theory is to sanction the existence of a strictly motionless ether while at the same time, dooming to failure any electromagnetic experiment designed to detect the existence of that ether.

B. THE PRINCIPLE OF SPECIAL RELATIVITY

8. *Einstein's basic postulate*

Classical dynamics was based on the following proposition:

I. *There exists an equivalence of all Galilean systems for the description of motion* (cf. footnote 46).

If we then admit *the validity of the Galilean transformation*, one of the consequences of the foregoing proposition is *the law of the addition of velocities in classical mechanics*. This law has the following possible corollary:

I′. *The velocity of light varies when it is related to two different Galilean systems.*

Yet the experimental results obtained in the field of classical electrodynamics lead to the following conclusion[47]:

II. *In vacuum, light propagates in an isotropic fashion, no matter what the motion of its source may be. Its velocity is equal to an absolute constant c in all Galilean systems.*

The earlier ether-theories sought to eliminate the incompatibility between propositions I′ and II by decomposing the velocity of light, with respect to the universal ether, into two terms:

The velocity of the ether contained in transparent bodies or dielectrics with respect to the universal ether;

The velocity of light with respect to the ether contained in transparent bodies.

The first term entailed a definition of an appropriate drag coefficient, while the second term was a constant. Thus, the combination of the two terms insured the equivalence of the special reference systems, but only of the first order.

On the other hand, the hypothesis of the contraction of lengths and dilatation of durations which was proposed by Lorentz involved equivalence in all orders, but only thanks to a sort of "appropriate distortion" that entered into the measurements made on systems in motion. To transpose an expression of Born's – the equivalence resulted from a sort of optical illusion.

In fact, there is no conflict between propositions I and II, but only between statements I′ and II. Indeed, the law of the addition of velocities in Classical Mechanics presupposes the validity of the Galilean transformation, which insures the invariance of the basic equation of dynamics in all Galilean systems. On the other hand, the isotropism and constancy of the velocity of light presuppose the validity of the Lorentz transformation, which insures the invariance of Maxwell's equations in all Galilean reference systems.

So we are ultimately forced to choose between these two types of transformations and to admit

1) either the continued validity of Newton's equations and of the Galilean transformation insuring their invariance. In that case, we must, in electrodynamics, assume the existence of new phenomena which find expression in the Lorentz-Poincaré equations (V-24) and (V-25), and which insure, by a sort of compensatory phenomenon, the invariance of

the Maxwell equations and the permanently elusive character of ether.

2) or the universal validity of the Lorentz-Poincaré equations. The Lorentz transformation entails the invariance of the Maxwell equations, but it requires the construction of a kinematics and dynamics that will be in harmony with it.

In this conflict between kinematics and optics, it is optics that has furnished the model to which Relativistic Mechanics conforms.[48] This choice is clearly brought out in *the postulates of Special Relativity*, which are none other than postulates I and II already formally incorporated in the theories of Lorentz and Poincaré:

I. *There exists a special equivalence of all Galilean systems. This equivalence is valid not only in dynamics, but throughout all of physics.*

II. *In vacuum light propagates isotropically. Its velocity is a universal constant c.*

The propositions follow, of course, from the formulas of Lorentz and Poincaré. But the originality of EINSTEIN's theory lies in the demonstration that these propositions are connected with a more precise analysis of the notions of space and time, and that they imply the general validity of Lorentz' formulas, which thus no longer express a mere appearance, but express instead the most fundamental properties of space-time.[49]

9. *Critique of the concept of simultaneity*

Pre-Einsteinian physics considers the notion of simultaneity as a self-evident truth. Now a truly complete appreciation of such simultaneity requires the presence in two different spots A and B, separated by a distance l, of synchronized clocks. To achieve or verify this synchronization one must of necessity use a signal. Electromagnetic signals are the most rapid, and the corrections imposed by their propagation-time are thus minimal.

1. If two points A and B both belong to one and the same system, assumed to be at rest, one cannot determine absolute simultaneities, but one can define a relative simultaneity in the following manner:

We say that two events taking place at A and B are simultaneous if the signals produced at A and B reach an observer M, situated midway between A and B, at the same time.[50] When the system AB moves in a uniform rectilinear manner, the same criterion remains valid, and the observers in this system disregard its motion. Now that is exactly what happens in an exchange of light-signals between observers within a given system, since the absolute motion of the system with respect to the ether cannot be observed.

2. Adhering to the convention just established, which determines simultaneity within any given Galilean system, it is easy to show that this simultaneity is not preserved in any other Galilean system.

Let us consider once more the classic example, thought up by Einstein, of a railway-track along which a train $A'B'$ is moving with a velocity v.

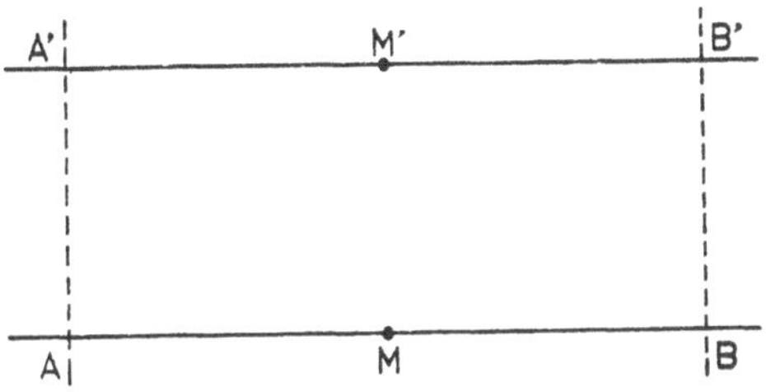

Fig. 20.

Let M' be the midpoint of the train $A'B'$, corresponding to the midpoint M of AB, when the signals are flashed (simultaneously for M) from A and B. It is clear that the observer M' moving towards B will receive the signal flashed from B sooner than that flashed from A. But, since the coincidence of the arrival of the signals at the midpoint of $A'B'$ is the only possible criterion of simultaneity, two events that are simultaneous for M will not be simultaneous for M'. Indeed, M or M' may, with equal right, affirm that his respective system is at rest, since no experiment can detect the relative motion of two different Galilean systems. Therefore, there is no such thing as absolute simultaneity.[51] This conclusion undercuts the hypothesis of universal time and the validity of the Galilean transformation.

10. *The Lorentz transformation*

By assuming that the speed of light is isotropic and equal to c in all reference systems (Postulate II), what we obtain is precisely the Lorentz transformation.

Let us examine a little more closely what is, according to Einstein, the situation with regard to the propagation of an electromagnetic perturbation in two Galilean reference systems, $S\,(OXYZ)$ and $S'\,(O'X'Y'Z')$.

Since the velocity of light is equal to c in both systems, the expressions

(V-27) $$ds^2 = -dx^2 - dy^2 - dz^2 + c^2dt^2$$

(V-28) $$ds'^2 = -dx'^2 - dy'^2 - dz'^2 + c^2dt'^2$$

are positive when we are dealing with the motion of a point-mass of velocity $v < c$. The expressions are zero when applied to the propagation of lightwaves. One will then have

(V-29) $$ds'^2 = f(xyzt)\,ds^2 .$$

It can be shown[52] that, on the basis of the reciprocity of the two reference systems, one simply has:

(V-30) $$f(xyzt) = k = 1 .$$

We are thus brought back to the problem of determining the change of coordinates that will satisfy

(V-31) $$ds'^2 \equiv ds^2 ,$$

i.e. those changes that transform the four-dimensional Euclidean space into itself. Now the way to solve that problem is known: *the orthogonal linear transformations of four-dimensional space* are the solutions.[53]

To simplify, let us here choose coordinate systems such that the translational velocity v of S' $(o'x'y'z')$ with respect to S $(oxyz)$ and the axes ox and $o'x'$ are parallel and oriented in the same sense. By virtue of symmetry, the linear orthogonal transformation we are seeking will have the following form:

(V-32) $$x' = g(v)(x - vt), \qquad y' = y, \qquad z' = z, \\ t' = h(v)t - l(v)x .$$

The identity (V–31) entails the following relations:

(V-33) $$\left(\frac{\partial x'}{\partial x}\right)^2 - c^2\left(\frac{\partial t'}{\partial x}\right)^2 = 1$$

(V-34) $$\frac{1}{c^2}\left(\frac{\partial x'}{\partial t}\right)^2 - \left(\frac{\partial t'}{\partial t}\right)^2 = -1$$

(V-35) $$\frac{\partial x'}{\partial x}\frac{\partial x'}{\partial t} - c^2\frac{\partial t'}{\partial x}\frac{\partial t'}{\partial t} = 0 .$$

By substituting the values of the partial derivatives obtained from (V-32), one will simply have:

$$g(v) = h(v) = \frac{\pm 1}{\sqrt{1-\beta^2}}, \tag{V-36}$$

$$l(v) = g(v)\frac{v}{c^2}, \tag{V-37}$$

expressions in which we must select the +-sign so that the two systems of axes coincide at the initial instant.

The transformation formulas – the very ones Lorentz had deduced from very different hypotheses – are thus as follows:

$$\text{(V-38)} \quad \begin{cases} x' = \dfrac{x - vt}{\sqrt{1-\beta^2}} & (1) \\ y' = y & (2) \\ z' = z & (3) \\ t' = \dfrac{t - \dfrac{\beta}{c}x}{\sqrt{1-\beta^2}} & (4) \end{cases}$$

Inversely, one will have:

$$\text{(V-39)} \quad \begin{cases} x = \dfrac{x' + vt'}{\sqrt{1-\beta^2}} & (1) \\ y = y' & (2) \\ z = z' & (3) \\ t = \dfrac{t' + \dfrac{\beta}{c}x'}{\sqrt{1-\beta^2}} & (4) \end{cases}$$

Formulas (V-38) and (V-39), valid in the special case where one of the axes of the coordinate systems S and S' is parallel to the relative velocity v, characterize the so-called *special Lorentz transformation*. The results obtained by Lorentz and Poincaré are easily derived from the transformations (V-38). We will call *Lorentzian systems* those reference systems which may be deduced from each other by means of the transformations (V-38), (V-39) and their generalizations.

11. *Consequences of the transformation formulas*

1) *The contraction of lengths.* A straight-line segment at rest in S' and parallel to the axis $o'x'$ has, in that system, a length of

$$l'_0 = x'_1 - x'_2 \tag{V-40}$$

In the system S, the length of the line-segment is obtained by determining the coordinates x_1 and x_2, corresponding to x'_1 and x'_2, *at the same instant for an observer in S.* One will then have, according to $(\text{V-38})_1$ and for $\Delta t = t_1 - t_2 = 0$:

$$l = x_1 - x_2 = (x'_1 - x'_2)\sqrt{1-\beta^2} = l'_0\sqrt{1-\beta^2} < l'_0 \tag{V-41}$$

l is the length that an observer in S attributes to the measuring-rod of S'. If l'_0 is the value of the standard unit of length, the observer in the system S finds that the measuring-rod attached to the S' system is too short.

Reciprocally, if l_0 is the length of a measuring-rod in S when its measurement is taken in its own system, then

$$l_0 = x_1 - x_2 \tag{V-42}$$

and an observer in S' attributes to that length at the same instant ($\Delta t' = 0$) the coordinates x'_1 and x'_2 which, in its own system, correspond to x_1 and x_2. According to $(\text{V-39})_1$, the measuring-rod in S' will then have the length:

$$l' = x'_1 - x'_2 = (x_1 - x_2)\sqrt{1-\beta^2} = l_0\sqrt{1-\beta^2} < l_0 . \tag{V-43}$$

In other words, the observer in S' will also find that the standard measuring unit of the system S is too short.

The value resulting from the measurement of the length of a straight-line segment is then maximum in the reference system attached to the segment (its "proper" system). In all other systems, this measurement will show a contraction proportionally involving $\sqrt{1-\beta^2}$. This contraction can no longer be interpreted as an effect of "ether-drift", that is, as the result of a real motion in absolute space. As a matter of fact, it constitutes a *reciprocal phenomenon*: if two observers at rest are supplied with measuring rods equal in every respect, a uniform rectilinear motion of the one with respect to the other leads the observers to notice a reciprocal shortening of their standard measure. The Lorentz contraction is a consequence of relative motion. It arises directly from the laws of transformation and does not necessitate any special hypothesis with regard to the structure of matter.[54]

2. *Dilatation of time-intervals.* In a similar manner, and according to formulas (V-39)$_4$, a time-interval $t'_1 - t'_2$ measured at one and the same point ($\Delta x' = x'_1 - x'_2 = 0$) of S' corresponds in S to a time-interval

$$t_1 - t_2 = \frac{t'_1 - t'_2}{\sqrt{1-\beta^2}} > t'_1 - t'_2 . \tag{V-44}$$

In other words, all phenomena of S' slow down in the proportion $\sqrt{1-\beta^2}$ for an observer in system S.

Reciprocally, a time-interval $t_1 - t_2$ measured at the same point $x = 0$ in S corresponds in S' to the time-interval:

$$t'_1 - t'_2 = \frac{t_1 - t_2}{\sqrt{1-\beta^2}} > t_1 - t_2 \tag{V-45}$$

according to formulas (V-38)$_4$. For an observer in system S', the phenomena of the S system slow down in a reciprocal way[55] in the relation $(1 - \beta^2)^{-\frac{1}{2}}$.

3) *A corollary of the contraction of lengths: the modification of angles and volumes.* Let us consider the straight line OM attached to system S having the origin $O\,(x_0, y_0)$ in the plane xOy and such that $\widehat{Ox, OM} = \alpha$. For every point in OM at the instant $\Delta t = 0$, we will have:

$$y - y_0 = (x - x_0) \tan \alpha . \tag{V-46}$$

In order to determine the slope of this line, an observer in the Lorentzian system S' evaluates the coordinates of $M\,(x, y)$ and $O\,(x_0, y_0)$ at the same instant t' of S' ($\Delta t' = 0$), an instant which naturally corresponds to different times of S.[56] According to the transformation formulas (V-39), we obtain for $\Delta t' = 0$:

$$x - x_0 = \frac{x' - x'_0}{\sqrt{1-\beta^2}}, \qquad y - y_0 = y' - y'_0 . \tag{V-47}$$

In the system S' the slope of the line will thus be:

$$(\text{V-48}) \qquad (\tan \alpha')_{\Delta t'=0} = \frac{y' - y'_0}{x' - x'_0} = \frac{\tan \alpha_{\Delta t=0}}{\sqrt{1-\beta^2}} > (\tan \alpha)_{\Delta t=0} .$$

It is minimal in its own proper system.

In a similar manner, let us seek the angle formed by a straight line OM bound to S and a straight line $O'N$ bound to S' (Figure 21). Assuming that

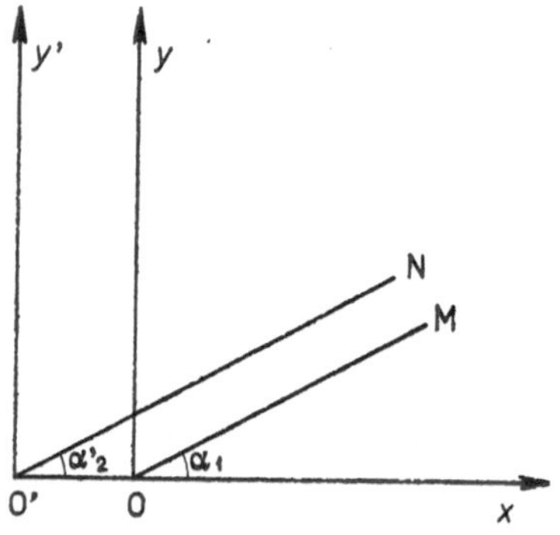

Fig. 21. The modification of angles.

these two lines are situated in the plane xOy, we put:

$$(\text{V-49}) \qquad y_1 = x_1 \tan \alpha_1 \quad \text{for } OM \text{ with respect to } S \quad (\Delta t = 0).$$

$$(\text{V-50}) \qquad y'_2 = x'_2 \tan \alpha'_2 \quad \text{for } O'N \text{ with respect to } S' \quad (\Delta t' = 0).$$

According to (V-48), an observer in the system S' will attribute to OM the slope

$$(\text{V-51}) \qquad (\tan \alpha'_1)_{\Delta t'=0} = \frac{(\tan \alpha_1)_{\Delta t=0}}{\sqrt{1-\beta^2}} .$$

If the slopes of the lines in their own respective systems are equal, that is, if $(\alpha'_2)_{\Delta t'=0} = (\alpha_1)_{\Delta t=0}$, we shall have:

$$(\text{V-52}) \qquad \begin{aligned} \frac{y'_2}{x'_2} &= (\tan \alpha'_2)_{\Delta t'=0} = (\tan \alpha_1)_{\Delta t=0} < \frac{\tan \alpha_{1\,\Delta t=0}}{\sqrt{1-\beta^2}} = \\ &= (\tan \alpha'_1)_{\Delta t'=0} = \frac{y'_1}{x'_1} . \end{aligned}$$

Similarly, in the system S

(V-53)
$$\frac{y_1}{x_1} = (\tan\alpha_1)_{\Delta t=0} = (\tan\alpha_2')_{\Delta t'=0} < \frac{(\tan\alpha_2')_{\Delta t'=0}}{\sqrt{1-\beta^2}} = = (\tan\alpha_2)_{\Delta t=0} = \frac{y_2}{x_2}.$$

The lines OM and $O'N$ which, within their own respective systems, form the same angle α with Ox, are not parallel in either S or S'.

The immediate result of the foregoing considerations is that the form of a body depends on the Lorentzian system to which we relate it. Thus, a sphere of radius R bound to a system S_0

(V-54)
$$x_0^2 + y_0^2 + z_0^2 = R^2$$

takes on the form of an ellipsoid in a Galilean system S moving at a uniform velocity v. We obtain, as a matter of fact, from (V-39):

(V-55)
$$\frac{x^2}{\sqrt{1-\beta^2}} + y^2 + z^2 = R^2.$$

In a general way, the volume of a body is maximal when it is measured in its proper system.

12. *Proper time*

"Proper time" is the term applied to the time τ measured by a clock bound to the reference system. An infinitesimally small interval of the proper time of S is:

(V-56)
$$\boxed{d\tau = dt\sqrt{1-\beta^2}}$$

that is

(V-57)
$$d\tau^2 = dt^2\left\{1 - \frac{1}{c^2}\left[\left(\frac{dx}{dt}\right)^2 + \left(\frac{dy}{dt}\right)^2 + \left(\frac{dz}{dt}\right)^2\right]\right\} = = \frac{1}{c^2}[c^2\,dt^2 - dx^2 - dy^2 - dz^2] = \frac{1}{c^2}ds^2.$$

The expression

(V-58)
$$ds^2 = c^2\,d\tau^2$$

being invariant under a Lorentz transformation (Cf. Section 3, above), the proper time $d\tau = ds/c$ is itself invariant.

13. *Geometrical representation of the Lorentz formulas*

One can make the symmetry of the coordinates of space and of time, as these coordinates appear in the Lorentz formulas, much clearer by means of the following representation.

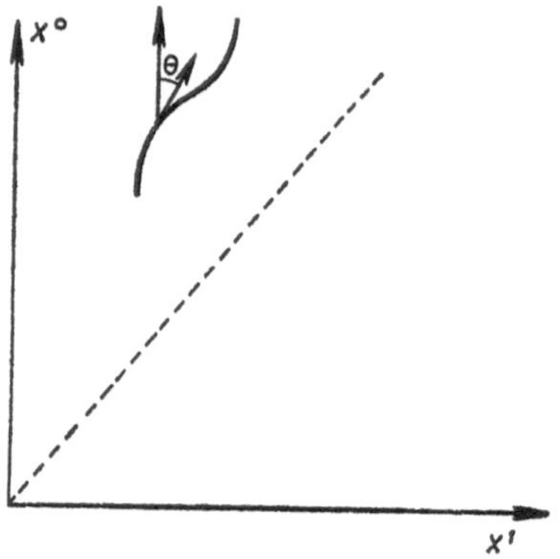

Fig. 22. Orientation of the trajectory of a point-mass.

Let us limit ourselves to the consideration of the spatial coordinates x^1 and x'^1 relative to the axes parallel to the direction of the motion. Formulas (V-38) and (V-39) applied to $x = x^1$ and $ct = x^0$ are written quite simply:

(V-59)
$$x'^1 = \frac{x^1 - \beta x^0}{\sqrt{1-\beta^2}} \qquad x^1 = \frac{x'^1 + \beta x'^0}{\sqrt{1-\beta^2}}$$
$$x'^0 = \frac{-\beta x^1 + x^0}{\sqrt{1-\beta^2}} \qquad x^0 = \frac{\beta x'^1 + x'^0}{\sqrt{1-\beta^2}}.$$

The motion of a point-mass is expressed by a world-curve $x^0 = f(x^1)$ (Figure 22). The tangent to this curve forms an angle θ with the time-axis such that:

(V-60)
$$\tan\theta = \frac{dx^1}{dx^0} = \frac{1}{c}\frac{dx}{dt} = \beta \leqslant 1 .$$

One will then always have: $\theta \leqslant \pi/4$. At the limits, one has a straight line with an inclination of $\pi/4$ and such that $\beta = 1$; this line corresponds to the trajectories of lightwaves.

Using formulas (V-59), one will obtain the position of the axes (x'^1, x'^0) of the system S by first noting that these axes have the same origin and by putting for the Ox'^0 axis: $x'^1 = 0$; that is, $x^1 = \beta x^0$; and for the Ox'^1 axis: $x'^0 = 0$; that is, $x^0 = \beta x^1$.

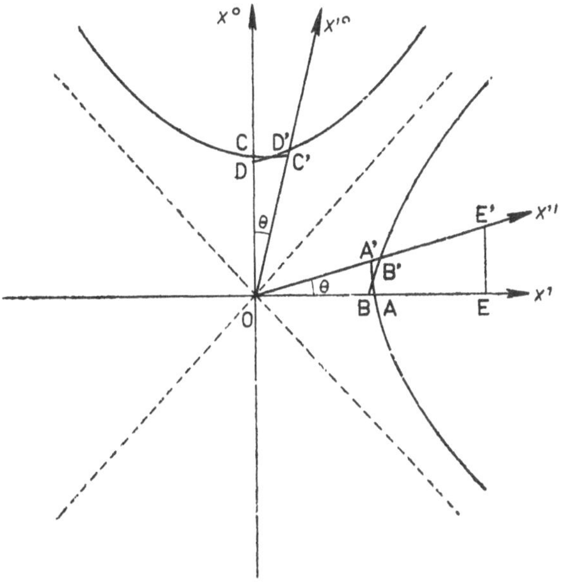

Fig. 23. Contraction of lengths and expansion of durations. (Space-time diagram.)

The Ox'^0 axis will then form with Ox^0 the angle θ, whose tangent is β. The Ox'^1 axis will also form with Ox^1 the angle θ, whose tangent is β. (Figure 23.)

1) *The relative character of simultaneity.* In the system $S\,(x^0, x^1)$, all the events situated along Ox^1 are simultaneous, but they are successive in the system $S'\,(x'^0, x'^1)$, since they are projected into different points of Ox'^0.

Conversely, the events that are simultaneous for S' (that is, situated along the axis Ox'^1) are no longer so for an observer S, since they are projected at different points along Ox^0. In particular, an event E' corresponding to $x'^0 = 0$ (or $t' = 0$) takes place at a corresponding point E of S at the end of the time

$$t = \frac{EE'}{c}.$$

2) *The contraction of lengths.* Let us draw the two conjugate hyperbolae:

$$(x^1)^2 - (x^0)^2 = 1 \qquad (x^1)^2 - (x^0)^2 = -1. \tag{V-61}$$

The first hyperbole cuts Ox^1 at the points A and A_1 ($x' = \pm 1$) and Ox'^1 at the points B' and B'_1 ($x'^0 = 0$, $x'^1 = \pm 1$), because these points correspond to the coordinates

$$x^1 = \frac{\pm 1}{\sqrt{1-\beta^2}} \qquad x^0 = \frac{\pm \beta}{\sqrt{1-\beta^2}}$$

which satisfy $(x^1)^2 - (x^0)^2 = 1$.

The second hyperbola cuts Ox^0 at the points C and $C_1(x^0 = \pm 1)$ and Ox'^0 at the points D' and D'_1 ($x'^1 = 0$, $x'^0 = \pm 1$), since these points correspond to the coordinates

$$x^1 = \frac{\pm \beta}{\sqrt{1-\beta^2}}, \qquad x^0 = \frac{\pm 1}{\sqrt{1-\beta^2}},$$

which satisfy $(x^1)^2 - (x^0)^2 = -1$.

If a standard measuring-rod represented by $OA = 1$ is fastened to the first reference system, its extremities describe the world-lines Ox^0 and AA' parallel to Ox^0. An observer in the second system records the simultaneous positions of the beginning and the end of this standard measuring rod. For $x'^0 = 0$, he then measures the length:

$$OA' < OB' = 1$$

and finds it less than that of his own standard measuring-rod OB'.

Inversely, the standard measure $OB' = 1$ attached to the second system S' (x'^0, x'^1) is such that its extremities describe the world-lines Ox'^0 and $B'B$ parallel to Ox'^0. An observer in the system S records the simultaneous positions of the beginning and the end of this measuring-rod for $x^0 = 0$. He thus measures the length:

$$OB < OA = 1$$

and likewise finds it less than that of his own standard measuring-rod OA.

3) *Dilatation of time intervals.* A clock fastened at $x^1 = 0$ in S is represented by a figurative point that traces, with the passage of time, the line Ox^0. When the clock-hand has made one entire revolution (which, we assume, corresponds to a time-unit), the figurative point is at $C(OC = 1)$. A second clock coinciding spatially with the first one at the instant

$x^0 = OC = 1$ is then represented in S' by the point C' of Ox'^0 ($x'^1 = x^1 = 0$) determined by the line CC' parallel to Ox^1 (CC' corresponding, in fact, to the instant $x^0 = 1$ in S). The segment OC' represents a time-interval such that $OC' < OD' = 1$.

The observer in S, noting that on the clock in S', placed at the same point in space as his own clock, the hand has not yet made a complete revolution (whereas the hand on his clock has made the complete circuit) concludes that the S' clock is slow.

But, conversely, the figurative point for the time indicated by a clock spatially attached to S' traces the line Ox'^0. The hand has made a complete revolution (one time unit) for the point D' which coincides spatially ($x^1 = x'^1 = 0$) with the point D of the system S (DD' being parallel to Ox'^1 corresponding to the instant $x'^0 = 1$ of S'). Now $OD < OC = 1$.

At the point C, the clock in S has not yet completed a revolution; therefore the observer bound to S' also concludes that the clocks of the system S are slow.

In other words, a moving clock is always slow with respect to a stationary one – which amounts to saying that the motion produces a dilation of the time-intervals. This conclusion, which seemed so paradoxical at first, has been elaborated upon by P. Langevin, who has given numerous illustrations of it.

In fact, the mutual contraction of lengths and reciprocal dilation of time-intervals become natural as soon as one admits the impossibility of defining absolute simultaneity. It is by implicitly maintaining such a possibility that one is led to *non-reciprocal modifications* of space and time for Galilean systems, which actually means negating *the principle of relativity*.

14. *Other expressions of the special Lorentz transformation*

1) *Real coordinates.* Relations (V-59) may be written in such a way as to make the symmetry of the x^1 and x^0 coordinates appear more clearly. Let us put:

(V-62) $$\boxed{\beta = \tanh \varphi}$$

that is:

(V-63) $$\cosh \varphi = \frac{1}{\sqrt{1-\beta^2}}, \qquad \sinh \varphi = \frac{\beta}{\sqrt{1-\beta^2}}$$

Formulas (V-59) are then written:

(V-64)
$$\boxed{\begin{aligned} x'^1 &= x^1 \cosh \varphi - x^0 \sinh \varphi \\ x^1 &= x'^1 \cosh \varphi + x'^0 \sinh \varphi \\ x'^0 &= -x^1 \sinh \varphi + x^0 \cosh \varphi \\ x^0 &= x'^1 \sinh \varphi + x'^0 \cosh \varphi . \end{aligned}}$$

2) *Imaginary coordinates.* If we now put, following Minkowski

(V-65) $$x^4 = ict$$

and

(V-66) $$\boxed{i\beta = \tan \psi}$$

one has

(V-67) $$\cos \psi = \frac{1}{\sqrt{1-\beta^2}} \qquad \sin \psi = \frac{i\beta}{\sqrt{1-\beta^2}}$$

and formulas (V-59) are further written as:

(V-68)
$$\boxed{\begin{aligned} x'^1 &= x^1 \cos \psi + x^4 \sin \psi \\ x^1 &= x'^1 \cos \psi - x'^4 \sin \psi \\ x'^4 &= -x^1 \sin \psi + x^4 \cos \psi \\ x^4 &= x'^1 \sin \psi + x'^4 \cos \psi . \end{aligned}}$$

These formulas represent, in the imaginary plane $(x^1\, O\, x^4)$, a rotation of the coordinate axes of the imaginary angle ψ. According to (V-62) and (V-66), one has

(V-69) $$\psi = i\varphi .$$

15. *The general Lorentz transformation. C. Møller's method*

The Lorentz transformation has as its purpose (Cf. Section 3, above) the determination of the changes in coordinates that allow us to pass from one Cartesian reference system S to another Cartesian system S' moving in a uniform and rectilinear manner with respect to the first one (Galilean systems), assuming the constancy of the velocity of light in each reference system or, what amounts to the same thing, the invariance of ds^2 (Lorentzian systems).

Up to now, we have always postulated that the relative velocity of the two Cartesian systems was parallel to the coordinate axes Ox and Ox', with the axes of S and S' likewise being parallel to each other. Whence we obtain formulas (V-38) and (V-39) or (V-64) and (V-68) which make up the *special Lorentz transformation.*

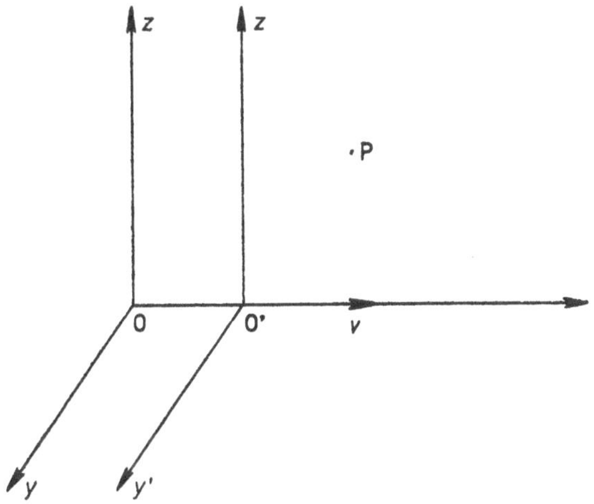

Fig. 24. The changing of the Galilean systems: the special transformation.

1) Let us consider once more the particular case in which the special transformation is valid, namely, the case that assumes that the relative velocity of the two reference systems S and S' is parallel to Ox and Ox' (Figure 24).

It is possible to group the four relations of (V-38) into two vector relations. In order to do this, we will designate the position of a point P in S and also in S' by putting

$$\text{(V-70)} \qquad \mathbf{r} = (x, y, z), \qquad \mathbf{r}' = (x', y', z')$$

and by considering $\mathbf{r}'$ and $\mathbf{r}$ as two related vectors in one and the same fixed coordinate system (cf. C. MØLLER [16], p. 41), $\mathbf{v}\,(v_x, o, o)$ being the velocity of S' with respect to S.

The formulas (V-38) are then deducible from the two vector relations

$$\text{(V-71)}_1 \qquad \mathbf{r}' = \mathbf{r} + \mathbf{v}\left[\left(\frac{\mathbf{r}\cdot\mathbf{v}}{v^2}\right)\left(\frac{1}{\sqrt{1-\beta^2}} - 1\right) - \frac{t}{\sqrt{1-\beta^2}}\right]$$

$$\text{(V-71)}_2 \qquad t' = \frac{t - \left(\dfrac{\mathbf{r}\cdot\mathbf{v}}{c^2}\right)}{\sqrt{1-\beta^2}}.$$

One obtains these formulas by successively giving **r** the values x, y and z, corresponding to v_x, o, o.

In a similar fashion, the inverse formulas (V-39) are deducible from the two vector relations:

$$\text{(V-72)}_1 \qquad \boxed{\mathbf{r} = \mathbf{r}' + \mathbf{v}'\left[\frac{(\mathbf{r}'\cdot\mathbf{v}')}{v^2}\left(\frac{1}{\sqrt{1-\beta^2}} - 1\right) - \frac{t'}{\sqrt{1-\beta^2}}\right]}$$

$$\text{(V-72)}_2 \qquad \boxed{t = \frac{t' - \left(\dfrac{\mathbf{r}'\cdot\mathbf{v}'}{c^2}\right)}{\sqrt{1-\beta^2}}}$$

in which $\mathbf{v}' = -\mathbf{v}$ represents the relative velocity of S with respect to S'. By successively giving $\mathbf{r}'$ the values x', y', z' corresponding to $\mathbf{v}' = -\mathbf{v}_x$, o, o, one obviously has the formulas (V-39) of the special transformation.

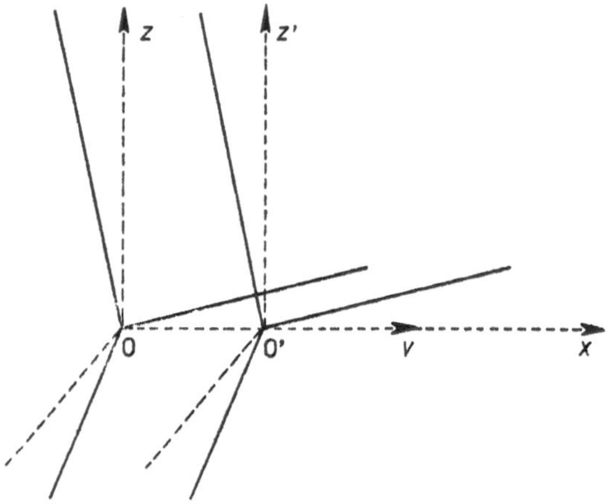

Fig. 25. The changing of a Galilean system: transformation without rotation.

2) Let us now transform the coordinate axes of S and S' by applying the same spatial rotation to each of them (Figure 25). In this case $\mathbf{r}, \mathbf{r}'$ and $\mathbf{v}, \mathbf{v}'$ are affected in the same way, and relations (V-71) and (V-72) continue to be valid. This time we have

$$\text{(V-73)} \qquad \mathbf{v} = (v_x, v_y, v_z) \quad \text{and} \quad \mathbf{v}' = (-v_x, -v_y, -v_z).$$

In this way we deduce, from the vector relations (V-71), the four following transformation formulas valid for the case of any arbitrary orientation of v with respect to the axes:

$$\text{(V-74)}_1 \qquad x' = x + \frac{\alpha v_x}{v^2}\left[v_x x + v_y y + v_z z - c^2 t\left(1 + \sqrt{1-\beta^2}\right)\right]$$

$$\text{(V-74)}_2 \qquad y' = y + \frac{\alpha v_y}{v^2}\left[v_x x + v_y y + v_z z - c^2 t\left(1 + \sqrt{1-\beta^2}\right)\right]$$

$$\text{(V-74)}_3 \qquad z' = z + \frac{\alpha v_z}{v^2}\left[v_x x + v_y y + v_z z - c^2 t\left(1 + \sqrt{1-\beta^2}\right)\right]$$

$$\text{(V-74)}_4 \qquad t' = t - \frac{1}{c^2\sqrt{1-\beta^2}}\left[v_x x + v_y y + v_z z - c^2 t\left(1 - \sqrt{1-\beta^2}\right)\right]$$

putting

$$\text{(V-75)} \qquad \boxed{\alpha = \frac{1}{\sqrt{1-\beta^2}} - 1.}$$

The inverse transformations are obtained easily from (V-74) by changing x, y, z to x', y', z' and inversely, and v_x, v_y, v_z to $-v_x, -v_y, -v_z$.

3) Let us now transform the coordinate axes of S and S', beginning with the initial position (Figure 24), by applying different spatial rotations to S and S'. It really amounts to the same thing if we apply a spatial rotation only to the coordinate axes of S', the initial position being defined by Figure 25. Let the rotation be D^{-1}. The orientation of the axes is then as indicated by the solid lines in Figure 26.

The last formula of transformation $(\text{V-71})_2$ is not modified, but $(\text{V-71})_1$ will be valid only on the condition that there be applied to the new vector $\mathbf{r}'$ of S_3' the inverse rotation D, which brings S_3' to the position S_2' corresponding to Figure 25.

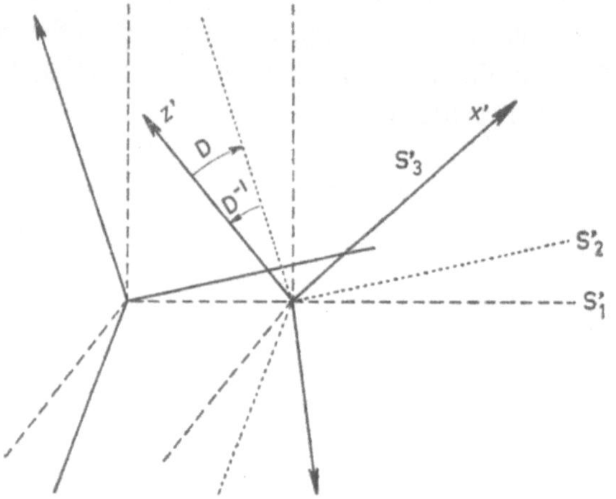

Fig. 26. The changing of a Galilean system: the general transformation.

One will then have

(V-76) $$D\mathbf{r}' = \mathbf{r} + \mathbf{v}\left[\frac{\alpha}{v^2}(\mathbf{r}\cdot\mathbf{v}) - \frac{t}{\sqrt{1-\beta^2}}\right].$$

On the other hand, the velocity $\mathbf{v}'$ of S with respect to S' becomes $D\mathbf{v}'$ in this inverse rotation. Now – as is the case in Figure 24 – it becomes equal to $-\mathbf{v}$:

(V-77) $$D\mathbf{v}' = -\mathbf{v}$$

By applying D^{-1} to (V-76) one finally has the transformation formulas

$(\text{V-78})_1$ $$\mathbf{r}' = D^{-1}\mathbf{r} - \mathbf{v}'\left[\frac{\alpha}{v^2}(\mathbf{r}\cdot\mathbf{v}) - \frac{t}{\sqrt{1-\beta^2}}\right]$$

$(\text{V-78})_2$ $$t' = \frac{t - \left(\dfrac{\mathbf{r}\cdot\mathbf{v}}{c^2}\right)}{\sqrt{1-\beta^2}}$$

with

(V-75) $$\alpha = \frac{1}{\sqrt{1-\beta^2}} - 1.$$

The inverse formulas would be obtained in a similar fashion by applying (starting with Figure 25) the rotation D^{-1} to the axes of S, and not to the axes of S'. One would thus have

$$(V\text{-}79)_1 \qquad \mathbf{r} = D^{-1}\mathbf{r}' - \mathbf{v}\left[\frac{\alpha}{v^2}(\mathbf{r}'\cdot\mathbf{v}') - \frac{t'}{\sqrt{1-\beta^2}}\right]$$

$$(V\text{-}79)_2 \qquad t = \frac{t' - \left(\dfrac{\mathbf{r}'\cdot\mathbf{v}'}{c^2}\right)}{\sqrt{1-\beta^2}}.$$

4) Finally, if we assume that the origins of the coordinate axes of S and S' do not coincide for $t = t' = 0$, we must substitute

$$(V\text{-}80) \qquad \mathbf{r}_1' = \mathbf{r}' + \mathbf{a}', \qquad t_1' = t' + \theta'$$

for $\mathbf{r}$ and t' ($\mathbf{a}'$ and θ' being constants). One will then have

$$(V\text{-}81) \qquad \Delta\mathbf{r}_1' = \Delta\mathbf{r}', \qquad \Delta t_1' = \Delta t'$$

and all the transformation formulas thus remain valid between the intervals $\Delta\mathbf{r}_1'$, $\Delta t_1'$ and $\Delta\mathbf{r}$, Δt.

In this way a transformation of coordinates from one Cartesian system to another involving

a) a special Lorentz transformation

b) a spatial rotation

c) a spatial translation and change of time-origin is described by formulas (V-79) and (V-80). The invariance of ds^2 for these formulas is easily verified

$$(V\text{-}82) \qquad ds^2 = c^2\,dt^2 - dr^2 = c^2\,dt'^2 - dr'^2 = ds'^2$$

Formulas (V-78) and (V-79) define a general Lorentz transformation: the passage from one Galilean system to another insures the invariance of ds^2 (V-82), since the orientation and relative velocity are entirely arbitrary. They determine the change of arbitrary Lorentzian systems.

Further on (Chapter VI, Section 7) the tensor formalism of the general Lorentz transformation will be taken up.

16. *Change of inertial system for a moving object. The clock paradox*[57]

The principles of Special Relativity expressed by the Lorentz transformation make it possible to compare the phenomena occurring in two Galilean systems: to every moving object there may be attached a proper Galilean system *to which the moving object remains permanently attached.* The Lorentz formulas permit us to pass from one proper system to another and demonstrate that there exists a complete reversibility in the description of observed effects – a reversibility that leads precisely to the concept of the Relativity of motion.

But such is no longer the case when we come to compare lengths and time intervals by means of two measuring-rods or two clocks, at least one of which does not remain attached to the same proper Galilean system.

Let us, for example, consider a measuring rod of length l_0 in its own Galilean system S. Let us now accelerate the motion of the rod very suddenly so that it acquires a velocity v with respect to S. If v is uniform, the new proper system is a system S' having the velocity v with respect to S. The length of the rod is then l_0' in S' and $l = l_0' \sqrt{1 - \beta^2}$ in S according to the Lorentz formulae. But of course, l (or l_0') cannot be compared to l_0. By bringing the rod to rest in the S system (for example, by again changing its acceleration) one will obtain a length l_{00} which may differ from l_0. As a matter of fact, l_{00} results from a double application of acceleration or, what amounts to the same thing, from a change of proper system that makes impossible the formation of any reciprocal observations.[58]

The comparison of the indications furnished by two clocks, one of which has been subjected to transformations of its proper system, leads to similar conclusions.

Let there be two clocks A and A' attached respectively to the proper Galilean systems S and S'. We know (V-44) that the indications of A', compared to those of A, show that A' is retarded in the A system:

(V-83)
$$(\Delta t)_S - (\Delta t')_{S(\Delta x' = 0)} = (\Delta t)_S - (\Delta t)_S \sqrt{1 - \beta^2} = \\ = (\Delta t)_S (1 - \sqrt{1 - \beta^2})$$

The phenomenon is reversible. In the system S' a retardation of A will be noted:

$$(\Delta t')_{S'} - (\Delta t)_{S'(\Delta x=0)} = (\Delta t')_{S'} - (\Delta t')_{S'}\sqrt{1-\beta^2} = = (\Delta t')_{S'}(1 - \sqrt{1-\beta^2}) \tag{V-84}$$

Let us now suppose that A and A' were originally in the same proper system S. A' is accelerated to the velocity v. A' is then attached, during the time $\Delta t'$, to the proper system S'. By giving an inverse acceleration to A', it is brought to rest in the original system S. If the durations of the accelerations are negligible, the comparison of A and A' show that A' (and not A) is retarded as indicated by (V-83).

But, in this case, the comparison between the indications of A and A' *is ultimately made in one and the same proper system*, namely S. This shows that the result of the experiment is not reversible. In a system permanently attached to A' (and consequently independent of S' at the outset and at the conclusion of the experiment) the final result (V-83) is obviously valid (since S' at that point coincides with S), and not the result indicated in (V-84).

Einstein himself pointed out this "paradox", which seems to make it possible to decide – contrary to the principles of Relativity – which clock remained motionless. In point of fact, however, this "paradox" is no longer within the framework of the Special theory, for it introduces accelerations that allow us to tell which clock is directly affected by them and likewise which clock changes its proper Galilean system during the course of the experiment.

This paradox applies especially to the so-called "traveller's experiment" of Langevin. In this case the slowing down of clock A' is expressed by the reduced aging of the traveller. Obviously, it is the appearance of accelerations through a change of proper Galilean systems at the instant of the traveller's departure and arrival that allows the phenomenon to manifest itself in a non-reciprocal guise.

These conclusions may be made clearer by recourse to the properties of the longitudinal Doppler effect.[59] Let us suppose that A sends out, in the direction of Ox, signals of frequency ν_0 in its own proper system S. For a moving object A' proceeding from O to O' with the velocity v, these signals will have a frequency[60]

$$\nu_a = \nu_0 \sqrt{\frac{1-\beta}{1+\beta}} < \nu_0 \tag{V-85}$$

in the proper system S'_a bound to A'. On the return trip (velocity $-v$) the frequency of the plane waves received by the moving object A′ will be

(V-86) $$\nu_r = \nu_0 \sqrt{\frac{1+\beta}{1-\beta}} > \nu_0$$

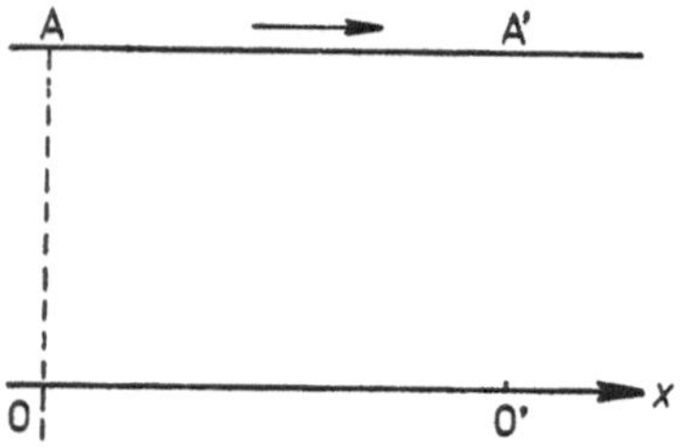

Fig. 27.

since the proper system S'_r of A' will then be another Galilean system.

If N is the number of signals sent out, and N_a and N_r the number of signals received:

(V-87) $$N = N_a + N_r, \qquad \frac{N_a}{\nu_a} = \frac{N_r}{\nu_r}$$

One then has:

(V-88) $$N_a = \frac{N}{1 + \dfrac{\nu_r}{\nu_a}} \qquad N_r = \frac{N}{1 + \dfrac{\nu_a}{\nu_r}}$$

By comparing (after two series of contrary accelerations) the time taken by A' to go from O to O', and then to return to O, one finds in the system S:

$$\Delta t = \frac{N}{\nu_0},$$

in the system S'_a and then in the system S'_r:

(V-89) $$\Delta t' = \frac{N_a}{\nu_a} + \frac{N_r}{\nu_r} = \frac{2N}{\nu_a + \nu_r} = \frac{N}{\nu_0}\sqrt{1-\beta^2} = \Delta t\sqrt{1-\beta^2}$$

The comparison is ultimately made in the system S, with which S'_r coincides when immobilized. So there is no reciprocity. The duration of the round-trip of A' is diminished by[61]

(V-90) $$\Delta t - \Delta t' = \Delta t(1 - \sqrt{1 - \beta^2}).$$

Since this question involves accelerations, it is related to the principles of General Relativity. And thus General Relativity appears, not merely as a mathematical generalization adding a more or less optional complement to the principles of the Special theory, but as an indispensible extension of that theory permitting the correct formulation of a certain number of problems that are raised by the kinematics and dynamics of Special Relativity, but which Special Relativity alone cannot solve.

NOTES

1. Various procedures have been proposed which are intended to measure the velocity of light along the "forward route" only and to reveal any possible anisotropy in the "forward-and-back" trajectories.
   Most of the indicated procedures are incorrect, for they implicitly presuppose the isotropy whose existence one seeks to confirm or disprove. Other methods, though theoretically correct, do not possess the necessary precision to verify this isotropy in actual fact.
   On this subject two sources that may be consulted are:
   O. Costa de Beauregard: 'De la mesure de la vitesse de la lumière sur un parcours aller simple', *Bull. Astron.* **15**, Fasc. 2 (1950) 159;
   H. Arzeliès [8], p. 64, and the collective bibliography on our particular subject on p. 67.
2. If l is the length of the trajectory, the "round-trip" time taken by light to cover this trajectory is, as a matter of fact:
   $$t = \frac{l}{c+v} + \frac{l}{c-v} = \frac{2lc}{c^2 - v^2}$$
   v being the absolute velocity that is sought.
   The average velocity actually observed is thus:
   $$c' = \frac{2l}{t} = \frac{c^2 - v^2}{c} = c\,(1 - \beta^2)$$
   and differs from c only in terms of the second order.
3. Ch. Doppler: *Abhand. Kgl. Böhmischen Gesell. Wiss.* (5) **2** (1841–1842) 465–482.
   The non-relativistic explanation of the Doppler effect is actually as follows: If a train of plane waves is propagated along Ox, starting from O at the initial instant, the number of waves reaching the point $P(x)$ seen in the system S will be
   $$\nu\left(t - \frac{x}{c}\right)$$
   at the instant t. The number of waves reaching the point Q (x') fixed in the system S' (S' coincides with S at the origin of the times and then moves in a rectilinear

and uniform manner with a velocity v along OX) will be:

$$\nu'\left(t - \frac{x'}{c}\right).$$

ν and ν' being the frequencies of each of the waves in the systems S and S'.

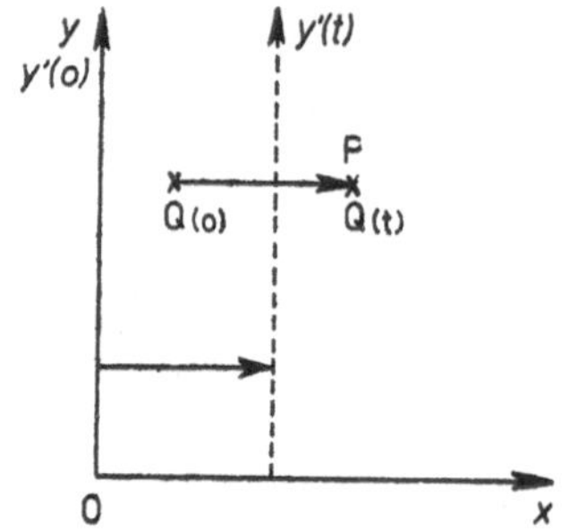

Fig. 13. The Doppler effect (propagation of plane waves along Ox).

If P coincides with Q at the instant t, we must have:

$$\nu\left(t - \frac{x}{c}\right) = \nu'\left(t - \frac{x'}{c}\right).$$

Now we pass from the x' coordinates to the x coordinates by means of the Galilean transformation:

$$x' = x - vt \qquad x = x' + vt$$

Whence

$$\nu\left(t - \frac{x' + vt}{c}\right) = \nu'\left(t - \frac{x'}{c}\right).$$

In particular for $x' = 0$, $\nu(1 - \beta) = \nu'$.
Let ν_0 be the frequency of the source at rest in a system S:
1) In the foregoing case, that is, if the observer at rest in the system S' changes his position, $\nu = \nu_0$. The frequency observed is $\nu' = \nu_0\,(1 - \beta_{obs})$.
2) If the source at rest in the system S' moves, $\nu' = \nu_0$, since the observer is motionless. The frequency observed is

$$\nu = \frac{\nu_0}{1 - \beta_{source}}$$

There is, thus, no reversibility in the frequencies observed as a result of the change in position of the observer $\nu_1 = \nu_0\,(1 - \beta)$, and as a result of the shifting of the source with the same velocity in the opposite direction

$$\nu_2 = \frac{\nu_0}{1 + \beta}.$$

But the difference is of the order β^2 and cannot, in practice, be used to determine an absolute motion.

$$\nu_2 - \nu_1 = \nu_0\left\{\frac{1}{1 + \beta} - (1 - \beta)\right\} = \nu_0\,\frac{\beta^2}{1 + \beta}.$$

3) In the general case where the observer and the source are both in motion

$$\nu = \nu_0\left(\frac{1-\beta_{\text{obs}}}{1-\beta_{\text{source}}}\right), \quad \nu - \nu_0 = \nu_0\left(\frac{\beta_{\text{source}}-\beta_{\text{obs}}}{1-\beta_{\text{source}}}\right).$$

If the source and the observer are moving with the same velocity ($\beta_s = \beta_{\text{obs}}$), there is no Doppler effect. That effect is produced only when $\beta_s \neq \beta_{\text{obs}}$, and the difference $\nu - \nu_0$ depends, in the first order, only on the relative source-observer motion ($\beta_s - \beta_{\text{obs}}$). Absolute velocities do not appear until the second order.

4. J. Bradley: *Phil. Trans.* **35** (1728) 637. The deviation of the light-rays does not depend on the presence of a refractive medium (a water-filled tube). Cf. G. B. Airy: *Proc. Roy. Soc. London* A **20** (1871) 35; **21** (1873) 121; *Phil. Mag.* **43** (1872) 310.
5. Doppler effects have been produced successfully with terrestrial sources by using, for example, canal-rays (J. Stark, 1906), but the measurement of second-order consequences by Doppler effects was not achieved until 1928 in Ives' experiments. S. Stark: *Ann. d. Phys.* **21** (1906) 40; J. Stark and K. Siegel: *Ann. d. Phys.* **21** (1906) 457; S. Stark, W. Hermann, and S. Kinoshita: *Ann. d. Phys.* **21** (1906) 462. H. E. Ives and G. R. Stillwell: *J. Opt. Soc. Amer.* **28** (1938) 215; cf. also p. 232.
6. Römer's measurement has as its object a kind of Doppler effect in which the frequency of the occultation of the satellites of Jupiter replaces the emission frequency of a certain wave. The principle of the Römerian measurement is as follows: Let us consider the successive observations of the occultations of one of Jupiter's satellites evaluated on earth at half-year intervals. At the time of the first observation, the sun, the earth, Jupiter, and its satellite occupy the positions S, T_1, J_1, σ_1; at the time of the second observation their positions are S, T_2, J_2, σ_2. In practical terms, J_1 is very close to J_2, since it takes Jupiter twelve years to accomplish one complete revolution around the sun. Likewise, the supplementary trajectory covered by the light rays at the time of the second occultation is very close to the diameter of the terrestrial trajectory $T_1T_2 = l$.

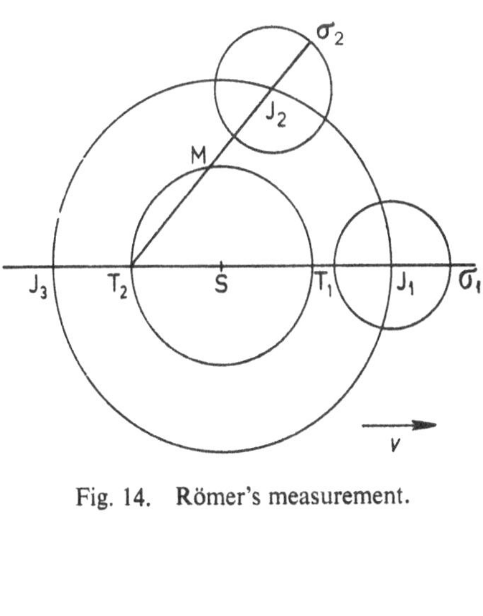

Fig. 14. Römer's measurement.

If the solar system is motionless in the ether, the time-lag in the second occultation would then be

$$t_1 = \frac{l}{c}.$$

If it moves with a velocity $\mathbf{v}$ (along T_2T_1) with respect to the ether, this lag becomes

$$t_2 = \frac{l}{c+v}.$$

These lags differ by

$$t_1 - t_2 = l\left(\frac{1}{c} - \frac{1}{c+v}\right) = \frac{lv}{c(c+v)} = \frac{l\beta^2}{v(1+\beta)} \simeq \frac{l\beta^2}{v}$$

that is, by a term in β^2.

In a like manner, if the measurement of two successive occultations is taken six years later, that is, when Jupiter is at J_3, the delay between the two occultations is

$$t_3 = \frac{l}{c-v}$$

if the solar system is moving with a velocity v. One still is confronted with the appearance of a second-order term:

$$t_2 - t_3 = l\left(\frac{1}{c-v} - \frac{1}{c+v}\right) = \frac{2lv}{c^2 - v^2} = \frac{2l}{v}\frac{\beta^2}{1-\beta^2}.$$

As Maxwell then remarked, the observation of the successive occultations of the satellites of Jupiter would theoretically permit us to determine the over-all motion of the solar system with respect to the ether. But the effects observed are only of the second order. And that is why the Römerian measurement achieves, to the first order, a correct determination of the velocity of light

$$c = \frac{l}{t_1}.$$

7. A much more radical hypothesis had been put forward by Stokes in 1845. He assumed that the ether is completely carried along by matter in motion. If the ether inside, and immediately surrounding, matter shares entirely that matter's motion, everything takes place as if that motion did not exist at all, as far as optical experiments are concerned. One could not then detect, in any order, any effect of ether-drift. In that form, Stokes' hypothesis ran into redhibitory difficulties in seeking to justify the constant direction of light-rays emitted by a star and the unchanged value of their velocity when these rays pass from motionless interstellar ether into the ether carried along by the earth. G. G. Stokes: *Phil. Mag.* **27** (1845) 9; *Math. and Phys. Papers* **1** (1880) 134.
8. Let us, nevertheless, point out an experiment due to Fizeau on the rotation of the plane of polarization of light-rays passing through a pile of mirrors. It seemed to produce a positive result, but, when it was again performed with much greater precision by Brace (1905) and Strasser (1907), the result was wholly negative.

8a. A. J. Fresnel: *Œuvres complètes*, Paris, 1866–1870, t. II, p. 627.

9. D. F. Arago: *Compt. Rend.* **8** (1839) 326; **36** (1853) 38. The first experiments

of Arago date from 1818. The most attainable publications, however, date from much later (1839 and 1853).

10. H. FIZEAU: *Compt. Rend.* **33** (1851) 349; *Ann. d. Phys. u. Chem.* **3** (1853) 457; A. A. MICHELSON and E. W. MORLEY: *Am. J. Sci.* **31** (1886) 377.
11. P. ZEEMAN: *Amst. Versl.* **23** (1914) 245; **24** (1915) 18.
12. M. HOEK: *Archives néerlandaises des Sciences exactes et naturelles* **3** (1868) 180.
13. Δt is of the order β^2. But the effect measured – a displacement of interference fringes, for example – is of the order $c\Delta t$, which is of the first order: ($\beta \simeq 1/c$). On the contrary, in Römer's experiment, for example, the measurement is directly concerned with $\Delta t \simeq \beta^2$.
    Michelson's experiment (cf. p. 105), like Fizeau's, also makes its measurements of quantities of the order $\Delta l = c\Delta t$. But the Δt are of the order

$$\frac{\beta^2}{c}$$

    (Cf. V-19); so the observed effect ($\sim\beta^2$) is of the second order.
14. A. J. FRESNEL: *Ann. de Chim. et de Phys.* **9** (1818) 57.
15. E. MASCART: *Ann. Sc. Ec. Norm. Sup.* **1** (1872) 157; **3** (1874) 363; *Traité d'optique.* Paris, 1893, Ch. XV, p. 38.
16. W. VELTMAN: *Astr. Nachr.* **75** (1870) 145; **76** (1870) 129; *Ann. d. Phys. u. Chem.* **150** (1873) 497.
17. A. POTIER: *Journ. Phys.* **3** (1874) 201; *Mémoires sur l'électricité et l'optique.* Paris, Gauthier-Villars, 1912, p. 292.
18. Experiments effected while studying the motion of a dielectric in an electric field: Röntgen, 1885; Eichenwald, 1903, Blondlot; or in a magnetic field: H. A. Wilson.
19. H. A. LORENTZ [5].
20. Cf. footnote 4 above.
21. More accurately stated: the electromagnetic field remains immobile, but the polarization current **P** and the magnetic induction M are carried along by matter, whence a change in the velocity of propagation and in the refraction index within a body in motion. (Cf. L. BLOCH [2], pp. 389 and 456.)
22. This can be shown by once more considering, along with BORN (*La Théorie de la relativité et ses bases physiques,* p. 200), the following example:
    Let us consider a dielectric moving along Ox with a velocity v and within which a plane monochromatic wave is propagated in the same direction. The electric field vector ($\mathbf{E}_y$) and the magnetic vector ($\mathbf{H}_z$) which characterize this wave are perpendicular to the direction of propagation. The displacement of the magnetic field determines a supplementary electrical displacement resulting from the polarization current **P** being carried along by matter. This displacement is directed along Oy and, according to the results of Wilson's experiment, its value is:

$$D = \varepsilon E' = (\varepsilon - 1)vH. \qquad (1)$$

    One thus becomes aware of the existence of a supplementary electric field:

$$E' = \frac{\varepsilon - 1}{\varepsilon} vH \qquad (2)$$

    and not $E' = vH$, as would be the case if the ether in the dielectric were completely

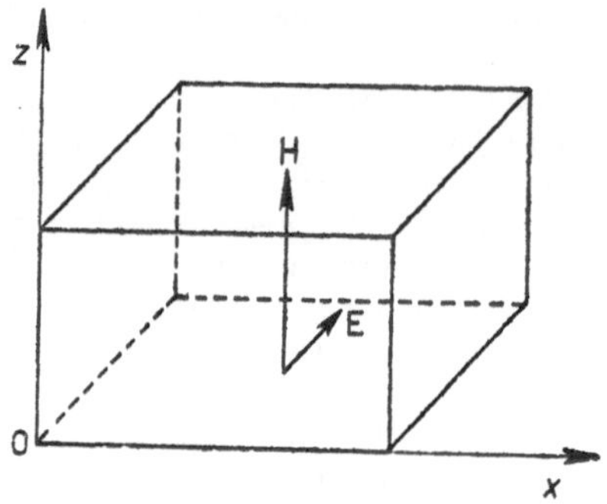

Fig. 17. Carrying along of a plane monochromatic wave by a dielectric in uniform motion

carried along by the matter in motion. H. A. Wilson's experiment with a revolving dielectric actually does confirm formula (2).

The value

$$\frac{\varepsilon - 1}{\varepsilon}$$

of the drag coefficient of the ether carried along by matter in motion, predicted by Maxwell's theory, coincides very well with the value Fresnel had proposed, but on the basis of far less convincing arguments. One actually has, according to Maxwell's theory (cf. III-72) $\varepsilon = n^2$.

Whence

$$\frac{\varepsilon - 1}{\varepsilon} = \frac{n^2 - 1}{n^2} = 1 - \frac{1}{n^2} = \alpha .$$

And that is indeed the value of the drag coefficient predicted by Fresnel. But in Lorentz' theory it is not the ether that is partially carried along, but the electrons contained in the matter. (Cf. the much more detailed calculations made by R. Becker [1], p. 290.)

23. Because, let us not forget, the deduction of the Maxwell equations from Lorentz' theory was achieved only for the first order – in keeping with the electrodynamics of bodies in motion. (Cf. Chapter IV, Section 6.)
24. A. A. Michelson: *Am. J. Sci.* **22** (1881) 20.
25. A. A. Michelson and E. W. Morley: *Am. J. Sci.* **34** (1887) 333.
26. With the aid of this interferometer, Michelson compared the length of the standard meter with the wave-length of the red cadmium line.

26a. With respect to a reference system bound to the laboratory, the light covers the same trajectory forward and backward $MM_1 = M_1M = l_1$, but on the forward trajectory its velocity is $c + v$, and on the backward, $c - v$. The time for the "round-trip"

$$t_1 = \frac{l_1}{c + v} + \frac{l_1}{c - v} = \frac{2l_1}{c} \frac{1}{1 - \beta^2}$$

is thus identical with the round-trip time in the reference system bound to the ether.

26b. With respect to a reference system bound to the laboratory, light covers the same distance $MM_2 = M_2M = l_2$ forward and back. Its velocity along MM_2 is the resultant of v along MM_1 and of c.

One thus has

$$c' = \sqrt{c^2 - v^2} = c\sqrt{1 - \beta^2}.$$

Whence the result that is identical with (V-14)

$$\bar{t}_2 = \frac{2l_2}{c'} = \frac{2l_2}{c}\frac{1}{\sqrt{1-\beta^2}}.$$

The difference $\overline{\Delta_1 t} = \bar{t}_2 - \bar{t}_1$ calculated in a reference system bound to the laboratory is identical to the difference in speed evaluated in a reference system bound to the ether. The fringe displacement resulting from the time-lag Δt should therefore have the value (V-20) in either reference system. This result was foreseeable in view of the absolute character of classical kinematic time: the time separating two events (emission from M and return to M of the wave-front) is the same, no matter in what Galilean reference system one evaluates it.

27. The purpose of this rotation is to eliminate any possible difference in length between the arms of the interferometer. (If $l_1 \equiv l_2$, the fringes observed before rotation would be due solely to the motion of the apparatus.)
28. R. J. KENNEDY: *Proc. Nat. Acad.* **12** (1926) 621.
29. K. K. ILLINGSWORTH: *Phys. Rev.* **30** (1927) 692.
30. A. PICCARD and E. STAHEL: *Naturwiss.* **14** (1926) 935; **16** (1928) 25.
31. G. JOOS: *Ann. d. Phys.* **7** (1930) 385.
32. Let us point out, nevertheless, one faintly positive result (and one which is rather different from what was expected), indicated by Miller. The results thus obtained seem to be superseded by subsequent experiments. (Cf. D. C. MILLER: *Rev. Mod. Phys.* **5** (1933) 203).
33. This experiment was designed to reveal the rotational effect, resulting from ether drift, that would be exerted on a suspended-plate condensor. (Cf. F. T. TROUTON and H. R. NOBLE: *Proc. Roy. Soc.* **72** (1903) 132.)
34. F. T. TROUTON and A. RANKINE: *Proc. Roy. Soc.* **80** (1908) 420.
35. C. T. CHASE: *Phys. Rev.* **30** (1927) 516.
36. R. TOMASHEK: *Ann. d. Phys.* **73** (1924) 105; **78** (1925) 743; **80** (1926) 509; **84** (1927) 161.
37. W. RITZ: *Ann. de Chim. et de Phys.* **13** (1908) 145.
38. W. DE SITTER: *Phys. Z.* **14** (1913) 429, 1267.
39. O. LODGE: *London Transaction* A **184** (1893) 727.
40. Cf. O. LODGE: *London Transaction* A **184** (1893) 727.
41. H. A. LORENTZ: *Amst. Verh. Acad. v. Wet.* **1** (1892) 74.
42. LORD RAYLEIGH: 'Does Motion through the Aether Cause Double Refraction?', *Phil. Mag.* **4** (1902) 678.
43. D. B. BRACE: 'On Double Refraction in Matter Moving Through the Aether', *Phil. Mag.* **6** (1904) 317.
44. F. T. TROUTON and A. O. RANKINE: 'On the Electrical Resistence of Moving Matter', *Proc. Roy. Soc.* **80** (1908) 420.
45. A. B. WOOD, G. A. TOMLISON, and L. ESSEX: 'The Effect of the Fitzgerald-Lorentz Contraction on the Frequency of Longitudinal Vibration of a Rod', *Proc. Roy. Soc.* **158** (1937) 606.
46. By "Galilean reference systems" we mean systems with rectilinear axes (Cartesian systems) moving with respect to each other in a rectilinear and uniform manner

(but not, of course, all deducible from each other by a Galilean transformation). These two definitions become identical only in the case of Newtonian Mechanics.

47. Let us note, with O. Costa de Beauregard, that experiments do not preclude the two following possibilities:
1. The value of the velocity of light depends on the absolute value (but not on the direction) of the ether-drift. In practical terms, this means that there might be a secular variation in the velocity of light.
2. The value of the velocity of light depends, with respect to the direction of the ether-drift, solely on the forward direction.
The first possibility – although quite improbable – is compatible with the principles of Special Relativity. The second is unverifiable. (Cf. O. COSTA DE BEAUREGARD [11], p. 15.)
Thus, in postulating the isotropism of the propagation of light only on the forward path, and assuming that the propagation is independent of the motion of the light-source (when, in fact, one can only verify an independence with respect to the movement of the source-observer ensemble), the principle of Special Relativity is not strictly imposed by experimental results. It simply constitutes the simplest hypothesis that permits interpretation of the experiment and leads to a coherent theory, all of whose predictions are experimentally verified.

48. A perfectly natural choice, when one considers that optics is the most precise and, to use O. Costa de Beauregard's expression, "the most geometrical and kinematic of the physical sciences". ([11], p. 15.) Let us note, moreover, that kinematics and dynamics are part of the physical sciences and thus must be influenced by the progress of the physical sciences. A *ne varietur* structure of those fields is thus entirely unthinkable.

49. A. EINSTEIN [12] and *Jahrbuch der Radioaktivität und Elektronik* **4** (1907) 411.

50. As Einstein remarks, to say that light takes the same time to cover *AM* as it does to cover *BM* is mere convention, an orientation procedure that tells us nothing about the properties of light. But in order to define absolute simultaneity, we would have to know that light actually does take the same time to cover *AM* as it does *BM*; in short, we would already have to be in possession of a means of measuring time (Einstein).

51. This conclusion had already been perceived by Henri Poincaré. But he did not go so far as to exclude the "theoretical" possibility of instantaneous signals, nor did he draw the logical conclusions of a purely physical definition of simultaneity. H. POINCARÉ: *La Valeur de la Science,* p. 35; and 'La mesure du temps', *Rev. Meta. et Morale* VI, 1, 28, p. 1.

52. Cf. e.g. J. CHAZY [19], t. II, p. 8.

53. The introduction of a four-dimensional space-time is proposed by H. POINCARÉ: *Rend. Pal.* **12** (1906) 129 and H. MINKOWSKI: 'Raum und Zeit' *Phys. Z.* **10** (1909) 104.

54. The contraction hypothesis, in the original form in which it appears in Fitzgerald's works, is the manifestation of the forces supposedly exerted by the ether on bodies in motion. These forces would presumably produce an absolute deformation, that is, one that is independent of any reference system used. Lorentz seeks to tie in these forces with much more general molecular interactions. The anisotropism exhibited by the contraction can, in the last analysis, be detected only as a result of corresponding modifications in duration and mass, but this anisotropism has an absolute character none the less.
In the light of Einstein's critique, the contraction no longer results from the play

of specific forces. It depends objectively (that is, independently of any observer) on whatever reference-system is adopted. It does not constitute an "appearance" for the reason that, since it is a reciprocal phenomenon, it cannot be contrasted with any contrary special "reality". The notion of length or dimension is thus objectively and essentially relative and results directly from the relative and irreducible character of simultaneity at a distance in two different Galilean systems. The important bibliography on this subject listed by H. ARZELIÈS ([8], p. 105) is especially noteworthy.

55. At this point one may bring up a very interesting discussion on the order of the temporal succession of two events and on the possibilities of causality (cf., for example, H. ARZELIÈS [8], p. 99).

1. Two events taking place simultaneously at two different points A and B of $S' \equiv S_0$, separated by a distance l_0, cannot be causally related. In order for that to be true, they would have to be separated by a time-interval

$$t_0 \geqslant \frac{l_0}{c}.$$

And the same holds true when the two events are related to a different Lorentzian system S. In fact, according to (V-39), two events that are simultaneous in $S' \equiv S_0$, that is, such that $t' = t_0 = 0$, will be, in S, separated by the intervals

$$l = \frac{l_0}{\sqrt{1-\beta^2}}, \qquad t = \frac{\beta l_0}{c\sqrt{1-\beta^2}}, \qquad \frac{l}{t} = \frac{c}{\beta} > c.$$

2. Two events $A(t'_A)$ and $B(t'_B)$ taking place at two different points in $S' \equiv S_0$ ($AB = l_0$) and such that $t_0 = t'_B - t'_A$ could be causally linked if

$$u_0 = \frac{l_0}{t_0} \leqslant c.$$

In the Lorentzian system S, one will still have

$$u = \frac{l}{t} = \frac{l_0 + vt_0}{t_0 + \dfrac{v}{c^2} l_0} = \frac{u_0 + v}{1 + \dfrac{vu_0}{c^2}} \leqslant c \quad \text{if} \quad u_0 \leqslant c.$$

A and B necessarily follow each other in the same order and present the possibility of being interpreted as cause and effect.

3. Two events $A(t'_A)$ and $B(t'_B)$ taking place at two different points of $S' \equiv S_0$ ($AB = l_0$) and such that $t_0 = t'_B - t'_A > 0$ cannot possibly be linked causally if

$$u_0 = \frac{l_0}{t_0} > c.$$

In the Lorentzian system S, one will have according to (V-39):

$$t = t_B - t_A = \frac{t_0 + \dfrac{v}{c^2} l_0}{\sqrt{1-\beta^2}} = t_0 \frac{1 + \dfrac{v}{c^2} u_0}{\sqrt{1-\beta^2}}.$$

It is possible to invert the course of events, that is, to assume

$$t<0 \quad \text{if} \quad \frac{1 + \frac{v}{c^2} u_0}{\sqrt{1-\beta^2}} < 0 \quad \text{or} \quad u_0(-v) > c^2$$

v being the velocity of S with respect to $S' \equiv S_0$. This condition is realizable since, in the proper system, the sequence of events presents no causal connection ($u_0 > c$ being a possibility).

Cf. [8] p. 101, the example proposed by Esclangon to illustrate this possibility.

56. One of the results of this situation is that the comparison of the measurements that must determine the slope of a straight line in "space" cannot have the absolute meaning that was conferred upon it by pre-relativistic theories. One is actually comparing data that cannot be simultaneous is S and S'. That is to say that, strictly speaking, the classical concept of a solid body cannot be retained. The results of all measurements are rigorously expressed in space-time diagrams relative to the two Lorentzian systems considered. Here we limit ourselves to the comparison of two "snapshots" (cf. J. L. Synge [18], p. 120) resulting from the cross-sections t = constant or t' = constant of space-time.
57. Cf. P. Langevin: 'L'évolution de l'espace et du temps', *Scientia* **10** (1911) 31. Also C. Møller [16], p. 258.
58. In this connection the problems of sudden stoppage and "take-off" may be examined. A few examples of them are to be found in W. H. MacCrea, 'The Fitzgerald-Lorentz contraction: some paradoxes and their resolution', *Proc. Roy. Dublin Soc.* 26 (1952) 27.
59. The longitudinal Doppler effect is due to the superposition of the non-relativistic Doppler effect $\nu = \nu_0 (1 \pm \beta \cos \theta)$, which is maximum when purely longitudinal ($\cos \theta = 1$; $\nu = \nu_0 [1 \pm \beta]$), and of the corrections required by relativity.
The transversal Döppler effect is purely relativistic, since the non-relativistic effect disappears in this case ($\cos \theta = 0$). It exhibits an immediate consequence of the phenomenon of the slowing-down of clocks. (Cf. Chapter X and also [8], p. 146.)
60. Cf. (X-15), p. 286.
61. The same conclusion is obtained if we assume that the source moves along with A. Cf. H. Arzeliès [8], p. 135.

THE FOUR-DIMENSIONAL FORMALISM OF SPECIAL RELATIVITY

1. *The pseudo-Euclidean universe of Special Relativity*

The basic postulate of Special Relativity – namely the constancy of the velocity of light in all Galilean reference systems – is expressed by the invariance of the basic quadratic form

$$ds^2 = c^2\,dt^2 - dx^2 - dy^2 - dz^2 . \tag{VI-1}$$

This invariance is insured under the general Lorentz transformation.

Expression (VI-1) characterizes the elementary interval of a four-dimensional Euclidean space related, as we are going to see, to a system of orthonormal rectilinear axes. The condition $ds^2 = 0$ defines, at the instant t and in a system of orthogonal axes, the light-wave limited at the initial instant from the origin of the coordinates. However, an Euclidean space for which the reduced form (VI-1) holds true is a pseudo-Euclidean space in the sense that every real, non-zero vector in that space does not have a norm, i.e., a length, which is necessarily defined as positive.

The changes of coordinates that permit the application of the principle of Special Relativity all apply to systems of orthonormal axes.

In a system of real axes, the normalization conditions naturally differ from those that would arise from a true Euclidean character of the space.

In order to establish a true Euclidean formalism, one may adopt the artifice of choosing *imaginary coordinates* (Cf. Chapter V, Section 7). The advantage of this procedure is that it once more makes it possible to give the reduced expression (VI-1) of ds^2 an *elliptical* form (the sum of four squares) and thus to avoid, as we will see (Cf. Chapter XIV, p. 419), the distinction between covariance and contravariance. The disadvantage resides in the introduction of coordinates whose imaginary character seems more or less tied in with the "natural" derivation of relativistic laws.

The use of *real coordinates* brings us back to a reduced *normal hyperbolic* form $(+\ -\ -\ -)$. It imposes the normalization condition that will be made clear by (VI-28) and obliges us to make a distinction between covariance and contravariance.

Under these conditions, we seem to gain very little by limiting ourselves to orthogonal axes satisfying (VI-28). In fact, this normalization condition really brings about very little significant simplification in the formalism. So it seems preferable in every way to assume a reference point in a *system of arbitrary rectilinear axes* (generally, oblique axes) without imposing upon it *a priori* the elliptical or hyperbolic "signature" of the ds^2, which depends on the chosen coordinates. A study of this method of reference is made in Chapter XIV-A (Mathematical Supplement).

Of course, the formalism relative to oblique axes is too broad for the description of the Lorentz transformations which apply exclusively to orthonormal systems, but it does include such systems as a special case. So we can obtain the Lorentz transformation if we are careful to make a proper choice of normalization conditions, depending on the type of coordinates used. These conditions limit the changes of oblique axes to transformations from one orthonormal system to another and lead to the four-dimensional formalism suitable to the expression of the Lorentz transformation.

This method – that is, the *a priori* definition of oblique axes – has the advantage of simultaneously introducing the various special cases corresponding to the various choices of coordinates and, consequently, to the various conditions of normalization. The resultant restrictions are quite obvious in each case.

In addition, it is easy to generalize the reference system of an oblique-axial Euclidean space (Chapter XIV-A) to Euclidean representations in any curvilinear system (Chapter XIV-B). One is thus led, without difficulty, to refer points in a non-Euclidean space to the systems of curvilinear coordinates which it then becomes necessary to introduce.[1] In this way one can state covariant laws, in a transformation of any arbitrary coordinates, in keeping with the principles of a Generalized Relativity.

Remark. Clearly, phenomena occurring in a four-dimensional Euclidean space are not necessarily described in a system of orthonormal axes. One can employ a representation using oblique axes (Chapter XIV-A) or even a system of any sort of curvilinear axes (XIV-B). In that case, it would, of course, also be necessary to introduce a distinction between covariance and contravariance and, if we are dealing with curvilinear coordinates, to make use of the notion of covariant differentiation. Nevertheless, the use of these axial systems, which may be convenient and even indispensable in solving certain problems, always remains *optional* as long as we are concerned with a Euclidean space. In other words, one can always come back to referring points of an extended region of that space to a system of orthonormal axes.

2. *Notational conventions*

1) *Indexes.* The *Greek indexes* μ, ν, ρ, σ take on the values 1, 2, 3, 4 when coordinates $x^1 = x$, $x^2 = y$, $x^3 = z$, $x^4 = ict$ are chosen, and 1, 2, 3, 0 when the real coordinates $x^1 = x$, $x^2 = y$, $x^3 = z$, $x^0 = ct$ are used.

The *Latin indexes* p, q, r, s take on the values 1, 2, 3.

2) *Summation convention.* We have adopted the following convention of summation:

The summation sign over an index is always understood when that index is repeated both above and below in the same expression.

This 'dummy' index has no special value other than simply indicating that we are concerned with a summation. Thus, in the coordinate system $x^\mu\,(x^1, x^2, x^3, x^0)$

$$A_\mu B^\mu = A_1 B^1 + A_2 B^2 + A_3 B^3 + A_0 B^0$$

$$A_p B^p = A_1 B^1 + A_2 B^2 + A_3 B^3$$

3) *Vector and tensor representation.* In a three-dimensional Euclidean space, we have represented a vector by the symbol **A**. The components of this vector are

$$A_x = A_1\,, \qquad A_y = A_2\,, \qquad A_z = A_3$$

or, in a shorter form, A_p with $p = 1, 2, 3$.

In a four-dimensional Euclidean space, we likewise use the symbol **A** to represent a vector. The components of this vector are $A_x = A_1$, $A_y = A_2$, $A_z = A_3$, $A_{ct} = A_0$ (or $A_{ict} = A_4$). In a shorter form, these components will be designated by A_μ with $\mu = 1, 2, 3, 0$ (or $\mu = 1, 2, 3, 4$). It should be noted that the three components A_p of a vector in three-dimensional space do not necessarily constitute the first three, or space, components of a four-vector. We will see that such is the case for the coordinates $x^\mu = (\mathbf{x}, x^0)$, but this is not true of the velocities.

3. *Reduced forms of the ds^2 in Special Relativity*

1) *Choice of imaginary coordinates.* a) If we put

(VI-2) $$\boxed{x^1 = ix \qquad x^2 = iy \qquad x^3 = iz \qquad x^4 = ct}$$

then (VI-1) will be written

(VI-3) $$ds^2 = (dx^1)^2 + (dx^2)^2 + (dx^3)^2 + (dx^4)^2 = \Sigma_\mu (dx^\mu)^2 \quad \mu = 1, 2, 3, 4$$

and represents the elementary interval of a four-dimensional Euclidean space with orthonormal rectilinear axes.

In fact, if we define a system of rectilinear axes by means of four unit-vectors $\mathbf{e}_\mu$ ($\mathbf{e}_1, \mathbf{e}_2, \mathbf{e}_3, \mathbf{e}_4$), we put

(VI-4) $$\mathbf{ds} = \mathbf{e}_1\, dx^1 + \mathbf{e}_2\, dx^2 + \mathbf{e}_3\, dx^3 + \mathbf{e}_4\, dx^4 .$$

The scalar product $\mathbf{ds} \cdot \mathbf{ds} = ds^2$ coincides with (VI-3) if the vectors $\mathbf{e}$ are: orthogonal

(VI-5) $$\mathbf{e}_\mu \cdot \mathbf{e}_\nu = 0 \qquad \text{for} \qquad \mu \neq \nu$$

normalized

(VI-6) $$\mathbf{e}_\mu^2 = 1 .$$

Conditions (VI-5) and (VI-6) may be grouped together to form the condition of orthonormalization

(VI-7) $$(\mathbf{e}_\mu \cdot \mathbf{e}_\nu) = \delta_{\mu\nu}$$

$\delta_{\mu\nu}$ being the Kronecker symbol,

(VI-8) $$\delta_{\mu\nu} = \begin{cases} 1 & \text{if} \quad \mu = \nu \\ 0 & \text{if} \quad \mu \neq \nu . \end{cases}$$

With the choice of coordinates (VI-2) and the introduction of orthonormal axes by (VI-7), the elliptical signature (VI-3) of the ds^2 is immediately deduced from the expression of the ds^2 in any system of oblique axes (cf. XIV-A)

(VI-9) $$ds^2 = g_{\mu\nu}\, dx^\mu\, dx^\nu ,$$

putting

(VI-10) $\boxed{g_{\mu\nu} = \delta_{\mu\nu}.}$

The conditions of orthonormality (VI-10) entail the equality of the covariant and contravariant components of a vector:

(VI-11) $A_\mu = g_{\mu\nu} A^\nu = \delta_{\mu\nu} A^\nu = A^\mu .$

Finally, the expressions for the *scalar product* (XIV-16) of two vectors and for the norm (XIV-18) – or *length* – of a vector are[2]:

(VI-12) $\mathbf{A} \cdot \mathbf{B} = g_{\mu\nu} A^\mu B^\nu = \Sigma_\mu A^\mu B^\mu$

(VI-13) $|A|^2 = g_{\mu\nu} A^\mu A^\nu = \Sigma_\mu (A^\mu)^2 .$

b) We will have recourse most frequently to

(VI-14) $\boxed{x^1 = x \qquad x^2 = y \qquad x^3 = z \qquad x^4 = ict}$

whence

(VI-15) $ds^2 = - \Sigma_\mu (dx^\mu)^2 , \qquad \mu = 1, 2, 3, 4 .$

This reduced form results from the use of rectilinear axes defined by the unit-vectors $\mathbf{e}_\mu$ orthonormalized according to

(VI-16) $(\mathbf{e}_\mu \cdot \mathbf{e}_\nu) = - \delta_{\mu\nu}$

instead of (VI-7). The reduced expression for ds^2 then results from the general form (VI-9) by choosing

(VI-17) $\boxed{g_{\mu\nu} = - \delta_{\mu\nu} .}$

In this system of axes the relationship between the covariant and contravariant components of a vector then becomes

(VI-18) $A_\mu = g_{\mu\nu} A^\nu = - \delta_{\mu\nu} A^\nu = - A^\mu$

and the expressions for the scalar product and the norm become, respectively:

(VI-19) $(\mathbf{A} \cdot \mathbf{B}) = g_{\mu\nu} A^\mu B^\nu = - \Sigma_\mu A^\mu B^\mu$

(VI-20) $|A|^2 = g_{\mu\nu} A^\mu A^\nu = - \Sigma_\mu (A^\mu)^2 .$

2) *Choice of real coordinates.* If one adopts real coordinates

(VI-21) $$\boxed{x^1 = x \qquad x^2 = y \qquad x^3 = z \qquad x^0 = ct}$$

then (VI-1) may be put in the following form:

(VI-22) $$ds^2 = (dx^0)^2 - \Sigma_p (dx^p)^2 , \qquad p = 1,2,3 .$$

This "reduced" form defines the normal hyperbolic signature $+ - - -$ of the ds^2. It corresponds to the choice of a system of rectilinear axes defined by the unit-vectors $\mathbf{e}_\mu$, $(\mathbf{e}_1, \mathbf{e}_2, \mathbf{e}_3, \mathbf{e}_0)$.

We still have

(VI-23) $$\mathbf{ds} = \mathbf{e}_1\, dx^1 + \mathbf{e}_2\, dx^2 + \mathbf{e}_3\, dx^3 + \mathbf{e}_0\, dx^0$$

and the scalar product $\mathbf{dsds} = ds^2$ is identical to (VI-22) if the vectors $\mathbf{e}_\mu$ are

a) orthogonal

(VI-24) $$(\mathbf{e}_\mu \cdot \mathbf{e}_\nu) = 0 \qquad \text{for} \qquad \mu \neq \nu$$

b) normalized by the conditions

(VI-25) $$\mathbf{e}_0^2 = 1 , \qquad \mathbf{e}_p^2 = -1 , \qquad p = 1,2,3 .$$

These conditions are brought together in the following formula

(VI-26) $$(\mathbf{e}_\mu \cdot \mathbf{e}_\nu) = \eta_{\mu\nu}$$

with

(VI-27) $$\eta_{\mu\nu} = \begin{vmatrix} -1 & & & \\ & -1 & & \\ & & -1 & \\ & & & +1 \end{vmatrix}$$

Conditions (VI-26) make it possible to express the ds^2 in the reduced form (VI-22), that is, in a system of real coordinates. (VI-22) is deducible from the general form (VI-9) if

(VI-28) $$\boxed{g_{\mu\nu} = \eta_{\mu\nu} .}$$

The covariant and contravariant components of a vector are then connected in the following manner

(VI-29) $$A_\mu = g_{\mu\nu} A^\nu = \eta_{\mu\nu} A^\nu$$

that is

(VI-30) $$A_p = \eta_{p\nu} A^\nu = -\delta_{p\nu} A^\nu = -A^p$$

(VI-31) $$A_0 = \eta_{0\nu} A^\nu = \delta_{0\nu} A^\nu = A^0 .$$

The scalar product of two vectors and the norm of a vector may also be expressed as follows:

(VI-32) $$\mathbf{A}\cdot\mathbf{B} = g_{\mu\nu} A^\mu B^\nu = A^0 B^0 - \Sigma_p A^p B^p \qquad p = 1, 2, 3$$

(VI-33) $$|A|^2 = g_{\mu\nu} A^\mu A^\nu = (A^0)^2 - \Sigma_p (A^p)^2 .$$

We verify, through the various expressions (VI-13), (VI-20) and (VI-33) of the norm of a *real* vector, that this norm is not necessarily defined as positive for every non-zero vector; the four-dimensional space of Special Relativity is *pseudo-Euclidean.*

4. *Space-like four-vectors. Time-like four-vectors. Isotropic four-vectors*

A four-vector **A** will be said to be *temporal, spatial* or *isotropic*, depending on whether its norm (or length) is positive, negative or zero. Let us choose a system of real coordinates. In a system orthonormalized by (VI-28), one will have, according to (VI-33)

(VI-34) $$|A|^2 = g_{\mu\nu} A^\mu A^\nu = (A^0)^2 - \Sigma_p (A^p)^2 \gtrless 0 .$$

A will be *temporal* if $|A|^2 > 0$ or $(A^0)^2 > \Sigma_p (A^p)^2$.
A will be *spatial* if $|A|^2 < 0$ or $(A_0)^2 < \Sigma_p (A^p)^2$.
A will be *isotropic* if $\Sigma_p (A^p)^2 = (A_0)^2$.

More especially, ds^2 *is positive for a particle of velocity* $v > c$. The tangents to the trajectory of a material particle then satisfy the relation $(dx^0)^2 - \Sigma_p (dx^p)^2 > 0$ and form temporal four-vectors. *They are inside the characteristic cone defined by* $ds^2 = 0$.

The trajectories satisfying $ds^2 = 0$ correspond to the light-rays $v = c$. The tangents to these trajectories are situated on the characteristic cone $ds^2 = 0$.

The characteristic cone $ds^2 = 0$ thus divides the four-dimensional space up into two regions: one ($ds^2 > 0$) includes all the temporal vectors and corresponds to the inside of the cone having its apex at the origin and its lines of generation along the isotropic vectors tangent to the trajectories of the light rays. This latter region is then further subdivided to form the

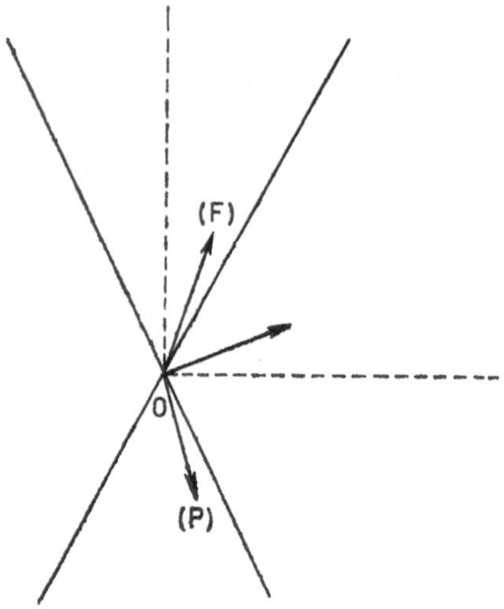

Fig. 28. The characteristic cone.

upper section (F) and the lower section P of the cone. In (F) all the A^0 components of a temporal four-vector are positive: this is the *region of the future*. In (P) these components are negative, and that is the *region of the past*.

The second region ($ds^2 < 0$) corresponds to the outside of the cone and includes all the spatial vectors; it is the "elsewhere" region.

5. *The invariance of the ds^2 under the displacement group in four-dimensional Euclidean space*

We now seek the transformations which, without altering the unit of length, allow us to pass from a system of axes orthonormal according to conditions (VI-10), (VI-17) or (VI-28), to another system likewise orthonormal according to the same conditions.

a) In the case where the introduction of imaginary coordinates imposes the choice of axes orthonormal according to (VI-10) ($g_{\mu\nu} = \delta_{\mu\nu}$) or even by the conditions (VI-17) ($g_{\mu\nu} = -\delta_{\mu\nu}$), the ds^2 has the reduced form (VI-3) or (VI-15). One must then insure the preservation of conditions

$$\text{(VI-35)} \qquad \Sigma_\rho (dx^\rho)^2 = \Sigma_\rho (dx'^\rho)^2 .$$

b) In the case where the choice of real coordinates ($x^0 = ct$) entails the normalization conditions (VI-28) ($g_{\mu\nu} = \eta_{\mu\nu}$), the reduced form of the ds^2 is given by (VI-22) and the invariance of the expression

$$- \Sigma_p (dx^p)^2 + (dx^0)^2 = - \Sigma_p (dx'^p)^2 + (dx'^0)^2 . \qquad \text{(VI-36)}$$

must be assured.

This invariance is achieved by the *displacement group in Euclidean* – or *pseudo-Euclidean – four-dimensional space.* This group includes:

1) *The translations* defined by

$$x'^\mu = x^\mu + a^\mu \qquad (a^\mu = \text{const}) . \qquad \text{(VI-37)}$$

Whence there results:

$$dx'^\mu = dx^\mu . \qquad \text{(VI-38)}$$

2) *The linear and orthogonal substitutions of Euclidean – or pseudo-Euclidean – four-dimensional space.* According to (XIV-23) these substitutions have always the form

$$\boxed{\begin{array}{ll} \mathbf{e}'_\mu = a^\nu_{\mu'} \mathbf{e}_\nu & \mathbf{e}_\mu = a^{\nu'}_\mu \mathbf{e}'_\nu \\ x'^\mu = a^{\mu'}_\nu x^\nu & x^\mu = a^\mu_{\nu'} x'^\nu \end{array}} \qquad \begin{array}{l}\text{(VI-39)}_1 \\ \text{(VI-39)}_2\end{array}$$

with (Cf. XIV-28)

$$\boxed{a^\rho_{\mu'} a^{\nu'}_\rho = a^{\rho'}_\mu a^\nu_{\rho'} = \delta^\nu_\mu .} \qquad \text{(VI-40)}$$

One must, in addition, uphold the orthogonality of the axes after the transformation. Whatever the norm may be, one will in this way achieve the invariance of the scalar product (cf. Chapter XIV)

$$g'_{\mu\nu} = (\mathbf{e}'_\mu \cdot \mathbf{e}'_\nu) = (\mathbf{e}_\mu \cdot \mathbf{e}_\nu) = g_{\mu\nu} \qquad \text{(VI-41)}$$

that is, the conditions (XIV-70)

$$\boxed{a^\rho_{\mu'} g_{\rho\lambda} = a^{\rho'}_\lambda g_{\mu\rho} .} \qquad \text{(VI-42)}$$

a) If the axes are orthonormalized by the conditions $g_{\mu\nu} = \delta_{\mu\nu}$ or even by the conditions $g_{\mu\nu} = - \delta_{\mu\nu}$, one will then have according to (VI-42)

$$a^\lambda_{\mu'} = a^{\mu'}_\lambda . \qquad \text{(VI-43)}$$

A transformation (VI-39) thus always entails, according to (VI-40) and (VI-43)

(VI-44) $$\Sigma_\rho a^\rho_{\mu'} a^\rho_{\nu'} = \delta_{\mu\nu}.$$

b) If the axes are orthonormal according to $g_{\mu\nu} = \eta_{\mu\nu}$, (VI-42) is expressed by:

(VI-45) $$a^q_{p'} = a^{p'}_q, \qquad a^0_{0'} = a^{0'}_0, \qquad a^0_{p'} = -a^{p'}_0, \qquad a^p_{0'} = -a^{0'}_p.$$

Remark. The choice $x^4 = ict$ necessitates that the coefficients $a^q_{p'}$, and $a^{4'}_4$, are real; whereas $a^4_{p'}$ and $a^p_{4'}$ are pure imaginaries. One will then have, according to (VI-40)

(VI-46) $$(a^4_{4'})^2 = 1 - \Sigma_p (a^p_{4'})^2 \geqslant 1.$$

On the contrary, putting $x^0 = ct$, all the coefficients $a^\rho_{\mu'}$ and $a^{\rho'}_\mu$ are real. One will thus have, according to (VI-40) and (VI-45)

(VI-47) $$(a^0_{0'})^2 = 1 + \Sigma_p (a^p_{0'})^2 \geqslant 1.$$

Likewise, in both cases, the inequality $(a^0_{0'})^2 \geqslant 1$ must entail

(VI-48) $$\boxed{a^0_{0'} \geqslant 1 \qquad (\text{or } a^4_{4'} \geqslant 1)}$$

excluding transformations of the type $(a^0_{0'})^2 \leqslant -1$. By choosing (VI-48), we confine ourselves to the transformations (VI-39), which form a group. Indeed, if such is the case, the passage from the x to the x', and then from the x' to the x'' must be of the same type as the passage from the x to the x''. One will have:

(VI-49) $$x'^\mu = a^{\mu'}_\nu x^\nu, \qquad x''^\mu = a^{\mu''}_{\nu'} x'^\nu$$

that is

(VI-50) $$x''^\mu = a^{\mu''}_{\nu'} a^{\nu'}_\rho x^\rho = a^{\mu''}_\rho x^\rho.$$

The coefficients

(VI-51) $$a^\mu_\rho = a^{\mu''}_{\nu'} a^{\nu'}_\rho$$

then share the same properties as the $a^{\mu''}_{\rho'}$ and the $a^{\mu'}_\rho$ do.

In order to form a group, the transformations (VI-49) must include as a special case the identical transformation, which forces the coefficient $a^0_{0'}$ to satisfy the inequality (VI-48).

This inequality shows that the sign of the component A^0 of a temporal four-vector ($A^2 > 0$) remains the same after a linear and orthogonal substitution of the preceding type. This sign, consequently, does not vary from one Galilean system to another.

In particular, in all the real Galilean systems, **dx** is time-like, since $ds^2 > 0$ if $v < c$. The component dx^0, which is positive in the region of the future, therefore keeps this sign in the change of axes defined above, that is, in a change of Galilean systems. This means that in all the Galilean systems having a physical meaning ($ds^2 > 0$) *time flows in the same direction.*

Thus, to the equivalence between the role of the spatial and temporal coordinates, which is expressed by the relations (VI-39), – to that equivalence the inequality $a^0_{0'} \gg 0$ adds the basic postulate of the irreversibility of time.

6. *The general Lorentz transformation and the special transformation*

We have just seen that the displacements group of four-dimensional Euclidean space retains the reduced form of the ds^2. The invariance of ds^2 under this group thus assures the special equivalence possessed by all Galilean systems and likewise, thanks to (VI-48), the direction of the flow of time. This group defines the general Lorentz transformation.

It is important to note that Lorentz transformations *without rotation* do not of themselves constitute a group when the velocities are randomly oriented with respect to the coordinate axes.[3] As a matter of fact, the composition of two Lorentz transformations *without rotation* $S_1 \rightarrow S_2 \rightarrow S_3$ actually does define a Lorentz transformation $S_1 \rightarrow S_3$, but the latter transformation *is not likewise without rotation.* It generally involves a rotation D of the spatial axes. In fact, one notices that if $S_1 \rightarrow S_2$ and $S_2 \rightarrow S_3$ are described by formulas of the type $(\text{V-71})_1$, $S_1 \rightarrow S_3$ depends on the expression $(\text{V-78})_1$. Moreover, if $\mathbf{v}_{(ij)}$ designates the velocity of S_i with respect to S_j

$$\mathbf{v}_{(12)} = -\mathbf{v}_{(21)}, \qquad \mathbf{v}_{(23)} = -\mathbf{v}_{(32)}$$
$$\text{but} \qquad D\mathbf{v}_{(31)} = -\mathbf{v}_{(13)}.$$

This phenomenon, called the *Thomas precession*[4], is an illustration, in kinematics, of the following property: if one of the transformations, $S_2 \rightarrow S_3$ for example, is an infinitesimal transformation, the rotation D is manifested by a precession of the axes of S_3 with respect to S_1, and the velocity of the precession is proportional to

$$\frac{[\mathbf{v}_{(12)} \wedge \dot{\mathbf{v}}_{(12)}]}{\mathbf{v}_{(12)}^2}.$$

The special Lorentz transformation is a special case of the transformation without rotation. It is connected with the case in which the axes of the selected systems are parallel, with the relative velocity parallel to one of them.

The general Lorentz transformation is equivalent to the ensemble of the three following transformations:

1) *A purely spatial displacement* involving the invariance of

$$d\sigma^2 = dx^2 + dy^2 + dz^2 \qquad \text{(VI-52)}$$

that is to say, involving a change of the origin of the axes of the space associated with a rotation of these axes.

2) *A change in the time origin.*

3) *A special Lorentz transformation* that is, a change of the system of reference arising in the special case of parallel axes oriented in the same direction, with one of them (ox) being directed along the path of the velocity $\mathbf{v}$.

As a matter of fact, one can always, by means of a four-variable translation, obtain one and the same origin for the systems $S\,(xyz)$ and $S'\,(x'y'z')$ and insure the coincidence of the time origins. A rotation of the spatial axes will thus make the axes ox and ox' coincide with the relative velocity $\mathbf{v}$ of the two systems of reference. A rotation of S' around ox' will then make oy' and oz' be parallel to oy and oz (and with the same direction). In this way we are brought back to the special Lorentz transformation, which is actually nothing more than the rotations which do not change the xoy and xoz planes (coinciding with $x'oy'$ and $x'oz'$), since these rotations leave intact the relations $y' = y$ and $z' = z$.

It is precisely the rotations defined in the imaginary plane x^1ox^4:

$$\begin{aligned} x'^1 &= x^1 \cos\psi + x^4 \sin\psi \\ x'^4 &= x^4 \cos\psi - x^1 \sin\psi \end{aligned} \qquad \text{(V-68)}$$

which enjoy this property. They are expressed, as we have seen, as functions of the imaginary angle ψ such that $\tan\psi = i\beta$.

In other words, one can always replace a general Lorentz transformation by a special transformation by subjecting the general transformation to a spatial displacement (translation-plus-rotation), leaving $d\sigma^2$ invariant, and to an arbitrary change of the time origin.

Of course, in the particular case of the special transformation, but only in that case, the composition of two transformations, $S_1 \to S_2 \to S_3$, without rotation, leads to a transformation of the same type.

7. *Expression of the coefficients in the general Lorentz transformation*

The general Lorentz transformation is defined by formulas (VI-39)

$$\text{(VI-39)}_1 \qquad \boxed{\mathbf{e}_\mu = a_\mu^{\nu'}\mathbf{e}'_\nu \qquad \mathbf{e}'_\mu = a_{\mu'}^{\nu}\mathbf{e}_\nu}$$

$$\text{(VI-39)}_2 \qquad \boxed{x^\mu = a_{\nu'}^{\mu}x'^\nu \qquad x'^\mu = a_\nu^{\mu'}x^\nu}$$

with

$$\text{(VI-40)} \qquad \boxed{a_\mu^{\rho'}a_{\rho'}^{\nu} = a_{\mu'}^{\rho}a_\rho^{\nu'} = \delta_\mu^\nu}$$

on condition that the following restrictions are taken into account

$$\text{(VI-42)} \qquad a_{\mu'}^{\rho}g_{\rho\nu} = a_\nu^{\rho'}g_{\mu\rho}$$

in a system of axes orthonormal according to the appropriate conditions.

a) If $g_{\mu\nu} = \pm\,\delta_{\mu\nu}$ (choice of *imaginary* coordinates) condition (VI-42) is written

$$\text{(VI-42)}_a \qquad a_{\mu'}^{\nu} = a_\nu^{\mu'}$$

and (VI-40) takes on the form

$$\text{(VI-40)}_a \qquad \textstyle\sum_\rho a_\mu^{\rho'}a_\nu^{\rho'} = \sum_\rho a_{\mu'}^{\rho}a_{\nu'}^{\rho} = \delta_{\mu\nu}.$$

b) If $g_{\mu\nu} = \eta_{\mu\nu}$ (choice of *real* coordinates), (VI-42) is identical to

$$a_{\mu'}^{\rho}\eta_{\rho\nu} = a_\nu^{\rho'}\eta_{\mu\rho}$$

that is

$$\text{(VI-42)}_b \qquad \boxed{a_{p'}^{q} = a_q^{p'}, \qquad a_{0'}^{0} = a_0^{0'}, \qquad a_{p'}^{0} = -\,a_0^{p'}, \qquad a_{0'}^{p} = -\,a_p^{0'}}$$

and (VI-40) may be written

$$\text{(VI-40)}_b \qquad \sum_r a^{r'}_p a^{r'}_q - a^{0'}_p a^{0'}_q = \delta_{pq},$$

$$-\sum_r a^{r'}_0 a^{r'}_0 + (a^{0'}_0)^2 = 1, \qquad -\sum_r a^{r'}_p a^{r'}_0 + a^{0'}_p a^{0'}_0 = 0.$$

In this section we will limit ourselves to the choice of *real coordinates* and, consequently, to conditions (VI-42)_b and (VI-40)_b.

Let us once more consider expression (V-78) of the general transformation of coordinates established in Chapter V. By writing out the components x^μ ($\mathbf{x} = x^p$, $x^0 = ct$) of the four-vector, we have further:

$$\text{(V-78)}_1 \qquad x'^p = D^{-1} x^p + D^{-1} v^p \left\{ \frac{\alpha}{v^2} (\mathbf{x} \cdot \mathbf{v}) - \frac{x^0}{c\sqrt{1-\beta^2}} \right\}$$

$$\text{(V-78)}_2 \qquad x'^0 = \frac{x^0 - \left(\dfrac{\mathbf{x} \cdot \mathbf{v}}{c} \right)}{\sqrt{1-\beta^2}}$$

with

$$\text{(VI-53)} \qquad \mathbf{v} = (v^p), \qquad v_p = -v^p, \qquad v^2 = \Sigma_p (v^p)^2$$

$$\text{(VI-54)} \qquad \mathbf{x} \cdot \mathbf{v} = \sum_r x^r v^r = -x^r v_r$$

and also

$$\text{(V-75)} \qquad \boxed{\alpha = \frac{1}{\sqrt{1-\beta^2}} - 1.}$$

The quantities $v^p = dx^p/dt$ do not constitute the space components of a four-vector, because dt is not an invariant. We will replace them by the components of the four-vector $u^\mu = dx^\mu/ds$ (ds being an invariant) such that

$$\text{(VI-55)} \qquad \begin{aligned} u^p &= \frac{dx^p}{ds} = \frac{dx^p}{dt}\frac{dt}{ds} = \frac{v^p}{c\sqrt{1-\beta^2}}, \\ u^0 &= \frac{dx^0}{ds} = c\frac{dt}{ds} = \frac{1}{\sqrt{1-\beta^2}}. \end{aligned}$$

$(\text{V-78})_1$ and $(\text{V-78})_2$ will be written thus:

$$(\text{VI-55})_1 \qquad x'^p = D^{-1} x^p - D^{-1} u^p \left\{ \alpha\left(\frac{1-\beta^2}{\beta^2}\right) x^r u_r + x^0 \right\}$$

$$(\text{VI-55})_2 \qquad x'^0 = x^0 u_0 + x^p u_p .$$

Let us represent by α_p^q the coefficients of the spatial rotation D^{-1} such that

$$(\text{VI-56}) \qquad D^{-1} x^p = \alpha_q^p x^q .$$

By comparison with $(\text{VI-39})_2$, which may also be written

$$(\text{VI-56})_1 \qquad x'^p = a_\nu^{p'} x^\nu = a_q^{p'} x^q + a_0^{p'} x^0$$

$$(\text{VI-56})_2 \qquad x'^0 = a_\nu^{0'} x^\nu = a_q^{0'} x^q + a_0^{0'} x^0 .$$

and by further comparison with (VI-55), one will then have

$$(\text{VI-57}) \qquad \boxed{a_q^{p'} = a_{p'}^{q} = \alpha_q^p - \alpha\left(\frac{1-\beta^2}{\beta^2}\right)\alpha_r^p u^r u_q = \alpha_q^p - \frac{\alpha}{v^2}\alpha_r^p v^r v_q}$$

$$(\text{VI-58}) \qquad \boxed{a_0^{p'} = -\,a_{p'}^{0} = -\,\alpha_r^p u^r , \qquad a_p^{0'} = -\,a_{0'}^{p} = u_p}$$

The preceding formulas (VI-57) and (VI-58) are summarized in the two following tables:

$$(\text{VI-59})_1 \qquad u_\mu^{\nu'} = \left|\begin{array}{cccc} \alpha_r^1\gamma_1^r & \alpha_r^2\gamma_1^r & \alpha_r^3\gamma_1^r & -u^1 \\ \alpha_r^1\gamma_2^r & \alpha_r^2\gamma_2^r & \alpha_r^3\gamma_2^r & -u^2 \\ \alpha_r^1\gamma_3^r & \alpha_r^2\gamma_3^r & \alpha_r^3\gamma_3^r & -u^3 \\ -\alpha_r^1 u^r & -\alpha_r^2 u^r & -\alpha_r^3 u^r & u^0 \end{array}\right|$$

$$(\text{VI-59})_2 \qquad a_{\mu'}^{\nu} = \left|\begin{array}{cccc} \alpha_r^1\gamma_1^r & \alpha_r^1\gamma_2^r & \alpha_r^1\gamma_3^r & \alpha_r^1 u^r \\ \alpha_r^2\gamma_1^r & \alpha_r^2\gamma_2^r & \alpha_r^2\gamma_3^r & \alpha_r^2 u^r \\ \alpha_r^3\gamma_1^r & \alpha_r^3\gamma_2^r & \alpha_r^3\gamma_3^r & \alpha_r^3 u^r \\ u^1 & u^2 & u^3 & u^0 \end{array}\right|$$

putting

$$(\text{VI-60}) \qquad \boxed{\begin{aligned} \gamma_p^r &= \delta_p^r + \frac{\alpha}{\beta^2}(1-\beta^2)\, u^r u^p = \\ &= \delta_p^r + \frac{\sqrt{1-\beta^2}}{\beta^2}\left(1-\sqrt{1-\beta^2}\right) u^r u^p \end{aligned}}$$

or

$$\boxed{\gamma_p^r = \delta_p^r + \frac{\alpha}{v^2} v^r v^p}$$

according to (VI-57), (VI-58) and (V-75). It should be noted that the subscript of the coefficients $a_\mu^{\nu'}$ or $a_{\mu'}^{\nu}$ always correspond to the lines of the matrices (VI-59) while the superscripts correspond to the columns.

Remark. In the less general case of a Lorentz transformation *without rotation* (cf. Chapter V, p. 149)

(VI-61) $$\alpha_p^q = \delta_p^q .$$

Formulas (VI-57) and (VI-58) are then written

(VI-62) $$\boxed{a_p^{q'} = a_{p'}^q = \delta_p^q + \frac{\alpha}{v^2} v^p v^q , \qquad a_{0'}^0 = a_0^{0'} = \frac{1}{\sqrt{1-\beta^2}} = u_0}$$

(VI-63) $$\boxed{\begin{aligned} a_0^{p'} &= -\, a_{p'}^0 = -\frac{v^p}{c\sqrt{1-\beta^2}} = -\, u^p , \\ a_p^{0'} &= -\, a_{0'}^p = -\frac{v^p}{c\sqrt{1-\beta^2}} = -\, u^p \end{aligned}}$$

that is

(VI-64)
$$a_\mu^{\nu'} = \begin{vmatrix} \gamma_1^1 & \gamma_1^2 & \gamma_1^3 & -u^1 \\ \gamma_2^1 & \gamma_2^2 & \gamma_2^3 & -u^2 \\ \gamma_3^1 & \gamma_3^2 & \gamma_3^3 & -u^3 \\ -u^1 & -u^2 & -u^3 & u^0 \end{vmatrix}$$

$$a_{\mu'}^{\nu} = \begin{vmatrix} \gamma_1^1 & \gamma_1^2 & \gamma_1^3 & u^1 \\ \gamma_2^1 & \gamma_2^2 & \gamma_2^3 & u^2 \\ \gamma_3^1 & \gamma_3^2 & \gamma_3^3 & u^3 \\ u^1 & u^2 & u^3 & u^0 \end{vmatrix}$$

8. *Application to the special Lorentz transformation*

In the even more particular case of a *special Lorentz transformation*

(VI-65) $$\mathbf{v} = v^1 , \qquad v^2 = v^3 = 0 .$$

The only non-zero coefficients of (VI-62) and (VI-63) are then

(VI-66)
$$\begin{aligned}
a_1^{1'} &= a_{1'}^1 = 1 + \alpha = \frac{1}{\sqrt{1-\beta^2}},\\
a_2^{2'} &= a_{2'}^2 = a_3^{3'} = a_{3'}^3 = 1\\
a_0^{1'} &= -\, a_{1'}^0 = \frac{-\beta}{\sqrt{1-\beta^2}},\\
a_1^{0'} &= -\, a_{0'}^1 = \frac{-\beta}{\sqrt{1-\beta^2}},\\
a_0^{0'} &= a_{0'}^0 = \frac{1}{\sqrt{1-\beta^2}}.
\end{aligned}$$

These values actually coincide with those that are directly deducible from the special transformation put, for example, into the form (V-64). Putting $\tanh\varphi = \beta$, one has

(VI-67)
$$a_\mu^{\nu'} = \begin{vmatrix} \cosh\varphi & 0 & 0 & -\sinh\varphi \\ 0 & 1 & 0 & 0 \\ 0 & 0 & 1 & 0 \\ -\sinh\varphi & 0 & 0 & \cosh\varphi \end{vmatrix}$$
$$a_{\mu'}^{\nu} = \begin{vmatrix} \cosh\varphi & 0 & 0 & \sinh\varphi \\ 0 & 1 & 0 & 0 \\ 0 & 0 & 1 & 0 \\ \sinh\varphi & 0 & 0 & \cosh\varphi \end{vmatrix}$$

Remark. By adopting the imaginary coordinates (VI-14) ($x^4 = ict$) and by putting $\tan\psi = i\beta$, formulas (V-68) concerning the special transformation would have led us to the following values for the coefficients $a_\mu^{\nu'}$ and $a_{\mu'}^{\nu}$:

(VI-68)
$$a_\mu^{\nu'} = \begin{vmatrix} \cos\psi & 0 & 0 & \sin\psi \\ 0 & 1 & 0 & 0 \\ 0 & 0 & 1 & 0 \\ -\sin\psi & 0 & 0 & \cos\psi \end{vmatrix}$$
$$a_{\mu'}^{\nu} = \begin{vmatrix} \cos\psi & 0 & 0 & -\sin\psi \\ 0 & 1 & 0 & 0 \\ 0 & 0 & 1 & 0 \\ \sin\psi & 0 & 0 & \cos\psi \end{vmatrix}$$

9. *Examples*

1) *Transformation of a vector* **A**. Let us choose a system of real coordinates. We have

$$A'^{\mu} = a^{\mu'}_{\nu} A^{\nu}, \qquad A'_{\mu} = a^{\nu}_{\mu'} A_{\nu} \tag{VI-69}$$

that is, for the contravariant components,

$$A'^{p} = a^{p'}_{q} A^{q} + a^{p'}_{0} A^{0} = \alpha^{p}_{r} \gamma^{r}_{q} A^{q} - \alpha^{p}_{r} u^{r} A^{0} \tag{VI-70$_1$}$$

$$A'^{0} = a^{0'}_{p} A^{p} + a^{0'}_{0} A^{0} = - \Sigma_{p} u^{p} A^{p} + u^{0} A^{0} \tag{VI-70$_2$}$$

and for the covariant components

$$A'_{p} = a^{q}_{p'} A_{q} + a^{0}_{p'} A_{0} = \Sigma_{q} \alpha^{p}_{r} \gamma^{r}_{q} A_{q} + \alpha^{p}_{r} u^{r} A_{0} \tag{VI-71$_1$}$$

$$A'_{0} = a^{p}_{0'} A_{p} + a^{0}_{0'} A_{0} = u^{p} A_{p} + u^{0} A_{0}\,. \tag{VI-71$_2$}$$

a) In particular, for the case of a Lorentz transformation *without rotation*, we obtain according to (VI-61):

$$A'^{p} = \left[\delta^{p}_{q} + \Sigma_{q} \frac{\alpha}{\beta^{2}} (1 - \beta^{2}) u^{p} u^{q} \right] A^{q} - u^{p} A^{0} \tag{VI-72$_1$}$$

$$A'^{0} = - \Sigma_{p} u^{p} A^{p} + u^{0} A^{0} \qquad \left(\alpha = \frac{1}{\sqrt{1 - \beta^{2}}} - 1 \right) \tag{VI-72$_2$}$$

$$A'_{p} = \left[\delta^{q}_{p} + \frac{\alpha}{\beta^{2}} (1 - \beta^{2}) u^{p} u^{q} \right] A_{q} + u^{p} A_{0} \tag{VI-73$_1$}$$

$$A'_{0} = u^{p} A_{p} + u^{0} A_{0}\,. \tag{VI-73$_2$}$$

b) Finally, in a *special transformation* such that

$$u^{p} = u^{1} = \frac{v^{1}}{c\sqrt{1 - \beta^{2}}} = \frac{\beta}{\sqrt{1 - \beta^{2}}}, \qquad u^{2} = u^{3} = 0\,. \tag{VI-74}$$

$$u^{0} = \frac{1}{\sqrt{1 - \beta^{2}}},$$

One will have, starting with (VI-72) and (VI-73):

$$\text{(VI-75)} \qquad A'^1 = \frac{A^1 - \beta A^0}{\sqrt{1-\beta^2}}, \; A'^2 = A^2, \; A'^3 = A^3, \; A'^0 = \frac{A^0 - \beta A^1}{\sqrt{1-\beta^2}}$$

$$\text{(VI-76)} \qquad A'_1 = \frac{A_1 + \beta A_0}{\sqrt{1-\beta^2}}, \; A'_2 = A_2, \; A'_3 = A_3, \; A'_0 = \frac{A_0 + \beta A_1}{\sqrt{1-\beta^2}}$$

Let us also note the inverse formulas

$$\text{(VI-77)} \qquad \begin{aligned} A^1 &= \frac{A'^1 + \beta A'^0}{\sqrt{1-\beta^2}}, \qquad A^2 = A'^2, \qquad A^3 = A'^3, \\ A^0 &= \frac{A'^0 + \beta A'^1}{\sqrt{1-\beta^2}} \end{aligned}$$

$$\text{(VI-78)} \qquad \begin{aligned} A_1 &= \frac{A'_1 - \beta A'_0}{\sqrt{1-\beta^2}}, \qquad A_2 = A'_2, \qquad A_3 = A'_3, \\ A_0 &= \frac{A'_0 - \beta A'_1}{\sqrt{1-\beta^2}}. \end{aligned}$$

Naturally (VI-75), (VI-76) and (VI-77), (VI-78) are directly deduced from the values (VI-66) or (VI-67) of the coefficients of the special transformation.[5]

2) *Transformation of an antisymmetric*[6] *tensor* $A^{\mu\nu}$. We still have:

$$\text{(VI-79)} \qquad A'^{\mu\nu} = a_\rho^{\mu'} a_\sigma^{\nu'} A^{\rho\sigma}, \qquad A'_{\mu\nu} = a_{\mu'}^{\rho} a_{\nu'}^{\sigma} A_{\rho\sigma}$$

that is

$$\text{(VI-80)}_1 \qquad A'^{pq} = \tfrac{1}{2}(a_r^{p'} a_s^{q'} - a_s^{p'} a_r^{q'})\, A^{rs} - (a_0^{p'} a_r^{q'} - a_r^{p'} a_0^{q'})\, A^{r0}$$

$$\text{(VI-80)}_2 \qquad A'^{p0} = \tfrac{1}{2}(a_r^{p'} a_s^{0'} - a_s^{p'} a_r^{0'})\, A^{rs} + (a_s^{p'} a_0^{0'} - a_0^{p'} a_r^{0'})\, A^{r0}$$

and, according to (VI-59)_1 and (VI-60):

$$\text{(VI-81)}_1 \qquad A'^{pq} = \alpha_m^p \alpha_n^q \Big[A^{mn} - \frac{\alpha}{\beta^2}(1-\beta^2)\, u_r (u^n A^{mr} + u^m A^{rn}) + \\ + A^{n0} u^m - A^{m0} u^n \Big]$$

(VI-81)$_2$
$$A'^{p0} = \alpha^p_m\Big[A^{ms}u_s + A^{m0}u_0 + + u^m u_r A^{r0} - \frac{\alpha}{\beta^2}(1-\beta^2)u^m u_r u_0 A^{r0}\Big]$$

And, for the covariant components:

(VI-81)$_3$
$$A'_{pq} = \tfrac{1}{2}(a^r_{p'}a^s_{q'} - a^s_{p'}a^r_{q'})\,A_{rs} + (a^r_{p'}a^0_{q'} - a^0_{p'}a^r_{q'})\,A_{r0}$$

(VI-81)$_4$
$$A'_{p0} = \tfrac{1}{2}(a^r_{p'}a^s_{0'} - a^s_{p'}a^r_{0'})\,A_{rs} + (a^r_{p'}a^0_{0'} - a^0_{p'}a^r_{0'})\,A_{r0}$$

that is, according to (VI-59)$_2$ and (VI-60):

(VI-82)$_1$
$$A'_{pq} = \sum_{rs}\alpha^p_r\alpha^q_s\Big[A_{rs} - (A_{r0}u_s - A_{s0}u_r) - - \frac{\alpha}{\beta^2}(1-\beta^2)(A_{rn}u_s - A_{sn}u_r)u^n\Big]$$

(VI-82)$_2$
$$A'_{p0} = \sum_r \alpha^p_r\Big[u^s A_{rs} + u^0 A_{r0} + u^m u_r A_{m0} - - \frac{\alpha}{\beta^2}(1-\beta^2)u_r u^m u^0 A_{m0}\Big].$$

a) In a transformation *without rotation* one will then have

(VI-83)$_1$
$$A'^{pq} = A^{pq} - \frac{\alpha}{\beta^2}(1-\beta^2)u_r(u^q A^{pr} - u^p A^{qr}) + + A^{q0}u^p - A^{p0}u^q$$

(VI-83)$_2$
$$A'^{p0} = A^{ps}u_s + A^{p0}u_0 + \frac{\alpha}{\beta^2}(1-\beta^2)u^p u_r A^{r0}$$

and

(VI-84)$_1$
$$A'_{pq} = A_{pq} - \frac{\alpha}{\beta^2}(1-\beta^2)u^s(u_q A_{ps} - u_p A_{qs}) + + (u_p A_{q0} - u_q A_{p0})$$

(VI-84)$_2$
$$A'_{p0} = A_{pr}u^r + A_{p0}u^0 + \frac{\alpha}{\beta^2}(1-\beta^2)u^r u_p A_{r0}\,.$$

b) In a *special transformation* such that

$$u^1 = u = \frac{\beta}{\sqrt{1-\beta^2}}, \qquad u^2 = u^3 = 0 \tag{VI-85}$$

one obtains

$$\begin{aligned} A'^{1q} &= \frac{A^{1q} - \beta A^{0q}}{\sqrt{1-\beta^2}}, & A'^{23} &= A^{23} \\ A'^{0p} &= \frac{A^{0p} - \beta A^{1p}}{\sqrt{1=\beta^2}}, & A'^{10} &= A^{10}, \qquad \text{for} \qquad p \neq 1 \end{aligned} \tag{VI-86}$$

and

$$\begin{aligned} A'_{1q} &= \frac{A_{1q} + \beta A_{0q}}{\sqrt{1-\beta^2}}, & A'_{23} &= A_{23} \\ A'_{0p} &= \frac{A_{0p} + \beta A_{1p}}{\sqrt{1-\beta^2}}, & A'_{01} &= A_{01}, \qquad p \neq 1 \end{aligned} \tag{VI-87}$$

10. *The addition of velocities and the general Lorentz transformation*

Let there be two Galilean systems S and S' and a moving point of velocity $\mathbf{v} = \mathbf{dx}/dt$ with respect to S, and of velocity $\mathbf{v}' = \mathbf{dx}'/dt'$ with respect to S'. To avoid confusion, let us here designate the velocity of S' with respect to S by $\mathbf{w}$. Let us put:

$$\beta = \frac{w}{c}. \tag{VI-88}$$

One still has

$$ds^2 = c^2\,dt^2 - \Sigma_p (dx^p)^2 = c^2\,dt'^2 - \Sigma_p (dx'^p)^2 . \tag{VI-89}$$

From this, there results:

$$1 - \frac{v^2}{c^2} = \left(\frac{dt'}{dt}\right)^2 \left(1 - \frac{v'^2}{c^2}\right) \tag{VI-90}$$

with

$$v^p = \frac{dx^p}{dt}, \quad v'^p = \frac{dx'^p}{dt}, \quad v^2 = \Sigma_p (v^p)^2, \quad v'^2 = \Sigma_p (v'^p)^2 . \tag{VI-91}$$

So

$$\frac{dt'}{dt} = \sqrt{\frac{1 - \frac{v^2}{c^2}}{1 - \frac{v'^2}{c^2}}}. \tag{VI-92}$$

Moreover, a transformation of coordinates

$$x'^{\mu} = a^{\mu'}_{\nu} x^{\nu}, \qquad x^{\mu} = a^{\mu}_{\nu'} x'^{\nu} \tag{VI-93}$$

is also written

$$x'^0 = a^{0'}_{p} x^p + a^{0'}_{0} x^0, \qquad x^0 = a^{0}_{p'} x'^p + a^{0}_{0'} x'^0 \tag{VI-94$_1$}$$

$$x'^p = a^{p'}_{q} x^q + a^{p'}_{0} x^0, \qquad x^p = a^{p}_{q'} x'^q + a^{p}_{0'} x'^0 . \tag{VI-94$_2$}$$

Whence

$$\frac{dx'^0}{dx^0} = \frac{dt'}{dt} = a^{0'}_{p} \frac{v^p}{c} + a^{0'}_{0}, \qquad \frac{dx^0}{dx'^0} = \frac{dt}{dt'} = a^{0}_{p'} \frac{v'^p}{c} + a^{0}_{0'} \tag{VI-95$_1$}$$

$$\frac{dx'^p}{dx^0} = a^{p'}_{q} \frac{v^q}{c} + a^{p'}_{0}, \qquad \frac{dx^p}{dx'^0} = a^{p}_{q'} \frac{v'^q}{c} + a^{p}_{0'} . \tag{VI-95$_2$}$$

We then obtain with (VI-92) and (VI-95)

$$\frac{dt'}{dt} = a^{0'}_{p} \frac{v^p}{c} + a^{0'}_{0} = \frac{1}{a^{0}_{p'} \frac{v'^p}{c} + a^{0}_{0}} = \sqrt{\frac{1 - \frac{v^2}{c^2}}{1 - \frac{v'^2}{c^2}}}. \tag{VI-96}$$

Further, one deduces from (VI-95)$_2$

$$\frac{v'^p}{c} \sqrt{\frac{1 - \frac{v^2}{c^2}}{1 - \frac{v'^2}{c^2}}} = a^{p'}_{q} \frac{v^q}{c} + a^{p'}_{0},$$

(VI-97)

$$\frac{v^p}{c} \sqrt{\frac{1 - \frac{v'^2}{c^2}}{1 - \frac{v^2}{c^2}}} = a^{p}_{q'} \frac{v'^q}{c} + a^{p}_{0'} .$$

11. *Application. Case where one of the systems is a proper system*

Let us now assume (case 1) that S' is a proper system of the point-mass. Under those conditions

$$\text{(VI-98)} \qquad \mathbf{v}'_1 = 0, \qquad \mathbf{v}_1 = \mathbf{w}$$

$\mathbf{w}$ being the velocity of S' with respect to S.
One will immediately have, as a result of (VI-96)

$$\text{(VI-99)} \qquad a^0_{0'} = \frac{1}{\sqrt{1-\beta^2}}, \qquad a^{0'}_p \frac{v^p_{(1)}}{c} = -\frac{\beta^2}{\sqrt{1-\beta^2}}.$$

Whence

$$\text{(VI-100)} \qquad a^{0'}_p = \frac{-v^p_{(1)}}{c\sqrt{1-\beta^2}} = -u^p$$

and, according to (VI-97)

$$\text{(VI-101)} \qquad \frac{v^p_{(1)}}{c\sqrt{1-\beta^2}} = a^p_{0'}$$

$$\text{(VI-102)} \qquad a^{p'}_q \frac{v^q_{(1)}}{c} = -a^{p'}_0.$$

On the contrary (case 2), let us now suppose that S is the proper system of the point-mass, in other words that

$$\text{(VI-103)} \qquad \mathbf{v}_{(2)} = 0, \quad \mathbf{v}'_{(2)} = \mathbf{w}' = -D^{-1}\mathbf{w} = -D^{-1}\mathbf{v}_{(1)} = -D^{-1}\mathbf{v},$$

$\mathbf{w}'$ being the velocity of S with respect to S'.
We thus obtain, proceeding from (VI-96):

$$\text{(VI-104)} \qquad \begin{aligned} a^{0'}_0 &= \frac{1}{\sqrt{1-\beta^2}}, \\ a^0_{p'} &= -\frac{v'^p_{(2)}}{c\sqrt{1-\beta^2}} = \frac{D^{-1}v^p}{c\sqrt{1-\beta^2}} = \frac{\alpha^p_q v^q}{c\sqrt{1-\beta^2}} = \alpha^p_q u^q \end{aligned}$$

and, proceeding from (VI-97),

$$\text{(VI-105)} \qquad a^{p'}_0 = \frac{v'^p_{(2)}}{c\sqrt{1-\beta^2}} = u'^p_{(2)} = -D^{-1}u^p = -\alpha^p_q u^q$$

and

$$\text{(VI-106)} \qquad a^p_{q'} \frac{v'^q}{c} = - a^p_{0'} .$$

In (VI-102) let us substitute the value (VI-105) of $a^{p'}_0$ and in (VI-106) the value (VI-101) of $a^p_{0'}$. We have

$$\text{(VI-107)} \qquad a^{p'}_q \frac{v^q}{c} = \alpha^p_q u^q$$

$$\text{(VI-108)} \qquad a^p_{q'} \frac{v'^q}{c} = - u^p .$$

These equations have the following solutions in a^p_q and $a^p_{q'}$:

$$\text{(VI-109)} \qquad a^{p'}_q = \alpha^p_q + \frac{\alpha}{v^2} \alpha^p_r v^r v^q$$

$$\text{(VI-110)} \qquad a^p_{q'} = \alpha^q_p + \frac{\alpha}{v^2} \alpha^q_r v^r v^p .$$

Indeed, the substitution of (VI-109) in (VI-107) entails

$$\text{(VI-111)} \qquad \begin{aligned} a^{p'}_q \frac{v^q}{c} &= \alpha^p_q \frac{v^q}{c} + \frac{\alpha}{v^2} \alpha^p_r v^r \left(\frac{v^2}{c}\right) = \\ &= (1+\alpha) \alpha^p_q \frac{v^q}{c} = \frac{\alpha^p_q v^q}{c\sqrt{1-\beta^2}} = \alpha^p_q u^q \end{aligned}$$

and the substitution of (VI-110) in (VI-108) has the result

$$\text{(VI-112)} \qquad \begin{aligned} a^p_{q'} \frac{v'^q}{c} &= \sum_q \left(\alpha^q_p \frac{v'^q}{c} + \frac{\alpha}{v^2} \alpha^q_r v^p v^r \frac{v'^q}{c} \right) \\ &= D^{-1} \frac{v'^p}{c} + \sum_r \frac{\alpha}{v^2} (D^{-1} v'^r) v^p \frac{v^r}{c} \\ &= - \frac{v^p}{c} - \frac{\alpha}{v^2} v^p \left(\frac{v^2}{c^2}\right) = \\ &= - \frac{v^p}{c}(1+\alpha) = - \frac{v^p}{c\sqrt{1-\beta^2}} = - u^p \end{aligned}$$

with

$$\alpha = \frac{1}{\sqrt{1-\beta^2}} - 1. \tag{V-75}$$

Relations (VI-107) and (VI-108) are thus satisfied identically.

Finally, the coefficients of the general Lorentz transformation have the values indicated by (VI-57), that is, by the tables (VI-59). These values had been obtained by having recourse to the results of Chapter V, that is, by having recourse to a generalization of the special transformation (Møller's method). Here they result (expressions VI-109 and VI-110) solely from the application of the transformation formulas to the special case where S (or S') coincides with the proper system. The general formulas of the addition of velocities necessarily leads to the solutions that are involved in the general Lorentz transformation.

NOTES

1. The reference-system of a non-Euclidean space must of necessity be set up within a system of curvilinear coordinates, if we are dealing with an extended region (cf. Chapter XV). Rectilinear coordinates may be used, but *only locally*.
2. In spite of this relation, the norm of a non-zero vector is not necessarily defined as positive (true Euclidean space), since the components of the vector are not all real

$$|A|^2 = \Sigma_p (A^p)^2 + (A^4)^2 = (A^4)^2 - (A_x)^2 - (A_y)^2 - (A_z)^2 \gtrless 0\,.$$

3. V. Lalan: *Compt. Rend.* **203** (1936) 1491; *Bull. Soc. Math. Fr.* **65** (1937) 98. A. Metz: *Compt. Rend.* **237** (1953) 29.
4. L. W. Thomas: *Phil. Mag.* **3** (1927) 1.
5. From (VI-68) one directly deduces the transformation formulas valid in a system of imaginary coordinates ($x^4 = ict$)

$$A'^1 = \frac{A^1 + i\beta A^4}{\sqrt{1-\beta^2}}, \qquad A'^2 = A^2, \qquad A'^3 = A^3, \qquad A'^4 = \frac{A^4 - i\beta A^1}{\sqrt{1-\beta^2}}$$

$$A_1' = \frac{A_1 + i\beta A_4}{\sqrt{1-\beta^2}}, \qquad A_2' = A_2, \qquad A_3' = A_3, \qquad A_4' = \frac{A_4 - i\beta A_1}{\sqrt{1-\beta^2}}$$

and conversely

$$A^1 = \frac{A'^1 - i\beta A'^4}{\sqrt{1-\beta^2}}, \qquad A^2 = A'^2, \qquad A^3 = A'^3, \qquad A^4 = \frac{A'^4 + i\beta A'^1}{\sqrt{1-\beta^2}}$$

$$A_1 = \frac{A_1' - i\beta A_4'}{\sqrt{1-\beta^2}}, \qquad A_2 = A_2', \qquad A_3 = A_3', \qquad A_4 = \frac{A_4' + i\beta A_1'}{\sqrt{1-\beta^2}}\,.$$

6. Cf. Chapter XIV, p. 423.

RELATIVISTIC KINEMATICS

A. RELATIVISTIC LAW OF ADDITION OF VELOCITIES

In what follows we continue to use a system of real coordinates

(VII-1) $$x^1 = x, \qquad x^2 = y, \qquad x^3 = z, \qquad x^0 = ct,$$

and we establish a reference system for a four-dimensional pseudo-Euclidean space by means of rectilinear axes defined by four unit-vectors $\mathbf{e}_\mu$ ($\mathbf{e}_1, \mathbf{e}_2, \mathbf{e}_3, \mathbf{e}_0$) orthonormal according to

(VII-2) $$g_{\mu\nu} = (\mathbf{e}_\mu \cdot \mathbf{e}_\nu) = \eta_{\mu\nu}$$

with

(VII-3) $$\eta_{\mu\nu} = \begin{vmatrix} -1 & & & \\ & -1 & & \\ & & -1 & \\ & & & 1 \end{vmatrix}.$$

In these circumstances, the elementary four-dimensional interval has the reduced form

(VII-4) $$ds^2 = c^2\,dt^2 - dx^2 - dy^2 - dz^2 = (dx^0)^2 - \Sigma_p (dx^p)^2, \quad (p = 1, 2, 3).$$

1. *The velocity four-vector*

The velocity of a point-mass

(VII-5) $$v^p = \frac{dx^p}{dt}$$

is not transformed in the same way as the space components of a vector (since dt is not invariant). For (VII-5) one substitutes the definition of a four-vector having the following components:

(VII-6) $$\boxed{\bar{u}^\mu = \frac{dx^\mu}{d\tau}, \qquad (\mu = 1, 2, 3, 0.)}$$

$d\tau$ representing the invariant differential of proper time.

The following definition is also used:

$$u^\mu = \frac{dx^\mu}{ds} = \frac{\bar{u}^\mu}{c} \tag{VII-7}$$

because, according to (V-58)

$$ds^2 = c^2\,d\tau^2\,. \tag{VII-8}$$

The four-vector u^μ is often called *the world-velocity.*
Moreover:

$$\begin{aligned} ds^2 &= c^2\,dt^2 - \Sigma_p(dx^p)^2 = c^2\,dt^2\left[1 - \frac{1}{c^2}\Sigma_p\left(\frac{dx^p}{dt}\right)^2\right] = \\ &= c^2\,dt^2(1-\beta^2) \end{aligned} \tag{VII-9}$$

Whence:

$$\frac{dt}{d\tau} = \frac{1}{\sqrt{1-\beta^2}}\,. \tag{VII-10}$$

By combining (VII-10) with (VII-7), one has

$$u^\mu = \frac{dx^\mu}{ds} = \frac{1}{c\sqrt{1-\beta^2}}\,\frac{dx^\mu}{dt}\,. \tag{VII-11}$$

The components of the world-velocity, u^p and u^0, are then expressed in the following manner as functions of v^p:

$$u^p = \frac{v^p}{c\sqrt{1-\beta^2}} \tag{VII-12$_1$}$$

$$u^0 = \frac{1}{\sqrt{1-\beta^2}}\,. \tag{VII-12$_2$}$$

These satisfy the normalization conditions

$$u_\mu u^\mu = (u^0)^2 - \Sigma_p(u^p)^2 = \frac{1 - \Sigma_p\left(\dfrac{v^p}{c}\right)^2}{1-\beta^2} = 1\,. \tag{VII-13}$$

2. *The modification of velocities in a Lorentz transformation*

Let us consider a particle in uniform rectilinear motion of velocity $\mathbf{v}$ with respect to a Galilean system S. Its world-velocity $u^\mu = dx^\mu/ds$ is expressed by means of (VII-12) as a function of the components $v^p = dx^p/dt$. If we pass from S to the Galilean system S' of velocity $\mathbf{w}$ with respect to S, ($\beta = w/c$), the u'^μ are transformed like the components of a four-vector

$$u'^{\mu} = a^{\mu}_{\nu'} u^{\nu} = a^{\mu'}_{q} u^{q} + a^{\mu'}_{0} u^{0} \qquad \begin{pmatrix} \mu, \nu = 1,2,3,0 \\ p, q = 1,2,3 \end{pmatrix}, \tag{VII-14}$$

and, conversely,

$$u^{\mu} = a^{\mu'}_{\nu} u'^{\nu} = a^{\mu}_{q'} u'^{q} + a^{\mu}_{0'} u'^{0}. \tag{VII-15}$$

We will then have for $\mu = p = 1, 2, 3$, and then for $\mu = 0$ – always keeping in mind (VII-12) –

$$\frac{v'^{p}}{c\sqrt{1-\frac{v'^{2}}{c^{2}}}} = a^{p'}_{q}\frac{v^{q}}{c\sqrt{1-\frac{v^{2}}{c^{2}}}} + a^{p'}_{0}\frac{1}{\sqrt{1-\frac{v^{2}}{c^{2}}}}. \tag{VII-16$_1$}$$

$$\frac{1}{\sqrt{1-\frac{v'^{2}}{c}}} = a^{0'}_{q}\frac{v^{q}}{c\sqrt{1-\frac{v^{2}}{c^{2}}}} + a^{p'}_{0}\frac{1}{\sqrt{1-\frac{v^{2}}{c^{2}}}}, \tag{VII-16$_2$}$$

and, conversely, by (VII-15)

$$\frac{v^{p}}{c\sqrt{1-\frac{v^{2}}{c^{2}}}} = a^{p}_{q'}\frac{v'^{q}}{c\sqrt{1-\frac{v'^{2}}{c^{2}}}} + a^{p}_{0'}\frac{1}{\sqrt{1-\frac{v'^{2}}{c^{2}}}}. \tag{VII-17$_1$}$$

$$\frac{1}{\sqrt{1-\frac{v^{2}}{v^{2}}}} = a^{0}_{q'}\frac{v'^{q}}{c\sqrt{1-\frac{v'^{2}}{c^{2}}}} + a^{0}_{0'}\frac{1}{\sqrt{1-\frac{v'^{2}}{c^{2}}}}. \tag{VII-17$_2$}$$

From the transformation formulas (VII-16)$_2$ and (VII-17)$_2$ we immediately obtain[1]

$$a^{0'}_{q}\frac{v^{q}}{c} + a^{0'}_{0} = \frac{1}{a^{0}_{q'}\frac{v'^{q}}{c} + a^{0}_{0'}} = \sqrt{\frac{1-\frac{v^{2}}{c^{2}}}{1-\frac{v'^{2}}{c^{2}}}}. \tag{VII-18}$$

and, by substitution in (VII-16)$_1$ and (VII-17)$_1$ we then obtain

$$\text{(VII-19)} \qquad \frac{v'^p}{c} = \frac{a_q^{p'} \dfrac{v^q}{c} + a_0^{p'}}{a_r^{0'} \dfrac{v^r}{c} + a_0^{0'}}$$

and

$$\text{(VII-20)} \qquad \frac{v^p}{c} = \frac{a_{q'}^{p} \dfrac{v'^q}{c} + a_{0'}^{p}}{a_{r'}^{0} \dfrac{v'^r}{c} + a_{0'}^{0}}.$$

3. *The Lorentz transformation and the general formula for the addition of velocities*

By applying the Lorentz transformation to the four-vector u^μ, we have determined

$$\text{(VII-21)} \qquad \mathbf{v}' = \varphi(\mathbf{v}, a_\mu^{\nu'})$$

$$\text{(VII-22)} \qquad \mathbf{v} = \varphi(\mathbf{v}', a_{\mu'}^{\nu})$$

The coefficients $a_\mu^{\nu'}$ and $a_{\mu'}^{\nu}$ appearing in (VII-21) and (VII-22) determine the passage from S' to S and vice-versa. In the special case where either one of the systems S or S' coincides with the proper system S_0 of the particle, these coefficients also determine the passage from S' to S_0 or from S to S_0. Now in this particular case we will have

$$\begin{aligned} &\text{either} \quad \mathbf{v}'_{(1)} = 0, \quad \mathbf{v}_{(1)} = \mathbf{w} \quad \text{if} \quad S' \equiv S_0 \\ &\text{or} \quad \mathbf{v}_{(2)} = 0, \quad \mathbf{v}'_2 = \mathbf{w}' = \\ &\qquad = -D^{-1}\mathbf{w} \quad \text{if} \quad S \equiv S_0 . \end{aligned}$$

It will thus be possible to determine the coefficients $a_{\mu'}^{\nu}$ and $a_\mu^{\nu'}$ as functions of the relative velocity $\mathbf{w}$ of the two reference systems if we adjust equations (VII-21) and (VII-22) in a particular way. That is what was done in Chapter VI. In this way we obtained:

$$\text{(VII-23)} \qquad a_\mu^{\nu'} = f_{(S' \equiv S_0)}(\mathbf{v}_{(1)}, \mathbf{v}'_{(1)}) = f'_0(\mathbf{v} = \mathbf{w}, \mathbf{v}'_{(1)} = 0)$$

$$\text{(VII-24)} \qquad a_{\mu'}^{\nu} = f_{(S \equiv S_0)}(\mathbf{v}_{(2)}, \mathbf{v}'_2) = f_0(\mathbf{v}'_2 = -D^{-1}\mathbf{w}, \mathbf{v}_2 = 0).$$

These are the values (VI-101) through (VI-110) of the coefficients of the general Lorentz transformation.

By substituting in (VII-21) and (VII-22), one thus has:

(VII-25) $$\mathbf{v}' = \varphi(\mathbf{v}, f_0'(\mathbf{w}))$$

(VII-26) $$\mathbf{v} = \varphi(\mathbf{v}', f_0(-D^{-1}\mathbf{w})).$$

Let us do this very thing by substituting the values (VI-59) and (VI-60), of the coefficients of the transformation, in (VII-19) and (VII-20).[2] We will have[3]

(VII-27) $$\frac{v'^p}{c} = \frac{\alpha_r^p \gamma_q^r \dfrac{v^q}{c} - \alpha_r^p u^r}{-\sum_m u^m \dfrac{v^m}{c} + u^0}$$

(VII-28) $$\frac{v^p}{c} = \frac{\sum_q \alpha_r^q \gamma_p^r \dfrac{v'^q}{c} + u^p}{\sum_m \alpha_s^m u^s \dfrac{v'^m}{c} + u^0}$$

with

(VII-29) $$\gamma_p^r = \delta_p^r + \frac{\sqrt{1-\beta^2}}{\beta^2}(1-\sqrt{1-\beta^2})\,u^r u^p = \\ = \delta_p^r + \frac{(1-\sqrt{1-\beta^2})\,w^r w^p}{v'^2\sqrt{1-\beta^2}}.$$

1) For the case of a Lorentz transformation without rotation, we must replace, in (VII-28) and (VII-27), the values (VI-62) and (VI-63) of the coefficients of transformation. We thus obtain:

(VII-30) $$\frac{v'^p}{c} = \frac{\left(\gamma_q^p \dfrac{v^q}{c} - u^p\right)\sqrt{1-\beta^2}}{1 - \sum_m \dfrac{v^m w^m}{c^2}} = \\ = \frac{\dfrac{v^p}{c}\sqrt{1-\beta^2} + \dfrac{w^p}{c}\left[\sum_q \dfrac{w^q v^q}{w^2}(1-\sqrt{1-\beta^2}) - 1\right]}{1 - \sum_n \dfrac{v^m w^m}{c^2}}$$

(VII-31)
$$\frac{v^p}{c} = \frac{\left(\sum_q \gamma_p^q \frac{v'^q}{c} + u^p\right)\sqrt{1-\beta^2}}{1 + \sum_m \frac{v'^m w^m}{c^2}} = \frac{\frac{v'^p}{c}\sqrt{1-\beta^2} + \frac{w^p}{c}\left[\sum_q \frac{w^q v'^q}{w^2}(1-\sqrt{1-\beta^2}) + 1\right]}{1 + \sum_m \frac{v'^m w^m}{c^2}}$$

In vector notation, (VII-30) and (VII-31) may also be written:

(VII-32)
$$\mathbf{v}' = \frac{\mathbf{v}\sqrt{1-\beta^2} + \mathbf{w}\left[\frac{\mathbf{v}\cdot\mathbf{w}}{w^2}(1-\sqrt{1-\beta^2}) - 1\right]}{1 - \frac{\mathbf{v}\mathbf{w}}{c^2}}$$

and

(VII-33)
$$\mathbf{v} = \frac{\mathbf{v}'\sqrt{1-\beta^2} + \mathbf{w}\left[\left(\frac{\mathbf{v}'\cdot\mathbf{w}}{w^2}\right)(1-\sqrt{1-\beta^2}) + 1\right]}{1 + \frac{\mathbf{v}'\cdot\mathbf{w}}{c^2}}.$$

2) Finally, in the more special case where the relative velocity of the two reference systems is still parallel to one of the axes (ox), which are themselves parallel to each other, one will have to substitute the values of the $a_\mu^{\nu'}$ and $a_{\mu'}^{\nu}$ relative to the special transformation in (VII-19) and (VII-20). One will then have, taking (VI-66) into account,

(VII-34)
$$v'_x = \frac{v_x - w}{1 - \frac{\beta}{c}v_x}, \qquad v'_y = \frac{v_y\sqrt{1-\beta^2}}{1 - \frac{\beta}{c}v_x}, \qquad v'_z = \frac{v_z\sqrt{1-\beta^2}}{1 - \frac{\beta}{c}v_x},$$
$$\left(\beta = \frac{w}{c}\right)$$

and conversely

(VII-35)
$$v_x = \frac{v'_x + w}{1 + \frac{\beta}{c}v'_x}, \qquad v_y = \frac{v'_y\sqrt{1-\beta^2}}{1 + \frac{\beta}{c}v'_x}, \qquad v_z = \frac{v'_z\sqrt{1-\beta^2}}{1 + \frac{\beta}{c}v'_x}.$$

These same formulas[4] would result immediately from (VII-32) and (VII-33) if we put

$$\text{(VII-36)} \qquad \mathbf{w} = w_x\,, \qquad w_y = w_z = 0\,.$$

4. *Length and direction of the velocity vector.*

a) Let us go back to the expressions $(\text{VI-59})_1$ and $(\text{VI-59})_2$, which determine the coefficients $a_q^{0'}$, $a_{q'}^{0}$, $a_0^{0'}$, and $a_{0'}^{0}$ of the transformation. By incorporating them into (VII-18) we obtain

$$\text{(VII-37)} \qquad -\Sigma_q \frac{u^q v^q}{c} + u^0 = \frac{1}{\Sigma_q \alpha_r^q u^r \dfrac{v'^q}{c} + u^0} = \sqrt{\frac{1 - \dfrac{v^2}{c^2}}{1 - \dfrac{v'^2}{c^2}}}\,.$$

Limiting ourselves to the case in which the Lorentz transformation is effected without rotation, (VII-37) may also be written

$$\text{(VII-38)} \qquad \frac{1 - \dfrac{\mathbf{v}\cdot\mathbf{w}}{c^2}}{\sqrt{1-\beta^2}} = \frac{\sqrt{1-\beta^2}}{1 + \dfrac{\mathbf{v}'\cdot\mathbf{w}}{c^2}} = \sqrt{\frac{1 - \dfrac{v^2}{c^2}}{1 - \dfrac{v'^2}{c^2}}}\,.$$

Let us square and multiply the two members of the first equality (VII-38) by each other. We obtain

$$\text{(VII-39)} \qquad \frac{1 + \dfrac{\mathbf{v}'\cdot\mathbf{w}}{c^2}}{1 - \dfrac{\mathbf{v}\cdot\mathbf{w}}{c^2}} = \frac{1 - \dfrac{v'^2}{c^2}}{1 - \dfrac{v^2}{c^2}} = \frac{1-\beta^2}{\left(1 - \dfrac{\mathbf{v}\cdot\mathbf{w}}{c^2}\right)^2} = \frac{\left(1 + \dfrac{\mathbf{v}'\cdot\mathbf{w}}{c^2}\right)^2}{1-\beta^2}\,.$$

From this we deduce

$$(\text{VII-40})_1 \qquad v^2 = c^2\left[1 - \frac{\left(1 - \dfrac{v'^2}{c^2}\right)\left(1-\beta^2\right)}{\left(1 + \dfrac{\mathbf{v}'\cdot\mathbf{w}}{c^2}\right)^2}\right]$$

and, reciprocally,

$$(\text{VII-40})_2 \qquad v'^2 = c^2\left[1 - \frac{\left(1 - \frac{v^2}{c^2}\right)\left(1 - \beta^2\right)}{\left(1 - \frac{\mathbf{v}\cdot\mathbf{w}}{c^2}\right)^2}\right].$$

These expressions can, of course, be just as easily deduced from (VII-32) and (VII-33).

Let us designate by θ the angle that the velocity $\mathbf{v}$ makes with ox and by θ' the angle that the velocity $\mathbf{v}'$ makes with ox'. Let us consider the special transformation and assume that ox is parallel to ox' and also to $\mathbf{w}$. One has

$$(\text{VII-41}) \qquad \tan\theta' = \frac{\sqrt{v_y'^2 + v_z'^2}}{v'}$$

and

$$(\text{VII-42}) \qquad \sin\theta' = \frac{\sqrt{v_y'^2 + v_z'^2}}{v'_x}, \qquad \cos\theta' = \frac{v'_x}{v'}.$$

One will thus obtain

$$(\text{VII-43}) \qquad v^2 = \frac{v'^2 + w^2 + 2v'w\cos\theta' - \left(\frac{v'w}{c}\sin\theta'\right)^2}{\left(1 + \frac{v'w}{c^2}\cos\theta'\right)^2}$$

and

$$(\text{VII-44}) \qquad v'^2 = \frac{v^2 + w^2 + 2vw\cos\theta - \left(\frac{v\cdot w}{c}\sin\theta\right)^2}{\left(1 - \frac{v\cdot w}{c^2}\cos\theta\right)^2}$$

b) Confining ourselves to the case of the special transformation, let us choose our coordinate axes in such a way that the velocity $\mathbf{v}'$ is in the plane xoy ($v_z' = 0$). According to (VII-35), $v_z = 0$ and $\mathbf{v}$ is also in the plane xoy.

If θ and θ' designate the angles made by the velocities $\mathbf{v}$ and $\mathbf{v}'$ with ox in this plane, the results we obtain by comparing relations (VII-41) and

(VII-42) to the transformation formulas (VII-34) are:

$$\tan\theta' = \frac{v_y'}{v_x'} = \frac{v_y\sqrt{1-\beta^2}}{v_x - w} = \frac{v\sqrt{1-\beta^2}\sin\theta}{v\cos\theta - w} \tag{VII-45}$$

$$v'\sin\theta' = v_y' = \frac{v_y\sqrt{1-\beta^2}}{1-\frac{\beta}{c}v_x}, \qquad v'\cos\theta' = \frac{v_x - w}{1-\frac{\beta}{c}v_x}. \tag{VII-46}$$

From (VII-45) we deduce

$$\tan\theta' = \frac{\sqrt{1-\beta^2}\sin\theta}{\cos\theta - \frac{w}{v}} \tag{VII-47}$$

and (VII-47) leads to

$$\begin{aligned} \sin\theta' &= \frac{\sqrt{1-\beta^2}\sin\theta}{\left(1+\frac{w^2}{v^2}-\frac{2w}{v}\cos\theta-\beta^2\sin^2\theta\right)^{\frac{1}{2}}} \\ \cos\theta' &= \frac{\cos\theta - \frac{w}{v}}{\left(1+\frac{w^2}{v^2}-\frac{2w}{v}\cos\theta-\beta^2\sin^2\theta\right)^{\frac{1}{2}}}. \end{aligned} \tag{VII-48}$$

Moreover, by squaring each relation (VII-46) and adding, we immediately obtain

$$v'^2 = \frac{v_y^2(1-\beta^2) + (v_x - w)^2}{\left(1-\frac{\beta}{c}v_x\right)^2} \tag{VII-49}$$

that is

$$v' = v\frac{\left[1+\frac{w^2}{v^2}-\frac{2w}{v}\cos\theta-\beta^2\sin^2\theta\right]^{\frac{1}{2}}}{1-\frac{\beta v}{c}\cos\theta}. \tag{VII-50}$$

We may further write (VII-50)

$$v' = v \frac{\left[\left(\frac{\beta c}{v} - \cos\theta\right)^2 - (1-\beta^2)\sin^2\theta\right]^{\frac{1}{2}}}{1 - \frac{\beta v}{c}\cos\theta}. \tag{VII-51}$$

Reciprocally, we will have the expression of v as a function of v' and θ':

$$v = v' \frac{\left[1 + \frac{w^2}{v'^2} + \frac{2w}{v'}\cos\theta' - \beta^2\sin^2\theta'\right]^{\frac{1}{2}}}{1 + \frac{\beta v'}{c}\cos\theta'}. \tag{VII-52}$$

5. *The limiting velocity*

The law of addition of velocities immediately presents us with the existence of a limiting velocity c.

If we add two velocities, both lower than c, the resultant velocity is likewise lower than c. In fact, if $v' < c$ and $w < c$, one will always have, according to (VII-40): $v < c$.

If we add two velocities, at least one of which is equal to c, the resultant velocity is still equal to c. In fact, for $v' = c$ or $w = c$ or $v' = w = c$, one obtains by (VII-40) a resultant velocity equal to c.

Of course, the existence of a limiting velocity arises only insofar as the Lorentz transformation is applicable, that is, if we are concerned with "velocity" in the ordinary sense and if the principles of Special Relativity are specifically respected.

a) The limiting velocity appears whenever the uniform motion of material bodies and, more generally, of the propagation of the various forms of energy, is involved. It does not, for example, concern the phase velocity of electromagnetic waves, which may be higher than c.[5] On the other hand, the group velocity corresponding to a displacement of energy always remains lower than c (cf. pp. 204 and 205).[6]

b) It assumes the effective definition of Galilean reference systems. Thus, it cannot result from the description of the motion of a galaxy in the reference system attached to another galaxy. Such a description gives rise to noteworthy difficulties concerning the very notion of absolute 'distance'[7] and absolute cosmic time. The relative velocity of two galaxies is proportional to their distance from each other (Hubble's law), in first approximation. But this Hubble's law is not valid for very distant galaxies.

But if we adhere to the principles and definitions of Special Relativity, a velocity is always a time-like four-vector and thus necessarily implies c as a limiting value.

6. *Asymmetry of the parts played by the "relative" velocity and the "coordinate displacement velocity"*

If we invert the roles of the relative velocity $\mathbf{v}'$ and the coordinate displacement velocity $\mathbf{w}$ – without modifying the magnitude or direction of $\mathbf{v}'$ or $\mathbf{w}$ – the direction of the resultant $\mathbf{v}$ is modified.

Let us assume, in order to simplify, that $\mathbf{v}'$ is parallel to oy ($v'_y = \mathbf{v}'$, $v'_x = v'_z = 0$). If $\mathbf{w}$ is directed along ox, one has, according to relations (VII-35)

$$v_x^{(1)} = w\,, \qquad v_y^{(1)} = v'\sqrt{1 - \frac{w^2}{c^2}}\,. \tag{VII-53}$$

If now $\mathbf{v}'$ (still directed along oy) is the coordinate displacement velocity of the systems S and S' and if $\mathbf{w}$ (still directed along ox) represents the relative velocity of the moving object with respect to S', this will amount to changing v' to w and ox to oy in formulas (VII-53). One thus obtains:

$$v_y^{(2)} = v'\,, \qquad v_x^{(2)} = w\sqrt{1 - \frac{v'^2}{c^2}}\,. \tag{VII-54}$$

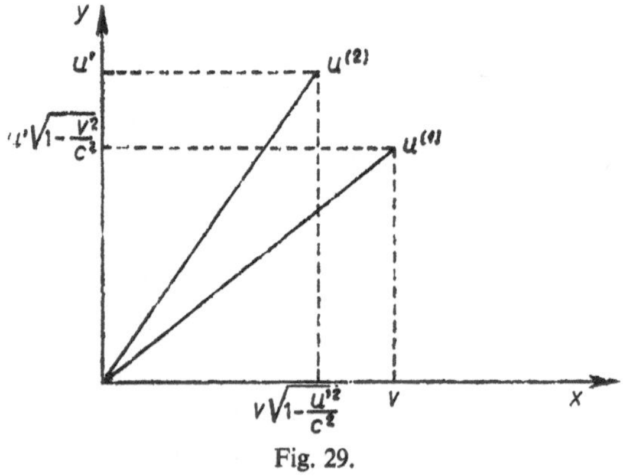

Fig. 29.

The absolute value of the resulting velocity

$$v^2 = v'^2 + w^2 - \frac{v'^2 w^2}{c^2} \tag{VII-55}$$

remains unchanged, but its direction is modified if the velocities to be added are not parallel.

7. The special case of the addition of parallel velocities

In the case of v', the relative velocity with respect to S', being itself parallel to the coordinate displacement velocity, formulas (VII-35) are simplified. One has, as a matter of fact

$$v'_y = v'_z = 0\,, \qquad v'_x = v' \tag{VII-56}$$

and (VII-35) is then written

$$v = \frac{v' + w}{1 + \dfrac{v'w}{c^2}}. \tag{VII-57}$$

If we put, as in (V-66)

$$\tan\psi = i\frac{v}{c}, \qquad \tan\psi_1 = \frac{iw}{c}, \qquad \tan\psi_2 = \frac{iv'}{c} \tag{VII-58}$$

we obtain, according to (VII-57)

$$\tan\psi = \frac{\tan\psi_1 + \tan\psi_2}{1 - \tan\psi_1 \tan\psi_2} = \tan(\psi_1 + \psi_2). \tag{VII-59}$$

Let us suppose that $\beta = w/c$ and $\beta' = v'/c$ are very small with respect to unity. The velocity resulting from the addition of the parallel velocities v' and w is approximately

$$v \simeq (w + v')(1 - \beta\beta'). \tag{VII-60}$$

It differs from the classical expression by a term of the order $\beta\beta'$.

In particular, if $v' = c/n$, $(n > 1)$, we obtain according to (VII-60) the closely related expression:

$$v \simeq \left(w + \frac{c}{n}\right)\left(1 - \frac{w}{nc}\right) \simeq \frac{c}{n} + w\left(1 - \frac{1}{n^2}\right) \tag{VII-61}$$

by neglecting the terms involving $1/c^2$ with respect to unity. We here obtain Fizeau's formula by applying the law of the addition of velocities to photons of velocity c/n, n being the index of the medium.

B. WAVE PROPAGATION AND RELATIVISTIC KINEMATICS

8. *Propagation of a plane wave in refractive media moving uniformly with respect to each other*

Let us suppose that a plane wave is propagated in a medium of index n, and let us call xOy the plane normal to the various planes of the wave. Let u and u' be the velocities of the wave-front[8], that is, the phase velocities, with respect to two Galilean reference systems S and S', which coincide at the instant $t = 0$. Let us designate the relative velocity of the two systems by w, with this velocity directed along Ox.

Let us consider the wave-front that passes the origin O at the instant $t = 0$. It reaches a point P at the end of the time

(VII-62) $$t_0 = \frac{PH}{u} = \frac{x\cos\theta + y\sin\theta}{u}$$

evaluated in the system S.

The number of waves reaching an observer at P at the end of time t is then

(VII-63) $$\nu(t - t_0) = \nu\left(t - \frac{x\cos\theta + y\sin\theta}{u}\right).$$

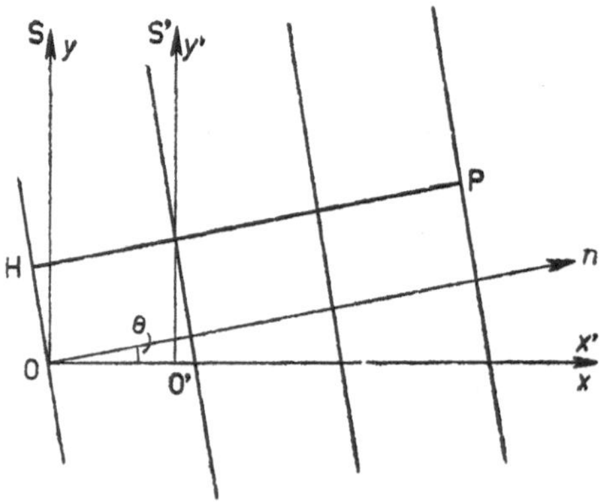

Fig. 30. The propagation of a plane wave in two Galilean systems S and S'.

This number cannot depend on the Galilean system chosen for the evaluation of the number. If we adopt the coordinate system S', the coordinates of the point P will be x', y'; time t will become t', but the number of waves is identical to the one already given:

$$\text{(VII-64)} \qquad \nu'(t' - t'_0) = \nu(t - t_0).$$

One will then have

$$\text{(VII-65)} \qquad \nu\left(t - \frac{x\cos\theta + y\sin\theta}{u}\right) = \nu'\left(t' - \frac{x'\cos\theta' + y'\sin\theta'}{u'}\right).$$

The substitution of the Lorentz formulas

$$\text{(VII-66)} \qquad x = \frac{x' + wt'}{\sqrt{1-\beta^2}}, \qquad y = y', \qquad t = \frac{t' + \dfrac{\beta}{c}x'}{\sqrt{1-\beta^2}}, \qquad \beta = \frac{w}{c}$$

in (VII-65), gives us an expression in which the coefficients of x', y', t' must cancel each other out in an identical manner. As a matter of fact, the equality between numbers of waves quite obviously does not depend on the selected point P nor on the time t under consideration. So we then have

$$\text{(VII-67)} \qquad \frac{\nu}{\sqrt{1-\beta^2}} - \frac{w\nu\cos\theta}{u\sqrt{1-\beta^2}} = \nu'.$$

$$\text{(VII-68)} \qquad \frac{\beta\nu}{c\sqrt{1-\beta^2}} - \frac{\nu\cos\theta}{u\sqrt{1-\beta^2}} = -\frac{\nu'\cos\theta'}{u'}.$$

$$\text{(VII-69)} \qquad \frac{\nu\sin\theta}{u} = \frac{\nu'\sin\theta'}{u'}.$$

From this we deduce, on one hand,

$$\text{(VII-70)} \qquad \tan\theta' = \frac{\sqrt{1-\beta^2}\sin\theta}{\cos\theta - \dfrac{\beta u}{c}}$$

that is

$$\text{(VII-71)} \qquad \sin\theta' = \frac{\sqrt{1-\beta^2}\sin\theta}{\sqrt{\left(\dfrac{\beta u}{c} - \cos\theta\right)^2 + (1-\beta^2)\sin^2\theta}}.$$

$$\cos\theta' = \frac{\cos\theta - \dfrac{\beta u}{c}}{\sqrt{\left(\dfrac{\beta u}{c} - \cos\theta\right)^2 + (1-\beta^2)\sin^2\theta}}, \tag{VII-72}$$

and, on the other

$$u' = \frac{u - \beta c\cos\theta}{\sqrt{\left(\dfrac{\beta u}{c} - \cos\theta\right)^2 + (1-\beta^2)\sin^2\theta}}. \tag{VII-73}$$

We will see, in Chapter X, that these relations express the laws of the relativistic Doppler effect and of the aberration phenomena.

It is interesting to note that relations (VII-70) and (VII-73) are identical with formulas (VII-47) and (VII-51) for the addition of the velocities of a particle, on the condition that we put

$$\frac{u}{c^2} = \frac{1}{v}. \tag{VII-74}$$

Thus, the transformation of the phase velocity u of a plane wave is deduced from the transformation of the velocity $v = c^2/u$ of an associated particle.[9] In the special case of $v = c$, the phase velocity u of the associated wave is

$$u = \frac{c^2}{v} = c. \tag{VII-75}$$

It has the same value as the velocity of the particle with which it is associated.

9. *Huygens' principle and Special Relativity*[10]

Let us now assume that a spherical wave with center O', which is also the origin of a system of Cartesian axes S', is propagated in a medium having an index n and at rest with respect to S'. In this medium, that is, in system S', the propagation velocity V' of the light-wave is equal to the phase velocity

$$V' = u' = \frac{c}{n}. \tag{VII-76}$$

At the instant t', the light-wave coincides, in S', with the sphere of radius $r' = u' t'$:

$$x'^2 + y'^2 + z'^2 - u'^2 t'^2 = 0. \tag{VII-77}$$

Let us now consider the reference system S coinciding with S' at the initial instant and moving with a constant velocity **w** with respect to S'. Let us choose the constant direction **w** as our $O\mathbf{x}'$ axis. The point (x', y', z', t') in the S' system has, in the S system, the coordinate (x, y, z, t) deduced from (x', y', z', t') by relations (VII-66). The wave-surface at the instant t is determined by the equation

$$\frac{(x - at)^2}{b} + y^2 + z^2 - bu'^2 t^2 = 0 \tag{VII-78}$$

in which we have put

$$a = w \frac{1 - \dfrac{u'^2}{c^2}}{1 - \dfrac{\beta^2 u'^2}{c^2}}, \qquad b = \frac{1 - \beta^2}{1 - \dfrac{\beta^2 u'^2}{c^2}} \qquad \left(\beta = \frac{w}{c}\right). \tag{VII-79}$$

(VII-78) is obtained by application of the transformations (VII-66) to (VII-77). It is the equation for an ellipsoid if the condition $n > 1$, that is $u' < c$, is fulfilled. And indeed, one still has in this case, by (VII-79)

$$0 < a < c, \qquad 0 < b < 1. \tag{VII-80}$$

Let us now examine the propagation of a wavelet originating at the point P_0' $(x_0', y_0', 0)$ on a wave-surface and at the instant t_0'. At the instant $t_0' + \Delta t'$ and in the system S', this wavelet is a tiny sphere whose intersection with the plane xoy is the circumference having as its equation:

$$(x' - x_0')^2 + (y' - y_0')^2 - u'^2 \Delta t'^2 = 0. \tag{VII-81}$$

Its center coincides with the starting point (x_0', y_0') of this elementary perturbation.

In the system S, the intersection of the wavelet originating at the same point P_0 $(x_0\, y_0)$ with the plane xoy is an ellipse described, according to (VII-78), by the equation:

$$f(x, y,) \equiv \frac{(x - x_0 - a\Delta t)^2}{b} + (y - y_0)^2 - bu'^2 \Delta t^2 = 0. \tag{VII-82}$$

The half-lengths of the axes of the ellipse are then $b\,u'\Delta t$, for the short axis parallel to Ox, and $\sqrt{b}\,u'\Delta t$ for the long axis lying along Oy. The wavelets are thus flattened out in the direction of the motion. On the other hand, their center $Q\,(x_0 + a\Delta t, y_0)$ does not coincide with the point P_0, where they originated, and which is the point that remains the center of

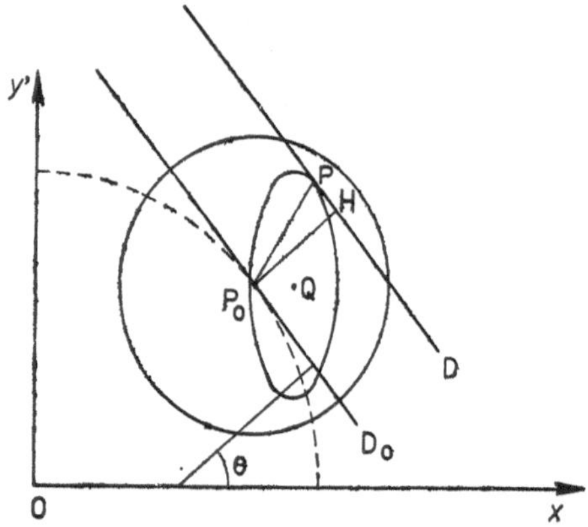

Fig. 31. The propagation of a spherical wave in two Galilean systems.

spherical wavelets in the system S'. The center of the elliptical wavelets moves with the velocity a in the direction of the motion.

Let us now suppose that a plane wave is propagated perpendicular to the plane xoy. In the system S, the intersection of xoy with the wave-front passing through the origin P_0 of the wavelets is a straight line D_0 whose normal makes an angle θ with ox.

In the system S, the equation of this straight line is

$$x_0 \cos\theta + y_0 \sin\theta = \text{const}. \tag{VII-83}$$

Now the angle that the normal to D_0 makes with the axis $ox = ox'$ is no longer θ but θ' when it is evaluated in the system S'. One actually has, by the inversion of (VII-70)

$$\tan\theta = \frac{\sqrt{1-\beta^2}\sin\theta'}{\cos\theta' + \dfrac{\beta u'}{c}}. \tag{VII-84}$$

Let us suppose that the straight line D_0 at the instant t passes through the point P_0 $(x_0 y_0)$ whence the wavelets originate. By applying Huygens' principle in system S', the straight line D' deduced from D_0 by a translation $u'\Delta t'$ (u' being the phase velocity in S') is the envelope of the family of circles (VII-81).

If Huygens' principle is compatible with the requirements of Relativity, the plane wave D, deduced from D_0 by a translation $u\Delta t$ (u being the phase velocity in S) must also be the envelope of the family of ellipses E determined by (VII-82).

Let P be the point where the ellipse E touches its envelope D. The distance P_0P is then, by definition, the product by Δt of the velocity of propagation V of the plane wave

(VII-85) $$P_0 P = V\Delta t$$

that is

(VII-86) $$x - x_0 = V_x \Delta t\,, \qquad y - y_0 = V_y \Delta t\,,$$

where x and y are the coordinates of the point P in S.

Moreover, the envelope of the family of ellipses is easily obtained from (VII-82) and (VII-83) by effecting a variation of the parameters x_0 and y_0. As a matter of fact, this envelope satisfies the equation

(VII-87) $$\frac{\partial f}{\partial x_0} \sin\theta - \frac{\partial f}{\partial y_0} \cos\theta = 0\,,$$

where $\partial f/\partial x_0$ and $\partial f/\partial y_0$ are deduced from equation (VII-82). In this way one obtains:

(VII-88) $$(x - x_0 - a\Delta t)\sin\theta - b(y - y_0)\cos\theta = 0\,.$$

Equations (VII-82), (VII-83) and (VII-88) give a parametric representation of the envelope of the family of ellipses E. By eliminating the parameters x_0, y_0 between the three relations, the equation for the envelope is obtained in the form:

(VII-89) $$x\cos\theta + y\sin\theta = \text{const} + \left[a + u'\sqrt{b^2 + b\tan^2\theta}\right]\Delta t\cos\theta\,.$$

According to (VII-89), the distance P_0H between two plane waves D_0 and D is then

(VII-90) $$P_0 H = \left[a + u'\sqrt{b^2 + b\tan^2\theta_3}\right]\Delta t\cos\theta\,.$$

If u is the phase velocity of the plane wave, that distance is also $u\Delta t$. By comparing with (VII-90) we deduce

$$\text{(VII-91)} \qquad [a + u'\sqrt{b^2 + b\tan^2\theta_3}]\cos\theta = u\,.$$

Let us replace the quantities a, b, $\cos\theta$ and $\tan^2\theta$ of this equation by their respective expressions derived from (VII-79) and (VII-84). One obtains

$$\text{(VII-92)} \qquad u = \frac{(u' + \beta c\cos\theta')}{\sqrt{\left(\frac{\beta u'}{c} + \cos\theta'\right)^2 + (1-\beta^2)\sin^2\theta'}},$$

that is the inverse of relation (VII-73).

In addition, the coordinates x_0 and y_0 both satisfy equations (VII-82) and (VII-88). And further, taking (VII-86) into account, these equations may be written:

$$\text{(VII-93)} \qquad (V_x - a)^2\frac{\Delta t^2}{b} + V_y^2 - bu'^2\Delta t^2 = 0\,,$$

$$\text{(VII-94)} \qquad (V_x - a)\Delta t\sin\theta - bV_y\Delta t\cos\theta = 0\,.$$

From this we conclude:

$$\text{(VII-95)} \qquad V_x = a + \frac{u'\sqrt{b}}{\sqrt{b + \tan^2\theta}}, \qquad V_y = \frac{u'\sqrt{b}\tan\theta}{\sqrt{b + \tan^2\theta}}.$$

Taking (VII-79), (VII-84) and (VII-76) all into account, we are able to obtain the following expressions, which define, according to Huygens' principle, the transformation $V \to V'$ of the velocities of propagation of light rays

$$\text{(VII-96)} \qquad V_x = \frac{V_x' + w}{1 + \frac{\beta V_x'}{c}}, \qquad V_y = \frac{\sqrt{1-\beta^2}\,V_y'}{1 + \frac{\beta V_x'}{c}}.$$

Relations (VII-96) are identical to (VII-35). Thus, the velocity of propagation V is transformed like the velocity of a particle. If a plane monochromatic wave is propagated in a moving medium S' of velocity w with respect to S and of index n, the transformation formulas concerning the velocities of propagation are identical to relations (VII-32) and (VII-33) in which $\mathbf{v}$ and $\mathbf{v}'$ represent the velocities of the particles. In these relations, $\mathbf{v}$ and $\mathbf{v}'$ will be replaced by V and V', taking into account the fact that

(VII-97) $$V' = \frac{c}{n}.$$

10. *Phase velocity and propagation velocity*[11]

In a refractive medium of index n, the value V of the velocity of propagation of a plane wave is generally different from the value u of the phase velocity.

a) In fact, according to (VII-73), the phase velocity u is expressed in the following way in terms of u' and θ:

(VII-98)

$$u = \frac{u'\left(1 - \frac{w^2}{c^2}\right)^{\frac{1}{2}}\left[1 - \frac{w^2}{c^2} + \frac{w^2}{c^2}\left(1 - \frac{u'^2}{c^2}\right)\sin^2\theta\right]^{\frac{1}{2}} + w\left(1 - \frac{u'^2}{c^2}\right)\cos\theta}{1 - \frac{w^2}{c^4}u'^2}.$$

If in the refractive medium that is carried along (system S') $u' = c/n$, one will then have:

(VII-99)

$$u = \frac{\frac{c}{n}\left(1 - \frac{w^2}{c^2}\right)^{\frac{1}{2}}\left[1 - \frac{w^2}{c^2} + \frac{w^2}{c^2}\left(1 - \frac{1}{n^2}\right)\sin^2\theta\right]^{\frac{1}{2}} + w\left(1 - \frac{1}{n^2}\right)\cos\theta}{1 - \frac{w^2}{c^2 n^2}}$$

and, neglecting the terms where w^2/c^2 appears before unity:

(VII-100) $$u \simeq \frac{c}{n} + w\left(1 - \frac{1}{n^2}\right)\cos\theta.$$

b) On the other hand, the propagation velocities V and V' of the plane wave are bound together, according to the preceding developments, in the way that the velocities v and v' of a particle are. In order to facilitate comparison with (VII-98), let us express V as a function of V' and θ

proceeding from relation (VII-50). We obtain

(VII-101)

$$V = \frac{V'\left(1-\frac{w^2}{c^2}\right)^{\frac{1}{2}}\left[1-\frac{w^2}{c^2}-\frac{w^2}{V'^2}\left(1-\frac{V'^2}{c^2}\right)\sin^2\theta\right]^{\frac{1}{2}}+w\left(1-\frac{V'^2}{c^2}\right)\cos\theta}{1-\frac{V'^2w^2}{c^4}-\frac{w^2}{c^2}\left(1-\frac{V'^2}{c^2}\right)\sin^2\theta}.$$

If, in the refractive medium carried along

$$V' = \frac{c}{n},$$

then there results from (VII-101)

(VII-102)

$$V = \frac{\frac{c}{n}\left(1-\frac{w^2}{c^2}\right)^{\frac{1}{2}}\left[1-\frac{w^2}{c^2}-\frac{w^2n^2}{c^2}\left(1-\frac{1}{n^2}\right)\sin^2\theta\right]^{\frac{1}{2}}+w\left(1-\frac{1}{n^2}\right)\cos\theta}{1-\frac{w^2}{c^2n^2}-\frac{w^2}{c^2}\left(1-\frac{1}{n^2}\right)\sin^2\theta}$$

and, neglecting the terms where w^2/c^2 is compared to one.

(VII-103) $$V \simeq \frac{c}{n} + w\left(1-\frac{1}{n^2}\right)\cos\theta.$$

The comparison of expressions (VII-98) and (VII-101) [or (VII-99) and (VII-102)] shows clearly that in any Galilean system the propagation velocity V has a value other than that of the phase velocity u.
Only the approximate expressions (VII-100) and (VII-103) (for $w^2/c^2 \ll 1$) coincide.

There exists, nevertheless, a rigorous identity between the expression of the phase velocity and the propagation velocity in the two following cases:

1) *When the propagation takes place in the vacuum* ($n = 1$), relation (VII-76) involves the following consequence

(VII-104) $$V' = u' = c.$$

In the system S' the phase velocity and the propagation velocity of light rays both have the value c. But in the case ($n = 1$), the transformations between the propagation velocities are identical with the transformations between the phase velocities. According to the principles of Special Relativity, the phase velocity is then equal to the velocity of propagation in the vacuum, and in all Galilean reference systems.

This property is directly verifiable by putting $n = 1$ in relations (VII-99) and (VII-102). One sees immediately that, for $n = 1$, these relations make

(VII-105) $$V = u = c$$

in any Galilean system whatever.

C. MØLLER has quite rightly pointed out the difference between this conclusion and the one that can be deduced from a theory based on the concept of absolute space.[12] In a theory of this latter sort, the phase velocity was assimilated with the propagation velocity in the special reference system bound to the motionless ether. In all other systems these velocities differed from each other. Special Relativity brings us to an entirely different conclusion by virtue of its intrinsic principle: in the vacuum light is propagated isotropically, with the velocity c in all Galilean systems. The wave-surfaces will then be spherical, no matter what system, S or S' is adopted.

2) *In a refractive medium of index n* if the normal to the plane wave is oriented in the same direction as the relative velocity w of the medium. In that case $\theta = 0$. One then deduces from (VII-73) or (VII-96) for V' (or u') $= c/n$

(VII-106) $$u = V = \frac{\dfrac{c}{n} \pm w}{1 \pm \dfrac{w}{nc}}$$

and by neglecting the terms involving $1/c^2 \ll 1$

(VII-107) $$u = V \simeq \left(\frac{c}{n} \pm w\right)\left(1 \mp \frac{w}{nc}\right) \simeq \frac{c}{n} \pm w\left(1 - \frac{1}{n^2}\right).$$

In this way we again arrive at Fizeau's formula. And if we actually consider Fizeau's experiment, we are indeed obliged to add two parallel velocities: one is the velocity of light in a transparent medium of index n, $V = c/n$; the other is the coordinate displacement velocity w of the medium.

In that case we get the approximate formula (VII-107) which leads precisely to the result verified by Fizeau.

According to Fresnel's theory, Fizeau's formula resulted from a partial ether-dragging; according to Lorentz' theory, it arose from a dragging of the waves (induction and polarization) in a motionless ether. Fizeau's formula proceeds far more directly from Einstein's theory, wherein it is deducible from simple kinematic considerations and does not require any structural hypothesis.[13]

It should be noted that the index n appearing in (VII-107) is the index $n(\nu')$ with reference to the frequency ν' of the plane wave in system S', that is, in the refractive medium travelling with a velocity w. This index differs from $n(\nu)$ relative to the frequency ν, that is in the observer's system S. In fact, from (VII-67) and (VII-68) one obtains

(VII-108) $$\nu' = \frac{\nu\sqrt{1-\beta^2}}{1 + \dfrac{w}{u'}\cos\theta'}$$

and if $u' = c/n$,

(VII-109) $$\nu' = \frac{\nu\sqrt{1-\beta^2}}{1 + n\beta\cos\theta'} \simeq \nu(1 - n\beta\cos\theta'),$$

neglecting the terms $\beta^2 = w^2/c^2$ with respect to unity.

From (VII-109) one deduces

(VII-110) $$n(\nu') = n(\nu) + \frac{dn}{d\nu}d\nu = n(\nu) - \frac{dn}{d\nu}\nu\beta n\cos\theta' = = n(\nu)\left[1 - \frac{dn}{d\nu}\nu\beta\cos\theta'\right]$$

(VII-111) $$\frac{1}{n(\nu')} = \frac{1}{n(\nu)}\left(1 + \frac{dn}{d\nu}\nu\beta\cos\theta'\right).$$

By substituting in (VII-107) one then obtains, within the limits of permissible approximation,

(VII-112) $$V = \frac{c}{n} + \left[1 - \frac{1}{n^2} + \frac{\nu}{n}\frac{dn}{d\nu}\right]w\cos\theta'.$$

This result of the theory was verified with the greatest precision by

ZEEMAN.[14] By measuring the velocity V of light in a moving quartz measuring rod, Zeeman succeeded in producing the dispersion effect that is expressed by (VII-112).

In Chapter X the explanation of the Doppler effect and aberration phenomena will be examined. That explanation also is an immediate consequence of the principles of relativistic kinematics.

NOTES

1. This relation was obtained in the preceding chapter by differentiating the Lorentz formula

$$\frac{dt'}{dt} = a_p{}^{0'}\frac{v^p}{c} + a_0{}^{0'} = \frac{1}{a_{p'}{}^0\dfrac{v'^p}{c} + a_{0'}{}^0} = \sqrt{\frac{1-\dfrac{v^2}{c^2}}{1-\dfrac{v'^2}{c^2}}}\,.$$

2. In all the formulas concerning the addition of velocities, care will be taken to use w to designate the relative velocity of two reference systems S and $S'(\beta = w/c)$. We do this in order to avoid any possible confusion with the notations v and v', which represent the velocities of the material point in S and S'.
3. Let us recall that

$$u^p = \frac{w^p}{c\sqrt{1-\beta^2}}, \quad u^0 = \frac{1}{\sqrt{1-\beta^2}}, \quad \left(\beta = \frac{w}{c}\right).$$

4. The formulas for the addition of velocities in the case of a Special Lorentz transformation are immediately deducible from the transformation formulas themselves. From

$$x' = \frac{x - wt}{\sqrt{1-\beta^2}}, \quad y' = y, \quad z' = z, \quad t' = \frac{t - \dfrac{\beta}{c}x}{\sqrt{1-\beta^2}} \quad \left(\beta = \frac{w}{c}\right)$$

we obtain

$$dt' = \frac{dt - \dfrac{\beta}{c}dx}{\sqrt{1-\beta^2}} \quad \text{that is to say} \quad \frac{dt'}{dt} = \frac{1 - \dfrac{\beta}{c}v_x}{\sqrt{1-\beta^2}}$$

and

$$v_x' = \frac{dx'}{dt'} = \frac{dx'}{dt}\frac{dt}{dt'} = \left(\frac{v_x - w}{\sqrt{1-\beta^2}}\right)\frac{\sqrt{1-\beta^2}}{1 - \dfrac{\beta}{c}v_x},$$

$$v_y' = \frac{dy'}{dt'} = \frac{dy'}{dt}\frac{dt}{dt'} = v_y\frac{\sqrt{1-\beta^2}}{1 - \dfrac{\beta}{c}v_x},$$

$$v_z' = \frac{dz'}{dt'} = \frac{dz'}{dt}\frac{dt}{dt'} = \frac{v_z\sqrt{1-\beta^2}}{1 - \dfrac{\beta}{c}v_x}$$

5. The phase velocity u which appears, for example, in the equation

$$\frac{1}{u^2}\frac{\partial^2\varphi}{\partial t^2} = \frac{\partial^2\varphi}{\partial x^2}$$

is indeed defined as a quantity *homogeneous with a velocity.* It obviously does not arise from definition (VII-5).
6. Cf. SOMMERFELD: *Phys. Z.* **8** (1907) 841; **33**, 413; *Ann. d. Phys.* **44** (1914) 177. L. BRILLOUIN: *Ann. Phys.* **44** (1914) 303; *Comptes Rendus du Congrès International de l'Électricité* **2** (1932) 753.
7. Cf. G. C. MAC VITTIE: *General Relativity and Cosmology.* New York, 1956, pp. 147–153.
8. The phase velocities are designated by u and u' as in Chapter III. One will have no trouble distinguishing the phase velocity u from the world four-velocity $\mathbf{u}$, which appears, from the practical point of view, only via its components $u^\mu = dx^\mu/ds$.
9. This property makes it possible to give a relativistic expression to the wave-particle association.
With every particle of constant velocity v there is associated a plane wave whose phase velocity is

$$u = \frac{c^2}{v},$$

that is, a plane wave of wave-length

$$\lambda = \frac{u}{\nu} = \frac{c^2}{v}\frac{h}{W} = \frac{h}{mv}$$

if the energy of a particle is defined by $W = h\nu = mc^2$. (Cf. Chapter VIII.) (L. DE BROGLIE, *Doctoral Thesis.*)
10. We here adopt the method and, to a great extent, the notation of C. MØLLER [16], p. 58.
11. By propagation velocity we mean the velocity V of a signal. This velocity may be either the velocity of a wave-front, or the group-velocity (or the propagation velocity of the amplitude of a wave-train, cf. p. 67), or even the energy velocity (or propagation velocity of the Poynting vector). In this tenth Section we consider the velocity of a wave-front, but in practical terms, the various definitions of the velocity of a signal are treated as one in most ordinary cases.
12. C. MØLLER [16], p. 61. If $u' = c$, $a = 0$ and $b = 1$. The Poynting vector, which measures the flow of energy-density, and which is consequently bound to the propagation velocity, is then normal to the plane wave, as is the phase velocity, in all inertial systems.
13. This divergence in the interpretation of the same formula and the same experiment illustrates Henri Poincaré's declaration: There are no crucial experiments in physics that have an absolute validity. The interpretation of these experiments depends on the underlying postulates (in the present instance, the postulates of kinematics) admitted throughout the domain of physics.
14. P. ZEEMAN: *Amst. Versl.* **23** (1913) 245; **24** (1915) 18.

RELATIVISTIC DYNAMICS

A. RELATIVISTIC DYNAMICS OF A POINT-MASS

1. *Momentum, energy and proper mass of a particle*

In non-relativistic mechanics, the impulse of a particle of velocity $\mathbf{v}\,(v^1, v^2, v^3)$ is defined by the expression

(VIII-1) $$\mathbf{p}_N = m_0 \mathbf{v}$$

in which m_0 is a constant characterizing the particle and defining its *inertial mass.*

In relativistic theory $\mathbf{p}_N$, as well as $\mathbf{v}$, cannot be made to resemble the spatial components of a vector, and we are obliged to substitute a covariant definition of momentum for (VIII-1).

In the system of real coordinates $x^\mu\,(x^1, x^2, x^3, x^0 = ct)$ we choose, as in (VI-28), the system of axes that is orthonormalized according to

(VIII-2) $$g_{\mu\nu} = (\mathbf{e}_\mu \cdot \mathbf{e}_\nu) = \eta_{\mu\nu}$$

which leads to the reduced form

(VIII-3) $$ds^2 = (dx^0)^2 - \Sigma_p (dx^p)^2 .$$

With the aid of the world-velocity

(VIII-4) $$\bar{u}^\mu = \frac{dx^\mu}{d\tau} \quad \text{or} \quad u^\mu = \frac{\bar{u}^\mu}{c} = \frac{dx^\mu}{ds} \qquad \left(u^0 = \frac{dt}{d\tau}\right),$$

we can then define a momentum four-vector by putting

(VIII-5) $$\boxed{P^\mu = m_0 \bar{u}^\mu = m_0 c u^\mu , \qquad (u^\mu u_\mu = 1)}$$

with

(VIII-6) $$\boxed{u^p = \frac{v^p}{c\sqrt{1-\beta^2}}, \qquad u^0 = \frac{1}{\sqrt{1-\beta^2}}, \qquad \left(\beta = \frac{v}{c}\right).}$$

m_0 represents a constant that is characteristic of the particle: its *proper mass.*

We will have, according to (VIII-5) and (VIII-6)

(VIII-5)$_1$
$$P^q = m_0 \bar{u}^q = \frac{m_0 v^q}{\sqrt{1-\beta^2}} = mv^q$$

(VIII-5)$_2$
$$P^0 = m_0 \bar{u}^0 = m_0 c u^0 = \frac{m_0 c}{\sqrt{1-\beta^2}} = mc$$

by putting

(VIII-7)
$$m = \frac{m_0}{\sqrt{1-\beta^2}}.$$

For low velocities ($\beta \ll 1$), m reduces to the constant m_0 and (VIII-5)$_1$ to the non-relativistic definition (VIII-1): m_0 represents the *proper mass* or *rest-mass of the particle.*

According to (VIII-5), the components P^μ satisfy[1]

(VIII-8)
$$P_\mu P^\mu = (P^0)^2 - \Sigma_q (P^q)^2 = \frac{m_0^2 c^2}{1-\beta^2}\left(1 - \Sigma_q \frac{(v^q)^2}{c^2}\right) = m_0^2 c^2 .$$

If

(VIII-9)
$$p^2 = \Sigma_q (p^q)^2 = \Sigma_q (P^q)^2 \qquad \text{with} \qquad \mathbf{p} = (P^1, P^2, P^3)$$

we then obtain

(VIII-10)
$$(P^0)^2 = p^2 + m_0^2 c^2 .$$

Let us put

(VIII-11)
$$\frac{W}{c} = P^0 = \frac{m_0 c}{\sqrt{1-\beta^2}} = mc$$

or

(VIII-12)
$$W = \frac{m_0 c^2}{\sqrt{1-\beta^2}} = mc^2 .$$

(VIII-10) may further be written

(VIII-13)
$$\frac{W^2}{c^2} = p^2 + m_0^2 c^2$$

We are going to see that W defined by (VIII-12) differs from the kinetic energy T of the particle by a constant rest-energy term

(VIII-14) $$W_0 = m_0 c^2 .$$

W_0 is the value of W in the proper system ($\beta = 0$).

2. *Minkowski force. The basic law of relativistic dynamics*

The Newtonian dynamics of a point-mass rests on the fundamental law

(VIII-15) $$\mathbf{f}_{(N)} = \frac{d\mathbf{p}_{(N)}}{dt} = m_0 \frac{d\mathbf{v}}{dt}.$$

From this one deduces

(VIII-16) $$\frac{d}{dt}(\tfrac{1}{2} m_0 v^2) = (\mathbf{f}_{(N)} \cdot \mathbf{v})$$

which expresses the conservation of energy: the variation dT of the kinetic energy

(VIII-17) $$T = \tfrac{1}{2} m_0 v^2$$

is equal to the work $\mathbf{f} \cdot \mathbf{v}\, dt = \mathbf{f} d\mathbf{l}$. of the external forces.
Law (VIII-15) is not relativistic, for $\mathbf{f}_{(N)}$ cannot represent the space components of a vector in a Lorentz transformation, since dt is not an invariant. In order to obtain a covariant law in a Lorentz transformation, the world-velocity will be substituted for $\mathbf{v}$ and the invariant differential of the proper time for dt. The force $\mathbf{F}$, or *Minkowski* force, is then the four-vector

(VIII-18) $$\boxed{\mathbf{F} = \frac{d\mathbf{P}}{d\tau} = m_0 \frac{d\bar{\mathbf{u}}}{d\tau}.}$$

Its components are

(VIII-19) $$F^\mu = \frac{dP^\mu}{d\tau} = m_0 \frac{d\bar{u}^\mu}{d\tau} = m_0 c \frac{du^\mu}{d\tau}$$

or

(VIII-20) $$F^\mu = m_0 c \frac{dx^\rho}{d\tau} \frac{du^\mu}{dx^\rho} = m_0 c^2 u^\rho \frac{du^\mu}{dx^\rho}.$$

By forming the scalar product $F_\mu u^\mu$, and by taking (VIII-5) into account (or by taking $\bar{u}^\mu \bar{u}_\mu = c^2$ into account), one thus has

(VIII-21) $$F_\mu u^\mu = 0 \qquad (\text{or } F_\mu \bar{u}^\mu = 0).$$

We obtain from (VIII-19) and (VIII-6) the following expressions for the F^μ components

(VIII-22)$_1$ $$F^p = m_0 c \frac{du^p}{d\tau} = m_0 \frac{dt}{d\tau} \frac{d}{dt} \frac{v^p}{\sqrt{1-\beta^2}} = \frac{m_0}{\sqrt{1-\beta^2}} \frac{d}{dt} \frac{v^p}{\sqrt{1-\beta^2}}$$

(VIII-22)$_2$ $$F^0 = m_0 c \frac{du^0}{d\tau} = m_0 c \frac{dt}{d\tau} \frac{d}{dt} \frac{1}{\sqrt{1-\beta^2}} = \frac{m_0 c}{\sqrt{1-\beta^2}} \frac{d}{dt} \frac{1}{\sqrt{1-\beta^2}}.$$

Let us represent by **f** the space vector $(f^1 f^2 f^3)$ whose components

(VIII-23) $$\mathbf{f}^q = m_0 \frac{d}{dt} \frac{v^q}{\sqrt{1-\beta^2}}$$

are reducible to the components of the Newtonian force **f** when β is negligible with respect to unity. By taking (VIII-23) and the definition (VIII-5)$_1$ into account, we have

(VIII-24) $$\boxed{\mathbf{f} = \frac{d\mathbf{p}}{dt}}$$

and, according to (VIII-22)$_1$

(VIII-25) $$\boxed{F^p = \frac{f^p}{\sqrt{1-\beta^2}}.}$$

On the other hand, according to (VIII-21), (VIII-6) and (VIII-25):

(VIII-26) $$F_0 u^0 = -F_p u^p = \frac{-f_p v^p}{c(1-\beta^2)}$$

whence

(VIII-27) $$\boxed{F_0 = \frac{-f_p v^p}{c\sqrt{1-\beta^2}} = \frac{\mathbf{f}\cdot\mathbf{v}}{c\sqrt{1-\beta^2}}} \qquad \mathbf{f} = (f^1, f^2, f^3).$$

By comparing it with the expression $(\text{VIII-22})_2$ of F^0, (VIII-27) may also be written

$$(\text{VIII-28}) \qquad \boxed{(\mathbf{f}\cdot\mathbf{v}) = \frac{d}{dt}\frac{m_0c^2}{\sqrt{1-\beta^2}} = \frac{dW}{dt}}$$

always keeping definition (VIII-12) in mind.

Let us now consider equation $(\text{VIII-5})_1$ such that

$$(\text{VIII-29}) \qquad \mathbf{p} = m\mathbf{v} \qquad \text{with} \qquad \mathbf{p} = (p^1, p^2, p^3) = (P^1, P^2, P^3)$$

According to this relation, (VIII-24) may also be written:

$$(\text{VIII-30}) \qquad \mathbf{f} = \frac{d\mathbf{p}}{dt} = \frac{dm}{dt}\cdot\mathbf{v} + m\frac{d\mathbf{v}}{dt}$$

But by substituting (VIII-7) in (VIII-28) one also obtains

$$(\text{VIII-31}) \qquad \frac{dm}{dt} = \frac{\mathbf{f}\cdot\mathbf{v}}{c^2}$$

and by substitution in (VIII-30)

$$(\text{VIII-32}) \qquad m\frac{d\mathbf{v}}{dt} = \mathbf{f} - \left(\frac{\mathbf{f}\cdot\mathbf{v}}{c^2}\right)\mathbf{v}.$$

Thus, the force acting on a particle does not in general have the same direction as the acceleration of that particle. Yet they do have the same direction if the force is parallel or perpendicular $[(\mathbf{f}\cdot\mathbf{v}) = 0]$ to the velocity.[2]

3. *Equivalence of mass and energy*

If we compare the relativistic formula

$$(\text{VIII-28}) \qquad (\mathbf{f}\cdot\mathbf{v}) = \frac{dW}{dt}$$

with results (VIII-16) valid in Newtonian dynamics, we are led to define the kinetic energy of a particle by putting:

$$(\text{VIII-33}) \qquad T = \frac{m_0c^2}{\sqrt{1-\beta^2}} + \text{const.}$$

Now, for low velocities ($\beta \ll 1$) this kinetic energy reduces to

(VIII-34) $$T = m_0c^2 + \tfrac{1}{2}m_0v^2 + \cdots + \text{const.}$$

So we then obtain the results (VIII-17) if,

(VIII-35) $$\text{const} = -\, m_0c^2 .$$

The kinetic energy of the particle is then

(VIII-36) $$\boxed{T = \frac{m_0c^2}{\sqrt{1-\beta^2}} - m_0c^2}$$

or according to (VIII-7), (VIII-12) and (VIII-14)

(VIII-37) $$T = (m - m_0)\,c^2 = W - W_0$$

(VIII-38) $$\boxed{m = m_0 + \frac{T}{c^2}.}$$

The expression

(VIII-12) $$W = \frac{m_0c^2}{\sqrt{1-\beta^2}} = mc^2$$

is thus equal, according to (VIII-37), to the kinetic energy of the particle increased by the constant term $W_0 = m_0c^2$, the rest-energy of the particle.

The rest-mass of the particle m_0 is equivalent to an energy W_0/c^2. Reciprocally, every proper energy W_0 is equivalent to a proper mass

(VIII-39) $$\boxed{m_0 = \frac{W_0}{c^2}}$$

and associates with the particle of velocity $\mathbf{v}$ the momentum

(VIII-40)$_1$ $$P^q = p^q = \frac{m_0v^q}{\sqrt{1-\beta^2}} = \frac{W_0}{c^2}\frac{v^q}{\sqrt{1-\beta^2}}$$

(VIII-40)$_2$ $$P^0 = \frac{m_0c}{\sqrt{1-\beta^2}} = \frac{W_0}{c}\frac{1}{\sqrt{1-\beta^2}}.$$

The principle of the *Equivalence of mass and energy* (Einstein, 1905) is likewise called the principle of the *Inertia of energy.*

In particular, if a free particle (v = constant) emits an energy-radiation $E = h\nu$, its mass will become m'. In the proper system ($\mathbf{p} = 0$), the principle of the conservation of energy

(VIII-41) $$W_0 = W_0' + E_0$$

will result in

(VIII-42) $$m_0 = m_0' + \frac{E_0}{c^2}$$

by taking into account (VIII-13) and the condition $\mathbf{p} = 0$.

There no longer exists, as there did in Newtonian dynamics, a principle of the conservation of mass distinct from the principle of the conservation of energy. The variation of the proper mass

(VIII-43) $$\Delta m_0 = m_0 - m_0' = \frac{E_0}{c^2}$$

which is equal to the energy of the radiation emitted, divided by c^2, is immediately deducible from the principle of the conservation of energy alone.

4. *Modification of velocities and basic quantities (momentum, energy, force) of dynamics in a Lorentz transformation*

1) *Momentum and energy.* To the world-velocity $\bar{u}^\mu = cu^\mu$ there corresponds, in the system S, the momentum-energy four-vector

(VIII-45) $$P^\mu = m_0 \bar{u}^\mu = m_0 c u^\mu .$$

In the system S', having the uniform velocity $\mathbf{w}$ with respect to S, one will then get

(VIII-44) $$P'^\mu = m_0 \bar{u}'^\mu = a_\nu^{\mu'} m_0 \bar{u}^\nu = a_\nu^{\mu'} P^\nu .$$

Conversely

(VIII-45) $$P^\mu = a_{\nu'}^\mu P'^\nu$$

One will then have

$$\text{(VIII-44)}_1 \qquad p'^q = a_r^{q'} p^r + a_0^{q'} \frac{W}{c}$$

$$\text{(VIII-44)}_2 \qquad \frac{W'}{c} = a_r^{o'} p^r + a_o^{o'} \frac{W}{c}.$$

Conversely

$$\text{(VIII-45)}_1 \qquad p^q = a_{r'}^q p'^r + a_{o'}^q \frac{W'}{c}$$

$$\text{(VIII-45)}_2 \qquad \frac{W}{c} = a_{r'}^o p'^r + a_{o'}^o \frac{W'}{c}.$$

In these formulas we shall substitute the coefficients $a_\mu^{\nu'}$ and $a_{\mu'}^\nu$ relative to the general Lorentz transformation $(\text{VI-57})_1$ and $(\text{VI-57})_2$, or to the Lorentz transformation without rotation (VI-62) and (VI-63), or even the special Lorentz transformation (VI-66). In particular, one will have for a Lorentz transformation without rotation

$$\text{(VIII-46)}_1 \qquad \mathbf{p}' = \mathbf{p} + \mathbf{w}\left\{\frac{\alpha}{\mathbf{w}^2}(\mathbf{p}\cdot\mathbf{w}) - \frac{W}{c^2\sqrt{1-\beta^2}}\right\}, \qquad \beta = \frac{w}{c}$$

$$\text{(VIII-46)}_2 \qquad W' = \frac{W - (\mathbf{p}\cdot\mathbf{w})}{\sqrt{1-\beta^2}}$$

putting, as was done for (V-75),

$$\alpha = \frac{1}{\sqrt{1-\beta^2}} - 1$$

and, in the case of a special transformation:

$$\text{(VIII-47)}_1 \qquad p'^1 = \frac{p^1 - \dfrac{W}{c^2} w}{\sqrt{1-\beta^2}}, \qquad p'^2 = p^2, \qquad p'^3 = p^3$$

$$\text{(VIII-47)}_2 \qquad W' = \frac{W - p^1 w}{\sqrt{1-\beta^2}}.$$

The inverse formulas are immediately deducible from (VIII-46), [or from (VIII-47)] by changing $\mathbf{p}$ to $\mathbf{p}'$, $\mathbf{v}$ to $\mathbf{v}'$ and $\mathbf{w}$ to $-\mathbf{w}$.

2) *Force.* On passing from S to S' the F^μ components of the Minkowski force become

$$F'^\mu = a_\nu^{\mu'} F^\nu = a_q^{\mu'} F^q + a_o^{\mu'} F^0 \tag{VIII-48}$$

and, taking (VIII-25) and (VIII-27) into account

$$\frac{f'^p}{\sqrt{1 - \frac{v'^2}{c^2}}} = a_q^{p'} \frac{f^q}{\sqrt{1 - \frac{v^2}{c^2}}} + a_o^{p'} \frac{(\mathbf{f} \cdot \mathbf{v})}{c\sqrt{1 - \frac{v^2}{c^2}}}. \tag{VIII-49}$$

Thus, according to (VI-96) or (VII-18):

$$f'^p = \frac{a_q^{p'} f^q + a_o^{p'} \left(\frac{\mathbf{f} \cdot \mathbf{v}}{c}\right)}{a_r^{o'} \frac{v^r}{c} + a_o^{o'}} \tag{VIII-50}$$

In particular, by substituting in (VIII-50) the values (VI-62) and (VI-63) of the coefficients $a_\mu^{\nu'}$ relative to a Lorentz transformation without rotation, one finds[3]:

$$\boxed{\mathbf{f}' = \left\{\mathbf{f} + \mathbf{w}\left[\frac{\alpha}{\mathbf{w}^2}(\mathbf{f} \cdot \mathbf{w}) - \frac{\mathbf{f} \cdot \mathbf{v}}{c^2\sqrt{1 - \beta^2}}\right]\right\} \frac{\sqrt{1 - \beta^2}}{1 - \left(\frac{\mathbf{w} \cdot \mathbf{v}}{c^2}\right)}} \tag{VIII-51}$$

always putting $\beta = w/c$.

Finally, in the case of a special Lorentz transformation, one will substitute the special values (VI-66) of the $a_\mu^{\nu'}$ in (VIII-50), or else put $\mathbf{w} = w^1$, $w^2 = w^3 = 0$ in (VIII-51). One thus has:

$$f'^1 = \frac{f^1 - \beta\left(\frac{\mathbf{f} \cdot \mathbf{v}}{c}\right)}{1 - \frac{\beta v^1}{c}},$$

$$f'^2 = \frac{f^2\sqrt{1 - \beta^2}}{1 - \frac{\beta v^1}{c}}, \qquad f'^3 = \frac{f^3\sqrt{1 - \beta^2}}{1 - \frac{\beta v^1}{c}}. \tag{VIII-52}$$

The inverse formulas will be deducible from

(VIII-53) $$F^{\mu} = a^{\mu}_{\nu'} F'^{\nu}$$

which leads us to

(VIII-54) $$f^{p} = \frac{a^{p}_{q'} f'^{q} + a^{p}_{o'} \left(\dfrac{\mathbf{f}' \cdot \mathbf{v}'}{c} \right)}{a^{o}_{r'} \dfrac{v'^{r}}{c} + a^{o}_{o'}}$$

The inverse formulas are likewise immediately derivable from (VIII-51) and (VIII-52) by changing **f** to **f**′, **v** to **v**′ and **w** to $-$ **w**.

5. *Systems of free particles*

1) *Energy, momentum, and proper mass of a system of free particles.* Let us consider a set of n non-interacting particles. Let us designate by **p** and w, respectively, the sum of the moments and of the energies of the n particles:

(VIII-55) $$\mathbf{p} = \Sigma_i p_{(i)}, \qquad W = \Sigma_i W_{(i)}.$$

According to relation (VIII-33), which may be applied to each particle

(VIII-56) $$W = \Sigma_i (T_{(i)} + m_{0(i)} c^2) = T + m_0 c^2$$

putting

(VIII-57) $$T = \Sigma_i T_{(i)}, \qquad m_0 = \Sigma_i m_{0(i)}.$$

We will call "*the proper system for the collection of the n particles*" the system S_0 such that [4]

(VIII-58) $$\mathbf{p}_{(0)} = 0.$$

If **v** is the uniform velocity of the center of gravity of the system of particles with respect to a Galilean system S (that is, the velocity of S_0 with respect to S) one will have, according to (VIII-45) and by transforming a system S' into S_0

(VIII-59)$_1$ $$p^{q} = a^{q}_{r'} p'^{r} + a^{q}_{o'} p'^{0} = a^{q}_{o'} \frac{W_0}{c}$$

(VIII-59)$_2$ $$p^{0} = a^{o}_{r'} p'^{r} + a^{o}_{o'} p'^{0} = a^{o}_{o'} \frac{W_0}{c}$$

for, in the system $S' \equiv S_0$, one still has

$$p'^r = p^{(0)r} = 0 , \qquad p'^0 = \frac{W_0}{c} .$$

Moreover, if S' is the proper system, that is if

$$\frac{dx'^r}{dx'^0} = \frac{v'^r}{c} = 0 ,$$

one still has, as in (VI-101)[5]

(VIII-60) $$a^q_{o'} = a^0_{0'} \frac{v^q}{c}$$

and, by substitution in (VIII-59),

(VIII-61) $$p^q = a^o_{o'} \frac{W_0}{c^2} v^q , \qquad W = a^o_{o'} W_0 .$$

One then puts

(VIII-62) $$M = a^o_{o'} \frac{W_0}{c^2}$$

and relations (VIII-61) will be written

(VIII-63) $$p^q = M v^q , \qquad W = M c^2 .$$

Now, one still has, according to (VI-100):

(VIII-64) $$a^o_{o'} = \frac{1}{\sqrt{1 - \beta^2}} .$$

From this results:

(VIII-65) $$M = \frac{W_0}{c^2 \sqrt{1 - \beta^2}} = \frac{W}{c^2} .$$

For the system of particles Σ_1, one will then define a proper mass

(VIII-66) $$M_0 = \frac{W_0}{c^2}$$

such that, according to (VIII-65),

(VIII-67) $$M = \frac{M_0}{\sqrt{1 - \beta^2}} .$$

Definitions (VIII-66) and (VIII-56) then give us

$$M_0 = m_0 + \frac{T_0}{c^2}. \tag{VIII-68}$$

Remarks. a) For a collection of *free* particles one always has

$$\mathbf{f}_{(i)} = \frac{d\mathbf{p}_{(i)}}{dt} \equiv 0 \tag{VIII-69}$$

If we assume that, in the systems S and $S' \equiv S_0$, the velocities of the various particles are small with respect to c, the proper time of each particle coincides approximately with the time defined for each system S and S'. One then has

$$\mathbf{p} = \Sigma \mathbf{p}_{(i)} = \text{const}. \tag{VIII-70}$$

In a like manner from VIII-57 we have,

$$M = \text{const} \qquad W = m_0 c^2 + T = \text{const}. \tag{VIII-71}$$

W represents the total energy H of the system of particles, an energy which is equal to the kinetic energy T increased by the sum of the proper energies of all the particles; $\mathbf{P}$ and M are the total impulse and total mass corresponding to the ensemble of the system of free particles.

b) One always has

$$\boxed{M_0 - m_0 = \frac{T_0}{c^2} > 0.} \qquad \text{or} \qquad \boxed{M_0 > \Sigma_i m_{0(i)}} \tag{VIII-72}$$

The rest-mass corresponding to the ensemble of the system is greater than the sum of the proper masses of the constituent particles. The difference results from the internal kinetic energy of the system, which is always positive.

c) From (VIII-63) and (VIII-67) there further results the relation

$$p^2 - \frac{W^2}{c^2} = -M_0^2 c^2. \tag{VIII-73}$$

2) *Collisions between systems of free particles. The equivalence of mass and energy.* Let us now suppose that a collision takes place between the set of free particles Σ_1 and another set of particles Σ_2.

Let $\mathbf{p}$ and $\mathbf{p}'$, W and W' be the momentum and energy of the set Σ_1, measured before the collision in two Galilean systems S and S′. Let $\mathbf{p} + \Delta\,\mathbf{p}$, $\mathbf{p}' + \Delta\,\mathbf{p}'$, $W + \Delta\,W$ and $W' + \Delta\,W'$ be the momentum and the energy of Σ_1 measured after the collision in the same Galilean systems. Since the Lorentz formulas are linear, one will obtain in passing from S to S':

$$\text{(VIII-74)}_1 \qquad \Delta \mathbf{p}^q = a^q_{r'} \Delta \mathbf{p}'^r + a^q_{o'} \frac{\Delta W'}{c}$$

$$\text{(VIII-74)}_2 \qquad \Delta \frac{W}{c} = a^o_{r'} \Delta \mathbf{p}'^r + a^o_{o'} \frac{\Delta W'}{c}\,.$$

If S' coincides with a proper system S_0 such that

$$\text{(VIII-75)} \qquad \Delta \mathbf{p}' = \Delta \mathbf{p}_{(0)} = 0$$

that is, if the momentum $\mathbf{p}' = \Sigma\, \mathbf{p}'_{(i)}$ remains unchanged in S_0 by the collision, one will have, as in (VIII-59)

$$\text{(VIII-76)}_1 \qquad \Delta p^q = a^q_{o'} \frac{\Delta W_0}{c} = a^o_{o'} \frac{\Delta W_0}{c^2} v^q = \Delta M \cdot v^q$$

$$\text{(VIII-76)}_2 \qquad \Delta W = a^o_{o'} \Delta W_0 = \Delta M \cdot c^2$$

putting

$$\text{(VIII-77)} \qquad \Delta M = a^o_{o'} \frac{\Delta W_0}{c^2} = \frac{\Delta W_0}{c^2 \sqrt{1 - \beta^2}} = \frac{\Delta W}{c^2}\,.$$

One will further have

$$\text{(VIII-78)} \qquad \Delta M = \frac{\Delta M_0}{\sqrt{1 - \beta^2}}$$

if

$$\text{(VIII-79)} \qquad M_0 = \frac{W_0}{c^2}\,.$$

Assuming the validity of the principle of conservation of energy and momentum, the set Σ_2 has its energy and momentum increased by ΔW and $\Delta \mathbf{p}$ respectively. This amounts to adding to this system a particle of proper mass ΔM_0 and having a velocity $\mathbf{v}$ with respect to S.

Now, one had, according to (VIII-74)

$$\text{(VIII-80)} \qquad (\Delta \mathbf{p})^2 - \left(\frac{\Delta W}{c}\right)^2 = (\Delta \mathbf{p}')^2 - \left(\frac{\Delta W'}{c}\right)^2,$$

by taking into account the relations (VI-45) between the coefficients $a^{\nu}_{\mu'}$. Passing over into the proper system ($\Delta \mathbf{p}' = 0, \Delta W' = \Delta W_0 = c^2 \Delta M_0$,) one will then still have

$$\text{(VIII-81)} \qquad (\Delta p)^2 - \left(\frac{\Delta W}{c}\right)^2 = -(\Delta M_0)^2 c^2 .$$

3) *Application to an annihilation process.* Let us assume that a particle of proper mass m_0 is susceptible to annihilation by liberating an energy W. Let us designate by S_0 the particle's proper system and by S another Galilean reference system. So far as the system Σ_1, reduced to a single particle, is concerned, one will have

in S_0

$$\text{(VIII-82)} \qquad \Delta p_{(0)} = 0 , \qquad \Delta W_0 = W_0$$

in S

$$\text{(VIII-83)} \qquad \Delta \mathbf{p} = \mathbf{p} = \frac{m_0 \mathbf{v}}{\sqrt{1-\beta^2}}, \qquad \Delta W = W = \frac{W_0}{\sqrt{1-\beta^2}} .$$

By substituting in (VIII-76)

$$\text{(VIII-84)}_1 \qquad \frac{m_0 v^q}{\sqrt{1-\beta^2}} = a^{o}_{o'} \frac{W_0}{c^2} v^q = \frac{W_0}{c^2\sqrt{1-\beta^2}} v^q$$

$$\text{(VIII-84)}_2 \qquad W = a^{o}_{o'} W_0 = \frac{W_0}{\sqrt{1-\beta^2}} .$$

These relations are satisfied identically if

$$\text{(VIII-85)} \qquad W_0 = m_0 c^2 .$$

Thus, in a relativistic process of annihilation producing an energy W, the principle of the conservation of energy necessarily leads us to attribute an internal energy $W_0 = m_0 c^2$ to the particle.

The results of "materialization" and "dematerialization" experiments entirely confirm these conclusions.

Positive electrons or positons[6] can form, with negative electrons or negatons, "pairs" capable of self-annihilation by producing radiation.[7] Conversely, electromagnetic radiation can disappear by forming a pair of electrons of opposite signs.[8] These phenomena, observed especially in cosmic rays, are predicted by Dirac's theory, which is a relativistic theory of electrons with spin.

If $E = h\nu$ is the energy of the radiation emitted in a process of materialization, the principle of the conservation of energy applied in the proper system S_0

(VIII-86) $$W_0 = W_0' + E_0 \qquad \text{with} \qquad W_0 = 2m_0c^2, \qquad W_0' = 0$$

leads to the relation

(VIII-87) $$2m_0c^2 = h\nu_0$$

between the proper mass of each particle and the proper frequency of the radiation.

6. *Systems of bound particles*

Let us now assume that the particles are subjected to mutual interactions. We relate the motion of the particles to Galilean systems S, S' such that the velocity of each particle when expressed in that system is small with respect to c. Under these circumstances, the proper time of a particle of the system coincides approximately with the time that characterizes the ensemble of the reference system, and the interacting forces are a function of the position of the particles. They are thus deduced from a potential V. Admitting this approximation, one will then have:

(VIII-88) $$\mathbf{f}_{(i)} = \frac{d}{dt}(m\mathbf{v})_{(i)} = -\frac{\partial V}{\partial \mathbf{x}_{(i)}}$$

and

(VIII-89) $$\Sigma_i \mathbf{f}_{(i)}\mathbf{v}_{(i)} = \frac{d\Sigma_i(\frac{1}{2}mv^2)_{(i)}}{dt} = -\frac{dV}{dt}$$

that is

(VIII-90) $$T + V = H = \text{const}.$$

If the particles are very far apart, interaction disappears and V must be constant. We will choose $V_\infty = 0$, so that $H = T_\infty$. Under these circumstances, *V is always negative if the particles are bound together*, that is, if $\mathbf{f} \cdot \mathbf{dx} = -dV > 0$. According to (VIII-90), the kinetic energy T is not, in general, constant, but rather depends on time.

Let us now define a reference system S_0 having the precedent properties (velocities $(v_i)_0 \ll c$) and such that

(VIII-91) $$\mathbf{p}_{(0)} = \Sigma_i \mathbf{p}_{(i)(0)} = 0.$$

In another Galilean system $S\,(v_i \ll c)$, we shall have, as in (VIII-59)

(VIII-92)$_1$ $$p^q = a^q_{o'} P'^0 = a^q_{o'} \frac{W_0}{c^2} = a^o_{o'} v^q \frac{W_0}{c^2} = \frac{W}{c^2} v^q = \mu v^q$$

(VIII-92)$_2$ $$\frac{W}{c} = a^o_{o'} P'^0 = a^o_{o'} \frac{W_0}{c} = \mu c$$

putting

(VIII-93) $$\mu = a^o_{o'} \frac{W_0}{c^2} = \frac{W_0}{c^2 \sqrt{1-\beta^2}} = \frac{W}{c^2},$$

that is

(VIII-94) $$\mu = \frac{\mu_0}{\sqrt{1-\beta^2}}$$

with

(VIII-95) $$\mu_0 = \frac{W_0}{c^2} = m_0 + \frac{T_0}{c^2}.$$

But here, the kinetic energy T (and especially T_0) *is not constant*; and the same is true of the expression μ_0 which appears in (VIII-95). Thus μ_0, μ and p are functions of time and cannot represent the total mass and the total momentum when we are dealing with a system of interacting particles.

On the other hand, we define a momentum Π relative to the ensemble of the system by substituting for the energy

(VIII-96) $$W_0 = m_0 c^2 + T_0,$$

the expression

(VIII-97) $$\Omega_0 = m_0 c^2 + T_0 + V_0 = m_0 c^2 + H_0,$$

that is, by replacing T_0 by H_0 and W_0 by Ω_0 in (VIII-95) and (VIII-92). In the case of bound particles, (VIII-97) is indeed constant according to (VIII-90), whereas (VIII-96) depends on time. In the case of free particles, ($V = 0$), Ω_0 reduces to W_0, which is then constant.

One will thus define in a manner analogous to (VIII-92) a total momentum

$$\text{(VIII-98)}_1 \qquad \Pi^q = a^o_{o'} v^q \frac{\Omega_0}{c^2} = \frac{(m_0 c^2 + H_0) v^q}{c^2 \sqrt{1-\beta^2}} = \frac{M_0 v}{\sqrt{1-\beta^2}} = M v^q$$

and an energy

$$\text{(VIII-98)}_2 \qquad \Omega = a^o_{o'} \Omega_0 = \frac{(m_0 c^2 + H_0)}{\sqrt{1-\beta^2}} = \frac{M_0 c^2}{\sqrt{1-\beta^2}} = M c^2$$

putting, in place of (VIII-95)

$$\text{(VIII-99)} \qquad M_0 = m_0 + \frac{H_0}{c^2},$$

M_0, M, Π^q and Ω are then constant.

1) *When we have a stable system.* In this case one always has

$$\text{(VIII-100)}_a \qquad H_0 = T_0 + V_0 < 0$$

And we must then supply the system of bound particles with an energy

$$\text{(VIII-101)} \qquad \Delta E = -H_0 > 0,$$

in order to convert it into its elements. One has, according to (VIII-99),

$$\text{(VIII-102)}_a \qquad \boxed{\Delta m = m_0 - M_0 = \frac{\Delta E}{c^2} > 0} \quad \text{or} \quad \boxed{M_0 < \Sigma_i m_{0(i)}.}$$

The proper mass corresponding to the ensemble of the system is less than the sum of the proper masses of all the particles. The difference represents the *mass defect of the system.*

This is the case of all *stable nuclei*: *the mass of the nucleus is less than the sum of the masses of the nucleons of which it is made up.*

2) *When we have an unstable system.* One then obtains

$$\text{(VIII-100)}_b \qquad H_0 = T_0 + V_0 > 0.$$

On conversion into its constituent elements, the system then furnishes an energy $E = H_0 > 0$. In this case

$$\text{(VIII-102)}_b \qquad \boxed{\Delta m = m_0 - M_0 = -\frac{\Delta E}{c^2} < 0} \quad \text{or} \quad \boxed{M_0 > \Sigma_i m_{0(i)}.}$$

Thus, for *unstable nuclei*, the *proper mass corresponding to the ensemble of the nucleus is greater than the sum of the proper masses of the nucleons*: the nucleus can then be broken down into its elements *to liberate the energy* $\Delta E/c^2$.

B. THE RELATIVISTIC DYNAMICS OF CONTINUOUS MEDIA

7. *The non-relativistic equations of a fluid in a system of orthogonal coordinates*

Let us consider a continuous medium whose material density is μ and whose velocity is $\mathbf{v}$ at the point $P(x)$. μ and $\mathbf{v}$ are connected by means of the continuity equation; that equation expresses the fact that the variation of the density in the volume $d\mathscr{V}$ is equal to the material flux crossing a closed surface dS surrounding this volume

$$\text{(VIII-103)} \qquad \int_{\mathscr{V}} \frac{\partial \mu}{\partial t} d\mathscr{V} = -\int_{S} \mu v_n \cdot dS = -\int_{\mathscr{V}} \operatorname{div}(\mu \mathbf{v})\, d\mathscr{V}$$

or

$$\text{(VIII-104)} \qquad \frac{\partial \mu}{\partial t} + \operatorname{div}(\mu \mathbf{v}) = 0 .$$

The mass element $dm = \mu d\mathscr{V}$ enclosed in the volume $\mathscr{V}$, is in equilibrium under the action of the following forces:

1) *Inertial force.*

$$\text{(VIII-105)} \qquad dm \cdot \gamma = \mu \gamma d\mathscr{V}$$

putting

$$\text{(VIII-106)} \qquad \gamma = (\gamma^1, \gamma^2, \gamma^3), \qquad \gamma^r = \frac{\partial v^r}{\partial t} .$$

2) *The resultant of the external forces* acting on the element $d\mathscr{V}$ of volume (mass forces): $f d\mathscr{V}$.

3) *The resultant of the surface forces* which represent the interactions (pressures or stresses) between the elements situated at various points of the surface dS: $-\mathbf{p}dS$. It can be shown[9] that the component $P^r dS$ of

the surface forces may also be written:

(VIII-107) $$P^r dS = p^{rq} d\sigma_q,$$

$d\sigma_q$ representing the components of a vector normal to the surface element dS.

The equilibrium conditions relative to the volume $\mathscr{V}$ will be obtained by cancelling out the total force and the moment corresponding to the elementary forces

(VIII-108) $$f^r \, d\mathscr{V} - p^{rq} d\sigma_q - \mu\gamma^r \, d\mathscr{V} = 0.$$

In a system of orthogonal coordinates one will then have

(VIII-109) $$\int_{\mathscr{V}} (f^r - \mu\gamma^r) \, d\mathscr{V} - \int_S p^{rq} d\sigma_q = 0$$

(VIII-110) $$\int_{\mathscr{V}} [x^s(f^r - \mu\gamma^r) - x^r(f^s - \mu\gamma^s)] \, d\mathscr{V} - \\ - \int_S (x^s p^{rq} - x^r p^{sq}) \, d\sigma_q = 0.$$

By transforming the surface integrals into volume integrals by means of Green's formula, one then obtains:

(VIII-111) $$f^r - \mu\gamma^r - \partial_q p^{rq} = 0$$

since (VIII-109) is valid no matter what the volume $\mathscr{V}$ may be. Taking (VIII-111) into account, we note that (VIII-110) is always satisfied if

(VIII-112) $$p^{rq} = p^{qr}.$$

8. *The relativistic equations of a continuous medium*

Let us select a system of rectangular coordinates fixed to the element $d\mathscr{V}$ under consideration (the proper system). One will then have

(VIII-113) $$v^q = 0,$$

but the derivatives of v^q are, in general, different from zero.

Then let us consider once more the non-relativistic equations of the dynamics of continuous media, that is (VIII-104) and (VIII-111). They refer to a momentum

(VIII-114) $$p^q = \mu v^q$$

such that

$$\frac{\partial p^q}{\partial t} = \mu \frac{\partial v^q}{\partial t} = \mu \gamma^q \tag{VIII-115}$$

according to the definition (VIII-106). These equations may, therefore, also be written

$$\frac{\partial \mu}{\partial t} + \partial_r p^r = 0 \tag{VIII-116}$$

$$\frac{\partial p^r}{\partial t} + \partial_q p^{rq} = f^r. \tag{VIII-117}$$

A relativistic theory will involve the appearance of a momentum p^r corresponding to all the forms of energy. Let us here limit ourselves to the purely material contributions, excluding, for example, any contribution of electromagnetic origin. The momentum vector will then include:

the precedent contribution $p^r = \mu v^r$

the energy flux produced by the interactions inside the medium. Now, for a displacement $v^r dt$, the work done by the surface forces is, according to (VIII-107)

$$- P^r dS \cdot v_r dt = - p^{rq} d\sigma_q v_r dt. \tag{VIII-118}$$

Through the surface dS there flows, then, an energy flux $- p^{rq} v_q$ which, in a relativistic theory, corresponds to a momentum $- 1/c^2 p^{rq} v_q$.

It is convenient, then, to replace the components p^r in the equations (VIII-116) and (VIII-117) by:

$$P^r = p^r - \frac{1}{c^2} p^{rq} v_q = \mu v^r - \frac{1}{c^2} p^{rq} v_q. \tag{VIII-119}$$

One thus obtains

$$\frac{\partial \mu}{\partial t} + \partial_r (\mu v^r - \frac{1}{c^2} p^{rq} v_q) = 0 \qquad (p, q, r = 1, 2, 3) \tag{VIII-120}$$

$$\frac{\partial}{\partial t} \mu v^r - \left(\frac{1}{c^2} p^{rq} v_q \right) + \partial_q p^{rq} = f^r. \tag{VIII-121}$$

We seek to establish the relativistic equations in the proper system, that is in a system such that

$$u^p = 0, \qquad u^0 = 1. \tag{VIII-122}$$

Now the u^μ are components of a normalized vector ($u^\mu u_\mu = 1$) and, in all reference systems

(VIII-123) $$u_\mu \partial_\lambda u^\mu = 0 .$$

In the proper system, one will then have, taking into account (VIII-123)

(VIII-124) $$\partial_\lambda u^p = \frac{1}{c} \partial_\lambda v^p , \qquad \partial_\lambda u^0 = 0 ,$$

We will then define a vector and a tensor whose components F^μ and $P^{\mu\nu}$ have, in the proper system, the following values:

(VIII-125) $$F^p = f^p , \qquad F^0 = 0 ,$$

(VIII-126) $$P^{pq} = P^{qp} , \qquad P^{po} = P^{op} = P^{oo} = 0 .$$

Thus, taking (VIII-122) into account:

(VIII-127) $$P^{\mu\nu} u_\nu \equiv 0 , \qquad F^\mu u_\mu = 0 .$$

We then note that the equations (VIII-120) and (VIII-121) may be grouped together to form the system

(VIII-128) $$\boxed{\partial_\mu (\mu c^2 u^\mu u^\rho + P^{\mu\rho}) = F^\rho} \qquad (\mu, \rho = 1, 2, 3, 0) .$$

Indeed, one obtains for $p = r = 1, 2, 3$

(VIII-129)$_1$ $$\partial_q (\mu c^2 u^q u^r + P^{qr}) + \partial_0 (\mu c^2 u^r + P^{or}) = f^r$$

and for $\rho = 0$

(VIII-129)$_2$ $$\partial_r (\mu c^2 u^r + P^{ro}) + \partial_0 (\mu c^2 + P^{oo}) = 0$$

that is, taking (VIII-122), (VIII-124) and (VIII-126) – which are the equations relative to the proper system – into account,

(VIII-130) $$\partial_q p^{qr} + \mu \frac{\partial v^r}{\partial t} + \partial_0 P^{or} = f^r$$

(VIII-131) $$\mu c \partial_r v^r + \partial_r P^{ro} + c\frac{\partial \mu}{\partial t} + \partial_0 P^{oo} = 0.$$

But according to (VIII-127),

(VIII-132) $$\partial_\lambda (P^{\mu\nu} u_\nu) = 0.$$

One will thus have in the proper system

(VIII-133) $$\partial_\lambda P^{\mu o} + \frac{P^{\mu q}}{c} \partial_\lambda v_q = 0,$$

that is

(VIII-134) $$\partial_\lambda P^{ro} = -\frac{p^{rq}}{c} \partial_\lambda v_q, \qquad \partial_\lambda P^{oo} = 0.$$

By substituting the expressions (VIII-134) in (VIII-130) and (VIII-131), we finally obtain the equations (VIII-120) and (VIII-121).

The equations (VIII-128) have been established within the proper system S_0. But their covariant form (covariant under a Lorentz transformation) assures their validity for all Galilean systems. So they represent the motion of a continuous medium in all the systems with orthonormal axes appearing in Special Relativity.

On the other hand, equations (VIII-120) and (VIII-121) are deduced from (VIII-128), if we operate within the proper system. In general, then, it is solely in the system S_0 that (VIII-121) and the continuity equation (VIII-120) are valid.

9. *The material energy-momentum tensor*

We will define the energy-momentum tensor of a material medium by putting

(VIII-135) $$\boxed{M^{\rho\sigma} = \mu_0 c^2 u^\rho u^\sigma + P^{\rho\sigma}}$$

$P^{\rho\sigma}$ refers to the interactions inside the medium and, according to (VIII-127), satisfies

(VIII-136) $$P^{\rho\sigma} u_\sigma \equiv 0, \qquad \partial_\lambda (P^{\rho\sigma} u_\sigma) = 0.$$

The motion of the fluid is then described by the equations

(VIII-137) $$F^\sigma = \partial_\rho M^{\rho\sigma}.$$

Taking into account (VIII-136) and the normalization conditions $u^\mu u_\mu = 1$, we can verify that the tensor $M^{\rho\sigma}$ always satisfies the following conditions

(VIII-138) $$M^{\rho\sigma}u_\sigma = \mu_0 c^2 u^\rho .$$

Moreover, we also obtain, by starting with (VIII-127) and (VIII-136)

(VIII-139) $$F^\sigma u_\sigma = u_\sigma \partial_\rho (\mu_0 c^2 u^\rho u^\sigma + P^{\rho\sigma}) = 0 .$$

Taking the normalization conditions into account

(VIII-140) $$u_\sigma u^\sigma = 1 , \qquad u_\sigma \frac{du^\sigma}{dx^\rho} = u^\sigma \frac{du_\sigma}{dx^\rho} = 0 ,$$

(VIII-139) may also be written

(VIII-141) $$\mu_0 c^2 \partial_\rho u^\rho + u_\sigma \partial_\rho P^{\rho\sigma} = 0 .$$

By substituting this result in (VIII-137) one then obtains:

(VIII-142) $$\begin{aligned} F^\sigma &= \partial_\rho (\mu_0 c^2 u^\rho u^\sigma + P^{\rho\sigma}) = \\ &= \mu_0 c^2 u^\rho \partial_\rho u^\sigma + (\delta^\sigma_\lambda - u^\sigma u_\lambda) \partial_\rho P^{\rho\lambda} . \end{aligned}$$

10. *The case of a perfect fluid*

A continuous medium is said to be a *perfect fluid* if the pressures p^{rq} are expressed as a function of a scalar p, the *internal pressure* of the perfect fluid.

If we adopt a system of rectilinear axes normalized according to:

(VIII-2) $$g_{\mu\nu} = (\mathbf{e}_\mu \cdot \mathbf{e}_\nu) = \eta_{\mu\nu} ,$$

we will have:

(VIII-143) $$p^{rs} = - p\eta^{rs} ,$$

the only non-zero p^{rs} being $p^{11} = p^{22} = p^{33} = p$.

According to (VIII-126), one will then have in the proper system:

(VIII-144) $$P^{rs} = p^{rs} = - p\eta^{rs} , \qquad P^{ro} = P^{or} = P^{oo} = 0 .$$

For all Galilean systems, the tensor $P^{\mu\nu}$ will then be defined by putting

(VIII-145) $$P^{\rho\sigma} = - p(\eta^{\rho\sigma} - u^\rho u^\sigma) ,$$

a relation identical to (VIII-144) in the proper system S_0. $P^{\mu\sigma}$ identically satisfies (VIII-136).

In the case of a perfect fluid, the energy-momentum tensor is then, according to (VIII-135) and (VIII-145),

(VIII-146) $$\boxed{M^{\rho\sigma} = (\mu_0 c^2 + p)\, u^\rho u^\sigma - p\eta^{\rho\sigma}.}$$

Remark. When the interactions produced inside the medium are negligible, the tensor $P^{\mu\nu}$ disappears, and the energy-momentum of the fluid is reduced to the single term

(VIII-147) $$M^{\rho\sigma} = \mu_0 c^2 u^\rho u^\sigma .$$

In particular, if we are dealing with a perfect fluid, the material tensor has the expression (VIII-147) when the internal pressure is negligible. According to (VIII-142), one then obtains, as in (VIII-20),

(VIII-148) $$F^\sigma = \mu_0 c^2 u^\rho \partial_\rho u^\sigma .$$

C. USE OF CURVILINEAR COORDINATES

In four-dimensional pseudo-Euclidean space (Minkowski space) one can always refer to dynamic or electromagnetic phenomena, involving an extended region, to a coordinate system orthonormalized according to the conditions (VIII-2). In this way we obtain the equations listed at the beginning of this chapter and of the next. These equations are covariant in a Lorentz transformation.

In this pseudo-Euclidean space it is also possible, and frequently convenient, to adopt a system of curvilinear coordinates. (The choice of polar coordinates is a familiar example.) Naturally, the changes of reference system no longer involve relations (VI-42), which characterize the Lorentz group and apply only to Galilean systems.

We are going to see (in Chapter XV) that a formalism valid in pseudo-Euclidean space, referred to any arbitrarily chosen system of coordinates, is also applicable, without significant modifications, in a Riemannian space. Nevertheless, in the latter case, and where one is concerned with an extended region of space, the use of curvilinear coordinates is necessary, whereas their use is merely possible as long as one is concerned with Euclidean space. And that is why the presentation of such a formalism often accompanies the study of non-Euclidean theories in which such a formalism is absolutely inevitable. Nevertheless, in this chapter we will give the expression for the equations of dynamics with reference to any curvilinear system defined in Euclidean space. Of course, in order to apply the principles of Special Relativity, that is to say the Lorentz transformation, to such a formalism, it is first necessary to rewrite it (as is always possible) in a system of orthonormal axes. In this way we again obtain the relations obtained in Sections A and B of this chapter.

11. *Trajectory of a material point expressed in any arbitrarily chosen system of coordinates*

In the Euclidean vector space, let us define any arbitrarily chosen curvilinear system of coordinates (y^μ). For every point there is defined a natural local system of reference $\mathbf{e}_\mu$ formed by the unit vectors tangent at y^μ. So it is convenient to replace the ordinary derivative, in all the foregoing definitions, by the covariant derivative defined in Chapter XIV, Section 10.

The velocity of a material point whose motion is a function of a parameter λ is determined by the four-vector

(VIII-149) $$\mathbf{u} = \frac{d\mathbf{M}}{d\lambda}$$

having the components

(VIII-150) $$u^\mu = \frac{dy^\mu}{d\lambda}.$$

The acceleration of the moving point

(VIII-151) $$\boldsymbol{\gamma} = \frac{d\mathbf{u}}{d\lambda}$$

has the components

(VIII-152) $$\gamma^\mu = \frac{\nabla u^\mu}{d\lambda} = u^\rho \nabla_\rho u^\mu$$

since (Cf. XIV-104) ∇u^μ represents the contravariant components of $d\mathbf{u}$

in the natural local system of reference defined by $\mathbf{e}_\mu$. One will then have

$$\gamma^\mu = \frac{1}{d\lambda}\left(du^\mu + \begin{Bmatrix}\mu\\ \nu\sigma\end{Bmatrix} u^\nu dy^\sigma\right) = \frac{du^\mu}{d\lambda} + \begin{Bmatrix}\mu\\ \nu\sigma\end{Bmatrix} u^\nu u^\sigma. \tag{VIII-153}$$

The point under consideration will describe a straight line in Euclidean space if its acceleration is always zero. For any system of coordinates, the equation of that straight line is then defined by the expression:

$$\frac{du^\mu}{d\lambda} + \begin{Bmatrix}\mu\\ \nu\sigma\end{Bmatrix} u^\nu u^\sigma = 0 \tag{VIII-154}$$

or

$$\frac{d^2 y^\mu}{d\lambda^2} + \begin{Bmatrix}\mu\\ \nu\sigma\end{Bmatrix} \frac{dy^\nu}{d\lambda}\frac{dy^\sigma}{d\lambda} = 0 \tag{VIII-155}$$

valid no matter what the adopted curvilinear system may be. One may choose for the parameter λ the abscissa s or the time t considered as a scalar parameter.

In a system of rectilinear axes (oblique or rectangular), the $g_{\mu\nu}$ are constant and the symbols $\{^{\rho}_{\mu\nu}\}$ vanish. The trajectories of the free material points referred to a system of rectilinear axes (x^μ) are then written

$$\frac{d^2 x^\mu}{d\lambda^2} = 0. \tag{VIII-156}$$

They define the rectilinear uniform motion of a moving point in a Galilean system

$$\frac{d^2 x^p}{dt^2} = 0. \tag{VIII-157}$$

In any coordinate system, the expressions of the rectilinear trajectories $(\gamma^\mu = 0)$

$$\boxed{u^\rho \nabla_\rho u^\mu = 0} \quad \text{or} \quad \boxed{\frac{d^2 y^\mu}{d\lambda^2} + \begin{Bmatrix}\mu\\ \nu\sigma\end{Bmatrix} \frac{dy^\nu}{d\lambda}\frac{dy^\sigma}{d\lambda} = 0} \tag{VIII-158}$$

show that the straight lines thus defined are also the geodesics of pseudo-Euclidean vector space.

12. *The basic law of the dynamics of a point*

If a material point is subjected to a force, the relativistic law (VIII-18)

$$\mathbf{F} = m_0 \frac{d\bar{\mathbf{u}}}{d\tau} = m_0 c^2 \frac{d\mathbf{u}}{ds},$$

written in Section 2 in a system of reduced Galilean coordinates, is still valid. But, if we are concerned with any curvilinear coordinates whatever, we must substitute the absolute differential ∇u^μ for the increments du^μ. One thus obtains

$$F^\mu = m_0 c^2 \frac{\nabla \mu^\mu}{ds} \tag{VIII-159}$$

that is

$$F^\mu = m_0 c^2 \frac{dy^\rho}{ds} \nabla_\rho u^\mu = m_0 c^2 u^\rho \nabla_\rho u^\mu . \tag{VIII-160}$$

(VIII-159) may also be written

$$F^\mu = m_0 c^2 \left(\frac{du^\mu}{ds} + \left\{ \begin{matrix} \mu \\ \nu\sigma \end{matrix} \right\} u^\nu \frac{dy^\sigma}{ds} \right) \tag{VIII-161}$$

or

$$\boxed{F^\mu = m_0 c^2 \left(\frac{d^2 y^\mu}{ds^2} + \left\{ \begin{matrix} \mu \\ \nu\sigma \end{matrix} \right\} \frac{dy^\nu}{ds} \frac{dy^\sigma}{ds} \right).} \tag{VIII-162}$$

Instead of describing the straight line defined by (VIII-158) in any system of coordinates, the material point subjected to the force **F** describes a trajectory determined by the expression (VIII-162). In particular, if we are dealing with a charged material point subjected to the action of an electromagnetic field, the F^μ components correspond to the Lorentz force defined by (IX-35).

13. *Motion of a homogeneous fluid. The matter tensor*

In (VIII-135) we introduced a symmetrical tensor of the second order called the *matter energy-momentum tensor*:

$$M^{\rho\sigma} = \mu_0 c^2 u^\rho u^\sigma + P^{\rho\sigma} . \tag{VIII-135}$$

μ_0 is the mass density of the homogeneous fluid and $P^{\rho\sigma}$ characterizes the interactions – pressures or stresses – inside the medium. Just as in (VIII-127) one still has

(VIII-163) $$P^{\rho\sigma}u_\sigma = 0\,.$$

Taking into account conditions $u^\sigma u_\sigma = 1$, we then deduce, as in (VIII-138)

(VIII-164) $$M^{\rho\sigma}u_\sigma = \mu_0 c^2 u^\rho\,.$$

If we are specifically dealing with a perfect fluid, the tensor $P^{\rho\sigma}$ has the following form

(VIII-165) $$P^{\rho\sigma} = p(u^\rho u^\sigma - g^{\rho\sigma})\,,$$

an expression that we deduce from (VIII-145) by changing $\eta^{\rho\sigma}$ (relative to a Galilean system) to $g^{\rho\sigma}$ (any curvilinear system). One thus obtains, according to (VIII-135)

(VIII-166) $$\boxed{M^{\rho\sigma} = (\mu_0 c^2 + p)u^\rho u^\sigma - pg^{\rho\sigma}\,.}$$

In the case of a perfect fluid, condition (VIII-163) is identically satisfied. since $P^{\rho\sigma}$ has the expression (VIII-165).

The F^μ components of the forces exerted on an element of the volume of a homogeneous fluid satisfy an equation similar to (VIII-137). By substituting ordinary derivatives for the covariant derivatives, one will consequently have

(VIII-167) $$\boxed{F^\mu = \nabla_\rho M^{\mu\rho}}$$

that is

(VIII-168) $$F^\mu = \mu_0 c^2 (u^\rho \nabla_\rho u^\mu + u^\mu \nabla_\rho u^\rho) + \nabla_\rho P^{\mu\rho}\,.$$

Now, just as in (VIII-127), the F^μ components of the four-vector **F** must satisfy the conditions of orthogonality:

(VIII-169) $$F^\mu u_\mu = 0\,.$$

Let us apply these conditions to (VIII-168) by taking into account the relations

(VIII-170) $$u_\mu u^\mu = 1\,, \qquad u^\mu \nabla_\rho u_\mu = u_\mu \nabla_\rho u^\mu = 0\,.$$

We obtain

(VIII-171) $$u_\mu \nabla_\rho P^{\mu\rho} + \mu_0 c^2 \nabla_\rho u^\rho = 0 .$$

Moreover, if we are dealing with a perfect gas, expression (VIII-165) of $P^{\mu\rho}$ also imposes

(VIII-172) $$u_\mu \nabla_\rho P^{\mu\rho} = u_\mu \nabla_\rho p (u^\mu u^\rho - g^{\mu\rho}) = p \nabla_\rho u^\rho .$$

In the case of a perfect gas, conditions (VIII-169) [and consequently (VIII-171) also] will then be satisfied if the continuity equation

(VIII-173) $$\nabla_\rho u^\rho = 0$$

is itself verified.

14. *Equations of conservation and equations of motion*

The motion of a homogeneous fluid can be characterized by two systems of equations.

1) *The conservation equations.* These are deduced from the preceding expressions by integrating the invariant density (cf. Chapter XIV, p. 434)

(VIII-174) $$\sqrt{-g}\, u_\mu F^\mu = \sqrt{-g}\, u^\mu \nabla_\rho M^\rho_\mu$$

over a four-dimensional element of volume

$$d\tau = dy^1 \wedge dy^2 \wedge dy^3 \wedge dy^0 .$$

One thus obtains:

(VIII-175) $$\int (u^\mu \nabla_\rho M^\rho_\mu) \sqrt{-g}\, d\tau = 0 .$$

2) *The equations of motion.* These are obtained with the help of the vector density formed by starting with

(VIII-176) $$\sqrt{-g}\, F^p = \sqrt{-g} \nabla_\rho M^{p\rho} .$$

We must then integrate over an element of spatial volume

$$d\mathscr{V} = dy^1 \wedge dy^2 \wedge dy^3 .$$

The equations of the motion of any homogeneous fluid will then be written

(VIII-177) $$\int F_p \sqrt{-g}\, d\mathscr{V} = \int (\nabla_\rho M^\rho_p) \sqrt{-g}\, d\mathscr{V} = 0 .$$

In order to transform this expression, we use the following identity applicable to every symmetrical tensor $M_p^\rho = g_{p\lambda} M^{\lambda\rho}$:

(VIII-178)
$$\sqrt{-g}\,\nabla_\rho M_p^\rho = \sqrt{-g}\left(\partial_\rho M_p^\rho - \left\{ {\sigma \atop p\rho} \right\} M_\sigma^\rho + \left\{ {\rho \atop \sigma\rho} \right\} M_p^\sigma \right)$$
$$= \partial_\rho \mathscr{M}_p^\rho - \frac{\sqrt{-g}}{2} M^{\lambda\rho} (\partial_p g_{\rho\lambda} + \partial_\rho g_{p\lambda} - \partial_\lambda g_{p\rho}) =$$
$$= \partial_\rho \mathscr{M}_p^\rho - \tfrac{1}{2} \mathscr{M}^{\lambda\rho} \partial_p g_{\lambda\rho}$$

with

(VIII-179)
$$\mathscr{M}_\mu^\rho = \sqrt{-g}\, M_\mu^\rho, \qquad \mathscr{M}^{\lambda\rho} = \sqrt{-g}\, M^{\lambda\rho}.$$

Taking (VIII-178) into account, the equations of motion (VIII-177) are then written

(VIII-180)
$$\int F_p \sqrt{-g}\, d\mathscr{V} = \int (\partial_\rho \mathscr{M}_p^\rho - \tfrac{1}{2} \mathscr{M}^{\lambda\rho} \partial_p g_{\lambda\rho})\, d\mathscr{V} = 0.$$

15. *A special case: the equations of conservation and of motion of a perfect fluid*

1) *The conservation equations.* Let us once more consider expression (VIII-166) for the components $M^{\rho\sigma}$ of the matter energy-momentum tensor in the case of a perfect fluid. We deduce from it

(VIII-181)
$$M_\mu^\rho = (\mu_0 c^2 + p)\, u_\mu u^\rho - p \delta_\mu^\rho.$$

The conservation equations (VIII-175) are then written:

(VIII-182)
$$\int \sqrt{-g} \left(\mu_0 + \frac{p}{c^2} \right) \nabla_\rho u^\rho \, d\tau = 0.$$

taking into account the conditions (VIII-170). The integral (VIII-182) then takes the form:

(VIII-183)
$$\int \sqrt{-g} \left(\mu_0 + \frac{p}{c^2} \right) \nabla_\rho u^\rho \, d\tau =$$
$$= \int \partial_\rho (\mu_0 \sqrt{-g}\, u^\rho)\, d\tau + \int \frac{p}{c^2} \partial_\rho (\sqrt{-g}\, u^\rho)\, d\tau = 0.$$

Let us integrate (VIII-183) on a world-tube generated by a line of current and limited by two spatial cross-sections Σ and Σ'. The first integral (VIII-183) is transformed into a surface integral (Green's Theorem). It thus represents the flux of the vector $\mu_0 \sqrt{-g}\, u^\rho$ through the closed surface bounded by the wall L of the world-tube and by the two cross-sections Σ and Σ'. Putting

(VIII-184) $$dS_\rho = \frac{\sqrt{-g}}{6}\varepsilon_{\rho\mu\nu\sigma}\, dy^\mu \wedge dy^\nu \wedge dy^\sigma ,$$

(VIII-183) may also be written

(VIII-185) $$\int\limits_{L+\Sigma-\Sigma'} \mu_0 \sqrt{-g}\, u^\rho\, dS_\rho + \int \frac{p}{c^2}\partial_\rho(\sqrt{-g}\, u^\rho)\, d\tau = 0 .$$

Now, integrating over the wall L generated by the lines of current supplies a zero contribution. If, further, the internal pressure of the fluid disappears ($p = 0$), the only contribution furnished by (VIII-185) is represented by

(VIII-186) $$m = m'$$

putting

(VIII-187) $$m = \int_\Sigma \mu_0 \sqrt{-g}\, u^\rho\, dS_\rho = \int_\Sigma \mu\, dS_0 ,$$

if

(VIII-188) $$\mu = \mu_0 \sqrt{-g}\, u^0 .$$

2) *The equations of motion.* The equations of motion are deduced from (VIII-180) by replacing M^ρ_p by the expression that may be immediately deduced from (VIII-181). In fact, taking into account (VIII-106)

(VIII-189)
$$\begin{aligned}\mathcal{M}^q_p &= \sqrt{-g}\left[c^2\left(\mu_0 + \frac{p}{c^2}\right)u_p u^q - p\delta^q_p\right]\\ &= \sqrt{-g}\left[c^2\left(\mu_0 + \frac{p}{c^2}\right)v_p \frac{v^q (u^0)^2}{c^2} - p\delta^q_p\right]\end{aligned}$$

(VIII-190)
$$\begin{aligned}\mathcal{M}^0_p &= \sqrt{-g}\, c^2\left(\mu_0 + \frac{p}{c^2}\right)u_p u^0 =\\ &= \sqrt{-g}\, c^2\left(\mu_0 + \frac{p}{c^2}\right)\frac{v_p (u^0)^2}{c} .\end{aligned}$$

By comparing (VIII-189) and (VIII-190):

$$\mathscr{M}^q_p = \mathscr{M}^0_p \frac{v^q}{c} - p\sqrt{-g}\,\delta^q_p. \tag{VIII-191}$$

One also obtains:

$$\partial_\rho \mathscr{M}^\rho_p = \partial_0 \mathscr{M}^0_p + \partial_q \mathscr{M}^q_p = \partial_0 \mathscr{M}^0_p + \frac{v^q}{c}\partial_q \mathscr{M}^0_p - \partial_p(\sqrt{-g}p) = d_0 \mathscr{M}^0_p - \partial_p(\sqrt{-g}p). \tag{VIII-192}$$

The equations of motion (VIII-180) then take on the following form:

$$\int F_p\sqrt{-g}\,d\mathscr{V} = \int d_0 \mathscr{M}^0_p\,d\mathscr{V} - \int \partial_p(\sqrt{-g}p)\,d\mathscr{V} - \tfrac{1}{2}\int \mathscr{M}^{\lambda\rho}\partial_p g_{\lambda\rho}\,d\mathscr{V} = 0. \tag{VIII-193}$$

The divergence term contributes nothing, and (VIII-193) is finally written

$$\boxed{\int F_p\sqrt{-g}\,d\mathscr{V} = \int \frac{d\mathscr{P}_p}{dt}\,d\mathscr{V} - \tfrac{1}{2}\int \mathscr{M}^{\lambda\rho}\partial_p g_{\lambda\rho}\,d\mathscr{V}} \tag{VIII-194}$$

putting

$$\mathscr{P}_p = \frac{1}{c}\mathscr{M}^0_p = \sqrt{-g}\left(\mu_0 + \frac{p}{c^2}\right)v_p(u^0)^2. \tag{VIII-195}$$

Thus, according to the equations of motion of a perfect fluid in any system of curvilinear coordinates, the quantity

$$\mathscr{P}_p = \frac{1}{c}\mathscr{M}^0_p$$

behaves like a momentum.

NOTES

1. One can verify the fact that the definition $\mathbf{p} = m\mathbf{v}$ with

$$m = \frac{m_0}{\sqrt{1-\beta^2}}$$

entails the conservation of the total momentum, taking the Lorentz formulas into account. In order to do this, one considers, for example, a collision process between two identical particles.

Reciprocally, it can be shown that the hypothesis of a total conservation of momentum defined by $\mathbf{p} = m(m_0, v)\,\mathbf{v}$, at the time of a collision process, entails the definition

$$m = \frac{m_0}{\sqrt{1-\beta^2}},$$

if the Lorentz formulas are accepted. Cf. C. MØLLER [16], p. 67; P. G. BERGMANN [9], p. 87.

2. This is the case of the motion of a charged particle in a constant magnetic field $\mathbf{H}$

$$\mathbf{f} = \frac{q}{c}[\mathbf{v} \wedge \mathbf{H}].$$

One can then apply the Newtonian law

$$\mathbf{f} = m\frac{d\mathbf{v}}{dt}.$$

3. (VIII-51) could also be obtained by forming

$$\mathbf{f}' = \frac{dp'}{dt'},$$

$\mathbf{p}'$ being determined as a function of $\mathbf{p}$ by the relation (VIII-44). In the right-hand member one will take into account

$$\frac{d}{dt'} = \frac{dt}{dt'}\frac{d}{dt},$$

with dt'/dt having the expression (VI-92).

4. This choice is always possible, for

$$p^2 - \frac{W^2}{c^2} < -\sum_i (m_{0(i)}c^2) < 0 \qquad \text{if} \qquad n > 1$$

In fact

$$\mathbf{p}^2 - \frac{W^2}{c^2} = \left(\mathbf{p} + \frac{W}{c}\right)\left(\mathbf{p} - \frac{W}{c}\right) = \Sigma_i\left(\mathbf{p}_{(i)} - \frac{W_{(i)}}{c}\right)\Sigma_j\left(\mathbf{p}_{(j)} + \frac{W_{(j)}}{c}\right).$$

One has further

$$\mathbf{p}_{(i)} - \frac{W_{(i)}}{c} < 0 \quad \text{for} \quad \left(\mathbf{p}_{(i)} - \frac{W_{(i)}}{c}\right)\left(\mathbf{p}_{(i)} + \frac{W_{(i)}}{c}\right) = -m_{0(i)}c^2 < 0.$$

And from this results

$$\Sigma_i\left(\mathbf{p}_{(i)} - \frac{W_{(i)}}{c}\right)\Sigma_j\left(\mathbf{p}_{(j)} + \frac{W_{(j)}}{c}\right) < \Sigma_i\left(\mathbf{p}_{(i)} - \frac{W_{(i)}}{c}\right) \times$$

$$\times\, \Sigma_i\left(\mathbf{p}_{(i)} + \frac{W_{(i)}}{c}\right) = \Sigma_i\mathbf{p}_{(i)}{}^2 - \frac{W_{(i)}{}^2}{c^2}$$

Finally

$$\mathbf{p}^2 - \frac{W^2}{c^2} < \Sigma_i\mathbf{p}_{(i)}{}^2 - \frac{W_{(i)}{}^2}{c^2} = -\Sigma_i(m_{0(i)}c^2) < 0.$$

5. One has, in fact,

$$dx^q = a_{r'}{}^q, dx'^r + a_{0'}{}^q, dx'^0, \qquad dx^0 = a_{r'}{}^0, dx'^r + a_{0'}{}^0, dx'^0$$

and, if

$$\frac{dx'^r}{dx'^0} = 0$$

(proper system)

$$\frac{dx^0}{dx'^0} = a_{0'}{}^0, \qquad \frac{dx^q}{dx'^0} = \frac{dx^q}{dx^0}\frac{dx^0}{dx'^0} = a_{0'}{}^0 \frac{v^q}{c} = a_{0'}{}^0$$

6. C. D. Anderson: *Science* **76** (1932) 238.
P. M. S. Blackett and G. P. S. Occhialini: *Proc. Roy. Soc.* A **139** (1933) 699.
7. P. A. M. Dirac: *The Principles of Quantum Mechanics*. 3d ed., Oxford, 1947, Par. 73.
8. C. D. Anderson and Neddermeyer: *Phys. Rev.* **43** (1933) 1034.
F. Rasetti, L. Meitner, and K. Philipp: *Naturwiss.* **21** (1933) 286.
I. Curie and F. Joliot: *Compt. Rend.* **196** (1933) 158.
9. Let us consider, for example, the elementary tetrahedron whose sides OBC, OCA, OAB, and ABC, are respectively $dS^{23} = \alpha_1 dS$, $dS^{31} = \alpha_2 dS$, $dS^{12} = \alpha_3 dS$, where α represents the unit vector normal to dS.
On the four sides of the tetrahedron, the surface forces are then:

$$\pi_{23}\, dS^{23}, \quad \pi_{31}\, dS^{31}, \quad \pi_{12}\, dS^{12}, \quad -PdS.$$

For this elementary tetrahedron, the conditions of equilibrium will then be as follows:

$$\pi_{23}\, dS^{23} + \pi_{31}\, dS^{31} + \pi_{12}\, dS^{12} - PdS + f\, d\mathscr{V} = \mu\gamma d\mathscr{V}.$$

As $d\mathscr{V}/dS$ tends towards zero, the tetrahedron becomes infinitesimally small and at the limit, one obtains:

$$\alpha_1\pi_{23} + \alpha_2\pi_{31} + \alpha_3\pi_{12} - P = 0,$$

that is

$$\mathbf{P}dS = \pi_{23}\, dS^{23} + \pi_{31}\, dS^{31} + \pi_{12}\, dS^{12} = \tfrac{1}{2}\, \varepsilon^{pqr}\, \pi_{pq}\, d\sigma_r,$$

putting

$$dS^{pq} = \varepsilon^{pqr}\, d\sigma_r.$$

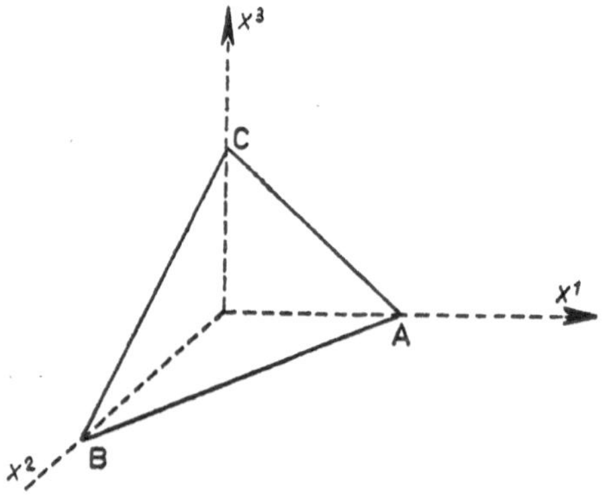

Fig. 32.

$\varepsilon^{pqr} = +1, -1$, or 0, depending on whether p, q, r represents an even or odd permutation of 1, 2, 3 – or on whether two of the indexes p, q, r are equal.
The components $P^r dS$ of $\mathbf{P}dS$ are thus linear functions of $d\sigma_r$ and may be written

$$P^r dS = p^{qr} d\sigma_q$$

putting

$$p^{rq} = \tfrac{1}{2}\, \varepsilon^{rps}\, \pi_{ps}{}^{q}\,.$$

Cf., for example, A. Lichnerowicz [35], p. 153 and also Bricard: *Le Calcul vectoriel*, p. 159.

RELATIVISTIC ELECTROMAGNETISM

A. THE COVARIANT FORM OF MAXWELL'S THEORY

1. *The electromagnetic field, a tensor of second rank*

Let us adopt the system of real coordinates

$$x^\mu (x^1 = x, x^2 = y, x^3 = z, x^0 = ct).$$

The electromagnetic equations will be written in a system of orthnormal axes with metric

$$g_{\mu\nu} = \eta_{\mu\nu}$$

that is

$$g_{pq} = g^{pq} = -\delta_{pq}, \quad g_{p0} = -g^{p0} = 0, \quad g_{00} = g^{00} = 1.$$

Let us reconsider equations (I) and (III) of Maxwell's theory:

$$\text{(I)} \qquad \boxed{\begin{array}{ll} \text{(a)} & \left\{ \begin{array}{l} \mathbf{curl}\,\mathbf{H} - \dfrac{1}{c}\dfrac{\partial \mathbf{D}}{\partial t} = \dfrac{4\pi\mathbf{I}}{c} \\[2ex] \mathbf{curl}\,\mathbf{E} + \dfrac{1}{c}\dfrac{\partial \mathbf{B}}{\partial t} = 0 \end{array} \right. \\ \text{(b)} & \end{array}}$$

$$\text{(III)} \qquad \boxed{\begin{array}{ll} \text{(a)} & \operatorname{div}\mathbf{D} = 4\pi\rho \\ \text{(b)} & \operatorname{div}\mathbf{B} = 0. \end{array}}$$

Using the notations

$$\text{(IX-1)} \qquad \partial_p = \frac{\partial}{\partial x^p} (p = 1, 2, 3), \qquad \partial_0 = \frac{\partial}{\partial x^0} = \frac{1}{c}\frac{\partial}{\partial t},$$

and introducing the current-density four-vector with components

$$\text{(IX-2)} \qquad \boxed{J_\mu = \left(-\frac{4\pi\mathbf{I}}{c}, \; 4\pi\rho\right),}$$

it becomes clear that systems (I) and (III) may likewise be written in the forms:

	(a)	(b)
(I)	$\partial_p H_q - \partial_q H_p - \partial_0 D_r = - J_r$	$\partial_p E_q - \partial_q E_p + \partial_0 B_r = 0$
(III)	$\sum_p \partial_p D_p = J_0$	$\sum_p \partial_p B_p = 0$

with $p, q, r = 1, 2, 3$, and the letters p, q, r forming, in (I), a cyclic permutation with respect to 1, 2, 3.

In a special Lorentz transformation:

$$x'^1 = \frac{x^1 - \beta x^0}{\sqrt{1-\beta^2}}, \quad x'^2 = x^2, \quad x'^3 = x^3, \quad x'^0 = \frac{x^0 - \beta x^1}{\sqrt{1-\beta^2}}, \tag{IX-3}$$

equations (I) and (III) preserve the same form, that is, they are still written[1]

(I)′	$\partial'_p H'_q - \partial'_q H'_p - \partial_0 D'_r = - J'^r$	$\partial'_p E'_q - \partial'_q E'_p + \partial'_0 B'_r = 0$
(III)′	$\sum_p \partial'_p D'_p = J'^0$	$\sum_p \partial'_p B'_p = 0$

if J (J_r, J_0) is transformed like the components of a four-vector

$$J'_1 = \frac{J_1 + \beta J_0}{\sqrt{1-\beta^2}}, \quad J'_2 = J_2, \quad J'_3 = J_3, \quad J'_0 = \frac{J_0 + \beta J_1}{\sqrt{1-\beta^2}}, \tag{IX-4}$$

and if the vectors **H**, **D** on one hand and **B**, **E** on the other are transformed in the following manner:

$$(\text{IX-5})_1 \qquad D'_1 = D_1, \quad D'_2 = \frac{D_2 - \beta H_3}{\sqrt{1-\beta^2}}, \quad D'_3 = \frac{D_3 + \beta H_2}{\sqrt{1-\beta^2}}$$

$$E'_1 = E_1, \quad E'_2 = \frac{E_2 - \beta B_3}{\sqrt{1-\beta^2}}, \quad E'_3 = \frac{E_3 + \beta B_2}{\sqrt{1-\beta^2}}$$

$$(\text{IX-5})_2 \qquad H'_1 = H_1, \quad H'_2 = \frac{H_2 + \beta D_3}{\sqrt{1-\beta^2}}, \quad H'_3 = \frac{H_3 - \beta D_2}{\sqrt{1-\beta^2}}$$

$$B'_1 = B_1, \quad B'_2 = \frac{B_2 + \beta E_3}{\sqrt{1-\beta^2}}, \quad B'_3 = \frac{B_3 - \beta E_2}{\sqrt{1-\beta^2}}.$$

It will be seen that the transformation formulas $(IX\text{-}5)_1$ and $(IX\text{-}5)_2$ coincide with expressions (VI-87) which define the transformation of an anti-symmetrical tensor of the second order. One has only to put

$$H_p = \varepsilon_{pqr} f^{qr}, \qquad D_p = f_{po} = - f^{po} \tag{IX-6}$$

that is

$$H_1 = f_{23} = f^{23}, \quad H_2 = f_{31} = f^{31}, \quad H_3 = f_{12} = f^{12}$$
$$D_1 = f_{10} = - f^{10}, \quad D_2 = f_{20} = - f^{20},$$
$$D_3 = f_{30} = - f^{30}$$

and likewise

$$B_p = \varepsilon_{pqr} \varphi^{qr}, \qquad E_p = \varphi_{po} = - \varphi^{po} \tag{IX-7}$$

that is

$$B_1 = \varphi_{23} = \varphi^{23}, \quad B_2 = \varphi_{31} = \varphi^{31}, \quad B_3 = \varphi_{12} = \varphi^{12}$$
$$E_1 = \varphi_{10} = - \varphi^{10}, \quad E_2 = \varphi_{20} = - \varphi^{20},$$
$$E_3 = \varphi_{30} = - \varphi^{30},$$

in order to identify (IX-5) with (VI-87). (In (IX-6) and (IX-7) p, q, r take on the values 1, 2, 3 and $\varepsilon_{pqr} = +1, -1$ depending on whether the indices form an even or odd permutation with respect to 1, 2, 3, – ε_{pqr} becoming zero if both indices are identical.)

Thus the components **H**, **D** of the magnetic field and of the electric displacement vector form the components $f_{\mu\nu}$ ($\mu, \nu = 1, 2, 3, 0$) of an anti-symmetrical tensor of the second rank. In a like manner, the ensemble of the components of the electrical field and the magnetic induction make up a single tensor $\varphi_{\mu\nu}$. Definitions (IX-6) and (IX-7) may further be written:

$$f_{\mu\nu} = \begin{vmatrix} 0 & H_3 & -H_2 & D_1 \\ -H_3 & 0 & H_1 & D_2 \\ H_2 & -H_1 & 0 & D_3 \\ -D_1 & -D_2 & -D_3 & 0 \end{vmatrix}, \quad \varphi_{\mu\nu} = \begin{vmatrix} 0 & B_3 & -B_2 & E_1 \\ -B_3 & 0 & B_1 & E_2 \\ B_2 & -B_1 & 0 & E_3 \\ -E_1 & -E_2 & -E_3 & 0 \end{vmatrix}. \tag{IX-8}$$

With these notations, equations (I) (a) and (III) (a) are written simply

$$\boxed{\partial_\rho f^{\mu\rho} = J^\mu} \qquad \mu = 1, 2, 3, 0. \tag{M_1}$$

Equations (I) (a) correspond to $\mu = p = 1, 2, 3$ and equations (III) (a) to $\mu = 0$.

On the other hand, the equations (I) (b) and (III) (b) have the following form:

$$(\mathrm{M}_2) \qquad \boxed{\partial_\rho \varphi_{\mu\nu} + \partial_\nu \varphi_{\rho\mu} + \partial_\mu \varphi_{\nu\rho} = 0}$$

(I) (b) occurs if one of the indices μ, ν, ρ has a zero value and (III) (b) if μ, ν, $\rho = 1, 2, 3$.

Finally, from (M_1) the *continuity equation* immediately follows

$$(\text{IX-9}) \qquad \boxed{\partial_\mu J^\mu = 0}$$

which is the conversion of (III′) into four-dimensional notation.

2. *The electromagnetic potential*

Let us once more consider the equations (IV) defining **E** and **B** as functions of an electromagnetic potential,

$$(\text{IV-a}) \qquad \mathbf{E} = -\,\mathbf{grad}\, V - \frac{1}{c}\frac{\partial \mathbf{A}}{\partial t}$$

$$(\text{IV-b}) \qquad \mathbf{B} = \mathbf{curl}\, \mathbf{A}\,.$$

By introducing the notations

$$(\text{IX-10}) \qquad \varphi_\mu = (\mathbf{A}, -V),$$

equations (IV) will be written

$$(\text{IX-11})_1 \qquad E_p = \partial_p \varphi_0 - \partial_0 \varphi_p$$

$$(\text{IX-11})_2 \qquad B_p = \partial_q \varphi_r - \partial_r \varphi_q \qquad p, q, r = 1, 2, 3.$$

p, q, r = cyclic permutation on 1, 2, 3.

Taking (IX-7) into account, these equations form part of the single system

$$(\mathrm{M}_3) \qquad \boxed{\varphi_{\mu\nu} = \partial_\mu \varphi_\nu - \partial_\nu \varphi_\mu}$$

equivalent to $(\text{IX-11})_1$ for μ or $\nu = 0$, and to $(\text{IX-11})_2$ for μ and $\nu \neq 0$.

The electromagnetic potential then forms a four-vector with the components φ_μ. It becomes immediately evident that, according to (M_3), this four-vector is defined only to the gradient of some arbitrary function.

In fact, if φ_μ is changed to

$$\bar{\varphi}'_\mu = \varphi_\mu - \partial_\mu \psi , \tag{IX-12}$$

the field $\varphi_{\mu\nu}$ is not altered. Transformation (IX-12), introduced by H. Weyl in order to characterize the variations in the scale of length, is called a *gauge transformation.* We say that the electromagnetic potential is defined except to a gauge transformation.[2]

3. *Maxwell's equations and the general Lorentz transformation*

Maxwell's equations may be put in either of two equivalent forms

$$\boxed{\begin{aligned} &\partial_\rho f^{\mu\rho} = J^\mu && (M_1)\\ &\partial_\rho \varphi_{\mu\nu} + \partial_\nu \varphi_{\rho\mu} + \partial_\mu \varphi_{\nu\rho} = 0 && (M_2)\end{aligned}}$$

or

$$\boxed{\begin{aligned} &\partial_\rho f^{\mu\rho} = J^\mu && (M_1)\\ &\varphi_{\mu\nu} = \partial_\mu \varphi_\nu - \partial_\nu \varphi_\mu . && (M_3)\end{aligned}}$$

Indeed, the definition (M_3) imposes the conditions (M_2) on the components of the field $\varphi_{\mu\nu}$.

From (M_3) we deduce

$$\partial^\mu \varphi_{\mu\nu} = \Box\, \varphi_\nu - \partial_\nu \partial^\mu \varphi_\mu \tag{IX-13}$$

with

$$\Box = \partial^\rho \partial_\rho = \eta^{\rho\sigma} \partial_\rho \partial_\sigma = \partial_0^2 - \sum_p \partial_p^2 .$$

If the electromagnetic potential satisfies, moreover, the *Lorentz condition* (VII) (cf. p. 57), that is, if

$$\partial^\mu \varphi_\mu = 0 , \tag{IX-14}$$

then we may write (IX-13)

$$\Box\, \varphi_\nu = \partial^\mu \varphi_{\mu\nu} . \tag{IX-15}$$

In the case of a medium that is only slightly dispersive and that has a dielectric constant ε and a magnetic permeability μ

$$\varepsilon\mu = 1, \qquad \varphi_{\mu\nu} = \mu f_{\mu\nu} = \frac{1}{\varepsilon} f_{\mu\nu} ,$$

(IX-16) $$\square\, \varphi_\nu = -\, \mu J_\nu .$$

Let us form $\square\, \varphi_{\mu\nu}$ from (M_3) and substitute (IX-16) in the equation thus obtained.

One has

(IX-17) $$\square\, \varphi_{\mu\nu} = -\, \mu(\partial_\mu J_\nu - \partial_\nu J_\mu) .$$

It is often useful to introduce a *dual* field when writing Maxwell's equations[2a]

(IX-18) $$\varphi^{\mu\nu *} = \tfrac{1}{2}\varepsilon^{\mu\nu\rho\sigma}\varphi_{\rho\sigma}, \qquad \overset{*}{\varphi}_{\mu\nu} = -\, \tfrac{1}{2}\varepsilon_{\mu\nu\rho\sigma}\varphi^{\rho\sigma}$$

with

(IX-19) $$\varepsilon_{\mu\nu\rho\sigma} = \varepsilon^{\mu\nu\rho\sigma} = \begin{cases} +1 & \text{if the indexes } \mu, \nu, \rho, \sigma \text{ form an even permutation of } 1, 2, 3, 0 \\ -1 & \text{if the indexes } \mu, \nu, \rho, \sigma \text{ form an odd permutation of } 1, 2, 3, 0 \\ 0 & \text{if two of the indexes are equal.} \end{cases}$$

We obtain

(IX-20) $$\begin{aligned} \varphi^{*23} &= \overset{*}{\varphi}_{23} = \varphi_{10} = -\, \varphi^{10}, & \varphi^{*10} &= -\, \overset{*}{\varphi}_{10} = \varphi_{23} = \varphi^{23} \\ \varphi^{*31} &= \overset{*}{\varphi}_{31} = \varphi_{20} = -\, \varphi^{20}, & \varphi^{*20} &= -\, \overset{*}{\varphi}_{20} = \varphi_{31} = \varphi^{31} \\ \varphi^{*12} &= \overset{*}{\varphi}_{12} = \varphi_{30} = -\, \varphi^{30}, & \varphi^{*30} &= -\, \overset{*}{\varphi}_{30} = \varphi_{12} = \varphi^{12} \end{aligned}$$

and equations (M_2) may be written in still another form

$(M_2)^*$ $$\boxed{\partial_\rho \varphi^{*\mu\rho} = 0 .}$$

When we perform a general Lorentz transformation

$$x^\mu = a^\mu_{\nu'} x'^\nu, \qquad x'^\mu = a^{\mu'}_\nu x^\nu,$$

the constant coefficients

$$a^{\mu'}_\nu = \frac{\partial x^{\mu'}}{\partial x^\nu}$$

and

$$a^\mu_{\nu'} = \frac{\partial x^\mu}{\partial x'^\nu}$$

satisfying the relations which characterize the Lorentz transformation, the quantities

$$\varphi_\mu, \qquad J_\mu \qquad \text{and} \qquad \partial_\mu = \frac{\partial}{\partial x^\mu} = \frac{\partial}{\partial x'^\lambda}\frac{\partial x'^\lambda}{\partial x^\mu} = a^{\lambda'}_\mu \partial'_\lambda$$

are transformed like the components of a four-vector. Furthermore, the fields and inductions $f^{\mu\nu}$ and $\varphi_{\mu\nu}$ are transformed according to (VI-86) and (VI-87), like the components of antisymmetrical tensors of the second rank:

$$f'_{\mu\nu} = \tfrac{1}{2}(a^{\rho}_{\mu'}a^{\sigma}_{\nu'} - a^{\sigma}_{\mu'}a^{\rho}_{\nu'})\,f_{\rho\sigma}, \quad f'^{\mu\nu} = \tfrac{1}{2}(a^{\mu'}_{\rho}a^{\nu'}_{\sigma} - a^{\mu'}_{\sigma}a^{\nu'}_{\rho})\,f^{\rho\sigma} \tag{IX-21}$$

$$\varphi'_{\mu\nu} = \tfrac{1}{2}(a^{\rho}_{\mu'}a^{\sigma}_{\nu'} - a^{\sigma}_{\mu'}a^{\rho}_{\nu'})\,\varphi_{\rho\sigma}, \quad \varphi'^{\mu\nu} = \tfrac{1}{2}(a^{\mu'}_{\rho}a^{\nu'}_{\sigma} - a^{\mu'}_{\sigma}a^{\nu'}_{\rho})\,\varphi^{\rho\sigma}. \tag{IX-22}$$

The covariance of Maxwell's equations (M_1), (M_2) [or $(M_2)^*$] and (M_3) is thus always assured.[3]

4. *The Lorentz electron theory. The energy-momentum tensor*

1) *The "microscopic" field and the current four-vector.* The equations of the Lorentz microscopic field

$$\boxed{\begin{aligned} &\mathbf{curl}\,\mathbf{h} - \frac{1}{c}\frac{\partial e}{\partial t} = \frac{4\pi}{c}\,\rho v & \text{(IX-23)} \\ &\mathbf{curl}\,\mathbf{e} + \frac{1}{c}\frac{\partial \boldsymbol{h}}{\partial t} = 0 & \text{(IX-24)} \end{aligned}}$$

$$\boxed{\begin{aligned} &\operatorname{div}\mathbf{e} = 4\pi\rho & \text{(IX-25)} \\ &\operatorname{div}\mathbf{h} = 0 & \text{(IX-26)} \end{aligned}}$$

may be immediately transformed into four-dimensional relativistic formalism by putting:

$$\varphi_{\mu\nu} = \begin{bmatrix} 0 & h_3 & -h_2 & e_1 \\ -h_3 & 0 & h_1 & e_2 \\ h_2 & -h_1 & 0 & e_3 \\ -e_1 & -e_2 & -e_3 & 0 \end{bmatrix} \tag{IX-27}$$

$$j^{\mu} = \left(\frac{4\pi}{c}\,\rho\mathbf{v}, 4\pi\rho\right) \quad \text{with} \quad v = (v^1, v^2, v^3). \tag{IX-28}$$

Equations (IX-23)–(IX-26) are then written

$$\boxed{\begin{aligned} &\partial_{\rho}\varphi^{\mu\rho} = j^{\mu} & (L_1) \\ &\partial_{\rho}\varphi_{\mu\nu} + \partial_{\nu}\varphi_{\rho\mu} + \partial_{\mu}\varphi_{\nu\rho} = 0. & (L_2) \end{aligned}}$$

(L_1) groups together the equations having a right-hand member (IX-23) and (IX-25), while (L_2) groups those equations without a right-hand member, namely (IX-24) and (IX-26).

From L_1 we obtain the continuity equation

$$\boxed{\partial_\mu j^\mu = 0.} \tag{IX-29}$$

In keeping with its definition (IX-28), the current j_μ is proportional to the world-velocity

$$\boxed{j_\mu = 4\pi\rho_0 u_\mu} \tag{IX-30}$$

ρ_0 being the density of charge in the proper coordinate system, that is the density of charge for an observer who is at rest with respect to the charges.

Indeed, by taking (IX-30) and (VII-12) into account, we shall have

$$j_p = 4\pi\rho_0 u_p = -\frac{4\pi\rho_0 v^p}{c\sqrt{1-\beta^2}} = -\frac{4\pi\rho}{c}v^p \tag{IX-31$_1$}$$

$$j_0 = 4\pi\rho_0 u_0 = \frac{4\pi\rho_0}{\sqrt{1-\beta^2}} = 4\pi\rho \tag{IX-31$_2$}$$

with

$$\boxed{\rho = \frac{\rho_0}{\sqrt{1-\beta^2}}.} \tag{IX-32}$$

The quantity of electricity contained in a volume $d\mathscr{V}$ will then be, according to (V-43) and (IX-32)

$$dq = \rho\, d\mathscr{V} = \frac{\rho_0}{\sqrt{1-\beta^2}} d\mathscr{V}_0 \sqrt{1-\beta^2} = \rho_0\, d\mathscr{V}_0 = dq_0. \tag{IX-33}$$

It is, thus, an *invariant*. In particular, the charge q of the electron remains the same in all Lorentzian reference systems.

2) *The Lorentz force.* The vector expression of the Lorentz force

$$\mathbf{f} = \rho\left(\mathbf{e} + \left[\frac{\mathbf{v}}{c}\wedge\mathbf{h}\right]\right) \tag{IX-34}$$

allows us to define the three spatial components f^p of the four-vector:

$$\boxed{f^\mu = \frac{1}{4\pi}\varphi^{\rho\mu} j_\rho.} \tag{IX-35}$$

One has, as a matter of fact:

(IX-36) $$f_p = \frac{1}{4\pi}(\varphi_{pq} j^q + \varphi_{p0} j^0)$$

and, according to (IX-27) and (IX-31)

(IX-37) $$f_p = \rho\left(e_p + \varphi_{pq}\frac{v^q}{c}\right) = \rho\left(e_p + \left[\frac{\mathbf{v}}{c}\wedge \mathbf{h}\right]_p\right).$$

The fourth component is further written:

(IX-38) $$f_0 = \frac{1}{4\pi}\varphi_{p0} j^p = \frac{\rho}{c}\varphi_{p0} v^p = \frac{\rho}{c}(\mathbf{v}\cdot\mathbf{e}).$$

According to (IX-37), it represents the power $\boldsymbol{f}\cdot \mathbf{dl}/dx^0$ developed by the electromagnetic forces.

3) *The energy-momentum or Maxwell tensor.* Let us consider the second-order tensor

(IX-39) $$\tau^\lambda_\mu = -\varphi_{\mu\rho}\varphi^{\lambda\rho} + \tfrac{1}{4}\delta^\lambda_\mu \varphi_{\rho\sigma}\varphi^{\rho\sigma}$$

or

(IX-40) $$\tau_{\mu\nu} = g_{\nu\lambda}\tau^\lambda_\mu = -\varphi_{\mu\rho}\varphi_\nu{}^\rho + \tfrac{1}{4} g_{\mu\nu}\varphi_{\rho\sigma}\varphi^{\rho\sigma}$$

with

(IX-41) $$g_{\mu\nu} = \eta_{\mu\nu} = \delta_{\mu\nu}(-1, -1, -1, +1).$$

This tensor is symmetrical

(IX-42) $$\tau_{\mu\nu} = \tau_{\nu\mu}$$

and, according to (IX-27), its components have the values summarized in the following table

(IX-43) $$\tau_{\mu\nu} = \begin{vmatrix} \tau_{11} & \tau_{12} & \tau_{13} & -4\pi s_1 \\ \tau_{21} & \tau_{22} & \tau_{23} & -4\pi s_2 \\ \tau_{31} & \tau_{32} & \tau_{33} & -4\pi s_3 \\ -4\pi s_1 & -4\pi s_2 & -4\pi s_3 & 4\pi w \end{vmatrix}$$

with

$$\boxed{\begin{aligned} &\text{(IX-44)} \quad \tau_{pq} = -(e_p e_q + h_p h_q) + \tfrac{1}{2}\delta_{pq}(\mathbf{e}^2 + \mathbf{h}^2) \\ &\text{(IX-45)} \quad s_p = \frac{1}{4\pi}[\boldsymbol{e} \wedge \boldsymbol{h}]_p \\ &\text{(IX-46)} \quad w = \frac{1}{8\pi}(\boldsymbol{e}^2 + \boldsymbol{h}^2). \end{aligned}}$$

The components τ_{po} of the tensor $\tau_{\mu\nu}$ are thus related to the *momentum density* of a distribution of charges and τ_{oo} expresses its *energy density*.

Let us take the divergence of (IX-39). We obtain:

(IX-47) $$\partial_\lambda \tau^\lambda_\mu = -(\partial_\lambda \varphi_{\mu\rho})\,\varphi^{\lambda\rho} - \varphi_{\mu\rho}\partial_\lambda \varphi^{\lambda\rho} + \tfrac{1}{4}\partial_\mu(\varphi_{\rho\sigma}\varphi^{\rho\sigma})$$

that is to say, when we take into account the antisymmetry in λ, ρ and the definition (L_1) of the current:

(IX-48) $$\partial_\lambda \tau^\lambda_\mu = -\tfrac{1}{2}(\partial_\lambda \varphi_{\mu\rho} + \partial_\rho \varphi_{\lambda\mu})\,\varphi^{\lambda\rho} + \varphi_{\mu\rho} j^\rho + \tfrac{1}{4}\partial_\mu(\varphi_{\rho\sigma}\varphi^{\rho\sigma}).$$

But, according to the second group of Maxwell's equations (M_2):

(IX-49) $$\partial_\lambda \tau^\lambda_\mu = \tfrac{1}{2}(\partial_\mu \varphi_{\rho\lambda})\,\varphi^{\lambda\rho} + \varphi_{\mu\rho} j^\rho + \tfrac{1}{4}\partial_\mu(\varphi_{\rho\sigma}\varphi^{\rho\sigma}).$$

Now

(IX-50) $$\begin{aligned} \tfrac{1}{2}(\partial_\mu \varphi_{\rho\lambda})\,\varphi^{\lambda\rho} &= \tfrac{1}{2}\eta^{\lambda\sigma}\eta^{\rho\tau}(\partial_\mu \varphi_{\rho\lambda})\,\varphi_{\sigma\tau} = \tfrac{1}{4}\eta^{\lambda\sigma}\eta^{\rho\tau}\partial_\mu(\varphi_{\rho\lambda}\varphi_{\sigma\tau}) = \\ &= \tfrac{1}{4}\partial_\mu(\varphi_{\rho\lambda}\varphi^{\lambda\rho}) = -\tfrac{1}{4}\partial_\mu(\varphi_{\rho\sigma}\varphi^{\rho\sigma}) \end{aligned}$$

and (IX-49) may then be written

(IX-51) $$\boxed{\partial_\lambda \tau^\lambda_\mu = \varphi_{\mu\rho} j^\rho.}$$

Going back to (IX-35), the right-hand member of (IX-51) gives the four-vector

(IX-52) $$4\pi f_\mu = \varphi_{\rho\mu} j^\rho$$

whose three spatial components represent the Lorentz force. One will then have:

(IX-53) $$\partial_\lambda \tau^\lambda_\mu = -4\pi f_\mu$$

and the fourth relation (IX-53) may then be written:

(IX-54) $$\partial_p \tau_0^p + \partial_0 \tau_0^0 = -4\pi f_0$$

that is, according to (IX-43) and (IX-38):

(IX-55) $$\boxed{\frac{1}{c}\frac{\partial w}{\partial t} + \operatorname{div} \mathbf{s} = -\rho\left(\frac{\mathbf{v}}{c}\cdot\mathbf{e}\right).}$$

This is the transposition into the Lorentz theory of Poynting's Theorem as established in (III-52). The relation (IX-53) thus represents the definition of the Lorentz force and also the condition of conservation that the electromagnetic energy must satisfy.

5. *Lorentz equations and Maxwell's equations*

Maxwell's equations may be deduced from Lorentz's microscopic equations by assuming that ponderable matter is at rest[4] (cf. Chapter IV, Section 5). So it is in a reference system bound to ponderable matter, that is, in the conducting or dielectric medium, that groups (I), (II), (III) and (V) of the Maxwellian equations are set up. And it is in this same special system that the relations $\mathbf{D} = \varepsilon\mathbf{E}$, $\mathbf{B} = \mu\mathbf{H}$ between inductions and fields are valid. Let us represent the quantities defined in this proper system by a zero index. The following equations, derived as in Chapter IV, will be valid in such a system

	(a)	(b)
(I)°	$\mathbf{curl}^{(0)}\mathbf{H}^{(0)} - \frac{1}{c}\frac{\partial \mathbf{D}^{(0)}}{\partial t} = \frac{4\pi}{c}\mathbf{I}^{(0)}$	$\mathbf{curl}^{(0)}\mathbf{E}^{(0)} + \frac{1}{c}\frac{\partial \mathbf{B}^{(0)}}{\partial t} = 0$
(III)°	$\operatorname{div}^{(0)}\mathbf{D}^{(0)} = 4\pi\rho^{(0)}$	$\operatorname{div}^{(0)}\mathbf{B}^{(0)} = 0$

(II)°
(1) $\mathbf{D}^{(0)} = \varepsilon\mathbf{E}^{(0)}$
(2) $\mathbf{B}^{(0)} = \mu\mathbf{H}^{(0)}$
(3) $\mathbf{I}^{(0)} = \sigma_c\mathbf{E}^{(0)}$.

In particular, relation (II)°-3 represents Ohm's law in a system bound to matter.

We have seen in the first section of this chapter that systems (I) and (III) are invariant in a Lorentz transformation, since the J_μ transform like the components of a vector and the (**H**, **D**) and (**E**, **B**) form two antisymmetrical tensors of the second order. We then put

(IX-56)
$$f_{\mu\nu}^{(0)} = \begin{vmatrix} 0 & H_3^{(0)} & -H_2^{(0)} & D_1^{(0)} \\ -H_3^{(0)} & 0 & H_1^{(0)} & D_2^{(0)} \\ H_2^{(0)} & -H_1^{(0)} & 0 & D_3^{(0)} \\ -D_1^{(0)} & -D_2^{(0)} & -D_3^{(0)} & 0 \end{vmatrix}$$
$$\varphi_{\mu\nu}^{(0)} = \begin{vmatrix} 0 & B_3^{(0)} & -B_2^{(0)} & E_1^{(0)} \\ -B_3^{(0)} & 0 & B_1^{(0)} & E_2^{(0)} \\ B_2^{(0)} & -B_1^{(0)} & 0 & E_3^{(0)} \\ -E_1^{(0)} & -E_2^{(0)} & -E_3^{(0)} & 0 \end{vmatrix}$$

(IX-57)
$$J_\mu^{(0)} = \left(-\frac{4\pi}{c} I^{(0)}, \qquad 4\pi\rho^{(0)} \right)$$

and systems (I)°, (III)° and (II)° may then be written

$(M_1)^{(\circ)}$
$$\partial_\rho^{(0)} f^{\mu\rho(0)} = J^{\mu(0)}$$

$(M_2)^{(\circ)}$
$$\partial_\rho^{(0)} \varphi_{\mu\nu}^{(0)} + \partial_\nu^{(0)} \varphi_{\rho\mu}^{(0)} + \partial_\mu^{(0)} \varphi_{\nu\rho}^{(0)} = 0$$

$(M_4)^{(\circ)}$
$$\begin{cases} (1) & f_{p0}^{(0)} = \varepsilon \varphi_{p0}^{(0)} \\ (2) & f_{pq}^{(0)} = \dfrac{1}{\mu} \varphi_{pq}^{(0)} \\ (3) & J_p^{(0)} = -\dfrac{4\pi\sigma}{c} \varphi_{p0}^{(0)}. \end{cases}$$

In a general Lorentz transformation one will obtain, beginning with $(M_1)^{(\circ)}$ and $(M_2)^{(\circ)}$, systems (M_1) and (M_2) valid in any Lorentzian system.[5]

(M_1)
$$\partial_\rho f^{\mu\rho} = J^\mu$$

(M_2)
$$\partial_\rho \varphi_{\mu\nu} + \partial_\nu \varphi_{\rho\mu} + \partial_\mu \varphi_{\nu\rho} = 0.$$

But, from their very form, it may be seen that equations $(M_4)^{(\circ)}$ are not invariant in a Lorentz transformation. Their expression, which is adapted solely to the proper system, cannot be transposed into an arbitrary Galilean system.

1) *The current four-vector.* The current four-vector J_μ appearing in equations (M_1) is deduced by means of a Lorentz transformation of the four-vector $J_\mu^{(0)}$ attached to ponderable matter. The components $\rho^{(0)}$ and $\mathbf{I}^{(0)}$ of $\mathbf{J}^{(0)}$ then represent the average values $\bar{\rho}$ and $\overline{\rho\mathbf{v}}$ of the density of

charge and the convection-current density defined by the microscopic theory.

Now, in a Lorentz transformation from the proper system, one always has

(IX-58) $$J_\mu = a_\mu^{\lambda(0)} J_\lambda^{(0)} = a_\mu^{0(0)} J_0^{(0)} + a_\mu^{p(0)} J_p^{(0)}.$$

More especially, in a Lorentz transformation without rotation in which the coefficients have the values (VI-62) and (VI-63), one has according to (IX-58)[6]

(IX-59)$_1$ $$J_p = J_p^{(0)} - v_p \left\{ J_r^{(0)} v^r \frac{\alpha}{v^2} - \frac{J_0^{(0)}}{c\sqrt{1-\beta^2}} \right\}$$

(IX-59)$_2$ $$J_0 = \frac{J_0^{(0)} - \dfrac{J_r^{(0)} v^r}{c}}{\sqrt{1-\beta^2}}$$

with

(IX-60) $$v^2 = \sum_r (v^r)^2, \qquad \beta^2 = \frac{v^2}{c^2}$$

(IX-61) $$\alpha = \frac{1}{\sqrt{1-\beta^2}} - 1.$$

If we refer to the expressions of the four-velocity, (IX-59) may be written[7]

(IX-62)$_1$ $$J_p = J_0^{(0)} u_p + J_p^{(0)} + \frac{\alpha u_p}{u_s u^s} (J_r^{(0)} u^r,$$

(IX-62)$_2$ $$J_0 = J_0^{(0)} u_0 - J_r^{(0)} u^r.$$

We note that (IX-62) may be put in the following form

(IX-63) $$J_p = J_p^{(0)} + \rho_1 u_p$$

on condition that we put:

(IX-64) $$\rho_1 = J_0^{(0)} + \frac{\alpha}{u_s u^s} J_r^{(0)} u^r.$$

But the expression (IX-63) is not covariant under a Lorentz transformation.[8]

On the other hand, we obtain a covariant re-transcription of (IX-62) by putting

(IX-65) $$\boxed{J_\mu = J_0^{(0)} u_\mu + I_\mu,}$$

since I_μ represents the components of the four-vector:

(IX-66) $$I_\mu = a_\mu^{r(0)} J_r^{(0)} = (a_p^{r(0)} J_r^{(0)}, -J_r^{(0)} u^r)$$

which, in the proper system ($u^r = 0$) reduces to $J_r^{(0)}$:

(IX-67) $$I_\mu^{(0)} = (J_r^{(0)}, 0).$$

Thus, *even in the case where the density of charge is zero in the proper system* ($J_0^{(0)} = 0$), *there exists in every Galilean system, a density of charge*

(IX-68) $$J_0 = I_0 = -J_r^{(0)} u^r$$

produced by the conduction current of value $J_r^{(0)}$ in the proper system.

Finally, if the conduction current is zero in the proper system ($J_r^{(0)} = 0$) – and that is the case for an insulator – one simply has

(IX-69) $$J_\mu = J_0^{(0)} u_\mu .$$

The current is then proportional to the world four-velocity.

2) *The relations between fields and inductions.* Whereas equations (I) and (III) are covariant, relations (II) do not keep the same form in a Lorentz transformation: $(M_4)^{(o)}$ is valid solely in the proper system.

In any Galilean system, we will then define the four-vectors:

(IX-70) $$\bar{E}_\mu = \varphi_{\mu\rho} u^\rho, \qquad \bar{B}_\mu = \overset{*}{\varphi}_{\mu\rho} u^\rho$$

(IX-71) $$\bar{D}_\mu = f_{\mu\rho} u^\rho, \qquad \bar{H}_\mu = \overset{*}{f}_{\mu\rho} u^\rho$$

which involve the tensors $\varphi_{\mu\rho}, f_{\mu\rho}$ defined in (IX-8) and the dual tensors (IX-18).

Taking into account expressions (IX-8) of $\varphi_{\mu\rho}$ and of $f_{\mu\rho}$, we note that the components $\bar{E}_\mu$, $\bar{B}_\mu$, $\bar{D}_\mu$ and $\bar{H}_\mu$ may be written:

(IX-72)
$$\bar{E}_\mu = \frac{1}{\sqrt{1-\beta^2}}\left(\tilde{\mathbf{E}}, \left(\tilde{\mathbf{E}}\cdot\frac{\mathbf{v}}{c}\right)\right),$$
$$\bar{B}_\mu = \frac{1}{\sqrt{1-\beta^2}}\left(\tilde{\mathbf{B}}, \left(\tilde{\mathbf{B}}\cdot\frac{\mathbf{v}}{c}\right)\right)$$

(IX-73)
$$\bar{D}_\mu = \frac{1}{\sqrt{1-\beta^2}}\left(\tilde{\mathbf{D}}, \left(\tilde{\mathbf{D}}\cdot\frac{\mathbf{v}}{c}\right)\right),$$
$$\bar{H}_\mu = \frac{1}{\sqrt{1-\beta^2}}\left(\tilde{\mathbf{H}}, \left(\tilde{\mathbf{H}}\cdot\frac{\mathbf{v}}{c}\right)\right)$$

putting

$$\tilde{\mathbf{E}} = \mathbf{E} + \left(\frac{\mathbf{v}}{c} \wedge \mathbf{B}\right), \qquad \tilde{\mathbf{B}} = \mathbf{B} + \left(\frac{\mathbf{v}}{c} \wedge \mathbf{E}\right) \tag{IX-74}$$

$$\tilde{\mathbf{D}} = \mathbf{D} + \left(\frac{\mathbf{v}}{c} \wedge \mathbf{D}\right), \qquad \tilde{\mathbf{H}} = \mathbf{H} + \left(\frac{\mathbf{v}}{c} \wedge \mathbf{H}\right). \tag{IX-75}$$

In the proper system ($\mathbf{v} = 0$), the four-vectors (IX-72) and (IX-73) then reduce to

$$\bar{E}_\mu = (\mathbf{E}^{(0)}, 0) \qquad \bar{B}_\mu = (\mathbf{B}^{(0)}, 0) \tag{IX-76}$$

$$\bar{D}_\mu = (\mathbf{D}^{(0)}, 0) \qquad \bar{H}_\mu = (\mathbf{H}^{(0)}, 0). \tag{IX-77}$$

Thus, in any Galilean system, the first two equations (M_4) have the following form:

$$\tilde{\mathbf{D}} = \varepsilon \tilde{\mathbf{E}}, \qquad \tilde{\mathbf{H}} = \frac{1}{\mu} \tilde{\mathbf{B}} \tag{IX-78}$$

that is

$$\bar{D}_\rho = \varepsilon \bar{E}_\rho, \qquad \bar{H}_\rho = \frac{1}{\mu} \bar{B}_\rho \tag{IX-79}$$

or

$$\boxed{f_{\rho\sigma} u^\sigma = \varepsilon \varphi_{\rho\sigma} u^\sigma} \tag{IX-80}$$

$$\boxed{\overset{*}{f}_{\rho\sigma} u^\sigma = \frac{1}{\mu} \overset{*}{\varphi}_{\rho\sigma} u^\sigma} \tag{IX-81}$$

with the further possibility of writing (IX-81):

$$f_{\mu\nu} u_\rho + f_{\rho\mu} u_\nu + f_{\nu\rho} u_\mu = \frac{1}{\mu} (\varphi_{\mu\nu} u_\rho + \varphi_{\rho\mu} u_\nu + \varphi_{\nu\rho} u_\mu). \tag{IX-82}$$

Finally, the third equation (M_4) is transformed into:

$$\boxed{J^\rho - u^\rho (J^\mu u_\mu) = - \frac{4\pi\sigma_c}{c} \varphi^{\rho\sigma} u_\sigma .} \tag{IX-83}$$

Indeed, the left-hand member of this equation represents the conduction current, that is the total current J^ρ less the convection current

$$u^\rho (J^\mu u_\mu) = u^\rho (J^\mu u_\mu)^{(0)} = u^\rho J^{0(0)} . \tag{IX-84}$$

The equation thus means that the conduction current $-(4\pi/c)\,I^\rho$ is proportional (with the factor $-(4\pi/c)\,\sigma_c$) to the field $\varphi^{\rho\lambda}\,u_\lambda = \bar{E}^\rho$: which is *the relativistic expression of Ohm's law.*

Going on to the vector notations and solving (IX-80) and (IX-82) with respect to **D** and **B** we obtain further:

(IX-85)
$$\mathbf{D} = \frac{1}{1+\varepsilon\mu\beta^2}\left\{\varepsilon(1+\beta^2)\mathbf{E} + (\varepsilon\mu - 1)\left[\left[\frac{\mathbf{v}}{c}\wedge\mathbf{H}\right] - \varepsilon\frac{\mathbf{v}}{c}\left(\frac{\mathbf{v}}{c}\cdot\mathbf{E}\right)\right]\right\}$$

(IX-86)
$$\mathbf{B} = \frac{1}{1+\varepsilon\mu\beta^2}\left\{\mu(1+\beta^2)\mathbf{H} - (\varepsilon\mu - 1)\left[\left[\frac{\mathbf{v}}{c}\wedge\mathbf{E}\right] - \mu\frac{\mathbf{v}}{c}\left(\frac{\mathbf{v}}{c}\cdot\mathbf{H}\right)\right]\right\}.$$

These relations constitute the generalization of the vector equations (II)° for an arbitrary Galilean system.

In a non-dispersive medium

(IX-87) $$\varepsilon\mu = 1$$

and, in this particular case, one has again

(IX-88) $$\mathbf{D} = \varepsilon\mathbf{E}, \qquad \mathbf{B} = \mu\mathbf{H}$$

in all Galilean coordinate systems.

Again using tensor notation, the field $f_{\mu\nu}$ may be expressed as a function of the field $\varphi_{\mu\nu}$ and conversely.[9] In order to do this, let us multiply (IX-82) by u^ρ and add. Taking into account the normalization conditions for the world-velocity

(VII-13) $$u_\rho u^\rho = 1$$

one obtains

(IX-89) $$f_{\mu\nu} + (f_{\rho\mu}u_\nu + f_{\nu\rho}u_\mu)\,u^\rho = \frac{1}{\mu}\varphi_{\mu\nu} + \frac{1}{\mu}(\varphi_{\rho\mu}u_\nu + \varphi_{\nu\rho}u_\mu)\,u^\rho$$

that is, according to (IX-82),

(IX-90) $$\boxed{f_{\mu\nu} = \frac{1}{\mu}\varphi_{\mu\nu} + \left(\frac{1-\varepsilon\mu}{\mu}\right)(\varphi_{\rho\mu}u_\nu + \varphi_{\nu\rho}u_\mu)\,u^\rho .}$$

Conversely,

(IX-91) $$\boxed{\varphi_{\mu\nu} = \mu f_{\mu\nu} - \left(\frac{1-\varepsilon\mu}{\varepsilon}\right)(f_{\rho\mu}u_\nu + f_{\nu\rho}u_\mu)\,u^\rho .}$$

In a non-dispersive medium ($\varepsilon\mu = 1$) we then obtain

(IX-92) $$f_{\mu\nu} = \varepsilon\varphi_{\mu\nu} = \frac{1}{\mu}\,\varphi_{\mu\nu}.$$

6. *The energy-momentum tensors*

Let us define the non-symmetrical tensor of the second order

(IX-93) $$(\tau_\mu{}^\nu)_M = -\,\varphi_{\mu\rho}f^{\nu\rho} + \tfrac{1}{4}\delta_\mu^\nu\varphi_{\rho\sigma}f^{\rho\sigma}$$

which, in vacuum ($f_{\mu\nu} = \varphi_{\mu\nu}$), reduces to Maxwell's symmetric tensor, By taking, as we did in (IX-48), the divergence of this tensor, we obtain, not (IX-51), but:

(IX-94) $$\partial_\nu(\tau_\mu{}^\nu)_M = \varphi_{\mu\rho}J^\rho + \tfrac{1}{4}(\varphi^{\rho\sigma}\partial_\mu f_{\rho\sigma} - f_{\rho\sigma}\partial_\mu\varphi^{\rho\sigma}).$$

Now, by taking (IX-90) into account, with ε and μ assumed to be constant,

(IX-95) $$\varphi^{\rho\sigma}\partial_\mu f_{\rho\sigma} - f_{\rho\sigma}\partial_\mu\varphi^{\rho\sigma} = 4\left(\frac{1-\varepsilon\mu}{\mu}\right)\varphi^{\rho\sigma}\varphi_{\sigma\lambda}u^\lambda\partial_\mu u_\rho .$$

Hence

(IX-96) $$\partial_\nu(\tau_\mu{}^\nu)_M = \varphi_{\mu\rho}J^\rho - \left(\frac{1-\varepsilon\mu}{\mu}\right)\varphi^{\sigma\rho}\varphi_{\sigma\lambda}u^\lambda\partial_\mu u_\rho .$$

If we are dealing with a medium that is only very slightly dispersive, ($\varepsilon\mu \sim 1$), we obtain, as in the case of Maxwell's tensor

(IX-97) $$\partial_\nu(\tau_\mu{}^\nu)_M = \varphi_{\mu\rho}J^\rho .$$

Let us write out the components of $(\tau_\mu{}^\nu)_M$ and put:

(IX-98) $$\begin{aligned}&(\tau_\mu{}^\rho)_M = g^{\rho\nu}(\tau_{\mu\nu})_M = \eta^{\rho\nu}(\tau_{\mu\nu})_M ,\\ &\eta^{\rho\nu} = \delta^{\rho\nu}(-1-1-1+1)\end{aligned}$$

with

$$(\text{IX-99}) \qquad (\tau_{\mu\nu})_M = \begin{vmatrix} \tau_{11} & \tau_{12} & \tau_{13} & -4\pi s_{10} \\ \tau_{21} & \tau_{22} & \tau_{23} & -4\pi s_{20} \\ \tau_{31} & \tau_{32} & \tau_{33} & -4\pi s_{30} \\ -4\pi s_{01} & -4\pi s_{02} & -4\pi s_{03} & 4\pi w \end{vmatrix}.$$

We will have, according to (IX-93):

$$(\text{IX-100}) \qquad \tau_{pq} = -(E_p D_q + H_p B_q) + \tfrac{1}{2}\delta_{pq}[(\mathbf{E}\cdot\mathbf{D}) + (\mathbf{H}\cdot\mathbf{B})]$$

$$(\text{IX-101}) \qquad s_{p0} = \frac{1}{4\pi}[\mathbf{D}\wedge\mathbf{B}]_p, \qquad s_{0p} = \frac{1}{4\pi}[\mathbf{E}\wedge\mathbf{H}]_p$$

$$(\text{IX-102}) \qquad W = \frac{1}{8\pi}(\mathbf{ED} + \mathbf{HB}).$$

The tensor (IX-93) is called *Minkowski's tensor.* It was proposed to substitute for it the second-rank tensor:

$$(\text{IX-103}) \qquad (\tau_\mu{}^\nu)_S = -\tfrac{1}{2}(\varphi_{\mu\rho} f^{\nu\rho} + f_{\mu\rho}\varphi^{\nu\rho}) + \tfrac{1}{4}\delta_\mu^\nu \varphi_{\rho\sigma} f^{\rho\sigma}.$$

obtained by symmetrization beginning with (IX-93).

We then obtain, instead of (IX-94):

$$(\text{IX-104}) \qquad \begin{aligned} \partial_\nu(\tau_\mu{}^\nu)_S = \varphi_{\mu\rho} J^\rho - \tfrac{1}{2}(f_{\mu\rho}\partial_\nu\varphi^{\nu\rho} - \varphi_{\mu\rho}\partial_\nu f^{\nu\rho}) - \\ - \tfrac{1}{4}(\partial_\nu f_{\mu\rho} + \partial_\rho f_{\nu\mu} + \partial_\mu f_{\rho\nu})\varphi^{\nu\rho}. \end{aligned}$$

If we are dealing with a slightly dispersive medium ($\varepsilon\mu \sim 1$)

$$(\text{IX-105}) \qquad (\tau_\mu{}^\nu)_M = (\tau_\mu{}^\nu)_S$$

and if ε and μ are, in addition, constant, one still has:

$$(\text{IX-106}) \qquad \partial_\nu(\tau_\mu{}^\nu)_S = \partial_\nu(\tau_\mu{}^\nu)_M = \partial_\nu\tau_\mu{}^\nu = \varphi_{\mu\rho} J^\rho.$$

7. *Use of arbitrary curvilinear coordinates*

Up to this point we have always described electromagnetic phenomena in terms of axes orthonormal according to the condition

$$g_{\mu\nu} = \eta_{\mu\nu}, \qquad \eta_{\mu\nu} = \delta_{\mu\nu}(-1, -1, -1, +1).$$

In this system of axes, the ds^2 has the reduced form (VI-22) which is invariant under a Lorentz transformation.

1) *Maxwell's equations.* Let us select, in Euclidean space, a system formed by any arbitrary curvilinear coordinates. Maxwell's equations valid in an

orthonormal system are generally no longer covariant in other systems. We can restore a covariant formalism by making use of arbitrary axes, as indicated in Chapter VI: the $g_{\mu\nu}$ are then variable from one point to another, and the ordinary derivatives must be replaced by the *covariant derivatives.*

One will then have, instead of (M_1), (M_2) and (M_3):

$$\text{(IX-107)} \qquad \boxed{\nabla_\rho f^{\mu\rho} = J^\mu}$$

and one of the two equivalent systems[10]

$$(M_2) \qquad \boxed{\partial_\rho \varphi_{\mu\nu} + \partial_\nu \varphi_{\rho\mu} + \partial_\mu \varphi_{\nu\rho} = 0}$$

or

$$(M_3) \qquad \boxed{\varphi_{\mu\nu} = \partial_\mu \varphi_\nu - \partial_\nu \varphi_\mu.}$$

It is convenient to substitute for the contravariant tensors $f^{\mu\nu}$ and J^μ the tensor densities

$$\text{(IX-108)} \qquad \mathscr{F}^{\mu\nu} = \sqrt{-g}\, f^{\mu\nu} \qquad (g = \det g_{\mu\nu})$$

$$\text{(IX-109)} \qquad \mathscr{J}^\mu = \sqrt{-g}\, J^\mu .$$

Equation (IX-107) will thus be written[11]

$$(\mathscr{M}_1) \qquad \partial_\rho \mathscr{F}^{\mu\rho} = \mathscr{J}^\mu .$$

Moreover, definition (IX-18) of the dual tensors will be replaced by:

$$\text{(IX-110)} \qquad \varphi^{\mu\nu *} = \frac{1}{2\sqrt{-g}} \varepsilon^{\mu\nu\rho\sigma} \varphi_{\rho\sigma}, \qquad \overset{*}{\varphi}_{\mu\nu} = -\frac{\sqrt{-g}}{2} \varepsilon_{\mu\nu\rho\sigma} \varphi^{\rho\sigma},$$

$\varepsilon_{\mu\nu\rho\sigma} = \varepsilon^{\mu\nu\rho\sigma}$ being the permutation symbols defined by (IX-19). System (M_2) may then be written

$$(M_3)^* \qquad \partial_\rho (\sqrt{-g}\, \varphi^{\mu\rho *}) = 0$$

or

$$\text{(IX-111)} \qquad \nabla_\rho \varphi^{\mu\rho *} = 0 .$$

Finally, Maxwell's equations are then equivalent to the two following systems:

$$(M_1) \qquad \nabla_\rho f^{\mu\rho} = J^\mu \qquad \text{or} \qquad \partial_\rho (\sqrt{-g}\, f^{\mu\rho}) = \mathscr{J}^\mu$$

$$(M_2)^* \qquad \nabla_\rho \varphi^{\mu\rho *} = 0 \qquad \text{or} \qquad \partial_\rho (\sqrt{-g}\, \varphi^{\mu\rho *}) = 0 .$$

From them we deduce the continuity equation

(IX-112) $$\nabla_\mu J^\mu = 0 .$$

Moreover, if $\varepsilon\mu = 1$, ε and μ being constant, one has according to (M_3)

(IX-113) $$\nabla^\nu \varphi_{\mu\nu} = \nabla^\nu \nabla_\mu \varphi_\nu - \nabla^\nu \nabla_\nu \varphi_\mu .$$

The left-hand member is still

$$\mu \nabla^\nu f_{\mu\nu} = \mu J_\mu$$

always taking (IX-92) and (IX-107) into account.

One thus has in Euclidean space[12]

(IX-114) $$\mu J_\mu = \nabla_\mu \nabla^\nu \varphi_\nu - \Box\, \varphi_\mu$$

or

(IX-115) $$\Box\, \varphi_\mu = -\, \mu J_\mu$$

if we postulate the Lorentz condition

(IX-116) $$\nabla^\nu \varphi_\nu = 0$$

putting

(IX-117) $$\Box = \nabla^\rho \nabla_\rho = g^{\rho\sigma} \nabla_\rho \nabla_\sigma .$$

By substituting (IX-115) in (M_3) there further results

(IX-118) $$\Box\, \varphi_{\mu\nu} = -\, \mu(\partial_\mu J_\nu - \partial_\nu J_\mu) .$$

2) *The trajectory of a charged material point.* We are familiar, following the principles of relativistic dynamics, with the equations

(VIII-159) $$F^\mu = m_0 c^2\, u^\rho \nabla_\rho u^\mu$$

which determine the trajectory of a material particle subjected to any given force. u^μ is the world-velocity with reference to any arbitrary system of coordinates

(IX-119) $$u^\mu = \frac{dy^\mu}{ds} .$$

(VIII-159) may also be written:

(IX-120) $$F^\mu = m_0 c^2 \left(\frac{d^2 y^\mu}{ds^2} + \begin{Bmatrix} \mu \\ \sigma\rho \end{Bmatrix} \frac{dy^\sigma}{ds} \frac{dy^\rho}{ds} \right) .$$

Now, if we are dealing with a charged particle, F^μ is reduced to the Lorentz force (IX-35):

$$\text{(IX-35)} \qquad f^\mu = \frac{1}{4\pi}\varphi^{\rho\mu} j_\rho$$

which can also be deduced from Maxwell's tensor

$$\text{(IX-39)} \qquad \tau^\nu_\mu = -\varphi_{\mu\rho}\varphi^{\nu\rho} + \tfrac{1}{4}\delta^\nu_\mu \varphi_{\rho\sigma}\varphi^{\rho\sigma}$$

by putting, as in (IX-53)

$$\text{(IX-121)} \qquad 4\pi f_\mu = -\nabla_\nu \tau^\nu_\mu .$$

Whence

$$\text{(IX-122)} \qquad m_0 c^2 \left(\frac{d^2 y^\mu}{ds^2} + \begin{Bmatrix} \mu \\ \sigma\rho \end{Bmatrix} \frac{dy^\sigma}{ds}\frac{dy^\rho}{ds} \right) = \frac{1}{4\pi}\varphi^{\rho\mu} j_\rho .$$

In vacuum, the current j_ρ is proportional to the world-velocity of the particle. According to (IX-30)

$$\text{(IX-30)} \qquad j_\rho = 4\pi\rho_0 u_\rho .$$

(IX-122) may also be written:

$$\text{(IX-123)} \qquad \frac{du^\mu}{ds} + \begin{Bmatrix} \mu \\ \sigma\rho \end{Bmatrix} u^\sigma u^\rho = \frac{\rho_0}{m_0 c^2}\varphi^{\rho\mu} u_\rho$$

or

$$\text{(IX-124)} \qquad \frac{Du^\mu}{ds} = \frac{du^\mu + \begin{Bmatrix} \mu \\ \sigma\rho \end{Bmatrix} u^\sigma dy^\rho}{ds} = \frac{\rho_0}{m_0 c^2}\varphi^{\rho\sigma} u_\sigma$$

that is

$$\text{(IX-125)} \qquad \boxed{u^\lambda \nabla_\lambda u^\mu = \frac{\rho_0}{m_0 c^2}\varphi^{\rho\mu} u_\rho .}$$

The equations in this section, which are valid in any coordinate system, are invariant in a general transformation of coordinates

$$y'^\mu = a^{\mu'}_\nu y^\nu , \qquad y^\mu = a^\mu_{\nu'} y'^\nu$$

with

$$a^{\rho'}_\mu a^\nu_{\rho'} = a^\rho_{\mu'} a^{\nu'}_\rho = \delta^\nu_\mu$$

the

$$a_\nu^{\mu'} = \frac{\partial y'^\mu}{\partial y^\nu} \qquad \text{and the} \qquad a_{\nu'}^{\mu} = \frac{\partial y^\mu}{\partial y'^\nu}$$

then being functions of the coordinates.

It should be recalled that, in a Euclidean space, one can always return to rectilinear axes orthonormal according to (VI-28). Maxwell's equations are then (M_1) (M_2) (M_3). Their covariance is assured by the Lorentz transformation in which the constant coefficients $a_\nu^{\mu'}$ and $a_{\nu'}^{\mu}$ are made to obey conditions $(\text{VI-42})_b$, which characterize the changes of Galilean reference systems.

B. EXTENSIONS OF MAXWELL'S THEORY

8. *The deduction of Maxwell's equations from a variational principle*

Let us consider, in any arbitrary system of curvilinear coordinates, the integral

(IX-126) $$\mathscr{A} = \int \mathscr{L}\, d\tau, \qquad d\tau = dy^1 \wedge dy^2 \wedge dy^3 \wedge dy^0,$$

formed on the basis of a scalar density[13]

(IX-127) $$\mathscr{L} = \sqrt{-g}L.$$

We will assume that L is an invariant depending on the potential φ_μ, either directly or through the electromagnetic field

(IX-128) $$\varphi_{\mu\nu} = \partial_\mu \varphi_\nu - \partial_\nu \varphi_\mu.$$

Let us increase the variables φ_μ by increments $\delta\varphi_\mu$ which become zero at the limit of the domain of integration.

Let us find the condition

(IX-129) $$\delta\mathscr{A} = 0$$

for every increment $\delta\varphi_\mu$.

Let us put

(IX-130) $$\mathscr{F}^{\mu\nu} = \frac{\partial \mathscr{L}}{\partial \varphi_{\mu\nu}}, \qquad \mathscr{J}^{\mu} = -\frac{\partial \mathscr{L}}{\partial \varphi_\mu}.$$

We obtain[14]

(IX-131)
$$\delta\mathscr{A} = \int \delta\mathscr{L}\,d\tau = \int \left(\frac{1}{2}\frac{\partial\mathscr{L}}{\partial\varphi_{\mu\nu}}\delta\varphi_{\mu\nu} + \frac{\partial\mathscr{L}}{\partial\varphi_{\mu}}\delta\varphi_{\mu}\right)d\tau = \\ = \int\left(\frac{1}{2}\mathscr{F}^{\mu\nu}\delta\varphi_{\mu\nu} - \mathscr{J}^{\mu}\delta\varphi_{\mu}\right)d\tau$$

and, by integrating by parts:

(IX-132)
$$\delta\mathscr{A} = \int[\tfrac{1}{2}\mathscr{F}^{\mu\nu}(\partial_\mu\delta\varphi_\nu - \partial_\nu\delta\varphi_\mu) - \mathscr{J}^{\mu}\delta\varphi_\mu]\,d\tau \\ = -\int\partial_\nu(\mathscr{F}^{\mu\nu}\delta\varphi_\mu)\,d\tau + \int(\partial_\nu\mathscr{F}^{\mu\nu} - \mathscr{J}^{\mu})\delta\varphi_\mu\,d\tau\,.$$

The first integrand of the right-hand member is a divergence. Its contribution is zero *because the $\delta\varphi_\mu$ are zero, by hypothesis, at the limit of the domain of integration.* Condition (IX-129) thus imposes

(IX-133)
$$\partial_\nu\mathscr{F}^{\mu\nu} = \mathscr{J}^{\mu}\,.$$

Thus, by postulating definition (IX-128) of the electromagnetic field, a principle of stationary action applied to the invariant density $\mathscr{L}$, whose form is not given explicitly, leads to the Maxwellian equations (IX-133). One can write further

(IX-134)
$$d\mathscr{L} = \tfrac{1}{2}\mathscr{F}^{\mu\nu}d\varphi_{\mu\nu} - \mathscr{J}^{\mu}d\varphi_\mu$$

still without giving $\mathscr{L}$ explicitly. (IX-134) then expresses the definitions (IX-130).

If $\varepsilon = \mu = 1$ ($f_{\mu\nu} = \varphi_{\mu\nu}$) and in the absence of charge or current, one will have to choose

(IX-135)
$$\boxed{L = \tfrac{1}{4}\,\varphi^{\mu\nu}\varphi_{\mu\nu}\,.}$$

L is thus reduced to the invariant $H^2 - E^2$ in an orthonormal system.

Beginning with an arbitrary density $\mathscr{L}(\varphi_{\mu\nu}, \varphi_\mu)$, we obtain (IX-133) that is, the group $(\mathscr{M}_1)$ of Maxwell's theory; but it is not possible to deduce any relation between induction and field which will characterize the electromagnetic theory we seek to construct. In order to obtain such relations, *$\mathscr{L}$ must be expressed as a function of the fields and the potentials.*

Thus, starting with (IX-135), we would obtain

(IX-136)
$$\mathscr{F}^{\mu\nu} = \frac{\partial\mathscr{L}}{\partial\varphi_{\mu\nu}} = \sqrt{-g}\,\varphi^{\mu\nu}$$

but the use of a variational principle does not necessarily lead to the linear relations between inductions and fields which Maxwell's theory assumes. A suitable choice of $\mathscr{L}$ may, on the contrary, lead to *non-linear* relations and thus eliminate certain difficulties inherent in Maxwell's theory. Such are the principles of the electromagnetic theories developed by Mie and by Born-Infeld.

9. *Mie's theory*[15]

In Maxwell's theory, the energy density of the electromagnetic field in the presence of matter is represented by the component τ_0^0 of Minkowski's tensor (IX-93). This tensor represents the purely electromagnetic actions and does not include the energy-momentum of the matter itself. General Relativity, as we shall see, does not modify this conception and leaves intact a dynamics and an electrodynamics that are independent of each other.

One can, on the other hand, deny the dualism and assume that the material and electromagnetic energies are not heterogeneous but rather both of electromagnetic origin. That is the basis of Mie's theory. This attempt thus has as its goal the justification of the existence and the characteristics of the charges by means of the properties of the field. The Lorentz electronic theory, by contrast, can explain the existence of charges only by bringing in non-electromagnetic forces. In order to avoid having recourse to such forces, Mie's theory proposes a non-dualistic conception of the field and of matter.

Mie assumes that the field is completely described by ten state quantities. To begin with, he chooses as his basic state quantities the inductions **D**, **B**, the vector-potential **A** and the electric density ρ. The other quantities in the theory, that is, the fields **E**, **H**, the scalar potential V and the current **I**, are functions of **D**, **B**, **A** and ρ. The relations of the theory are then obtained by postulating the following evolutionary equations for the basic quantities:

(IX-137) $$\frac{1}{c}\frac{\partial \mathbf{D}}{\partial t} - \mathbf{curl}\,\mathbf{H} = -\frac{4\pi}{c}\mathbf{I}$$

(IX-138) $$\frac{1}{c}\frac{\partial \mathbf{B}}{\partial t} + \mathbf{curl}\,\mathbf{E} = 0$$

(IX-139) $$\frac{1}{c}\frac{\partial \mathbf{A}}{\partial t} + \mathbf{grad}\,V = -\mathbf{E}$$

(IX-140) $$\frac{1}{c}\frac{\partial \rho}{\partial t} + \operatorname{div}\frac{\mathbf{I}}{c} = 0 .$$

The relations are identical with relations (I), (IVa) and (III) of Maxwell's theory, except that in the latter theory we assume that (IX-139) and (IX-138) are deduced from the equations $\mathbf{B} = \mathbf{curl}\,\mathbf{A}$, $\operatorname{div}\mathbf{D} = 4\pi\rho$ – neither of which occurs here, since the quantities $\mathbf{B}$, $\mathbf{D}$, $\mathbf{A}$ and ρ are assumed to be independent.

In Maxwell's theory, equation (IX-97) relative to the conservation of energy in only slightly dispersive media is expressed, for the zero value of the μ index by the vector relation:

(IX-141) $$\mathbf{H}\cdot\frac{1}{c}\frac{\partial \mathbf{B}}{\partial t} + \mathbf{E}\cdot\frac{1}{c}\frac{\partial \mathbf{D}}{\partial t} + \operatorname{div}[\mathbf{E}\wedge\mathbf{H}] = -\frac{4\pi}{c}(\mathbf{I}\cdot\mathbf{E}).$$

In Mie's theory one obtains the same equation by scalar multiplication of (IX-141) by $\mathbf{E}$, of (IX-138) by $\mathbf{H}$ and then by adding the two results. Further, by scalar multiplication of (IX-139) by $4\pi\mathbf{I}/c$, and of (IX-140) by $4\pi V$, and then adding, one has:

(IX-142) $$4\pi\left(\frac{\mathbf{I}}{c^2}\frac{\partial \mathbf{A}}{\partial t} + \frac{V}{c}\frac{\partial \rho}{\partial t}\right) + \frac{4\pi}{c}\operatorname{div}(\mathbf{I}\cdot V) = -\frac{4\pi}{c}(\mathbf{I}\cdot\mathbf{E}).$$

Now, the interpretation of (IX-141) is very different here from what it is in Maxwell's theory. The term $-4\pi\mathbf{I}\cdot\mathbf{E}/c$ which indicated the production of non-electromagnetic energy arising from the variation in energy of the field, cannot have a physical significance in Mie's theory, since in it every energy is treated as electromagnetic in origin. On the contrary, by eliminating (IX-142) from (IX-141) we obtain a relation in which the term $-(4\pi/c)(\mathbf{IE})$ does not occur. It is that relation which has a physical meaning and which expresses the conservation of electromagnetic energy:

(IX-143) $$\frac{1}{c}\frac{\partial W}{\partial t} + \operatorname{div}\boldsymbol{\Sigma} = 0$$

with

(IX-144) $$dW = \mathbf{E}\,d\mathbf{D} + \mathbf{H}\,d\mathbf{B} - 4\pi\left(\frac{\mathbf{I}}{c}\,d\mathbf{A} + V\,d\rho\right)$$

(IX-145) $$\boldsymbol{\Sigma} = [\mathbf{E}\wedge\mathbf{H}] - \frac{4\pi}{c}(\mathbf{I}\cdot V).$$

So it all comes back to determining W or, what amounts to the same thing:

(IX-146) $$L = -(\mathbf{E}\cdot\mathbf{D}) + 4\pi\rho V + W\,,$$

starting with the relation

(IX-147) $$\begin{aligned} dL &= \mathbf{H}\,d\mathbf{B} - \mathbf{D}\,d\mathbf{E} + 4\pi\left(\rho\,dV - \frac{\mathbf{I}}{c}d\mathbf{A}\right) = \\ &= \tfrac{1}{2} f^{\mu\nu}\,d\varphi_{\mu\nu} - J^{\mu}\,d\varphi_{\mu}\,, \end{aligned}$$

with the definitions (IX-8) and (IX-2) being duly taken into account.

Thus, the deduction of the field equations from the scalar density

(IX-148) $$d\mathscr{L} = d\sqrt{-g}\,L(\varphi_{\mu\nu}, \varphi_{\mu}) = \tfrac{1}{2}\mathscr{F}^{\mu\nu}\,d\varphi_{\mu\nu} - \mathscr{J}^{\mu}\,d\varphi_{\mu}$$

the form of which is not explicitly stated, is tantamount to the form set forth in the preceding section. The relations deduced from (IX-148) by a principle of stationary action are the relations valid "outside the charges". In Mie's theory it is assumed that it is never rigorously possible to place one's self outside the charges. The hypothesis $J_{\mu} = 0$ constitutes a limiting case. So it is necessary to modify the invariant (IX-148) and replace it by

(IX-149) $$\boxed{L_m = \tfrac{1}{4}\,\varphi_{\mu\nu}\varphi^{\mu\nu} - f(\pm\sqrt{\varphi_{\mu}\varphi^{\mu}})}$$

f being an invariant function of the electromagnetic potential.

A special case. A spherically symmetrical static solution. In the static case, the equations (IX-137) through (IX-140) permit the following solution:

(IX-150) $$\mathbf{E} = -\,\mathbf{grad}\,V\,, \qquad \mathbf{H} = \mathbf{I} = 0\,.$$

According to (IX-130) and (IX-150) one will then have

$$\frac{\partial\mathscr{L}}{\partial\varphi_p} = \frac{\partial\mathscr{L}}{\partial\varphi_{pq}} = 0\,(p, q = 1, 2, 3)\,,$$

and the invariant L_m will be reduced, in the static case, to the following expression:

(IX-151) $$L_m = -\tfrac{1}{2}\sum_r \varphi_{r0}\varphi_{r0} - f(V)\,, \qquad \varphi_0 = -V\,.$$

The definitions (IX-130) applied to (IX-151) then give

(IX-152) $$f^{p0} = \frac{\partial L_m}{\partial \varphi_{p0}} = \varphi^{p0}$$

(IX-153) $$j^0 = 4\pi\rho = -\frac{\partial L_m}{\partial \varphi_0} = f'(\varphi_0) = -f'(V).$$

The equations (IX-137)–(IX-140), which here are reduced to div $\mathbf{D} = 4\pi\rho$, are then written according to (IX-152), (IX-150) and (IX-153)

(IX-154) $$\operatorname{div}\mathbf{E} = -\operatorname{div}\mathbf{grad}\, V = -f'(V).$$

Now, in the case where a field is created by a source having spherical symmetry, one has, in a system of polar coordinates

(IX-155) $$\operatorname{div}\mathbf{grad}\, V = \frac{1}{r^2}\frac{\partial}{\partial r}\left(r^2 \frac{\partial V}{\partial r}\right) = f'(V).$$

So we must seek finite and continuous solutions of (IX-154), solutions such that $V \to 0$ when $r \to \infty$. The charge of the particle is then obtained by integration over the whole space of the charge density[16]

(IX-156) $$\begin{aligned} q &= \iiint \rho\sqrt{-g}\cdot dr\, d\theta\, d\varphi = \frac{-1}{4\pi}\iiint r^2 f'(V)\sin\theta\, dr\, d\theta\, d\varphi \\ &= -\int_0^\infty r^2 f'(V)\, dr = \int \frac{\partial}{\partial r}\left(r^2\frac{\partial V}{\partial r}\right) dr = \left[r^2 \frac{\partial V}{\partial r}\right]_0^\infty. \end{aligned}$$

Moreover, the mass of the particle results from the integration over the whole space of the energy density:

(IX-157) $$m = \frac{1}{c^2}\iiint W\sqrt{-g}\cdot dr\, d\theta\, d\varphi = \frac{4\pi}{c^2}\int_0^\infty W r^2\, dr$$

with, according to (IX-146), (IX-149) and (IX-153):

(IX-158) $$W = \frac{E^2}{2} - f(V) + Vf'(V).$$

One will then have:

$$m = \frac{4\pi}{c^2} \int r^2 \left[\frac{1}{2} \left(\frac{\partial V}{\partial r} \right)^2 - f(V) + V f'(V) \right] dr. \tag{IX-159}$$

Integrating by parts and taking (IX-155) into account

$$\begin{aligned} \int r^2 \left(\frac{\partial V}{\partial r} \right)^2 dr &= \iint \left[\frac{\partial}{\partial r} \left(r^2 V \frac{\partial V}{\partial r} \right) - V \frac{\partial}{\partial r} \left(r^2 \frac{\partial V}{\partial r} \right) \right] dr = \\ &= - \int V r^2 f'(V) \, dr + \int \frac{\partial}{\partial r} \left(r^2 V \frac{\partial V}{\partial r} \right). \end{aligned} \tag{IX-160}$$

If we suppose that V is zero at the limit of the domain of integration, the latter integral gives no contribution.
Whence

$$m = \frac{4\pi}{c^2} \int r^2 \left[\frac{V}{2} f'(V) - f(V) \right] dr. \tag{IX-161}$$

The mass and the charge of a particle, which is the source of the field, would then be completely expressed, in the static case, as a function of the scalar potential V. According to Mie's conceptions, the electron is not a particle localized at a point. It remains coextensive with the whole space. It is the proper choice of the function $f(V)$ that makes possible the definition of the finite space integrals, which may also represent the charge and the mass.

One of the great objections to this theory concerns the role assigned to the electromagnetic potential. According to the usual hypotheses, the electromagnetic potential is defined apart from a gradient and does not have a physical meaning, as the field does. Moreover, expression (IX-149) of L_m is rather artificial, and the choice of f remains very arbitrary.

The Born-Infeld theory seeks to overcome these important difficulties by expressing the characteristics of the sources as a function of the fields and not of the potentials. The function of action then deviates even more radically from the Maxwellian invariant and does not merely introduce supplementary terms, as the function f does. Nevertheless, the natural choice of a suitable function of action and the disappearance of arbitrary hypotheses connected here with the choice of $f(V)$ are modifications that are initiated in the Born-Infeld theory but much more satisfactorily achieved in the unitary theory of Einstein-Schrödinger.

10. *The theory of M. Born and L. Infeld*[17]

1) Let us define a system of arbitrary coordinates

$$x^\mu (x^1 = x, \quad x^2 = y, \quad x^3 = z, \quad x^0 = ct)$$

and let us call $g_{\mu\nu}$ the metric tensor. We will assume that the basic quantities of the theory are the six components of the electromagnetic field $\varphi_{\mu\nu}$ (and not the ten components $\varphi_{\mu\nu}$, φ_μ as in Mie's theory). We designate the contravariant components of the field by

(IX-162) $$\varphi^{\mu\nu} = g^{\mu\rho} g^{\nu\sigma} \varphi_{\rho\sigma}$$

and we define the dual field by the usual notations

(IX-163) $$\varphi^{\mu\nu *} = \frac{1}{2\sqrt{-g}} \varepsilon^{\mu\nu\rho\sigma} \varphi_{\rho\sigma}, \qquad \overset{*}{\varphi}_{\mu\nu} = -\frac{\sqrt{-g}}{2} \varepsilon_{\mu\nu\rho\sigma} \varphi^{\rho\sigma},$$
$$(\varphi^{\mu\nu *} = g^{\mu\rho} g^{\nu\sigma} \overset{*}{\varphi}_{\rho\sigma})$$

$\varepsilon^{\mu\nu\rho\sigma} = \pm 1, 0$ being the usual Levi-Civita symbols.

We will then introduce, following Born and Infeld, a function of action that is the scalar density

(IX-164) $$\mathscr{L}_B = (\sqrt{-\pi} - \sqrt{-g})$$

g designating the determinant formed by the components $g_{\mu\nu}$ of the metric tensor and π the determinant whose elements are the components of the tensor:

(IX-165) $$\pi_{\mu\nu} = g_{\mu\nu} + \varphi_{\mu\nu}.$$

One can calculate π as a function of the determinants g and φ. If

(IX-166) $$g g^{\mu\nu} = \text{minor}\, g_{\mu\nu},$$

the $g^{\mu\nu}$ designate the usual contravariant components of the metric tensor ($g_{\mu\rho} g^{\nu\rho} = \delta^\nu_\mu$). One then obtains by a simple calculation of determinants:

(IX-167) $$\pi = g + \varphi + \frac{g}{2} g^{\mu\rho} g^{\nu\sigma} \varphi_{\mu\nu} \varphi_{\rho\sigma}.$$

(IX-164) may then be written

(IX-168) $$\boxed{\mathscr{L}_B = \sqrt{-g}(L_B - 1)}$$

putting[18]

(IX-169) $$\boxed{L_B = \left(1 + \frac{\varphi}{g} + \frac{1}{2} g^{\mu\rho} g^{\nu\sigma} \varphi_{\mu\nu} \varphi_{\rho\sigma}\right)^{\frac{1}{2}}}$$

It may be noted that L_B is expressed entirely as a function of the two basic invariants of Maxwell's theory.

(IX-170) $$F = H^2 - E^2 = \tfrac{1}{2} \varphi_{\mu\nu} \varphi^{\mu\nu} = \tfrac{1}{2} g^{\mu\rho} g^{\nu\sigma} \varphi_{\mu\nu} \varphi_{\rho\sigma}$$

(IX-171) $$G = (\mathbf{E} \cdot \mathbf{H}) = \tfrac{1}{4} \varphi_{\mu\nu} \varphi^{\mu\nu*} = \frac{1}{8\sqrt{-g}} \varepsilon^{\mu\nu\rho\sigma} \varphi_{\mu\nu} \varphi_{\rho\sigma} = \sqrt{\frac{\varphi}{-g}}.$$

The action function L_B may thus also be written

(IX-172) $$L_B = (1 + F - G^2)^{\frac{1}{2}}.$$

So the conjugate field can be defined:

(IX-173) $$\sqrt{-g} f^{\mu\nu} = \frac{\partial \mathscr{L}_B}{\partial \varphi_{\mu\nu}} = \sqrt{-g} \left(2 \frac{\partial L}{\partial F} \varphi^{\mu\nu} + \frac{\partial L}{\partial G} \varphi^{\mu\nu*}\right)$$

that is

(IX-174) $$\boxed{f^{\mu\nu} = \frac{\varphi^{\mu\nu} - G\varphi^{\mu\nu*}}{L}.}$$

The two conjugate fields $f^{\mu\nu}$ and $\varphi_{\mu\nu}$ are then connected by the non-linear relations (IX-174).

2) Let us suppose that the basic field $\varphi_{\mu\nu}$ satisfies the relation:

(IX-175) $$\varphi_{\mu\nu\rho} \equiv \partial_\mu \varphi_{\nu\rho} + \partial_\rho \varphi_{\mu\nu} + \partial_\nu \varphi_{\rho\mu} = 0$$

or

(IX-176) $$\partial_\rho (\sqrt{-g}\, \varphi^{\mu\rho*}) = 0.$$

The field $\varphi_{\mu\nu}$ is then deduced from a potential

(IX-177) $$\varphi_{\mu\nu} = \partial_\mu \varphi_\nu - \partial_\nu \varphi_\mu.$$

A variational principle applied to $\mathscr{L}_B$ leads to results very close to those explained in Section 8 of the present chapter. They are, in fact, applicable to the density $\mathscr{L}_B$, a function of the potential φ_μ through the intermediary of the field $\varphi_{\mu\nu}$. If

(IX-178) $$\sqrt{-g}\, f^{\mu\nu} = \frac{\partial \mathscr{L}_B}{\partial \varphi_{\mu\nu}}$$

one will have, instead of (IX-133),

(IX-179) $$\partial_\rho (\sqrt{-g}\, f^{\mu\rho}) = 0\,.$$

The Maxwell equations (IX-133) are replaced by the equations (IX-179) which differ from them by the disappearance of the right-hand member, since the potential φ_μ does not explicitly occur in the action function. But above all, the form (IX-172) of L_B imposes the relations (IX-174) between the inductions and the fields – relations which, unlike the Maxwell hypotheses, are no longer linear.

3) The current density is defined by the four-vector

(IX-180) $$\mathscr{J}^\mu = \partial_\rho (\sqrt{-g}\, \varphi^{\mu\rho})\,.$$

Now, according to (IX-179),

(IX-181) $$\partial_\rho (\sqrt{-g}\, f^{\mu\rho}) = \partial_\rho \left[\sqrt{-g}\left(2\frac{\partial L}{\partial F}\varphi^{\mu\rho} + \frac{\partial L}{\partial G}\varphi^{\mu\rho *}\right)\right] = 0\,.$$

Taking (IX-181) into account, one will then have:

(IX-182) $$-j^\mu = \frac{1}{2\dfrac{\partial L}{\partial F}}\left[2\,\varphi^{\mu\rho}\partial_\rho\left(\frac{\partial L}{\partial F}\right) + \varphi^{\mu\rho *}\partial_\rho\left(\frac{\partial L}{\partial G}\right)\right].$$

4) *A special case. A spherically symmetrical static field.* Let us suppose that the electromagnetic field is produced by a static distribution of charges and possesses, in addition, a spherical symmetry. It is convenient to use the system of polar coordinates

(IX-183) $$y^1 = r\,, \qquad y^2 = \theta\,, \qquad y^3 = \varphi\,, \qquad y^0 = ct\,.$$

One thus obtains

(IX-184) $$ds^2 = -\,dr^2 - r^2(d\theta^2 + \sin^2\theta\, d\varphi^2) + c^2\, dt^2$$

which is deduced from the general expression

(IX-185) $$ds^2 = g_{\mu\nu}\, dy^\mu\, dy^\nu\,,$$

putting

(IX-186)
$$g_{11} = \frac{1}{g^{11}} = -1, \qquad g_{22} = \frac{1}{g^{22}} = -r^2,$$
$$g_{33} = \frac{1}{g^{33}} = -r^2 \sin^2\theta, \qquad g_{00} = \frac{1}{g^{00}} = 1,$$

(IX-187)
$$\sqrt{-g} = r^2 \sin\theta .$$

In this coordinate system, the static field $\varphi_{\mu\nu}$ is reduced to the components φ_{p0} and the induction $f^{\mu\nu}$ to the components $f^{\mu 0}$. In addition, the hypothesis of sphericity limits these components to φ_{10} and f^{10}. One will then have, according to (IX-179) and (IX-187):

(IX-188)
$$\partial_1(r^2 f^{01}) = \partial_1(r^2 g^{00} g^{11} f_{01}) = \partial_1(r^2 f_{10}) = 0 .$$

Whence

(IX-189)
$$f_{10} = g_{11} g_{00} f^{10} = -f^{10} = \frac{k}{r^2}$$

k being a constant of integration.

In this particular case, the non-linear relations (IX-174) may be written

(IX-190)
$$f^{10} = \frac{\varphi^{10}}{\sqrt{1 + \varphi^{10}\varphi_{10}}}, \qquad G = 0 .$$

One will also obtain

(IX-191)
$$f^{10} f_{10} = \frac{\varphi^{10}\varphi_{10}}{1 + \varphi^{10}\varphi_{10}}$$

and, consequently

(IX-192)
$$\varphi_{10}\varphi^{10} = \frac{f^{10} f_{10}}{1 - f^{10} f_{10}},$$

that is

(IX-193)
$$\varphi_{10} = -\varphi^{10} = \frac{f^{01}}{\sqrt{1 - f^{01} f_{01}}} = \frac{k}{r^2} \frac{1}{\sqrt{1 + \dfrac{k^2}{r^4}}}$$

taking (IX-189) into account.

Let us then designate by b the ratio between the expression of the fields in conventional units (that is **E**, **B**, **H**, **D**) and their expression $(\varphi_{\mu\nu}, f^{\mu\nu})$ in natural units:

(IX-194)
$$\mathbf{E} = b(\varphi_{10}, \varphi_{20}, \varphi_{30}), \qquad \mathbf{B} = b(\varphi_{23}, \varphi_{31}, \varphi_{12})$$

(IX-195) $$\mathbf{D} = b(f_{10}, f_{20}, f_{30}), \qquad \mathbf{H} = b(f_{23}, f_{31}, f_{12}).$$

So we arrive at, according to (IX-189) and (IX-193):

(IX-196) $$D_r = bf_{10} = b\frac{k}{r^2}$$

(IX-197) $$E_r = b\varphi_{10} = \frac{bk}{r^2}\frac{1}{\sqrt{1+\frac{k^2}{r^4}}}$$

and, putting

(IX-198) $$kb = q, \qquad b = \frac{q}{r_0^2},$$

one obtains the following expressions for the radial components

(IX-199) $$D_r = \frac{q}{r^2}, \qquad E_r = \frac{q}{r_0^2}\frac{1}{\sqrt{1+\left(\frac{r}{r_0}\right)^4}}.$$

The field E_r remains finite at the origin ($r = 0$). It is then equal to the value

(IX-200) $$b = \frac{q}{r_0^2} = (E_r)_{r=0}.$$

The constant b represents the "absolute field".

One can thus interpret the sources either as point singularities responsible for an induction D_r that is infinite at the origin, or as a continuous distribution throughout the space – a distribution described by the field E_r that is finite at the origin. Obviously, it is the latter interpretation that expresses most clearly the absence of the field-particle dualism. The negation of a heterogeneity between the field and its sources is thus one of the essential results of Born's theory – a result permitted by the non-linearity of the field-induction relations, which was itself introduced by the form (IX-165) or (IX-172) of the Lagrangian of the theory.

In this theory a charge is, in principle, coextensive with the whole space. Its characteristics will be entirely determined as a function of the field (and not of the potential, as assumed in Mie's theory). And the vector density $\mathscr{J}^{\mu}$ defined by (IX-180) allows us to express a free density of charge and a free current as a function, not of the coordinates, but of the field.

More especially, one will be able to define a charge of the particle by integrating over the whole space – including the origin – the free density $\mathscr{J}^0$. The finite value of the field $\mathbf{E}$ at the origin makes it possible to obtain an integral of finite value that can represent the charge q.

In fact, if we are concerned with a static field possessing spherical symmetry, (IX-182) will be reduced to the following expression:

$$-j^0 = \frac{1}{\sqrt{1+\frac{k^2}{r^4}}}\varphi^{01}\cdot\partial_1\left(\sqrt{1+\frac{k^2}{r^4}}\right) =$$

(IX-201)
$$= \frac{k}{r^2}\frac{1}{1+\frac{k^2}{r^4}}\frac{\partial}{\partial r}\left(\sqrt{1+\frac{k^2}{r^4}}\right),$$

taking into account (IX-186), (IX-187) and (IX-183). The equation (IX-201) may also be written

(IX-202)
$$j^0 = \frac{2k^3}{r^7\left(1+\frac{k^2}{r^4}\right)^{\frac{3}{2}}} = \frac{2}{r\left(1+\frac{r^4}{k^2}\right)^{\frac{3}{2}}}$$

and the density $\rho = b\mathscr{J}^0$ has the following value:

(IX-203)
$$\rho = \frac{b}{4\pi}\mathscr{J}^0 = \frac{b}{4\pi}\sqrt{-g}\,j^0 = \frac{qr^2\sin\theta}{r_0^2\cdot 2\pi r\left(1+\frac{r^4}{r_0^4}\right)^{\frac{3}{2}}}.$$

By integrating *over the whole space*, without excluding a singular region surrounding the origin, one will then have

(IX-204)
$$\int \rho\,dr\,d\theta\,d\varphi = \int_0^\infty\int_0^\pi\int_0^{2\pi} \rho\,dr\,d\theta\,d\varphi = \frac{2q}{r_0^2}\int_0^\infty \frac{r^2\,dr}{\frac{r}{r_0}\left(1+\frac{r^4}{r_0^4}\right)^{\frac{3}{2}}}$$
$$= q\int_0^{\frac{1}{2}\pi}\cos\psi\,d\psi = q$$

putting

(IX-205) $$tg\psi = \frac{r^2}{r_0^2}.$$

The distribution of charge defined by the free density ρ then leads to a finite expression, the charge of the particle, by integrating over the whole space. This result – and the conceptions of charge that it involves – is here tied in with the non-linearity of the field-induction relations.

NOTES

1. By way of an exercise, let us verify that the two Maxwell groups are invariant under a Lorentz transformation. One has, according to (IX-3),

$$\partial_1 = \frac{\partial_1' - \beta\partial_0'}{\sqrt{1-\beta^2}}, \qquad \partial_2 = \partial_2', \qquad \partial_3 = \partial_3', \qquad \partial_0 = \frac{\partial_0' - \beta\partial_1'}{\sqrt{1-\beta^2}}$$

because

$$\partial_\mu = \frac{\partial}{\partial x^\mu} = \frac{\partial}{\partial x'^\lambda}\frac{\partial x'^\lambda}{\partial x^\mu} = \frac{\partial x'^\lambda}{\partial x^\mu}\partial_\lambda'.$$

The two groups will thus be expanded as follows:

(a)

$$\text{(I)}\begin{cases} \partial_2' H_3 - \partial_3' H_2 - \left(\dfrac{\partial_0' - \beta\partial_1'}{\sqrt{1-\beta^2}}\right) D_1 = -J_1 \\ \partial_3' H_1 - \left(\dfrac{\partial_1' - \beta\partial_0'}{\sqrt{1-\beta^2}}\right) H_3 - \left(\dfrac{\partial_0' - \beta\partial_1'}{\sqrt{1-\beta^2}}\right) D_2 = -J_2 \\ \left(\dfrac{\partial_1' - \beta\partial_0'}{\sqrt{1-\beta^2}}\right) H_2 - \partial_2' H_1 - \left(\dfrac{\partial_0' - \beta\partial_1'}{\sqrt{1-\beta^2}}\right) D_3 = -J_3 \end{cases}$$

$$\text{(III)}\left\{ \left(\dfrac{\partial_1' - \beta\partial_0'}{\sqrt{1-\beta^2}}\right) D_1 + \partial_2' D_2 + \partial_3' D_3 = J_0 \right.$$

(b)

$$\text{(I)}\begin{cases} \partial_2' E_3 - \partial_3' E_2 + \left(\dfrac{\partial_0' - \beta\partial_1'}{\sqrt{1-\beta^2}}\right) B_1 = 0 \\ \partial_3' E_1 - \left(\dfrac{\partial_1' - \beta\partial_0'}{\sqrt{1-\beta^2}}\right) E_3 + \left(\dfrac{\partial_0' - \beta\partial_1'}{\sqrt{1-\beta^2}}\right) B_2 = 0 \\ \left(\dfrac{\partial_1' - \beta\partial_0'}{\sqrt{1-\beta^2}}\right) E_2 - \partial_2' E_1 + \left(\dfrac{\partial_0' - \beta\partial_1'}{\sqrt{1-\beta^2}}\right) B_3 = 0 \end{cases}$$

$$\text{(III)}\left\{ \left(\dfrac{\partial_1' - \beta\partial_0'}{\sqrt{1-\beta^2}}\right) B_1 + \partial_2' B_2 + \partial_3' B_3 = 0. \right.$$

If we then put (IX-5) or, conversely:

$$\left\{\begin{array}{lll} D_1 = D_1', & D_2 = \dfrac{D_2' + \beta H_3'}{\sqrt{1-\beta^2}}, & D_3 = \dfrac{D_3' - \beta H_2'}{\sqrt{1-\beta^2}} \\ H_1 = H_1', & H_2 = \dfrac{H_2' - \beta D_3'}{\sqrt{1-\beta_2}}, & H_3 = \dfrac{H_3' + \beta D_2'}{\sqrt{1-\beta^2}} \end{array}\right.$$

$$\left\{\begin{array}{lll} E_1 = E_1', & E_2 = \dfrac{E_2' + \beta B_3'}{\sqrt{1-\beta^2}}, & E_3 = \dfrac{E_3' - \beta B_2'}{\sqrt{1-\beta^2}} \\ B_1 = B_1', & B_2 = \dfrac{B_2' - \beta E_3'}{\sqrt{1-\beta^2}}, & B_3 = \dfrac{B_3' + \beta E_2'}{\sqrt{1-\beta^2}} \end{array}\right.$$

we note, on substituting in (I) and (III), that these equations may also be written:

(a′)

$$(\mathrm{I})\left\{\begin{array}{l} \partial_2'H_3' - \partial_3'H_2' - \partial_0'D_1' + \beta\Sigma_p\partial_p'D_p' = -J_1\sqrt{1-\beta^2} \\ \partial_3'H_1' - \partial_1'H_3' - \partial_0'D_2' = -J_2 \\ \partial_1'H_2' - \partial_2'H_1' - \partial_0'D_3' = -J_3 \end{array}\right.$$

$$(\mathrm{III})\ \{\ \Sigma_p\partial_p'D_p' + \beta[(\partial_2'H_3' - \partial_3'H_2') - \partial_0'D_1'] = J_0\sqrt{1-\beta^2}$$

(b′)

$$(\mathrm{I})\left\{\begin{array}{l} \partial_2'E_3' - \partial_3'E_2' + \partial_0'B_1' - \beta\Sigma_p\partial_p'B_p' = 0 \\ \partial_3'E_1' - \partial_1'E_3' + \partial_0'B_2' = 0 \\ \partial_1'E_2' - \partial_2'E_1' + \partial_0'B_3' = 0 \end{array}\right.$$

$$(\mathrm{III})\ \{\ \Sigma_p\partial_p'B_p' - \beta[(\partial_2'E_3' - \partial_3'E_2') + \partial_0'B_1'] = 0.$$

By comparing (a′)(I) and (a′)(III), (b′)(I) and (b′)(III), one finally obtains (I)′ and (III)′ (p. 245), on condition that we postulate (IX-4).

2. The Lorentz condition $\partial^\mu\varphi_\mu = \partial^\mu\varphi_\mu = 0$ fixes the gauge by imposing on ψ the restriction $\Box\,\psi = 0$.

2a. As a matter of fact, the definitions of a dual tensor should be written

$$\varphi^{\mu\nu *} = \frac{1}{2\sqrt{-g}}\,\varepsilon^{\mu\nu\rho\sigma}\,\varphi_{\rho\sigma}$$

$$\varphi_{\mu\nu}^{\ *} = -\frac{\sqrt{-g}}{2}\,\varepsilon_{\mu\nu\rho\sigma}\,\varphi^{\rho\sigma}$$

with Levi-Civita's symbols

$$\eta^{\mu\nu\rho\sigma} = \frac{\varepsilon^{\mu\nu\rho\sigma}}{\sqrt{-g}}, \qquad \eta_{\mu\nu\rho\sigma} = -\sqrt{-g}\,\varepsilon_{\mu\nu\rho\sigma}$$

In the coordinate system adopted $\sqrt{-g} = \sqrt{-\eta} = 1$, η being the determinant formed from the components of the Minkowski metric. Thus, by choosing this metric, one obtains definitions (IX-18).

3. The covariance of Maxwell's equations in a Lorentz transformation has been pointed out by H. Poincaré: *Compt. Rend.* **140** (1905) 891; *Rend. Pal.* **21** (1906) 129; A. Einstein: *Ann. d. Phys.* **17** (1905) 891; H. Minkowski: *Gött. Nach.* 1908, 53; *Math. Ann.* **68** (1910) 472.

4. H. A. LORENTZ [5].
5. H. MINKOWSKI: *Gött. Nach.* 1908, 53; *Math. Ann.* **68** (1910) 472.
6. One has, in fact

$$a_0{}^{0(0)} = u_0 = \frac{1}{\sqrt{1-\beta^2}}, \quad a_p{}^{0(0)} = a_0{}^{p(0)} = u_p$$

$$a_p{}^{r(0)} = a_r{}^{p(0)} = \delta_p{}^r + \frac{\alpha u_p}{u_s u^s} u^r$$

values which immediately lead to expressions $(\text{IX-59})_1$ or $(\text{IX-59})_2$.

7. Of course, in the case of a special Lorentz transformation $u = u_1$, and (IX-59) is reduced to

$$J_1 = \frac{J_1{}^{(0)} - \beta J_0{}^{(0)}}{\sqrt{1-\beta^2}}, \quad J_2 = J_2{}^{(0)}, \quad J_3 = J_3{}^{(0)}, \quad J_0 = \frac{J_0{}^{(0)} - \beta J_1{}^{(0)}}{\sqrt{1-\beta^2}}$$

8. Cf. C. MØLLER [16], p. 197.
9. PHAM MAU QUAN: *J. Rat. Mech. Anal.* **5** (1956) 473.
10. As a matter of fact, one can replace the covariant derivatives by the ordinary derivatives in the expressions (M_2) and (M_3) by reason of the symmetry of the coefficients of affine connexion

$$\nabla_\rho\varphi_{\mu\nu} + \nabla_\nu\varphi_{\rho\mu} + \nabla_\mu\varphi_{\nu\rho} \equiv \partial_\rho\varphi_{\mu\nu} + \partial_\nu\varphi_{\rho\mu} + \partial_\mu\varphi_{\nu\rho}$$

$$+ \left\{ {\sigma \atop \mu\rho} \right\} \varphi_{\sigma\nu} + \left\{ {\sigma \atop \nu\rho} \right\} \varphi_{\mu\sigma} + \left\{ {\sigma \atop \rho\nu} \right\} \varphi_{\sigma\mu} + \left\{ {\sigma \atop \mu\nu} \right\} \varphi_{\rho\sigma} + \left\{ {\sigma \atop \rho\mu} \right\} \varphi_{\nu\sigma} + \left\{ {\sigma \atop \nu\mu} \right\} \varphi_{\sigma\rho} .$$

The bracketed terms { } cancel out two by two when we take into account their symmetry with respect to the two subindices and the antisymmetry of the $\varphi_{\mu\nu}$. Likewise

$$\varphi_{\mu\nu} = \nabla_\mu\varphi_\nu - \nabla_\nu\varphi_\mu = \partial_\mu\varphi_\nu - \partial_\nu\varphi_\mu + \left\{ {\sigma \atop \nu\mu} \right\} \varphi_\sigma - \left\{ {\sigma \atop \mu\nu} \right\} \varphi_\sigma = \partial_\mu\varphi_\nu - \partial_\nu\varphi_\mu .$$

11. One has (cf. XIV-132)

$$\partial_\rho\mathcal{F}^{\mu\rho} = \partial_\rho(\sqrt{-g}\, f^{\mu\rho}) = \sqrt{-g}\left(\partial_\rho f^{\mu\rho} + f^{\mu\rho}\frac{\partial_\rho\sqrt{-g}}{\sqrt{-g}}\right) =$$

$$= \sqrt{-g}\left(\partial_\rho f^{\rho\mu} + \left\{ {\rho \atop \sigma\rho} \right\} f^{\mu\sigma}\right).$$

In addition,

$$\nabla_\rho f^{\mu\rho} \equiv \partial_\rho f^{\mu\rho} + \left\{ {\mu \atop \sigma\rho} \right\} f^{\sigma\rho} + \left\{ {\rho \atop \sigma\rho} \right\} f^{\mu\sigma} .$$

The second term of the right-hand member is zero for the case of an antisymmetric tensor. Whence

$$\nabla_\rho f^{\mu\rho} \equiv \partial_\rho f^{\mu\rho} + \left\{ {\rho \atop \sigma\rho} \right\} f^{\mu\sigma} .$$

One will then have

$$\partial_\rho\mathcal{F}^{\mu\rho} = \sqrt{-g}\,\nabla_\rho f^{\mu\rho} = \sqrt{-g}\, J^\mu = \mathcal{J}^\mu .$$

12. In a Euclidean space – where the curvature is zero – one can, as a matter of fact, permute the order of the covariant derivations

$$\nabla_\nu \nabla_\mu \varphi_\rho = \nabla_\mu \nabla_\nu \varphi_\rho \,.$$

(Cf. XV-135.)

13. The appearance of a scalar density assures the invariance of $\mathscr{L} d\tau = L \sqrt{-g} d\tau$, $\sqrt{-g} d\tau$ here being an invariant (Cf. XIV-128). $\mathscr{L}$ reduces to L in an orthonormal system.
14. The variables appearing in the variation $\delta \mathscr{L}$ are not independent, for $\varphi_{\mu\nu} = -\varphi_{\nu\mu}$. Whence the introduction of the factor $\frac{1}{2}$.
15. G. Mie: *Ann. d. Phys.* **37** (1912) 511; **39** (1912) 1; **40** (1913) 1.
16. Indeed, in a system of spherical coordinates

$$ds^2 = -dr^2 - r^2(d\theta^2 + \sin^2\theta \, d\varphi^2) + c^2 dt^2$$

$$g_{11} = -g_{00} = -1 \,, \qquad g_{22} \sin^2\theta = g_{33} = -r^2 \sin^2\theta \,,$$

$$\text{whence} \qquad \sqrt{-g} = r^2 \sin\theta \,.$$

17. M. Born: *Proc. Roy. Soc.* A **143** (1934) 410; *Ann. Inst. H. Poincaré* 1937. M. Born and L. Infeld: *Proc. Roy. Soc.* A **144** (1934) 425.
18. If the fields are weak ($\varphi_{\mu\nu} \ll 1$), one obtains

$$L_B \simeq 1 + \frac{1}{4} \varphi_{\mu\nu}\varphi^{\mu\nu} \qquad \text{and} \qquad \mathscr{L}_B = \frac{\sqrt{-g}}{4} \varphi_{\mu\nu}\varphi^{\mu\nu}$$

that is, the expression introduced, in the vacuum case, into Maxwell's theory.

CHAPTER X

THE EXPERIMENTAL VERIFICATIONS OF SPECIAL RELATIVITY

The experimental verifications of Special Relativity do not comprise merely those experiments expressly designed to demonstrate the validity of the principles of that theory. It likewise includes verification of deductions immediately resulting from Special Relativity. At the present moment, this verification is consequently linked with a very large portion of classical or quantum physics. The principles of Special Relativity have, in fact, served as the basis for the construction or modification of numerous theories whose consequences, when put to the experimental test, also serve as tests of the initial hypotheses.

One of the most famous of these constructions is the quantum and relativistic theory of the electron formulated by Dirac in 1932. The principles of that theory are applicable to any high-velocity particle having a spin 1/2. The relativistic formalism of the theory involves *ipso facto* the prediction of a proper magnetic moment – or spin – of the electron, which non-relativistic theories had already postulated in a rather arbitrary way. In a relativistic theory, effects resulting from spin automatically find a place, and experimental verification of these effects constitute an indirect confirmation of Special Relativity. Specifically, measurements have actually been made of the fine structure of the hydrogen line H_α or deuterium D_α. The values experimentally obtained are in satisfactory agreement with the theoretical curves expressing the distribution of intensities according to the relativistic theory of fine structure.

A modified Dirac theory may allow the construction of a relativistic theory of particles having any spin, either integer or half-integer. Taking a rather different approach, the quantum theory of fields has achieved a satisfactory relativistic formulation in the works of Schwinger, Feynman and Dyson. Quantum electrodynamics is a special case and constitutes one of the extensions of relativistic electrodynamics.

We will not here examine the coherency nor the experimental results of the developments resulting more or less directly from Special Relativity.

We will limit ourselves to the study of some of the immediate experimental proofs of its basic principles. Some of these experimental proofs have already been referred to in preceding chapters. We will confine ourselves to a more detailed examination of some of the others.

A. THE RETARDATION OF MOVING CLOCKS

The proper time $\Delta\tau$, measured by a clock fixed within a reference-system S', is related to the time t of an observer in another Galilean system S by the formulas (V-44):

$$\Delta t = \frac{\Delta\tau}{\sqrt{1-\beta^2}} > \Delta\tau . \tag{X-1}$$

The duration $\Delta\tau$ of an event, measured in the proper system, is thus always less than in another system; the clocks in a moving system slow down for an observer in another Galilean system.

1. *The theory of the Doppler effect and the slowing-down of clocks*

Let us consider the clock formed by an atom vibrating with a proper frequency (i.e., a frequency measured by the reference system of the atom) ν_0. In all other reference systems this frequency will have the value:

$$\nu = \nu_0\sqrt{1-\beta^2} < \nu_0 \tag{X-2}$$

according to (X-1).

To the observer, therefore, there is a diminution in the frequency, that is to say, an increase in the wave-length of the radiation emitted by an atom in motion with respect to the spectrograph. The spectrograph, thus, must indicate a displacement towards the red band of the spectrum given off by the atom.

But superimposed upon the variation in frequency just noted by (X-2) – (a variation of the second order, since β^2 occurs in it) – is a classical Doppler effect (cf. p. 196). This effect is of the first order and thus conceals the relativistic slowing down (X-2).

If θ is the angle made by the direction of the motion of the source with the direction of the observer, the Doppler effect predicted by the pre-relativistic theory will change the frequency ν_0 to

$$\nu = \frac{\nu_0}{1 - \beta \cos \theta}. \tag{X-3}$$

The superimposing of the relativistic slowing-down (X-2) on the classical Doppler effect (X-3) then leads to the following expression for the effect observed:

$$\boxed{\nu = \frac{\nu_0 \sqrt{1 - \beta^2}}{1 - \beta \cos \theta}.} \tag{X-4}$$

The relativistic correction (X-2) is referred to as the *transverse Doppler effect*. As a matter of fact, if $\theta = \pi/2$, the classical Doppler effect disappears, and only (X-2) remains.

Expression (X-4) follows quite simply from a totally relativistic theory of the Doppler effect. It is immediately obtained by applying the formulas given in Chapter VII, p. 197, for the particular case where $n = 1$, that is, where we assume that the propagation of the waves takes place in a vacuum (or approximately, in air).

Let us suppose that the source moves along $O\mathbf{x}$ and that we observe the light sent out in the direction $\mathbf{on}$ $(\widehat{\mathbf{x}\, O\, \mathbf{n}} = \theta)$. (Cf. Figure 30, p. 196.) Let us consider an observer P in the system $S\,(Ox, Oy)$. His coordinates are (x, y) in the system S; they are (x', y') in system S' $(O'x', O'y')$, which is fixed to the source in motion. We assume that S', which changes its position by Ox sliding along $O'x'$ with a velocity w, coincides at the origin with system S.

The time required by the first wave-front to go from O to P is

$$t_1 = \frac{l}{c} = \frac{x \cos \theta + y \sin \theta}{c}. \tag{X-5}$$

The number of waves arriving at P at the instant t of system S is thus

$$\nu(t - t_1).$$

This number is unchanged no matter in what reference system it is measured. One will thus have:

$$\nu\left(t - \frac{l}{c}\right) = \nu'\left(t' - \frac{l'}{c}\right) \tag{X-6}$$

where ν' is the wave frequency in system S' fixed to the source. Consequently

$$\nu' = \nu_0 . \tag{X-7}$$

In addition, l'/c is the time taken by the wave OM to reach P – the time being evaluated here in system S'. l'/c can be expressed in a manner similar to (X-5) as a function of the coordinates x', y' of the point P in S':

$$\frac{l'}{c} = \frac{x'\cos\theta' + y'\sin\theta'}{c} \tag{X-8}$$

where θ' remains to be determined.

By substitution in (X-6) one then obtains:

$$\nu\left(t - \frac{x\cos\theta + y\sin\theta}{c}\right) = \nu'\left(t' - \frac{x'\cos\theta' + y'\sin\theta'}{c}\right). \tag{X-9}$$

Substituting the Lorentz formulas in (X-9):

$$x = \frac{x' + vt'}{\sqrt{1-\beta^2}}, \qquad y = y', \qquad t = \frac{t' + \frac{\beta}{c}x'}{\sqrt{1-\beta^2}} \tag{X-10}$$

one is then led to an expression in which the coefficients of x', y', t' must cancel out identically, because ν' and θ' are obviously independent of the position of the chosen point P.

So we arrive at:

$$\frac{\nu(1-\beta\cos\theta)}{\sqrt{1-\beta^2}} = \nu', \qquad \frac{\nu(\beta - \cos\theta)}{\sqrt{1-\beta^2}} = -\nu'\cos\theta' \tag{X-11}$$

$$\nu\sin\theta = \nu'\sin\theta'.$$

These are relations (VII-67), (VII-68) and (VII-69) of Chapter VII written for $n = 1$, that is, for $u = u' = c$. They are reducible to:

$$\nu' = \frac{\nu(1 - \beta \cos\theta)}{\sqrt{1-\beta^2}} \tag{X-12}$$

$$\tan\theta' = \frac{\sqrt{1-\beta^2}\sin\theta}{\cos\theta - \beta} \tag{X-13}$$

that is

$$\cos\theta' = \frac{\cos\theta - \beta}{1 - \beta\cos\theta}, \qquad \sin\theta' = \frac{\sqrt{1-\beta^2}\sin\theta}{1-\beta\cos\theta}.$$

But $\nu' = \nu_0$, the proper frequency of the atom. One then has

$$\nu = \frac{\nu_0\sqrt{1-\beta^2}}{1-\beta\cos\theta}. \tag{X-14}$$

The change of ν_0 to ν expresses the *Doppler effect in relativistic terms.*

The change of θ' to θ arises from *aberration phenomena*; a modification of the direction of the light-rays is produced as a result of the relative source-observer motion.

If we make our observation in the direction of the motion of the source (the longitudinal Doppler effect), then $\theta = 0$. So we deduce from (X-11)

$$\theta' = 0, \qquad \nu = \nu_0\sqrt{\frac{1+\beta}{1-\beta}}. \tag{X-15}$$

In this case no aberration phenomena exist.

If we make our observation in a direction perpendicular to the motion of the source (transverse Doppler effect), then and thus $\theta = \pi/2$

$$\cos\theta' = -\beta, \qquad \nu = \nu_0\sqrt{1-\beta^2}. \tag{X-16}$$

The Doppler effect is then due solely to the relativistic phenomenon of the slowing-down of clocks.

2. *Ives and Stillwell's experiments (1941)*[1]

The comparison of (X-15) and (X-16) shows that the non-relativistic

Doppler effect is of the first order, whereas the correction imposed by the slowing-down of clocks is of the second order. One can, of course, eliminate first-order effects by making the observation in a direction perpendicular to the motion of the source. But in that case, any error in the evaluation of angle θ completely conceals the occurrence of terms involving β^2 and reduces the verification of Einstein's formula to a mere illusion.

Ives and Stillwell's experiments consist of making two simultaneous observations, in opposite directions, of the radiation emitted by the same source. One thus observes the frequencies:

$$(X\text{-}17) \qquad \nu_1 = \frac{\nu_0\sqrt{1-\beta^2}}{1-\beta\cos\theta}, \qquad \nu_2 = \frac{\nu_0\sqrt{1-\beta^2}}{1+\beta\cos\theta}.$$

In practice, the chosen angle of observation is a very small one. So the wavelengths λ_1 and λ_2 of the radiation being observed satisfy the following relations:

$$(X\text{-}18) \qquad \frac{\lambda_1+\lambda_2}{2} = \frac{\lambda_0(1-\beta\cos\theta)}{2\sqrt{1-\beta^2}} + \frac{\lambda_0(1+\beta\cos\theta)}{2\sqrt{1-\beta^2}} = \frac{\lambda_0}{\sqrt{1-\beta^2}}.$$

The difference between the initial wave-length and the wave-length corresponding to the center of gravity of the two beams obtained is then

$$(X\text{-}19) \qquad \Delta_2\lambda = \frac{\lambda_1+\lambda_2}{2} - \lambda_0 \simeq \lambda_0\frac{\beta^2}{2}.$$

On the other hand, the observation of the radiation emitted in a direction θ at a very small angle from the direction of the source, makes it possible to measure the first-order effects:

$$(X\text{-}20) \qquad \Delta_1\lambda \simeq \lambda_0\beta.$$

The experiment is to compare the variations $\Delta_1\lambda$ and $\Delta_2\lambda$.

Ives and Stillwell use positive ion sources modified by BATHO and DEMPSTER.[2] These sources make it possible to obtain monokinetic atoms that produce a λ_0 band that is fine enough to allow $\Delta_2\lambda$ to be discernible. The hydrogen molecules ionized by the electrons emitted by a filament are accelerated by a large difference in potential (going as high as 40 000 volts) set up between two electrodes D and E situated very close together.

The hydrogen pressure is so low that no collisions or modifications of charge take place in the small interspace *DE*. (The low pressure is achieved by immersing a carbon trap in liquid air.) The H_2 and H_3 ionized molecules separate, producing neutral atoms. By this method it is possible to obtain monokinetic atoms, whose radiation is then observed (bands in the Balmer series).

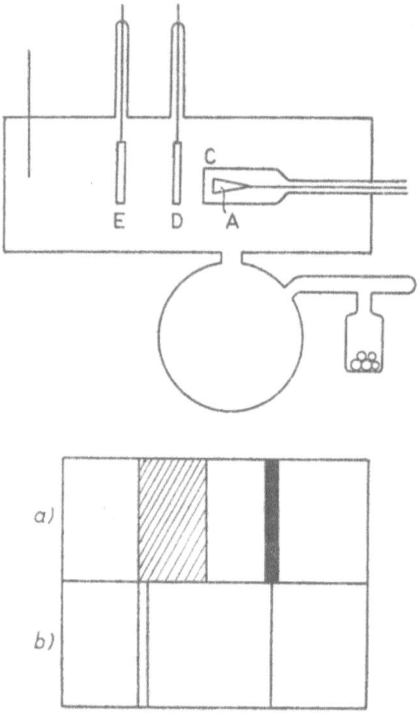

Fig. 33. The Ives-Stillwell experiment.

In ordinary ion-source apparatuses the atoms are not monokinetic, and the bands produced are greatly widened by Doppler effects of the first order. The comparison between the stationary band and the displaced one would have to be made in figures of type a). With the Ives-Stillwell apparatus, on the other hand, bands fine enough to allow detection of a displacement $\Delta_2\lambda$ of the center of gravity (a displacement of the second order) are obtainable. In Figure 33-b the double line is produced by the H_2 and H_3 molecules that have undergone acceleration.

The radiation is observed along a line making an angle of 7° with the axis of propagation of the channel rays. This radiation is received by the slit of a spectrograph placed at the curvature-center of a small concave mirror M. The axis of this mirror coincides with the direction of observation. Thus the radiation emitted by the atom passes in both directions along the straight lines joining a point in the mirror to the spectrograph slit. So one is in a position to apply formula (X-18).

When the accelerating potential is applied, the Doppler effect of the first order produces a displacement $\Delta_1\lambda = 20$ Å in the band $H_\beta = 4861$ Å. For example, the accelerating potential of 20 000 volts originally used by Ives and Stillwell corresponded to a displacement of 2 mm with the apparatus used. According to (X-20)

$$\beta \# \frac{20}{5000} = 0.004\,. \tag{X-21}$$

The second-order effect to be expected ought to have the calculated value

$$\Delta_2\lambda = \frac{\lambda_0}{2}\beta^2 = \frac{\Delta_1\lambda}{2}\beta = \frac{20}{2} \times 0.004 = 0.04\,\text{Å}\,, \tag{X-22}$$

and would bring about a displacement of

$$\frac{2 \times 0.04}{20} = 0.004\,\text{mm}\,.$$

Now that is precisely the order of magnitude of the half-width of the H_β bands observed; the fine structure did not appear. So it was legitimate to ask whether the effect observed was not due to a difference in the relative intensities of the unseparated components of the H_β band. In order to answer this objection, Ives and Stillwell repeated the experiments, raising the accelerating potential all the way to 43 000 volts.

With all these (and even a few other) precautions, the agreement is excellent between $\Delta_2\lambda$, calculated according to (X-19), and the values observed experimentally for various accelerating voltages – that is, for various values of β, with a maximum value of $\beta = 0.007$. The phenomena observed thus agree with the values predicted by the Einstein formula.

3. *The mean lifetime of mesons*[3]

The mesons that have been discovered in cosmic rays are charged or neutral particles with a mass that is midway between the mass of an electron and that of a proton. The μ-meson (with a mass two hundred times that of an electron) disintegrates, after a mean life τ, into an electron plus neutrino (a neutral particle of zero mass). This disintegration has been observed in photographs obtained in a Wilson cloud-chamber (WILLIAMS and ROBERTS, 1940)[4] or by means of counters (AUGER, MAZE, CHAMINADE, FRÉON, 1941)[5].

Counters, in fact, make it possible to measure the mean life-time τ_0 of a meson at rest. In order to do this, the mesons are caught in a metallic mass. One can, by means of counters, record the arrival of the incident meson and the departure of the disintegration electron. In practical terms, the trigger of the arrival-counter is delayed in order to insure its synchronization with the departure-counter. If Δt is the time interval between the arrival of a meson and its disintegration, it is possible to obtain the number $N(\Delta t)$ of mesons which disintegrate during a time t.

One thus has:

$$\text{(X-23)} \qquad y = -\frac{\Delta t}{\tau_0} + \text{const} \qquad \text{putting} \qquad y = \log \frac{N(\Delta t)}{t}.$$

The slope of the line $y = -\Delta t/\tau_0$ then makes it possible to measure the mean life of a meson at rest. In this way values have been obtained ranging between $\tau_0 = 2.15 \pm 0.07 \cdot 10^{-6}$ seconds (NERENSON and ROSSI, 1943)[6] and $\tau_0 = 2.7 \pm 0.5 \cdot 10^{-6}$ seconds (CACCIAPUOTI and RICCIONI, 1941)[7]. The mean life of the meson is thus approximately

$$\text{(X-24)} \qquad \tau_0 \# 2.2 \cdot 10^{-6} \text{ sec.}$$

In the higher atmosphere, the meson moves with an average velocity close to c, and it can cover several kilometers before disintegrating. It is therefore necessary to assume that the mean life of mesons in the higher atmosphere is far greater than their mean life at rest, in view of the fact that the higher atmosphere makes possible such a long trajectory. A mean life of $\tau_0 = 2.2 \cdot 10^{-6}$ seconds would indeed correspond to a trajectory of length

$$\text{(X-25)} \qquad L = v \cdot \tau_0 \simeq c \cdot \tau_0 \simeq 3 \cdot 10^8 \cdot 2.2 \cdot 10^{-6} = 600 \,\text{m}.$$

In fact, τ_0 is the duration of the mean-life of a meson when that duration is measured in the meson's proper system. In all other systems it has the much higher value of

$$\tau = \frac{\tau_0}{\sqrt{1-\beta^2}} \tag{X-26}$$

and corresponds to an average free trajectory:

$$L = \tau v \simeq \frac{\tau_0 c}{\sqrt{1-\beta^2}} = W \frac{\tau_0}{m_0 c} \tag{X-27}$$

putting

$$W = \frac{W_0}{\sqrt{1-\beta^2}} = \frac{m_0 c^2}{\sqrt{1-\beta^2}}. \tag{X-28}$$

One will then have

$$\frac{L}{W} = \frac{\tau_0}{m_0 c} = \text{const.} \tag{X-29}$$

The verification of this formula was achieved by ROSSI and HALL (1941).[8] In order to do this, they made measurements of L which they applied to mesons with a well-determined energy W. For $W = (5 \pm 0.7) \cdot 10^8$ eV. one finds

$$L = (4.5 \pm 0.6)\, 10^5 \text{ cm} \tag{X-30}$$

whence one deduces

$$\tau_0 = 2.4 \pm 0.3 \cdot 10^{-6} \text{ sec}, \tag{X-31}$$

if the mass of the meson is two hundred times that of an electron.[9] Now the energy $W = 5.10^8$ eV corresponds, according to (X-27), to

$$\sqrt{1-\beta^2} = \frac{\tau_0 c}{L} = \frac{2.4 \cdot 10^{-6} \cdot 3 \cdot 10^{10}}{4.5 \cdot 10^5} \tag{X-32}$$

which means that $\beta = 0.99$.

So the law of the slowing-down of clocks has been verified over a wide range of velocities, varying from $\beta = \Delta\lambda/\lambda \simeq 1/250 \simeq 0.004$ (Ives and Stillwell's experiment) to $\beta = 0.99$.

B. THE VARIATION OF MASS WITH VELOCITY

4. *The motion of a charged particle in an electromagnetic field*

The motion of a charged particle in a field of force is described by the equations (VIII-24) $\mathbf{f} = \mathbf{dp}/dt$.

If we are dealing with a charged particle subjected to the action of an electromagnetic field, the force **f** is to be identified with the Lorentz force. According to (VIII-25), (IX-35), (IX-30) and (IX-32) one has, in fact,

$$F^p = \frac{f^p}{\sqrt{1-\beta^2}} = \frac{1}{\sqrt{1-\beta^2}}\frac{1}{4\pi}\varphi^{\rho p}j_\rho = \rho\varphi^{\rho p}u_\rho . \tag{X-33}$$

Now, according to (VII-12)

$$u^\rho = \frac{v^\rho}{c\sqrt{1-\beta^2}}, \qquad \beta^2 = \frac{v^2}{c^2} = \Sigma_p \frac{(v^p)^2}{c^2} \tag{X-34}$$

putting

$$u^\rho = \frac{dx^\rho}{ds}, \qquad v^\rho = \frac{dx^\rho}{dt}. \tag{X-35}$$

(X-33) can then be written

$$f^p = \frac{\rho}{c}\varphi^{\rho p}v_\rho . \tag{X-36}$$

If the force **f** is produced by the action of an electromagnetic field on a charged particle, the equations of motion (X-36) then have the following form:

$$\frac{d}{dt}\frac{m_0 v^p}{\sqrt{1-\beta^2}} = \frac{q}{c}\varphi^{\rho p}v_\rho . \tag{X-37}$$

Let us multiply (X-37) by v_p and sum up over p. One then obtains

$$v_p\frac{d}{dt}\frac{m_0 v^p}{\sqrt{1-\beta^2}} = \frac{q}{c}v_p\varphi^{\rho p}v_\rho = \frac{q}{c}v_p\varphi^{0p}v_0 . \tag{X-38}$$

Now

$$v_p v^p = -\Sigma_p (v^p)^2 = -c^2\beta^2 \tag{X-39}$$

and

$$\varphi^{p0} = \partial^p\varphi^0 - \partial^0\varphi^p . \tag{X-40}$$

Equation (X-38) is then written

$$\frac{d}{dt}\left(\frac{m_0c^2\beta^2}{\sqrt{1-\beta^2}}\right)-\frac{m_0c^2}{2\sqrt{1-\beta^2}}\frac{d\beta^2}{dt}=q\left(v_p\partial^p\varphi^0\right)-q\left(v_p\partial^0\varphi^p\right) \tag{X-41}$$

or

$$m_0c^2\frac{d}{dt}\frac{1}{\sqrt{1-\beta^2}}=q\left(\frac{d\varphi^0}{dt}-\frac{\partial\varphi^0}{\partial t}\right)-\frac{q}{c}v_p\frac{\partial\varphi^p}{\partial t}. \tag{X-42}$$

Indeed, we still have

$$\frac{d\varphi^0}{dt}=\frac{\partial\varphi^0}{\partial t}+v^p\frac{\partial\varphi^0}{\partial x^p}. \tag{X-43}$$

Finally, the motion of a charged particle is described by the following equation

$$\frac{d}{dt}\left(\frac{m_0c^2}{\sqrt{1-\beta^2}}-q\varphi^0\right)=-q\left(\frac{\partial\varphi^0}{\partial t}+\frac{v_p}{c}\frac{\partial\varphi^p}{\partial t}\right). \tag{X-44}$$

Let us suppose that the particle possesses an initial velocity very close to zero and is accelerated by an electrostatic potential V, for example, as in the so-called Van de Graaf accelerator. From (X-44) one immediately obtains

$$\frac{m_0c^2}{\sqrt{1-\beta^2}}-q\varphi^0=\text{const}. \tag{X-45}$$

that is, according to our initial assumptions,

$$m_0c^2+qV=\frac{m_0c^2}{\sqrt{1-\beta^2}}. \tag{X-46}$$

From this it is quite easy to deduce

$$v=\frac{\sqrt{\dfrac{2qV}{m_0}\left(1+\dfrac{qV}{2m_0c^2}\right)}}{1+\dfrac{qV}{m_0c^2}}. \tag{X-47}$$

5. *The deviation of charged particles subjected to the action of parallel electric and magnetic fields perpendicular to the initial velocity of the particles*[10]

The prediction of a variation in the mass with respect to the velocity

$$m = \frac{m_0}{\sqrt{1-\beta^2}} \tag{X-48}$$

was already a result of the Lorentz theory of the electron. Rayleigh and Brace's experiment – which was intended to detect ether-drift – was, in fact, to demonstrate the influence of the contraction of lengths upon the refraction index of a transparent body in motion. The negative result of that experiment might thus be interpreted by assuming that the variations in mass, required by (X-48), succeeded in compensating for the effects of the contraction.

Now, that hypothesis (X-48) arises immediately from the theory of the deformable electron that Lorentz had substituted for the conception of the rigid electron proposed by Abraham. The experiments designed to

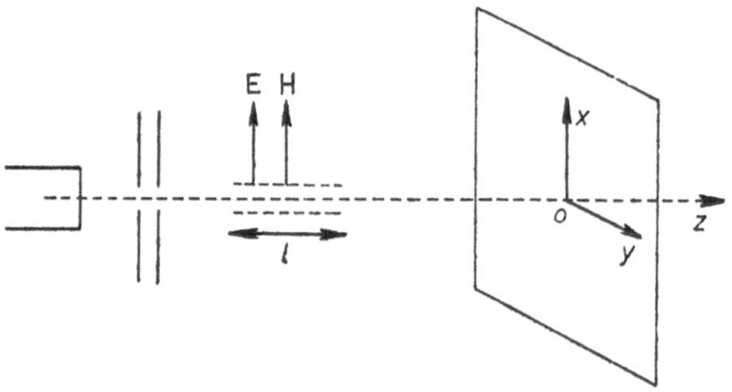

Fig. 34. The deviation of an electron beam by parallel electric and magnetic fields.

verify (X-48) thus seemed at first to make possible a choice between these two conceptions of the electron. Most of these experiments[11] consist of subjecting a beam of cathode rays, laterally limited by means of a diaphragm, to an electric field **E** and a magnetic field **H** parallel to each other and perpendicular to the initial velocity of the bundle (cf. Figure 34).

In the absence of a field, the cathode rays impinge at O. The field **E** produces a deviation

$$x = \frac{1}{2}\frac{e}{m}E\frac{l^2}{v^2},$$

l being the length along which the fields **E** and **H** exert their action. The field **H** causes a deviation perpendicular to the plane determined by **H** and by the direction of the bundle:

$$y = \frac{1}{2}\frac{e}{mc}H\frac{l^2}{v}.$$

The association of the two fields thus produces, for a certain value of v, deviations x and y such that

(X-49) $$\frac{y^2}{x} = \frac{1}{2}\frac{e}{m}\frac{H^2}{E}\frac{l^2}{c^2}.$$

The particles having the same e/m but varying in velocity have their

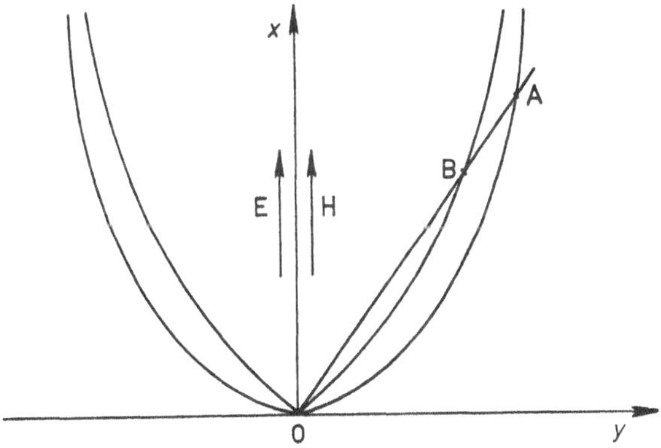

Fig. 35. Distribution of points of impact of particles having the same e/m.

points of impact distributed along a parabola corresponding to

(X-50) $$\frac{y^2}{x} = \frac{e}{m}\frac{H^2}{E}\frac{l^2}{2c^2} = \text{const} \quad \text{with} \quad \begin{cases} x = \dfrac{e}{2m}E\dfrac{l^2}{v^2} \\ y = \dfrac{e}{2mc}H\dfrac{l^2}{v}. \end{cases}$$

Taking into account a possible variation of mass with respect to velocity, the points of impact of the particles having the same e/m are distributed, no longer along a parabola, but along a fourth-degree curve obtained by eliminating v from the two equations

$$(X\text{-}51)_1 \qquad x = \frac{eE}{2m} \frac{l^2}{v^2} \sqrt{1-\beta^2}$$

$$(X\text{-}51)_2 \qquad y = \frac{eE}{2mc} \frac{l^2}{v} \sqrt{1-\beta^2}.$$

One then obtains

$$(X\text{-}52) \qquad \frac{y^2}{x} = \left(\frac{y^2}{x}\right)_{\text{parab.}} \cdot \sqrt{1-\beta^2}.$$

These curves are no longer tangent to Oy at the origin. This origin is reached, moreover, not for v infinite, as in non-relativistic theory, but for $v = c$. One has:

$$(X\text{-}53) \qquad \tan\alpha = \left(\frac{y}{x}\right)_{v \to c} = \frac{H}{E},$$

α designating the angle made by Ox with the tangent to the curve at O.

On the other hand, the points of impact of particles having different masses but identical velocity, are always distributed along the straight lines

$$(X\text{-}54) \qquad \frac{y}{x} = \frac{H}{E} \frac{v}{c}$$

issuing from the origin.

When mass varies with velocity, the points of impact for every particle of a given $\beta = v/c$ will be at the intersection of the straight line (X-54) corresponding to that particular value of β and of the classical parabola, on condition, according to (X-52), that the y^2/x thus obtained be contracted in the ratio $\sqrt{1-\beta^2}$. In this way, one passes from the point A to the point B.

In actual experiments, Guye and Lavanchy [12] produced constant deviations of two beams of cathode rays of different velocities by modifying – depending upon the beam to which they were applied – the electric and magnetic fields employed. It is then possible to deduce the ratio of the masses m and m' from the ratio of the fields. Guye and Lavanchy's measurements were made on electrons of such velocity that $0.22 < \beta < 0.49$.

These measurements were perfected by NACKEN in 1935.[13] He employed electrons having an energy close to 200 kV, corresponding to values of $\beta \simeq 0.7$.

It is found that formula (X-48) is very satisfactorily verified, whereas the non-relativistic calculation based on Abraham's hypothesis cannot account for the experimental results. The variation of mass with velocity thus seems to be a well-established fact.

This result is brought out most clearly by methods of accelerating heavy particles (protons, deuterons, α-particles) by means of a magnetic field. Cyclotrons make it possible to accelerate these charged particles. Under the influence of a magnetic field orthogonal to their velocity, the particles describe a circular trajectory. Their frequency is constant if m is constant[14], and the radius of the circle described ($v = r\omega$) increases with each impulse. At the very high velocities attained by the particles ($\beta = 0.145$ for deuterons of 20 MeV) v decreases as a result of the increase of $m(v)$. From this there arises a "braking" effect that can be eliminated by synchronizing the accelerating field with the motion of the charged particle, either by modulation of the intensity of the field that increases when m increases (synchrotron), or by modulation of the frequency of the accelerating field, since the frequency of the accelerations diminishes as m increases (synchrocyclotron).

6. *The elastic collision of two particles*

Let us situate the elastic collision of two particles, having the same rest-mass m_0, in a system S (the laboratory system). One of these particles P_0 is at 0 in system S; the other, P_1, has a uniform velocity $\mathbf{v}_1$. After the collision, the two particles follow from point O, the paths OP_1' and OP_2' with corresponding velocities $\mathbf{v}_1'$ and $\mathbf{v}_2'$ with respect to S.

Let us select the Cartesian axes Ox and Oy in the plane OP_1, OP_1' by orienting these axes in such a way that Ox coincides with OP_1' – that is, with $\mathbf{v}_1'$. According to the principle of the conservation of momentum, the velocity $\mathbf{v}_2'$ is also in the plane xOy. The velocities $\mathbf{v}_1'$, $\mathbf{v}_1'$ and $\mathbf{v}_2'$ correspond to the momenta $\mathbf{p}_1$, $\mathbf{p}_1'$ and $\mathbf{p}_2'$.

Let φ and θ be the angles formed by OP_1 and OP_2' with Ox; θ is then

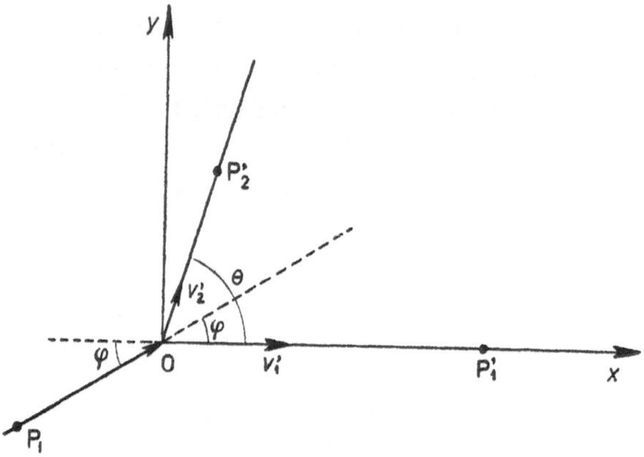

Fig. 36. The elastic collision of two particles.

also the angle of the fork OP_1', OP_2' defined by the trajectories of the two particles after the collision. According to the principle of the conservation of momentum, one will obtain by projection along Ox and Oy

(X-55) $$p_1 \cos \varphi = p_1' + p_2' \cos \theta$$

(X-56) $$p_1 \sin \varphi = p_2' \sin \theta .$$

From this there results

(X-57) $$2\, p_1' p_2' \cos \theta = p_1^2 - p_1'^2 - p_2'^2 .$$

Furthermore, the conservation of energy requires that

(X-58) $$m_1 + m_0 = m_1' + m_2' .$$

We must take into account the relation

(X-59) $$\frac{W^2}{c^2} = p^2 + m_0^2 c^2 \qquad \text{or} \qquad \frac{p^2}{c^2} = m^2 - m_0^2$$

which, substituted in (X-57), leads to the expression:

(X-60) $$\frac{2}{c^2} p_1' p_2' \cos \theta = (m_1^2 - m_0^2) - (m_1'^2 + m_2'^2 - 2\, m_0^2) =$$
$$= m_1^2 + m_0^2 - m_1'^2 - m_2'^2 .$$

Now, according to (X-58)

(X-61) $$m_1 = m_1' + m_2' - m_0$$

whence

$$\frac{2}{c^2} p_1' p_2' \cos\theta = 2(m_0^2 + m_1' m_2' - m_1' m_0 - m_2' m_0) \tag{X-62}$$

that is

$$\frac{p_1' p_2'}{c^2} \cos\theta = (m_2' - m_0)(m_1' - m_0). \tag{X-63}$$

a) In non-relativistic mechanics, one always has $m_0 = m_1' = m_2'$. So

$$\cos\theta = 0 \qquad (\text{if } p_1' p_2' \neq 0) \qquad \text{and} \qquad \theta = \frac{\pi}{2}. \tag{X-64}$$

After the collision, the trajectories of the particles form a right angle.

b) In relativistic mechanics, the trajectories of the particles form an angle θ such that:

$$\cos\theta = c^2 \frac{(m_2' - m_0)(m_1' - m_0)}{p_1' p_2'} \tag{X-65}$$

or according to (X-59):

$$\cos\theta = \frac{(m_2' - m_0)(m_1' - m_0)}{\sqrt{(m_1'^2 - m_0^2)(m_2'^2 - m_0^2)}} =$$

$$= \sqrt{\frac{(m_2' - m_0)(m_1' - m_0)}{(m_2' + m_0)(m_1' + m_0)}}. \tag{X-66}$$

Now if $p_1', p_2' \neq 0$ one should have

$$m_2' = \frac{m_0}{\sqrt{1 - \beta_2^2}} > m_0, \qquad m_1' = \frac{m_0}{\sqrt{1 - \beta_1^2}} > m_0 \tag{X-67}$$

that is

$$(m_2' - m_0)(m_1' - m_0) > 0 \tag{X-68}$$

whence

$$\cos\theta > 0, \qquad 0 < \theta < \frac{\pi}{2}. \tag{X-69}$$

The angle θ formed by the particles after the relativistic collision is, thus, always an acute angle.

The results obtained may be put in a slightly different form by considering the angles φ and ψ formed by OP_1' and OP_2' with the incident trajectory OP_1. In that way one obtains

(X-70) $$\theta = \varphi + \psi$$

and

(X-71) $$\tan\varphi\tan\psi = \tan\varphi\tan(\theta - \varphi) = \frac{\tan\varphi\tan\theta - \tan^2\varphi}{\tan\varphi + \tan\theta\tan^2\varphi}.$$

Now, according to (X-66), (X-55), (X-56) and (X-59)

(X-72) $$\tan^2\theta = \frac{2m_0(m_1' + m_2')}{(m_2' - m_0)(m_1' - m_0)}$$

(X-73) $$\tan^2\varphi = \frac{\sin^2\theta}{\left(\frac{p_1'}{p_2'} + \cos\theta\right)^2} = \frac{2m_0(m_2' - m_0)}{(m_1' - m_0)(m_1' + m_2')}.$$

By substitution in (X-71) and by taking (X-58) into account, one then obtains

(X-74) $$\tan\varphi\tan\psi = \frac{2m_0}{m_1' + m_2'} = \frac{2m_0}{m_0 + m_1}.$$

If the particles P_0 and P_1 have the same rest-mass, one then has

(X-75) $$m_1 = \frac{m_0}{\sqrt{1 - \frac{v_1^2}{c^2}}} \qquad \left(\beta = \frac{v_1}{c}\right),$$

that is

(X-76) $$\tan\varphi\tan\psi = \frac{2\sqrt{1-\beta^2}}{1 + \sqrt{1-\beta^2}}.$$

In the Newtonian approximation ($\beta \to 0$), the result (X-69) does indeed recur:

(X-77) $$\tan\theta = \tan(\varphi + \psi) \to \infty \qquad \text{whence} \qquad \theta = \varphi + \psi \to \frac{\pi}{2}.$$

These results agree very well with the experimental results.

So long as the incident particle's velocity remains low with respect to c, one finds, in keeping with the conclusions of Newtonian Mechanics, trajectories OP_1' and OP_2' perpendicular to each other after the collision.

Those are exactly the results obtained in a Wilson cloud-chamber when an α-particle happens to hit a helium nucleus.

However, when the incident particle's velocity is not negligible with respect to c, the predictions of Relativistic Mechanics are experimentally verified. If one observes, for example, in a Wilson cloud-chamber, the collision between a high-velocity electron and an electron at rest, the angle of the fork made by the trajectories of the two particles after the collision is less than $\pi/2$.

F. C. CHAMPION's experiments[15] (1932) with incident particles of different velocities, have made it possible to systematically verify the formula (X-69) by direct measurements of θ and φ. These results were confirmed by especially remarkable photographic observations of Wilson cloud-chambers. One of these photographs[16] (F. JOLIOT) shows the collision of an incident electron corresponding to $\beta = 0.968$ with an electron at rest; the angle of the fork is 60°. Another photograph (L. LEPRINCE-RINGUET)[17] is likewise of a relativistic electron-electron collision and involves an incident electron such that $\beta = 0.93$; the angle of the fork is here 72°.

7. *The Compton effect*

Let us now seek to determine the characteristics of the collison between an electron initially at rest and a photon of energy

(X-78) $$E = h\nu\,.$$

We cannot apply to the photon those relativistic formulas which involve terms in $1/\sqrt{1-\beta^2}$ because $\beta_{\text{photon}} = 1$. However, the relation

(X-79) $$\frac{W^2}{c^2} = p^2 + \mu_0^2 c^2$$

is still valid for the case of a photon whose proper mass μ_0 is assumed to be zero. One also obtains for a radiation $\gamma(\mu_0 = 0)$

(X-80) $$p = \frac{W}{c} = \frac{h\nu}{c}\,.$$

Let us assume that the incident photon moves along a path MM' parallel to the Ox axis. After collision, the photon goes off in the direction $O\gamma$, while the electron, which was motionless at E before the collision, subsequently follows the straight line $E\varepsilon$.

Let us designate by ν, W and $\mathbf{p}$ the frequency, energy and momentum, respectively, of the incident photon, and by ν', W', $\mathbf{p}'$ these same quantities after collision. If m_0 and $\mathbf{v}$ represent the proper mass and the velocity of the electron, the conditions of conservation of energy and momentum will be expressed in the following relations

$$W + m_0c^2 = W' + \frac{m_0c^2}{\sqrt{1-\beta^2}} \qquad \left(\beta = \frac{v}{c}\right) \tag{X-81}$$

$$\mathbf{p} = \mathbf{p}' + \frac{m_0\mathbf{v}}{\sqrt{1-\beta^2}}. \tag{X-82}$$

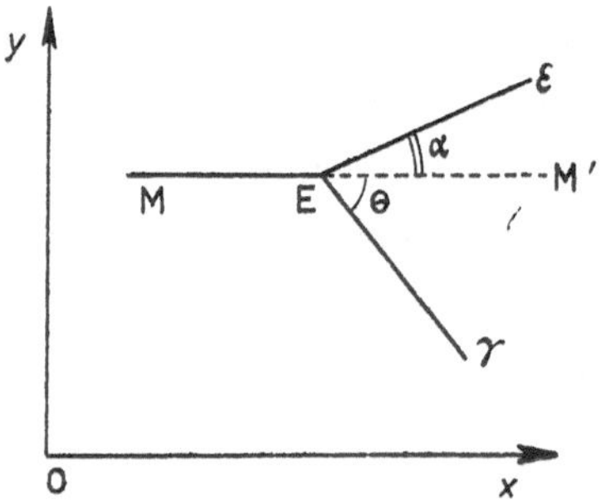

Fig. 37. Collision of photon and electron.

Let us project $\mathbf{p}$, $\mathbf{p}'$ and $\mathbf{v}$ along Ox and Oy. According to Figure 37, (X-82) may be decomposed into two equations:

$$p = p'\cos\theta + \frac{m_0v}{\sqrt{1-\beta^2}}\cos\alpha \tag{X-83$_1$}$$

$$0 = -\,p'\sin\theta + \frac{m_0v}{\sqrt{1-\beta^2}}\sin\alpha\,. \tag{X-83$_2$}$$

In keeping with (X-83)$_1$ and (X-83)$_2$, let us form $\sin^2\alpha + \cos^2\alpha$ and then replace p and p' in the equation thus obtained by their values (X-80). One thus obtains:

$$\frac{m_0^2v^2c^2}{1-\beta^2} = h^2(\nu^2 + \nu'^2 - 2\,\nu\nu'\cos\theta)\,. \tag{X-84}$$

In addition, (X-81) may also be written:

$$\frac{m_0 c^2}{\sqrt{1-\beta^2}} = h(\nu - \nu') + m_0 c^2 . \tag{X-85}$$

Let us square the preceding equation and subtract (X-84) from the result thus obtaining

$$m_0^2 c^4 = -2h^2 \nu\nu'(1 - \cos\theta) + m_0 c^2 \left[m_0 c^2 + 2h(\nu - \nu')\right] \tag{X-86}$$

that is

$$2h\nu\nu' \sin^2\frac{\theta}{2} = m_0 c^2 (\nu - \nu') . \tag{X-87}$$

By replacing ν and ν' by c/λ and c/λ' one then obtains:

$$\lambda' - \lambda = \frac{2h}{m_0 c} \sin^2\frac{\theta}{2} . \tag{X-88}$$

The change in wave-length produced by the Compton effect will be maximal if $\theta = \pi$; and the photon is then scattered in the backward direction. Its new wave-length in this case will be:

$$\lambda' = \lambda + \frac{2h}{m_0 c} . \tag{X-89}$$

If $\theta = \pi/2$, that is, if the photon is scattered at right angle, it undergoes a change in wave-length:

$$\Delta\lambda = \frac{h}{m_0 c} \tag{X-90}$$

equal to the Compton wave-length.

The adaptation of quantum theories to relativistic formalism supplied numerous applications, which constitute confirmation of the principles of both quantum and relativity theory. In order not to stray from the field in which classical theories apply, we have here merely indicated one of the simplest of these proofs: the Compton effect. The determination of the fine structure of the hydrogen-atom spectrum (Sommerfeld), the construction and elaboration of a Relativistic Wave Mechanics introducing *ipso facto* the spin of the electron (Dirac), and, finally, the development of a covariant formulation of quantum field-theory – all these constitute so many remarkable and fruitful extensions of Special Relativity.

C. THE EQUIVALENCE OF MASS AND ENERGY

8. *Mass defect and nuclear energy*

The relativistic expression of the energy of a system of bound particles leads to the following conclusion (cf. p. 223):

A *stable* system of bound particles possesses a rest mass M_0 less than the sum of the masses $m_0 = \Sigma_i(m_i)_0$ of its constituents. This diminution of mass

$$\Delta m = \Sigma_i(m_i)_0 - M_0 = \frac{\Delta E}{c^2} > 0 \tag{X-91}$$

corresponds to the *binding energy* ΔE existing among the constituents, and that is the energy that would have to be supplied to the system to resolve it into its elements.

On the other hand, if we are dealing with an *unstable* system of bound particles

$$\Delta m = \Sigma_i(m_i)_0 - M_0 = -\frac{\Delta E}{c^2} < 0\,. \tag{X-92}$$

The system breaks down into its elements in such a way that $\Sigma_i(m_i)_0 < M_0$. It thus liberates an energy ΔE.

The diminution of mass has been experimentally shown to exist by the use of nuclear structures in which the binding energies between nucleons are extremely high. By applying the methods of mass spectrography to the more stable nuclei, that is, nuclei with particularly high binding energies, it has been possible to prove the correctness of

$$\Delta m = \Sigma_i(m_i)_0 - M_0 > 0\,. \tag{X-93}$$

A simple example is that of the deuteron ${}^1_2D$, nucleus of heavy hydrogen.[18] Its mass, in the unit system in which the oxygen atom has a mass of 16, is

$$M_0 = 2.01417\,.$$

Now the deuteron is made up of a proton ($m_p = 1.00757$) and a neutron ($m_n = 1.00893$). So one has

$$\Delta m = \Sigma m_i - M_0 = 0.00233 \text{ mass units}$$

that is

$$\Delta m = 0.0387 \cdot 10^{-25} \text{ g}.$$

9. *The balance of energy and momentum in nuclear reactions*

The validity of the relation

$$\Delta E = \Delta m \cdot c^2 \qquad \text{(X-94)}$$

is assured by measurements made on nuclear reactions which transform a system having a certain diminution of mass into a new system with a different diminution in mass. The loss of mass in the course of the nuclear reaction will have to be equivalent to the energy released by that reaction.

a) We know (COCKCROFT and WALTON, 1932)[19] that lithium ${}^7_3\text{Li}$ bombarded by very fast protons is transformed into an unstable system that produces two alpha-particles on disintegration:

$${}^7_3\text{Li} + {}^1_1\text{H} \quad \rightarrow \quad {}^4_2\text{He} + {}^4_2\text{He}. \qquad \text{(X-95)}$$

Mass spectrography makes it possible to determine the values of the masses corresponding to ${}^7_3\text{Li}$, ${}^1_1\text{H}$ and ${}^4_2\text{He}$. By putting the mass of $O^{16} = 16.0000$ (i.e., using the physical scale), we obtain[20]

$$\begin{aligned} \Delta m &= 7.0166 + 1.0076 - (2 \times 4.0028) = \\ &= 0.0186 \text{ mass units} \end{aligned} \qquad \text{(X-96)}$$

corresponding to

$$\Delta m = 0.309 \cdot 10^{-25} \text{ g} \qquad \text{(X-97)}$$

$$\Delta m \cdot c^2 = 27.7 \cdot 10^{-6} \text{ erg}. \qquad \text{(X-98)}$$

This energy must represent the difference between the kinetic energy of the α-particles produced in the course of the reaction and the kinetic energy of the incident proton. Now the experimental determining of these kinetic energies produces the following result:

$$\Delta E = 17.28 \pm 0.03 \text{ MeV} = (27.6 \pm 0.05) \cdot 10^{-6} \text{ erg}. \qquad \text{(X-99)}$$

The comparison of (X-98) and (X-99) constitutes an excellent verification of (X-94).

b) Let us suppose that a nucleus N_0, at rest in system S, is bombarded with high-speed particles P_1. The transformation of the unstable system thus formed into a residual nucleus N_3 is accompanied by the emission of a very light particle P_2. The reaction is of the following type

(X-100) $$N_0 + P_1 \quad \rightarrow \quad N_3 + P_2 .$$

In order to verify (X-94) it will be necessary:

to measure the loss in mass, during the course of the reaction, by means of mass spectrography;

to determine the energy E released during the nuclear reaction. This energy is equal to the difference between the kinetic energies of the system after and before the reaction. The principles of the conservation of energy and the conservation of momentum are in fact written

(X-101) $$E = T_2 + T_3 - T_1 \qquad (T_0 = 0)$$

(X-102) $$\mathbf{p}_3 = \mathbf{p}_1 - \mathbf{p}_2 \qquad (p_0 = 0),$$

where T_0, T_1, T_2 and T_3 designate the kinetic energies, and $\mathbf{p}_0$, $\mathbf{p}_1$, $\mathbf{p}_2$, and $\mathbf{p}_3$ the momenta, of N_0, P_1, P_2, and N_3.

Let us assume that the residual nucleus N_3 is heavy enough and has a sufficiently low velocity for its kinetic energy to be calculated by applying the laws of Classical Mechanics. One will have

(X-103) $$p^2 = m^2 v^2 = 2mT .$$

In addition, if θ is the angle formed by the incident particle P_1 and the ejected particle P_2, (X-102) can be written

(X-104) $$p_3^2 = p_1^2 + p_2^2 - 2\, p_1 p_2 \cos\theta$$

that is, according to (X-103),

(X-105) $$m_3 T_3 = m_1 T_1 + m_2 T_2 - 2\sqrt{m_1 T_1 m_2 T_2} \cos\theta .$$

The most frequently observed particles are those ejected with $\theta = \pi/2$.

In those circumstances

$$T_3 = \frac{m_1 T_1 + m_2 T_2}{m_3} \tag{X-106}$$

and, by substitution in (X-101)

$$\begin{aligned} E &= T_2 + \frac{m_1 T_1 + m_2 T_2}{m_3} - T_1 = \\ &= \frac{(m_3 + m_2)}{m_3} T_2 - \frac{(m_3 - m_1)}{m_3} T_1 . \end{aligned} \tag{X-107}$$

In order to determine E, it sufficies to measure the energies T_1 and T_2 of the incident particle and the ejected particle. If we are concerned with a charged particle, the measurement of T_2 is obtained from the average range of the particle in a given substance. As a matter of fact, from previous study of radioactive emissions, the average range-energy relationship is known for protons and deutons whose energy can be measured directly by the method of magnetic deflections. The energy of the emitted neutrons is determined in a different way: one deduces it from the average range of the recoil proton produced by the bombardment of a hydrogenous substance by the emitted neutrons.

The results obtained in this way agree very well with relation (X-94). An accuracy often greater than 1% has been achieved in a great number of reactions of the most varied sort.

c) Finally, experiments on the pair production of oppositely charged particles having an energy $2m_0c^2$, starting from an electromagnetic radiation of energy $E_0 = h\nu_0$, along with experiments on the inverse phenomena of pair annihilation $2m_0c^2 \rightarrow h\nu_0$, make it possible to give meaning to the equivalence relation (X-94) in the case where the mass is entirely the result of the radiation or, contrariwise, entirely disappears as a result of the radiation.

PROBLEM

In cosmic radiation we are confronted with a charged particle whose velocity is almost that of light.

a) Calculate the components of the electromagnetic field **E**, **H** created by the motion of this charge.

b) Show that the field thus obtained is analogous to the field that would be produced by a short monochromatic wave-train.

(Cf. P. G. BERGMANN [9], p. 138).

Solution: a) We first find the expression for the field in the proper system S′ attached to the particle.

This is a purely electrostatic field

$$\varphi'^{po} = \partial'^{p}\left(\frac{q}{r'}\right) = \frac{qx'^{p}}{r'^{3}} \qquad (r'^{2} = \Sigma_{p}(x'^{p})^{2}). \tag{1}$$

We then pass over into system S of the laboratory; its velocity is $-\mathbf{v}$ with respect to S'. If $\mathbf{v}$ is the axis Ox, formulas (IX-5) are written:

$$\begin{aligned} &\varphi^{10} = \varphi'^{10}, \qquad \varphi^{20} = \frac{\varphi'^{20}}{\sqrt{1-\beta^{2}}}, \qquad \varphi^{30} = \frac{\varphi'^{30}}{\sqrt{1-\beta^{2}}} \\ &\varphi^{23} = 0, \qquad \varphi^{31} = -\frac{\beta\varphi'^{20}}{\sqrt{1-\beta^{2}}}, \qquad \varphi^{12} = \frac{\beta\varphi'^{30}}{\sqrt{1-\beta^{2}}}. \end{aligned} \tag{2}$$

The formulas of the special transformation then allow us to express φ^{po} for any Galilean system

$$\begin{aligned} \varphi'^{10} &= q\left[\frac{(x-vt)^{2}}{(1-\beta^{2})} + y^{2} + z^{2}\right]^{-\frac{3}{2}} \frac{x-vt}{\sqrt{1-\beta^{2}}}, \\ \varphi'^{20} &= qy\left[\frac{(x-vt)^{2}}{1-\beta^{2}} + y^{2} + z^{2}\right]^{-\frac{3}{2}}, \\ \varphi'^{30} &= qz\left[\frac{(x-vt)^{2}}{1-\beta^{2}} + y^{2} + z^{2}\right]^{-\frac{3}{2}}. \end{aligned} \tag{3}$$

By substitution in (2) we finally have

$$\begin{aligned} &\varphi^{10} = q\frac{(x-vt)}{\rho^{3}}, \qquad \varphi^{20} = \frac{qy}{\rho^{3}}, \qquad \varphi^{30} = \frac{qz}{\rho^{3}}, \\ &\varphi^{23} = 0, \qquad \varphi'^{31} = -\frac{\beta qy}{\rho^{3}}, \qquad \varphi'^{12} = \frac{\beta qz}{\rho^{3}}, \end{aligned} \tag{4}$$

with

$$\rho = (1 - \beta^2)^{\frac{1}{2}} \left(\frac{(x - vt)^2}{1 - \beta^2} + y^2 + z^2 \right)^{\frac{1}{2}}. \tag{5}$$

b) If $v = c$, $|\mathbf{E}| = \sqrt{\Sigma_p E_p{}^2}$ is maximum for $t_0 = x/v$.
For $t = t_0$, $|\mathbf{E}| = |\mathbf{H}|$, $\mathbf{E}$ is directed along Ox and is perpendicular to $\mathbf{H}$.

NOTES

1. H. E. Ives and G. R. Stillwell: *J. Opt. Soc. Amer.* **28** (1938) 215.
   H. E. Ives: *J. Opt. Soc. Amer.* **31** (1941) 369.
   R. Lennuier: *Revue Scientifique* **85** (1947) 740.
2. H. F. Batho and A. J. Dempster: *Astr. Journ.* **75** (1932) 34.
3. Cf. also R. Lennuier: *Revue Scientifique* **12** (1947) 740.
4. Williams and Roberts: *Nature* **145** (1940) 102.
5. P. Auger and Maze: *Compt. Rend.* **213** (1941) 381.
   Maze and Chaminade: *Compt. Rend.* **214** (1942) 266.
   Chaminade, Fréon and Maze: *Compt. Rend.* **218** (1944) 402.
6. Nereson and Rossi: *Phys. Rev.* **64** (1943) 199.
7. Cacciapuoti and Riccioni: *Ricerca Sc.* **12** (1941) 874.
8. Rossi and Hall: *Phys. Rev.* **59** (1941) 223.
9. L. Leprince-Ringuet and S. Gorodetzky: *Compt. Rend.* **213** (1941) 756.
10. Cf. W. Gerlach: *Handbuch der Physik*, 1926, 61–82.
11. E.g. W. Kaufmann: *Gött. Nachr. Math. Nat. Klasse*, 1901, 143.
    A. H. Bücherer: *Verh. der Deutschen Phys. Ges.* **6** (1908) 688.
    G. Neumann: *Ann. d. Phys.* **45** (1914) 529.
    Ch. E. Guye and Chr. Lavanchy: *Arch. Sc. Phys. Nat. Genève* **41** (1916) 353 and 441.
    W. Gerlach: *Handbuch der Physik*, 1926, p. 61.
12. Ch. E. Guye and Chr. Lavanchy: *Arch. Sc. Phys. Nat. Genève* **41** (1916) 286, 353 and 441.
13. M. Nacken: *Ann. d. Phys.* **25** (1935) 313.
14. Indeed, according to (IX-34):

$$\mathbf{f} = m\gamma = \frac{e}{c}[\mathbf{v} \wedge \mathbf{h}] \text{ or } m\omega^2 r = \frac{e}{c}\omega r H \text{ or } \nu = \frac{\omega}{2\pi} = \frac{eH}{2\pi mc}.$$

15. F. C. Champion: *Proc. Roy. Soc.* A **136** (1932) 630.
16. P. Curie: *Radioactivité.* t. I. Paris, Pl. XVI.
17. L. Leprince-Ringuet: *Thèse.* Paris, 1936, Pl. VI.
18. H. A. Bethe: *Elementary Nuclear Theory*, pp. 23–25.
19. J. D. Cockcroft and G. T. S. Walton: *Proc. Roy. Soc.* A **137** (1932) 229.
20. K. T. Bainbridge and E. B. Jordan: *Phys. Rev.* **51** (1937) 384.
    H. Bethe and M. S. Livingstone: *Rev. Mod. Phys.* **9** (1939) 370.
21. N. M. Smith: *Phys. Rev.* **56** (1939) 548.

PART THREE

GENERAL RELATIVITY

GENERAL RELATIVITY

A. THE NEWTONIAN LAW OF GRAVITATION

1. *The Newtonian law of gravitation and observational data*

Newton's law gives us an expression of the force of attraction exerted between two bodies. By assuming that these bodies are reduced to two material points separated by the distance r, this force, which is inversely proportional to r^2, depends on the two constants M and M' characterizing each of the two bodies involved

(XI-1) $$\boxed{F = -K\frac{MM'}{r^2}.}$$

Each constant M, M' is the *gravitational mass* of the body to which it refers; K is a universal constant depending on the unit system chosen for the expression of M and M'.

1) *The disagreement between Newton's law and observation.* The application of Newton's law to the experience was an almost universal success. Poincaré, for example, could write that Celestial Mechanics had no other object than the verification of the Newtonian law of gravity. In the midst of a host of striking proofs, the only disagreement worthy of note seemed to arise in mid-nineteenth century, in connection with the large planets.[1]

Around 1850, Le Verrier, taking up Laplace's work once more, studies the motion of the then-known planets of our solar system. He points out that the shift of the perihelion of Mercury, in particular, presents an advance of thirty-eight arc-seconds per century more than the value predicted by the Newtonian tables. The work of Newcomb (around 1880), based on more numerous and more accurate observations, confirm the existence of the anomaly noted by Le Verrier, but give as the value of the advance in the perihelion of Mercury 42.9″ of the arc per century. Moreover, Newcomb points out two other probable points of disagreement with the Newtonian theory, to wit, the fact that the advance of the perihelion of Mars differs by 8″ of arc per century from the calculated advance and is, therefore, three times greater than any probable error, and secondly, the advance of the node of Venus is equal to 10″ of arc per century and thus five times greater than the probable error.[2]

But other than those three points of disagreement, all of which concern the motion of the large planets, the discrepancies between observation and Newton's theory are quite debatable. Of them, the two principal ones have to do with the motion of the moon and the motion of Encke's comet.

The moon's motion does indeed present a secular acceleration effect, which had already been pointed out by Halley (1693). But it could be explained with a good deal of plausibility by the hypothesis of a variation in the eccentricity of the earth (Laplace) or else by the hypothesis of a slowing-down of the earth's rotation under the influence of the tides.

As for the motion of Encke's comet, it too is subject to variable accelerations that have been interpreted in terms of the intermittent presence of a meteoric current (Baklund).

Summing up, the noteworthy and unexplained points of disagreement between the predictions of the Newtonian tables and observations are those concerning the motion of the large planets, and particularly the advance of the perihelion of Mercury.

2) *"Newtonian" explanations of these discrepancies.* In order to interpret the gap between the Newtonian predictions and observation, various hypotheses have been resorted to, which may be qualified as "Newtonian", in the sense that they do not modify the basic law of action at a distance.

The asteroid ring. Le Verrier was led to postulate the existence of an inner-Mercurial planet, which would explain the advance in the perihelion of Mercury. But according to the characteristics that such a planet would have had, it ought to have been observable. So one is led, rather, to assume the existence of a ring of tiny inter-Mercurial planets. But even if this hypothesis explains the anomalies in the motion of Mars, it cannot simultaneously explain the discrepancies observed in connection with both Venus and Mars.

The non-sphericity of the sun or of the solar corona. A slight flattening of the photosphere would suffice to explain the advance in the perihelion of Mercury. Unfortunately, the comparison of the polar and equatorial diameters of the sun (measured by Auwers in 1832) seems quite definitely to contradict this hypothesis. In addition, such a flattening, if it existed, would involve a secular retardation of the node of Mercury just about equal to the advance of its perihelion, and this effect has not been observed.

Zodiacal light and Seeliger's hypothesis. The existence of zodiacal light indicates that the sun is surrounded by a cloud of diffuse matter having the shape of a bi-convex lens. This diffuse matter extends outwards, with an ever-decreasing density, beyond the earth's orbit. The plane of the ecliptic is the plane of its symmetry. Would not the presence of this diffuse matter suffice to produce the advance in the perihelion of Mercury?

This hypothesis was taken up by Seeliger, but it cannot simultaneously explain the discrepancy in the motions of the large planets, unless one further stipulates a certain distribution of the masses producing zodiacal light. This distribution, which would be a very inaccurate description at best, is highly arbitrary. It is, moreover, equivalent in part to the hypothesis of a ring of inter-Mercurial planets, and it is just as gratuitous. So the "Newtonian" hypotheses seeking to explain the three basic discrepancies seem at once inadequate and arbitrary.

3) *The "Non-Newtonian" laws of gravity.* One can try to explain the discrepancies between Newton's law and experience by slightly modifying that law in such a way that it will "produce" the experimental results.

Hall's law. The first of these non-Newtonian laws was proposed by Hall (1895). It consists of substituting for the Newtonian law of action at a distance the expression

$$F(r) = -K\frac{MM'}{r^N}. \qquad \text{(XI-2)}$$

And one does in fact find that there is an advance or a retardation in the perihelion, depending on whether N is greater than or equal to 2.[3]

The value of N deduced from the observed data concerning the perihelion of Mercury would be $N = 2.00000016$. But unfortunately, if the same coefficient K is kept, this law cannot be applied to the motion of the moon.

Newton's law with a corrective term. One might suppose that a corrective term varying with $1/r^n$ ($n = 3$, 4 or 5) has to be added to the Newtonian term $1/r^2$.

So the following form of the law of gravity has been proposed:

$$\text{(XI-3)} \qquad F = -K\frac{MM'}{r^2}\left(1 + \frac{\alpha}{r^n}\right)$$

α being positive so that there is an advance – and not a retardation – of the perihelion. But it is easy to see that no value of α can lead to a result that is simultaneously acceptable in the case of Mercury, the other planets, and the advance of the perigee of the moon.

A special case of (XI-3) is the law proposed by Decombes:

$$\text{(XI-4)} \qquad F(r) = -K\frac{MM'}{r^2}\left(1 + \frac{\alpha}{r^3}\right)$$

It was connected, according to its originator, with electric activity, with the coefficient α being a function of the mass and radius of the planet, as well as of electric induction.

And lastly, let us mention a different form of the law of gravity

$$\text{(XI-5)} \qquad F(r) = -K\frac{MM'}{r^2}e^{-\alpha r}.$$

This expression, thought up by Laplace, reminds us of the modification of Coulomb forces that was deduced from Yukawa's potential in the case of nuclear interactions.

Introduction of a non-Euclidean space. Finally, one can suppose that Newton's law applies to a space that is elliptical or spherical but, in any case, non-Euclidean. In that case the distance r between the attracting masses must be replaced by its expression in terms of a spherical or elliptical space. In a spherical space of radius R, this distance becomes

$$R \arcsin\frac{r}{R}$$

and one has

$$\text{(XI-6)} \qquad F(r) = -\frac{KMM'}{R^2\left(\arcsin\dfrac{r}{R}\right)^2} \simeq -\frac{KMM'}{r^2}\left(1 + \frac{r^2}{3R^2}\right).$$

But such a hypothesis can scarcely be given serious consideration, for it would necessitate assigning an impossible value to the radius R of the spherical space, if we are to come up with a correct evaluation of the advance of the perihelion.

2. *The gravitational potential and its properties. The equivalence of gravitational mass and inertial mass*

The Newtonian law of action at a distance

$$F = -K\frac{MM'}{r^2} \tag{XI-1}$$

involves, in this form, the "gravitational" *masses* M and M' of the two bodies present. These heavy masses play the role of charges in Coulomb's law.

On the other hand, the fundamental law of dynamics

$$\mathbf{F} = m\boldsymbol{\gamma} \tag{XI-7}$$

introduces a constant m characterizing the test-body. m expresses the *resistance to acceleration* of this test-body, in other words, its *inertial mass.*

We know that in vacuum all bodies fall with the same velocity. By comparing (XI-1) and (XI-7) we note that this property, experimentally observed, leads to the following consequence: the acceleration $\gamma = -M/m\,(K(M'/r^2))$ is independent of the nature of the test-body. Thus we are led to *postulate* that the gravitational mass M and the inertial mass m are equal, except for a multiplicative constant that is the same for all bodies. If we put

$$\frac{M}{m} = C, \tag{XI-8}$$

Newton's law (XI-1) is then written

$$F = -KC^2\frac{mm'}{r^2} = -G\frac{mm'}{r^2} \quad \text{with} \quad G = KC^2. \tag{XI-9}$$

If, with Newton, one postulates the *identity* of gravitational mass and inertial mass

(XI-10) $$C = 1\,, \quad M = m\,, \quad K = G \quad \text{whence} \quad F = -\,G\frac{MM'}{r^2}\,.$$

This is the convention usually adopted.[4] Anyhow, the Newtonian law of attraction takes on the following form

(XI-11) $$\boxed{F = -\,G\frac{mm'}{r^2}}$$

as a function of the inertial masses m and m'. It may further be written

(XI-12) $$\boxed{\mathbf{F} = m\,\mathbf{grad}\,U}$$

with

(XI-13) $$\boxed{U = G\frac{m'}{r}}\,.$$

G is the *Newtonian gravitational constant.* Its numerical value, evaluated according to (XI-11), is

(XI-14) $$\boxed{G = 6.664 \cdot 10^{-8}\,\mathrm{cm}^3\,\mathrm{g}^{-1}\,\mathrm{sec}^{-2}\,.}$$

U is called the Newtonian gravitational potential. The gradient of that potential is equal to the acceleration of the test-body produced by the gravitational forces – an acceleration that is independent of the nature, and consequently of the mass, of the test-body:

(XI-15) $$\gamma = \mathbf{grad}\,U\,.$$

The Newtonian hypothesis (XI-8) of the equality between gravitational mass and inertial mass thus leads one, by that very fact, to define an acceleration γ *independent of the test-body*:

Theoretically, this result is most remarkable.[5] Practically, it is based on the results of classical experiments with falling bodies. These experiments, which justify the Newtonian hypothesis (XI-8), were originally rather crude, and verifications of the identity of gravitational mass and inertial mass have been performed again and again in various contexts.

Completed by Newton's own experiments, and then by Bessel's on the oscillations of pendulums, these verifications were then performed in a very different way by Eötvös, Zeeman and Southerns. Southerns' experiments[6] make use of the radioactive properties of uranium oxide, which exhibits a large mass defect. Yet, the most conclusive results concerning the identity between gravitational mass and inertial mass are those arising from the work of Eötvös[7] and Zeeman[8]. Let us indicate briefly the principle of their experiments, the conclusions of which form the very basis of Einstein's conception of gravitation.[9]

Two bodies A_1 and A_2 of slightly different gravitational masses M_1 and M_2 and of inertial masses m_1 and m_2 are placed at the extremities of the beam of a torsion balance. Each of the bodies is subjected:
first, to the weight directed towards the center of the earth and proportional, by definition, to the gravitational mass. Its intensity is

$$\text{(XI-16)} \qquad \mathbf{F}_1 = M_1\gamma , \qquad \mathbf{F}_2 = M_2\gamma$$

and secondly, to the centrifugal force resulting from the earth's rotation about its own axis. This force f_1, proportional to the inertial mass m_1, is directed along the perpendicular HA_1 toward the axis of the terrestrial rotation. If ω is the angular velocity of rotation, and φ the latitude of the observation point

$$\text{(XI-17)} \qquad f_1 = m_1\omega^2 A_1 H = m_1\omega^2 R \cos\varphi \qquad f_2 = m_2\omega^2 R \cos\varphi .$$

The forces $\mathbf{F}$ and $\mathbf{f}$ are balanced by the torque of the fibre. One finds that the resultant moment of these forces produces a twisting of the fibre and is measured in terms of a rotation α of the beam. By interchanging the positions of the two bodies A_1 and A_2 the rotation becomes α'. The

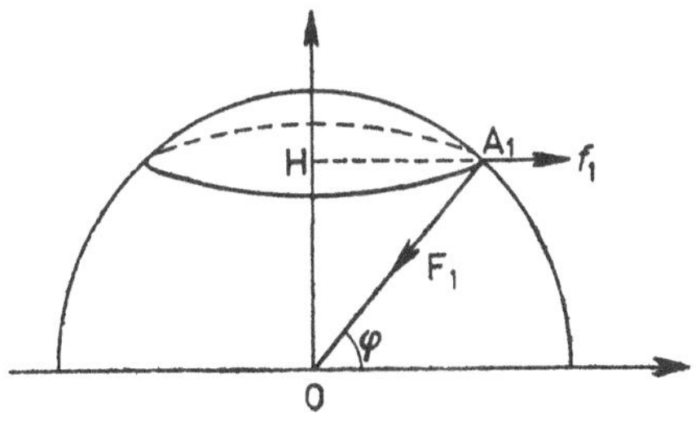

Fig. 38. The Eötvös-Zeeman experiment.

difference $\alpha - \alpha'$ then makes it possible to measure the difference

$$\frac{M_1}{m_1} - \frac{M_2}{m_2}.$$

Now all the experiments that have been performed according to this principle have demonstrated, within a very good approximation, the angles α and α' to be equal. The ratio M/m is thus constant, no matter which body is chosen and no matter what the orientation of the body is with respect to the earth.

3. *Poisson's law*

If we consider masses m_i contained in the volume $\mathscr{V}$ limited by a surface S, the application of Gauss' theorem leads us to the following result, which is analogous to the one we postulated in electrostatics.[10]

The flux of force through a surface S is proportional to the sum of the masses contained within the volume $\mathscr{V}$.

$$\text{(XI-18)} \qquad \int_S \gamma_n \, dS = 4\pi G \Sigma m_i$$

n being the normal to the surface-element dS.

Indeed,

$$\int_S \gamma_n \, dS = \int_S |\gamma| \cos \widehat{(\gamma, \mathbf{n})} \, dS = \int_S |\gamma| \, dS_n = \int_\omega |\gamma| \, r^2 \, d\omega,$$

dS_n being the projection of dS on the plane perpendicular to γ, and $d\omega$ the solid angle of the cone with base dS.

Now, according to (XI-13) and (XI-15), that is, according to Newton's law

$$|\gamma| = |\mathbf{grad}\, U| = G\frac{m_i}{r^2}.$$

Consequently

$$\int_S \gamma_n \, dS = G m_i \int_\omega d\omega = 4\pi \, G m_i .$$

For a continuous distribution of density μ per unit of volume, one has likewise

$$\text{(XI-19)} \qquad \int_S \gamma_n \, dS = 4\pi G \int_{\mathscr{V}} \mu \, d\mathscr{V}.$$

One thus obtains

(XI-20) $$\int_{\mathscr{V}} \operatorname{div} \gamma \, d\mathscr{V} = 4\pi G \int_{\mathscr{V}} \mu \, d\mathscr{V} .$$

From this we deduce the local relation:

(XI-21) $$\operatorname{div} \gamma = 4\pi G \mu$$

or, according to (XI-15)

(XI-22) $$\operatorname{div} \mathbf{grad}\, U = 4\pi G \mu$$

that is

(XI-23) $$\boxed{\Delta U = 4\pi G \mu} \qquad \text{with} \qquad \Delta = \sum_p \frac{\partial^2}{(\partial x^p)^2}, \qquad p = 1, 2, 3.$$

That is *Poisson's law*. Taking into account the definition of U, it is equivalent to the Newtonian law of action at a distance. And like the Newtonian law, it is invariant in a Galilean transformation, but not in a Lorentz transformation.

4. *Newton's law and the principle of Special Relativity*

Since Newton's law does not satisfy relativistic requirements, it seems natural to look for the form of a gravitational law that is covariant in a Lorentz transformation. Newton's law would be an approximate expression of that covariant law. Now the formulation of a relativistic law of gravity is far from being obvious. The model that classical electrodynamics might furnish – in a form suggested by Lorentz's electronic theory, for example – cannot be easily transposed into an explanation of the interaction of masses.

More specifically, every relativistic generalization of Poisson's law (XI-23) would lead one to substitute the d'Alembertian

$$\square \equiv \frac{1}{c^2} \frac{\partial^2}{\partial t^2} - \Delta ,$$

for the Laplacian

$$\Delta \equiv \Sigma_p \frac{\partial^2}{(\partial x^p)^2} .$$

In a system of rectangular coordinates, one would then have

$$\eta^{\rho\sigma}\partial_\rho\partial_\sigma U = 4\pi G\mu \qquad (\rho,\sigma = 1,2,3,0). \tag{XI-24}$$

In electrostatics, Gauss's theorem produces the appearance of a density of charge ρ, which plays a role analogous to μ. But ρ is the time component of a four-vector j^μ; and the relativistic expression of Gauss's theorem may also be obtained by substituting the four-vector potential A^μ for the electrostatic potential V. The conclusions of the relativistic dynamics of continuous media are very different. The density μ appears neither as a scalar nor as the component of a four-vector. According to Special Relativity, μc^2 is associated with an energy W, that is (cf. (VIII-147)) to the component M_{00} of a symmetrical tensor of the second order $M_{\mu\nu}$.

The gravitational potential ought then also to be represented by a tensor of the second order having the Newtonian potential U as one of its components. We will see that such is the conclusion resulting from General Relativity.

As a matter of fact, the relativistic expression of a law of gravity has been deduced by Einstein by means of an extension of the principle of Relativity. This expression does not, therefore, result from a correction of an already existing law, instead, it constitutes the natural extension of the main ideas of Special Relativity.

B. THE PRINCIPLE OF EQUIVALENCE AND THE INTRODUCTION OF A NON-EUCLIDEAN UNIVERSE

A Relativity principle expresses an equivalence of the reference systems used in describing physical phenomena and in stating the laws these systems obey. The Special Theory limits this equivalence to inertial systems defined by Galilean frames of reference; the General Theory extends it to accelerated systems. A postulate of equivalence or General Relativity thus makes it possible to absorb the effects produced by the so-called "fictitious" forces, that is, by the forces appropriate to accelerated systems; but such an equivalence involves, as we will see, the appearance of a non-Euclidean geometry.

Nevertheless this equivalence, generalized but still limited to inertial forces, would leave out the possible geometrization of the so-called "real" forces, and especially the gravitational ones. Indeed, the Principle of Equivalence postulated by the first extension of the Relativity Principle (1911) is a principle of a local assimilation of gravitational and inertial forces. This local equivalence in its turn makes it possible to give full significance to a Generalized Principle of Relativity (1916): the assimilation of the accelerated and inertial systems, that is, the absorption of the inertial forces by a non-Euclidean structure of space-time, then involves the reduction of the gravitational forces to a local non-Euclidean structure. The law of gravity is expressed by the existence of structural conditions.

So the principle of equivalence is formed by stratification. On one hand, the reduction of gravitational forces to inertial forces partially suppresses the distinction between real and fictitious forces and makes it possible locally to reduce the appearance of attraction phenomena to the occurrence of properly determined accelerations. On the other hand, the postulate of an equivalence between inertial and accelerated systems – and so, between inertial systems and systems endowed with gravitational forces – constitutes a principle of General Relativity that permits a new interpretation of these forces. Of course, the Lorentz force and the nuclear forces still have a phenomenological interpretation in General Relativity, that is, an interpretation foreign to all geometrization. It will be the role of the unified theories to seek the total geometrical reduction that a completely generalized Relativity Principle, as well as a perfect theory of the pure field, would require.

5. *Accelerated reference systems and "fictitious" inertial forces. The limits of the principle of Special Relativity*

The principle of Special Relativity is expressed by the covariance of physical laws under a Lorentz transformation. It assumes the impossibility of detecting the uniform rectilinear motion of a reference system by any experimental means.

This impossibility obviously does not extend to accelerated motions.[11] On the contrary, they can be shown to exist either by means of a mechanical experiment (the Foucault pendulum) or of an optical experiment (the Harress and Sagnac experiment).

1) *The Foucault pendulum.* If a pendulum oscillated without friction at the Pole, its plane of oscillation would pivot through 360° every 24 hours

in a direction opposite to that of the earth's rotation. This experiment, whose precise results depend upon the latitude at which it is performed, makes it possible to demonstrate the rotation of the earth on its axis. So an experiment in mechanics does reveal the rotation of the reference system to which it is attached.

2) *The experiments of Harress, Sagnac, Pogany.* These experiments constitute the "optical" counterpart" to the Foucault-pendulum experiment. Their object is to reveal the rotational motion of a disc by means of an optical experiment. Two light-beams emanating from the same source are separated at M by a semi-transparent mirror set at an angle of 45°. They cover, in opposite directions, an optical circuit that is carried along by the uniform rotation of the disc. This circuit is set up in glass prisms (HARRESS, 1912)[12] or in a water-filled tube fastened to the rotating disc (Pogany). Still another method (SAGNAC, 1913)[13] is the use of a series of mirrors placed on the edge of the rotating disc; in this way the beams of light describe in air a polygonal circuit tending towards a circumference that delimits a surface $\mathscr{S}$. In any case, an interference apparatus makes it possible to determine the difference between the lengths of time required by the two beams travelling in opposite directions to cover the same path. The light-source and the interferometer are

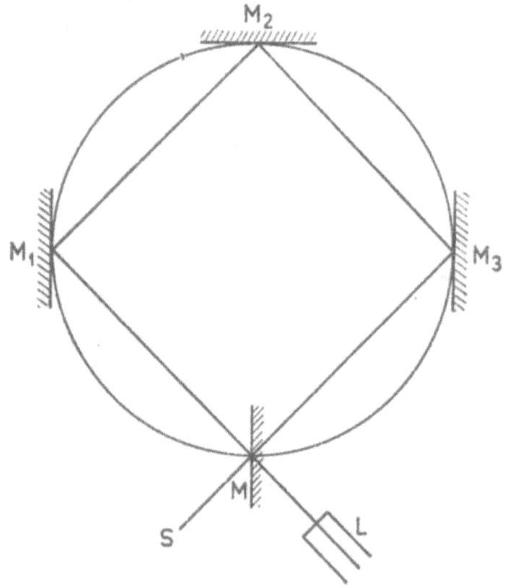

Fig. 39. Experiment of Sagnac.

fastened to the rotating disc and, along with the mirrors, make up a single rotating system. The observation of light-fringes shows that, in the proper system of the rotating disc, the light-ray propagated in the direction of the rotation of the disc makes a complete circuit in slightly less time than the ray moving in the opposite direction. If ω is the constant angular velocity of the disc, the difference in the circuit-times within the proper system of the rotating disc is

(XI-25) $$\Delta t = 4\frac{\omega \mathscr{S}}{c^2}.$$

From these experiments involving rotating systems we may draw the following conclusion:

The existence of accelerated motions seems to lead to the possibility of defining absolute motions. In the absence of reference-points supplied by other solid bodies, it would be necessary to admit that these absolute motions – for example, the rotation of the earth in the Foucault-pendulum experiment – take place with reference to an empty frame: the absolute space.

But in the light of the criticism resulting from Special Relativity, this conclusion is not very satisfactory, and one cannot help asking if these absolute motions are not necessarily bound to the presence of other masses, that is, to the existence of distant heavenly bodies. That will, in fact, be Mach's opinion.

6. *The local equivalence of gravitational and inertial forces*

1) *The forerunners of Einstein: Hertz and Mach.* Newton had admitted the distinction between *real forces*, which depend upon the physical properties of the system producing them, and *fictitious forces* resulting from the choice of a particular accelerated reference-system.

The inertial forces (centrifugal force, Coriolis' force) have a very special property: they produce accelerations that are independent of the characteristics of the body – and, in particular, of its mass. And that is why they are referred to as *fictitious*, for it seems possible to eliminate them by a judicious choice of reference system. Absolute space thus represents the privileged reference system that could absorb the fictitious inertial forces, artificially introduced in order to take into account the accelerated motion of the reference system being used. In absolute space only the *real* forces would remain, and the laws of physics would therefore be reduced to their natural expression. The notion of absolute space thus guarantees both the validity of the inertial principle and the possibility of distinguishing between the two kinds of forces.

Taking the opposite stand, Hertz, and subsequently Mach, eliminated any recourse to the idea of absolute space by seeking to justify the introduction of inertial forces by means of other considerations.

Hertz sought to reduce the electric and magnetic actions at a distance to contact actions. And he was to attempt the application of the same procedure to gravitational forces. Now the rectilinear motion of free masses is determined by the principle of inertia. According to Hertz, the different motions resulting from the intervention of inertial forces arise from connections with other masses. These connections determine the trajectories according to Gauss' principle of least constraint; the motion actually occurring is the one that differs least from a uniform rectilinear motion. It corresponds, not to the absence of forces, but to the absence of hidden masses.

Mach's criticisms attribute the special character of inertial systems to the intervention of distant masses whose influence we are unable to eliminate. In the absence of distant heavenly bodies – if, for example, the earth were alone in space – all reference systems would be equivalent and would constitute inertial systems. And thus, in that ideal situation, it would be impossible to detect the rotation of the Foucault pendulum. It is here that appears the principle of a possible equivalence between the fictitious inertial forces, on one hand, and the real gravitational forces arising from the influence of distant heavenly bodies, on the other. This equivalence will be the basis of Einstein's theory.

2) *Statement of a principle of local equivalence between inertial forces and gravitational forces.*[14] Einstein's criticisms do, in fact, show that the distinction between fictitious inertial forces and gravitational forces is illusory when we consider only a limited region of space-time. This conclusion results from a special property of gravitational forces: like inertial forces, they bestow on a test-body an acceleration that is independent of that body's mass. *So it is the equivalence between gravitational mass and inertial mass* (an equivalence that has been experimentally proved) *that removes any basis whatever for a local distinction between inertial and gravitational forces.*

Under these circumstances, it may be foreseen that the gravitational forces, like the inertial ones, may be modified and possibly eliminated by a careful choice of reference system. Let us cite the classical example of a ball that is released inside an elevator in free fall. The ball remains motionless with respect to the elevator, that is, always at the same distance from the floor. If the elevator is subjected to an acceleration greater than g (the acceleration due to gravity), the ball will stick to the ceiling. If the acceleration is less than g the ball will "fall" to the floor. The choice of a suitable accelerated system (the elevator, in the present instance), that is, the appearance of inertial forces thus modifies, and may even eliminate, the effects of gravity for an observer in that system. In other words, no experiment performed inside an accelerated system can detect the motion of that system, since it amounts to the same thing whether we assume that the system is motionless or whether we modify the value of the weight at that place. *So one cannot, within a limited region of space, distinguish the fictitious inertial forces from the real gravitational forces. A local equivalence subsists between them.*

In an extended region this equivalence is partially suppressed. For example, the existence of a gravitational field is shown, when the region is large enough, by a convergence of the lines of force. One cannot completely eliminate a gravitational field and leave only an inertial field. Without the complete dissociation of one field from the other, one could only affirm that their ensemble is not totally inertial.

So we state a principle of local equivalence in the following terms:

In a limited region of space, a gravitational field is equivalent to the field of force created by an accelerated motion (inertial field). No local experiment can make it possible to distinguish the one field from the other.

This statement is a principle of *Generalized Relativity*. According to Special Relativity, an equivalence exists only between Galilean systems; it is the notion of velocity that is relative. The foregoing statement postulates, more generally, a local equivalence between accelerated systems, through the introduction of gravitational forces or, as we will see, through a modification of geometry. In this way the notion of acceleration also becomes relative.

7. *Introduction of a non-Euclidean universe*

From 1913 on, Einstein was of the opinion that an equivalence of gravitational and inertial forces must entail a modification of the geometry. He was thus led to postulate the existence of a non-Euclidean universe. Yet, the expression of a law of gravity in terms of structural conditions satisfying a Riemannian space (1917) cannot be derived from the 1911 premises by any rigorous deduction. This expression is the product of a most attractive intuition, which permits the organization of results thus far obtained.

Before setting forth the Riemannian theory of gravitation, we are going to show, in this section and the next, how one is led to construct a non-Euclidean theory, that is, how one is led, first by the local reduction of gravitational to inertial forces and then by the postulate of a generalized equivalence between all accelerated systems, to the introduction of a non-Euclidean geometry.

Without going outside the framework of Newtonian mechanics, one can show that the effects of a force **F** may be expressed in terms of a geometrical structure.[15]

The statement of the inertial principle is based on the notion of equipollent Galilean systems.

In the case of a material point moving in a field of force **F**, it is still possible to preserve the form of the inertial principle, provided we relate the motion of the material point to a suitably accelerated reference system.

Indeed, let $\mathbf{v}$ and $\mathbf{v} + \mathbf{F}\, dt/m$ be the velocities at the instants t and $t + dt$ with respect to a fixed system S_0, where m is the mass of the material point. Then relate this motion to two coordinate systems S and S' equipollent to S_0 and moving with velocities $\mathbf{u}$ and $\mathbf{u}'$ with respect to it. The velocity of the moving material point $\mathbf{v} - \mathbf{u}$ with respect to $S(\mathbf{t})$ will be equal to its velocity $\mathbf{v} + (\mathbf{F}/m)\, dt - \mathbf{u}'$ with respect to $S'(\mathrm{t} + dt)$ if:

$$\mathbf{u}' - \mathbf{u} = \frac{\mathbf{F}}{m}\, dt. \qquad \text{(XI-26)}$$

In this case the inertial principle continues to be valid, since the components of the velocity of the moving body when related to the two proper equipollent systems are constant. However, in the statement of this principle we must change the meaning of the word equipollent: two reference systems with origins separated only by an infinitesimal distance will be called equipollent if they are made up of two parallel trihedrons in the usual geometrical sense and, moreover, are moving with respect to one another with a uniform rectilinear motion of velocity $\mathbf{F}\,dt/m$.

In the general case of a non-uniform gravitational field, the equipollence of two Galilean systems defined step by step depends on the path that, in space-time, joins the two origins. "Rigorously speaking", writes E. Cartan, "one has, by way of an explanation, simply established a linguistic convention, but this very fact illustrates the importance in the development of science, of linguistic conventions judiciously chosen."

Nevertheless, Newtonian mechanics requires a definition of equipollence incompatible with the principles of Special Relativity. The conditions imposed by Special Relativity are expressed in an altogether different way with respect to space and time. Indeed, if $\mathbf{e}_p$ ($\mathbf{e}_1$, $\mathbf{e}_2$, $\mathbf{e}_3$) are the unit vectors of rectilinear spatial axes, and if $\mathbf{e}_0$ is the unit vector of the time axis, the usual similarity is revealed by the disappearance of the three increments of $d\mathbf{e}_p$. One will then have the relations:

(XI-27) $$\mathbf{d}\mathbf{e}_p = 0\,, \qquad \mathbf{d}\mathbf{e}_0 = F^p \mathbf{e}_p\,dt\,.$$

They do not remain invariant under a relativistic transformation which mixes the two types of increment. The equipollence that made it possible to state a generalized inertial principle within the framework of Newtonian mechanics cannot be made to harmonize with the principle of invariance. If, with Einstein, we make invariance a basic principle, the law of gravity will have to be modified in such a way that its form remains unchanged in any Galilean system. That law turns out to be a relation between magnitudes characterizing a non-Euclidean universe. We are going to see that it actually involves only the curvature of space-time. It replaces the gravitational field by the data of the world-lines of the various material points, that is, by the geodesics of the space-time being considered. This procedure amounts to substituting a kinematic study for a dynamic study, but this kinematics contains the equivalent of the notion of force in the geometry which this kinematics presupposes.

By replacing the inertial forces, and consequently the gravitational forces, with a modification of the geometry, we postulate the existence of a non-Euclidean manifold in which the particles, which we have assumed to be subjected to some force, behave like free particles. According to the inertial principle, their trajectories must generalize Euclidean straight lines. But, on a curved surface, the shortest distances between two points are geodesics. The connection with the hidden masses of the Hertzian theory, the influence of the fixed stars in Mach's theory are expressed then, in General Relativity, by a more complex structure of space-time, which obliges the particles to describe geodesics in a non-Euclidean manifold. The equivalence of inertial and gravitational forces has its roots in the geometrical structure of the universe – the effect of material bodies being, not to generate forces, but to curve the universe. An Euclidean universe is one in which no matter exists.

Thus, the introduction of a non-Euclidean universe permits an extension of the principle of Relativity to accelerated systems defined by any curvilinear frame of reference. In other words, the covariance of the equations of Physics must be assured, not only under a Lorentz transformation, but under any transformation of coordinates whatsoever.

Naturally, it would be possible to define arbitrary systems of coordinates and arbitrary transformations in an Euclidean space. But then it would be possible to establish an equivalence valid throughout space. The equivalence between an accelerated and our inertial system must have, on the contrary, a *purely local significance.* And that is the state of affairs in a non-Euclidean space wherein this equivalence amounts to *fusing a small region of an affine manifold with the Euclidean space tangent to one of its points.*

And finally, the introduction of a non-Euclidean space, and notably of a Riemannian space, makes it possible to clear up the significance *and the limits* of the principle of equivalence.

8. *Study of a special case: the problem of the rotating disc*

Let us consider two coaxial discs S and S_0, and let us suppose that S is in motion, with respect to S_0, with a uniform rotational angular velocity ω – around the common axis. S_0 is a Galilean system represented, for example, by the laboratory. *We shall assume that the measurements taken on S and S_0 by an observer of S_0 supplied, consequently, with a measuring-rod attached to S_0, lead that observer to define an Euclidean geometry.*

Let us try to compare the measurements of length and time made on *S and S_0*.

1) *The geometry on a revolving solid body. The measurement of lengths.* The postulate of Special Relativity and its expression by means of Lorentz' formulas no longer applies, in principle, in the case of a rotating

disc, which does not constitute an inertial system. And yet one can extend the postulate in the following manner:

Measuring instruments – measuring rods and clocks – fastened to the rotating disc S undergo deformations from centrifugal force. According to the basic principles of Special Relativity, there is, in fact, no such thing as a perfectly rigid solid. And so these forces modify the standards of length and time whose values, with respect to S, will be defined *by assuming that all the corrections necessitated by the forces belonging to the accelerated systems have been made.*

In these circumstances, we will suppose that, at a given instant, the relationship between the lengths dl and dl_0 of the measuring rods fastened to S and S_0 *is the same as that which would exist between the lengths of the measuring rods of S' and S_0, S' being the Galilean system bound to the measuring rod dl at the instant being considered.* This amounts to the assumption that, after correction of the effects resulting from the accelerations, the measuring-rods fastened to the rotaling disc *are subject solely to a Lorentz contraction.*

Let us choose a system of polar coordinates (r, θ). The distance between two infinitesimally separated points (r, θ) and $(r + dr, \theta + d\theta)$ of the system S, measured by the standard measuring-rod of system S_0, is still

$$d\sigma^2 = dr^2 + r^2\, d\theta^2 \tag{XI-28}$$

for an observer in system S_0.

Now, for this observer in system S_0, the standard measuring-rod dl of S when oriented in any radial direction ($v = 0$) will still have unit-length. If, on the other hand, it is oriented along a perpendicular to the radius at a point P such that $OP = r$, the standard rod has a velocity $v = \omega r$ and undergoes a contraction that, for S_0, reduces its rest-length $dl = dl_0$ to $dl_0 \sqrt{1 - (\omega^2 r^2/c^2)}$. Similarly, for an observer in S_0, the distance between the two points (r, θ) and $(r + dr, \theta + d\theta)$ *measured by the standard measuring-rods bound to the accelerated system S*, becomes

$$d\sigma^2 = dr^2 + \frac{r^2\, d\theta^2}{1 - \dfrac{\omega^2 r^2}{c^2}}. \tag{XI-29}$$

More especially, a circumference

$$r = \text{const} \tag{XI-30}$$

which, measured by the Galilean standard-lengths of S_0 ($\omega = 0$) has the length

(XI-31) $$s_0 = \int ds_0 = r \int_0^{2\pi} d\theta = 2\pi r,$$

does not have the same length when the measurements are made with standards attached to an accelerated system. The measurements taken in that way lead us to attribute to the circumference a length

(XI-32) $$s = \int d\sigma = \frac{r}{\sqrt{1 - \frac{\omega^2 r^2}{c^2}}} \int_0^{2\pi} d\theta = \frac{s_0}{\sqrt{1 - \frac{\omega^2 r^2}{c^2}}} > s_0$$

and a surface area,

(XI-33) $$\mathcal{S} = \int_0^{2\pi}\int_0^{r} \frac{r d\theta}{\sqrt{1 - \frac{\omega^2 r^2}{c^2}}} \cdot dr = \frac{2\pi c^2}{\omega^2}\left(1 - \sqrt{1 - \frac{\omega^2 r^2}{c^2}}\right)$$

that is

(XI-34) $$\mathcal{S} \# \pi r^2 \left(1 + \frac{\omega^2 r^2}{4c^2}\right)$$

if the velocity $v = r\omega$ is very small in comparison with c.

These results [(XI-29), and (XI-33)] arise from any measurement performed with standards attached to an accelerated system. Now these standards are the ones that an observer, attached to the rotating disc S, would naturally choose. Thus, for this observer, *the geometry that he would be naturally led to construct, by means of the standards appropriate to his system, is not an Euclidean geometry.*[16]

Thus, for an observer in S, the relation between the circumference of a circle in S and its diameter is greater than π.

(XI-35) $$\frac{s}{2r} = \frac{s_0}{2r\sqrt{1 - \frac{\omega^2 r^2}{c^2}}} = \frac{\pi}{\sqrt{1 - \frac{\omega^2 r^2}{c^2}}} > \pi .$$

The natural geometry of the disc is not Euclidean and becomes increasingly non-Euclidean as the point P is moved farther and farther from the center.

Geodesics.[17] The geometry of S is defined by the two-dimensional spatial element

(XI-36) $$d\sigma^2 = g_{ab}\, dy^a\, dy^b\,, \qquad a, b = 1, 2\,.$$

By selecting the coordinates

(XI-37) $$y^1 = r\,, \qquad y^2 = \theta$$

we have, according to (XI-29)

(XI-38) $$g_{11} = 1\,, \qquad g_{22} = \frac{r^2}{1 - \dfrac{r^2\omega^2}{c^2}}\,, \qquad g_{12} = g_{21} = 0\,.$$

A free material point should follow the geodesics of system S. According to (XV-154), these geodesics are the curves

(XI-39) $$\frac{d^2 y^c}{d\sigma^2} + \begin{Bmatrix} c \\ ab \end{Bmatrix} \frac{dy^a}{d\sigma} \frac{dy^b}{d\sigma} = 0\,.$$

It is easy to calculate, beginning with the g_{ab} defined by (XI-38), the Christoffel symbols:

(XI-40) $$\begin{Bmatrix} c \\ ab \end{Bmatrix} = \tfrac{1}{2}\, g^{cd}(\partial_a g_{bd} + \partial_b g_{ad} - \partial_d g_{ab})\,, \qquad a, b, c, d = 1, 2\,.$$

Indeed, one deduces from (XI-38)

$$g^{11} = \frac{1}{g} \text{cofactor of } g_{11} = \frac{g_{22}}{g} = 1\,,$$

(XI-41) $$g^{22} = \frac{1}{g} \text{cofactor of } g_{22} = \frac{g_{11}}{g} = \frac{1 - \dfrac{r^2\omega^2}{c^2}}{r^2}$$

$$g^{12} = g^{21} = \frac{1}{g} \text{cofactor of } g_{12} = 0$$

and consequently, the values of the only non-zero symbols are:

(XI-42) $$\begin{Bmatrix} 1 \\ 22 \end{Bmatrix} = -\tfrac{1}{2}\, g^{11} \partial_1 g_{22} = \frac{-r}{\left(1 - \dfrac{r^2\omega^2}{c^2}\right)^2}$$

$$\begin{Bmatrix} 2 \\ 12 \end{Bmatrix} = \begin{Bmatrix} 2 \\ 21 \end{Bmatrix} = \tfrac{1}{2}\, g^{22} \partial_1 g_{22} = \frac{1}{r\left(1 - \dfrac{r^2\omega^2}{c^2}\right)}\,.$$

By substitution in (XI-39) one then obtains
For $c = 1$

$$\text{(XI-39)}_1 \qquad \frac{d^2r}{d\sigma^2} - \frac{r}{\left(1-\frac{r^2\omega^2}{c^2}\right)^2}\left(\frac{d\theta}{d\sigma}\right)^2 = 0.$$

For $c = 2$

$$\text{(XI-39)}_2 \qquad \frac{d^2\theta}{d\sigma^2} + \frac{2}{r\left(1-\frac{r^2\omega^2}{c^2}\right)}\frac{dr}{d\sigma}\frac{d\theta}{d\sigma} = 0$$

or

$$\text{(XI-43)} \qquad \frac{d}{d\sigma}\left(\frac{r^2}{1-\frac{r^2\omega^2}{c^2}}\frac{d\theta}{d\sigma}\right) = 0.$$

We deduce from (XI-43)

$$\text{(XI-44)} \qquad \frac{r^2}{1-\frac{r^2\omega^2}{c^2}}\frac{d\theta}{d\sigma} = K$$

which, substituted into (XI-29), gives us

$$\text{(XI-45)} \qquad \left(\frac{dr}{d\sigma}\right)^2 = 1 - \frac{r^2}{1-\frac{r^2\omega^2}{c^2}}\left(\frac{d\theta}{d\sigma}\right)^2 = 1 - \frac{K^2}{r^2}\left(1-\frac{r^2\omega^2}{c^2}\right)$$

or

$$\text{(XI-46)} \qquad \frac{dr}{d\theta}\frac{d\theta}{d\sigma} = \frac{dr}{d\theta}\frac{K}{r^2}\left(1-\frac{r^2\omega^2}{c^2}\right) = \pm\sqrt{1-\frac{K^2}{r^2}\left(1-\frac{r^2\omega^2}{c^2}\right)}.$$

If $K = 0$

$$\frac{dr}{d\sigma} = 1, \qquad \frac{d\theta}{d\sigma} = 0,$$

The curves θ = constant, that is, the radii of S, are thus the geodesics of the disc.

If $K \neq 0$, the differential equation of the geodesics will be written

$$\text{(XI-47)} \qquad \frac{dr}{d\theta} = \pm\frac{1}{K}\frac{r^2}{1-\frac{r^2\omega^2}{c^2}}\sqrt{1+\frac{K^2\omega^2}{c^2}-\frac{K^2}{r^2}}.$$

Let us put

(XI-48) $$\rho = \frac{r}{K}\sqrt{1 + \frac{\omega^2 K^2}{c^2}}$$

and then (XI-47) may also be written

(XI-49) $$\frac{1}{\rho^2\sqrt{1 - \frac{1}{\rho^2}}}\frac{d\rho}{d\theta} = \pm 1 + \frac{\frac{K^2\omega^2}{c^2}}{1 + \frac{K^2\omega^2}{c^2}}\frac{1}{\sqrt{1 - \frac{1}{\rho^2}}}\frac{d\rho}{d\theta}$$

and, by integrating

(XI-50) $$\operatorname{Arc}\cos\frac{1}{\rho} = \pm(\theta - \theta_0) + \frac{\frac{K^2\omega^2}{c^2}}{1 + \frac{K^2\omega^2}{c^2}}\sqrt{\rho^2 - 1}\,.$$

By choosing the origin such that $\theta_0 = 0$, we will thus have

(XI-51) $$\theta = \pm\operatorname{Arc}\cos\frac{a}{r} \mp \frac{a\omega^2}{c^2}\sqrt{r^2 - a^2}$$

with

(XI-52) $$a = \frac{K}{\sqrt{1 + \frac{K^2\omega^2}{c^2}}}\,.$$

In particular, if $K = 0$, one has according to (XI-44), $\theta =$ constant: the radii of S are geodesics.

It is readily seen that the sum of the angles of a curvilinear triangle formed by three geodesics lies between zero and π. As a matter of fact, we know that the angle φ formed by two curves defined by $M(y^a)$, $M + dM(y^a + dy^a)$ and by $M(y^a)$ and $M + \delta M(y^a + \delta y^a)$ is given by[18]

(XI-53) $$\cos\varphi = \frac{g_{ab}\,dy^a\,\delta y^b}{d\sigma\,\delta\sigma}, \qquad a, b = 1, 2$$

with

(XI-54) $$d\sigma^2 = g_{ab}\,dy^a\,dy^b, \quad \delta\sigma^2 = g_{ab}\,\delta y^a\,\delta y^b,$$

Here the parameters chosen are $y^1 = r$, $y^2 = \theta$. Taking (XI-38) into account, one will then have

$$\text{(XI-55)} \qquad \cos\varphi = \frac{dr}{d\sigma}\frac{\delta r}{\delta\sigma} + \frac{r^2}{1 - \dfrac{\omega^2 r^2}{c^2}} \frac{d\theta}{d\sigma}\frac{\delta\theta}{\delta\sigma}$$

and according to (XI-44) and (XI-45):

$$\text{(XI-56)} \qquad \cos\varphi = \sqrt{1 + \frac{K_1^2\omega^2}{c^2} - \frac{K_1^2}{r^2}}\,\sqrt{1 + \frac{K_2^2\omega^2}{c^2} - \frac{K_2^2}{r^2}} + K_1K_2\frac{1 - \dfrac{r^2\omega^2}{c^2}}{r^2}$$

where K_1 and K_2 are the values of K corresponding respectively to $(dr/d\sigma)\,(d\theta/d\sigma)$ and $(\delta r/\delta\sigma)\,(\delta\theta/\delta\sigma)$.

Let us consider (cf. Figure 40) a geodesic triangle OHA, HA being a geodesic normal to the radius vector, and the point A being on the limiting circle of radius $R = c/\omega$.

Let us calculate the values of K_1, K_2, K_3 relative to the geodesics OH, HA and OA. We will have:
for OH and OA, $d\theta/d\sigma = 0$. According to (XI-44) we then obtain

$$K_1 = K_3 = 0\,.$$

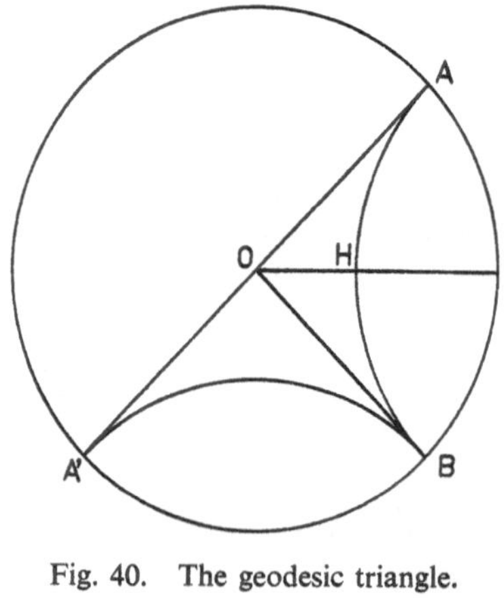

Fig. 40. The geodesic triangle.

For HA, $dr/d\sigma = 0$ at the point H. And so, according to (XI-43),

$$K_2 = \frac{r_0}{\sqrt{1 - (\omega^2 r_0^2/c^2)}} \qquad \text{with} \qquad r_0 = OH\,.$$

By substituting this value of K_2 in (XI-44), one will then obtain at the point H for the geodesic HA

$$\left(\frac{d\theta}{d\sigma}\right)_H = K_2 \frac{1 - \dfrac{r_0^2\omega^2}{c^2}}{r_0^2} = \frac{\sqrt{1 - \dfrac{\omega^2 r_0^2}{c^2}}}{r_0}\,.$$

We can now apply (XI-55) at $H\,(r_0, \theta = 0)$ and then at $A\,(r = c/\omega, \theta)$, in order to determine at those two points, with the help of the values for K_1, K_2, K_3, the angles φ_H and φ_A of the geodesic triangle. One obtains at the point H

$$\left(r = r_0\,, \qquad K_1 = 0\,, \qquad K_2 = \frac{r_0}{\sqrt{1 - \dfrac{\omega^2 r_0^2}{c^2}}} \right)$$

(XI-57) $\qquad \cos\varphi_H = 0 \qquad \text{whence} \qquad \varphi_H = \dfrac{\pi}{2}$

and at the point A

$$\left(r = \frac{c}{\omega}\,, \qquad K_1 = \frac{-r_0}{\sqrt{1 - \dfrac{\omega^2 r_0^2}{c^2}}}\,, \qquad K_3 = 0 \right)$$

(XI-58) $\qquad \cos\varphi_A = 1 \qquad \text{whence} \qquad \varphi_A = 0\,.$

Finally, at the point 0 where the geometry is Euclidean

(XI-59) $\qquad \varphi_0 < \dfrac{\pi}{2}\,.$

From this there results

(XI-60) $\qquad \varphi_{OHA} = \varphi_0 + \varphi_H + \varphi_A < \pi\,.$

The sum of the angles of a triangle is thus always included between zero and π.[19]

2) *Time measurement. Local time.* Let us compare the time measurements made in S and S_0. Two clocks H and H_0, attached to S and S_0 respectively are synchronized at the instant – which we choose as the origin – when they coincide spatially. Let t and t_0 be the indications supplied by H and H_0 at a later instant. Then a postulate analogous to the one we have applied to the measurement of lengths will be as follows:

The relation of the times t and t_0 *is identical with the relation that would result from the comparison of the indications supplied by two clocks H'* and H_0, *H' being a clock attached to the inertial system S', whose uniform velocity would be that of H at the instant considered.* This amounts to assuming that the time t indicated by H is deduced from t_0 by applying the Lorentz formula

(XI-61) $$t = t_0 \sqrt{1 - \frac{\omega^2 r^2}{c^2}}.$$

Let us suppose that the clock H returns to H_0 and is suddenly stopped. The observers in S and S_0 will both notice that clock H is slow. This is a special case of the clock-paradox. Observer S_0 will attribute this slowness to the accelerations H has undergone in the course of its motion; whereas, for observer S, clock H has been constantly at rest. He will be led to suppose that a gravitational field exists in system S, which, for him, is at rest. This gravitational field is deduced from the potential $U = -\omega^2 r^2/2$ and has the effect of modifying the running of the clocks in such a way that

(XI-62) $$t = t_0 \sqrt{1 + \frac{2U}{c^2}}.$$

The time t is variable in S and depends on the position of clock H. All the clocks equidistant from the axis will show the same time t. The time t defined by (XI-61) is said to be the *local time* of S.[20]

In defining the velocity of light by means of local time, one easily sees that this velocity is not constant in system S. In fact, in the inertial system S_0 a light-wave is such that

(XI-63) $$ds_0^2 = -dx_0^2 - dy_0^2 - dz_0^2 + c^2 dt_0^2 = 0$$

that is, by using cylindrical coordinates ($x_0 = r_0 \cos\theta_0, y_0 = r_0 \sin\theta_0, z_0$).

(XI-64) $$ds_0^2 = -dr_0^2 - r_0^2 d\theta_0^2 - dz_0^2 + c^2 dt_0^2 = 0.$$

Now the system S_0 is rotating with respect to S with an angular velocity $-\omega$. Whence

$$r = r_0, \qquad \theta = \theta_0 - \omega t_0, \qquad z = z_0. \tag{XI-65}$$

Substituting in (XI-64), this condition is then written

$$\begin{aligned} &- dr^2 - r^2\, d\theta^2 - dz^2 \mp 2\,\omega r^2\, d\theta\, dt_0 + \\ &+ c^2 \left(1 - \frac{r^2\omega^2}{c^2}\right) dt_0^2 = 0 \end{aligned} \tag{XI-66}$$

and according to (XI-61)

$$- dr^2 - r^2\, d\theta^2 - dz^2 \mp \frac{2\,\omega r^2\, d\theta\, dt}{\sqrt{1 - \dfrac{r^2\omega^2}{c^2}}} + c^2\, dt^2 = 0 \tag{XI-67}$$

or

$$d\sigma_e^2 \pm \frac{2\,\omega r^2}{\sqrt{1 - \dfrac{r^2\omega^2}{c^2}}}\, d\theta\, dt - c^2\, dt^2 = 0\,. \tag{XI-68}$$

$d\sigma_e^2$ is the Euclidean spatial element:

$$d\sigma_e^2 = dr^2 + r^2\, d\theta^2 + dz^2\,. \tag{XI-69}$$

In this system S, the velocity of light would then be defined by

$$V = \frac{d\sigma_e}{dt} \tag{XI-70}$$

as a function of local time. So one would then have, according to (XI-68),

$$V^2 = c^2 \mp \frac{2\,\omega r^2}{\sqrt{1 - \dfrac{r^2\omega^2}{c^2}}} \frac{d\theta}{dt}\,. \tag{XI-71}$$

Natural time. The Euclidean spatial element $d\sigma_e$ appears in expression (XI-68). The natural geometry of the disc is, however, defined by a non-

Euclidean spatial element

$$d\sigma^2 = dr^2 + \frac{r^2 d\theta^2}{1 - \frac{r^2\omega^2}{c^2}} + dz^2 . \tag{XI-72}$$

By substituting (XI-72) in (XI-66) one would thus obtain

$$- d\sigma^2 + c^2\, d\tau^2 = 0 \tag{XI-73}$$

with

$$d\tau = \sqrt{1 - \frac{r^2\omega^2}{c^2}} \left(dt_0 - \frac{\omega r^2\, d\theta}{c^2 \left(1 - \frac{r^2\omega^2}{c^2}\right)} \right). \tag{XI-74}$$

τ defines a *natural time* bound up in the natural geometry of the disc. By using this time τ, the velocity of light is still c. One can then regulate the clocks on S by an exchange of light-signals according to the initial indications of a clock H_1. At the initial instant, let us synchronize H_1 with the clock of S_0 coinciding spatially with H_1. A clock H on S, regulated according to H_1 by an exchange of light-signals, will indicate a time τ which, according to (XI-74), depends on the path that must be followed in order to get from H_1 to H. We obtain it by integration of (XI-74), assuming that $dt_0 = 0$. In particular, if we pass from H_1 to H by describing a circle of radius r around the origin

$$\begin{aligned} \tau_0 &= \frac{\pm r^2\omega}{c^2 \sqrt{1 - \frac{r^2\omega^2}{c^2}}} \int_0^{2\pi} d\theta = \frac{\pm 2\pi\omega r^2}{c^2 \sqrt{1 - \frac{\omega^2 r^2}{c^2}}} = \\ &= \pm \frac{2\omega\mathcal{S}}{c^2 \sqrt{1 - \frac{\omega^2 r^2}{c^2}}} \end{aligned} \tag{XI-75}$$

where $\mathcal{S}$ designates the area of the circle of radius r described by the signal.

If $v = \omega r$ is small with respect to c

$$\tau_0 = \pm \frac{2\omega\mathcal{S}}{c^2} . \tag{XI-76}$$

If H is regulated by H_1 by means of light-signals whose trajectories form circles q times around the origin, the indications on H and H_1 differ by $q\tau_0 = \pm 2\omega q\mathscr{S}/c^2$. At one and the same point, the natural time is then defined to within $q\tau_0$.[21]

If two light-rays going in opposite directions, describe a polygonal trajectory that is close to a circle of radius r, the time-difference of the two circuits after return to the point of departure is

$$\text{(XI-77)} \qquad \Delta\tau = \frac{4\omega\mathscr{S}}{c^2}.$$

And that is, in fact, the result that appeared in an interference experiment carried out by means of light propagated in air (Sagnac's experiment).

If one considers, not the propagation of photons with velocity c, but the motion of two bodies with velocities $+V$ and $-V$ with respect to the rotating disc, the same result is obtained: the time-difference of the circuits, independent of V, is still[22]

$$\text{(XI-77)} \qquad \Delta\tau = \frac{4\omega\mathscr{S}}{c^2}.$$

More especially, if one is concerned with the propagation of light rays in glass prisms (Harress' experiment) or in a tube filled with water and fastened to the rotating disc (Pogany's experiment), $V = c/n$ does not arise. An interference experiment brings out the result (XI-77).[23]

C. EINSTEIN'S LAW OF GRAVITATION

As a result of the very principles of General Relativity, gravitational forces should be absorbed locally by the existence of an appropriately-chosen non-Euclidean space time.[24] Einstein suggests that what is involved is a four-dimensional Riemannian space. Such a space differs from the Euclidean variety in that it exhibits a curvature that is represented by the Riemann-Christoffel tensor $G^{\rho}_{\mu\nu\sigma}$ (cf. XV-110).

Riemannian space is entirely determined by its metric $g_{\mu\nu}$. In fact, its affine connection $\Gamma^{\rho}_{\mu\nu}$ is identical with the Christoffel symbols at every point

$$\text{(XI-78)} \qquad \Gamma^{\rho}_{\mu\nu} = \left\{ {\rho \atop \mu\nu} \right\} = \tfrac{1}{2} g^{\rho\sigma} (\partial_\mu g_{\nu\sigma} + \partial_\nu g_{\mu\sigma} - \partial_\sigma g_{\mu\nu}),$$

and its curvature

(XI-79) $$G^{\rho}_{\mu\nu\sigma} = \partial_\sigma \begin{Bmatrix} \rho \\ \mu\nu \end{Bmatrix} - \partial_\nu \begin{Bmatrix} \rho \\ \mu\sigma \end{Bmatrix} + \begin{Bmatrix} \lambda \\ \mu\nu \end{Bmatrix} \begin{Bmatrix} \rho \\ \lambda\sigma \end{Bmatrix} - \begin{Bmatrix} \lambda \\ \mu\sigma \end{Bmatrix} \begin{Bmatrix} \rho \\ \lambda\nu \end{Bmatrix}$$

is expressed as a function of the $g_{\mu\nu}$ and their derivatives of the two first orders. One will thus have all the characteristics of Riemannian space by taking as the elementary interval

(XI-80) $$ds^2 = g_{\mu\nu}\, dy^\mu\, dy^\nu .$$

9. *The law of gravitation outside matter*

The law of gravitation outside matter will be expressed by structural conditions, that is, by restrictions imposed on the curvature of the universe, which is the only feature distinguishing Riemannian from Euclidean space.

These conditions may be obtained by setting equal to zero certain linear combinations formed with the help of the components of the curvature tensor. Einstein requires that the components of this tensor satisfy the ten following conditions:

(XI-81) $$G^{\rho}_{\mu\nu\rho} = G^{1}_{\mu\nu 1} + G^{2}_{\mu\nu 2} + G^{3}_{\mu\nu 3} + G^{0}_{\mu\nu 0} = 0 .$$

The contracted curvature tensor $G_{\mu\nu} = G^{\rho}_{\mu\nu\rho}$ is called the *Ricci tensor.* According to (XI-79) it has the following expression:

(XI-82) $$\begin{aligned} G_{\mu\nu} = G^{\rho}_{\mu\nu\rho} = \partial_\rho \begin{Bmatrix} \rho \\ \mu\nu \end{Bmatrix} - \partial_\nu \begin{Bmatrix} \rho \\ \mu\rho \end{Bmatrix} + \\ + \begin{Bmatrix} \lambda \\ \mu\nu \end{Bmatrix} \begin{Bmatrix} \rho \\ \lambda\rho \end{Bmatrix} - \begin{Bmatrix} \lambda \\ \mu\rho \end{Bmatrix} \begin{Bmatrix} \rho \\ \lambda\nu \end{Bmatrix} . \end{aligned}$$

So the Einstein hypothesis amounts to assuming that outside matter, the law of gravitation is obtained by setting equal to zero the components of the Ricci tensor:

(XI-83) $$\boxed{G_{\mu\nu} = 0 .}$$

A priori, it seems that these conditions are still too restrictive. Indeed, they do represent ten differential equations between the ten $g_{\mu\nu}$. So one could theoretically determine the $g_{\mu\nu}$ completely – which would impose the choice of a reference system. This conclusion would be paradoxical, since the choice of the reference system must, of course, remain arbitrary.[25] Nevertheless, it is possible to adopt conditions (XI-83) because of the existence of the following four identities among the $G_{\mu\nu}$:

(XI-84) $$\nabla_\rho (G^\rho_\mu - \tfrac{1}{2}\delta^\rho_\mu G) \equiv 0$$

by putting

(XI-85) $$G^\rho_\mu = g^{\rho\sigma} G_{\mu\sigma}, \qquad G = g^{\mu\nu} G_{\mu\nu}.$$

Thus, the number of independent equations resulting from (XI-83) is $10 - 4 = 6$, which leaves the desired arbitrariness for the choice of the reference system.

In more general terms, let us suppose that the equations of gravitation are expressed by the conditions

(XI-86) $$\boxed{S^\rho_\mu = 0}$$

S^ρ_μ being a tensor $g^{\rho\nu} S_{\mu\nu}$, solely a function of the $g_{\mu\nu}$ and of their derivatives of the two first orders. Let us also suppose that the ten conditions (XI-86) are reduced to six, thus taking care of the arbitrary character of the reference system by making the S^ρ_μ satisfy the conservation equations:

(XI-87) $$\boxed{\nabla_\rho S^\rho_\mu \equiv 0.}$$

E. Cartan has shown that the only S^ρ_μ tensor that satisfies all the requirements is the one having the form[26]

(XI-88) $$\boxed{S^\rho_\mu = G^\rho_\mu - \tfrac{1}{2}\delta^\rho_\mu G}$$

$G_{\mu\nu}$ being the Ricci tensor determined by (XI-82). Condition (XI-86) results in $G = 0$ and may be further written

(XI-83) $$G_{\mu\nu} = 0$$

These are the equations of gravitation outside matter and in the absence of any electromagnetic field.

10. *The law of gravitation inside matter or in the presence of an electromagnetic field*

The presence of matter or of an electromagnetic field is expressed by the appearance of an energy-momentum tensor T^{ν}_{μ} that is conserved:

(XI-89) $$\nabla_\rho T^\rho_\mu = 0 .$$

In this case, the law of gravitation will consist of counter-balancing the effects of the conserved tensor S^ρ_μ, of geometrical origin, by those of the tensor T^ρ_μ, likewise conserved but not of geometrical origin, since it arises from matter (material tensor), from an electromagnetic field, or from any other non-gravitational contribution of energy.

The law of gravity will then be written:

(XI-90) $$\boxed{S^\rho_\mu = \chi T^\rho_\mu}$$

χ is a constant connected with the gravitational constant G.
The identities (XI-87) lead to (XI-89), thus ensuring the divergence-free character of the tensor T^ρ_μ.

Going back to expression (XI-88) of S^ρ_μ, one will then have:

(XI-91) $$G^\rho_\mu - \tfrac{1}{2}\delta^\rho_\mu G = \chi T^\rho_\mu$$

or

(XI-92) $$G_{\mu\nu} - \tfrac{1}{2}g_{\mu\nu}G = \chi T_{\mu\nu} .$$

11. *The trajectories of a particle subjected to a gravitational field are the geodesics of a Riemannian space*

If the presence of matter made up, for example, of neutral particles gives up no electromagnetic (or caloric, etc.) contribution, the tensor T^ρ_μ reduces to the material tensor M^ρ_μ. According to expression (VIII-166) applied to a perfect gas, and where, besides, $p = 0$, one can write:

(XI-93) $$M^\rho_\mu = \mu_0 c^2 u_\mu u^\rho .$$

If, on the other hand, the sources are made up of charged particles, there exists an electromagnetic field. Assuming that this field obeys Maxwell's equations, its energy-momentum is determined by the Maxwell tensor:

(XI-94) $$\tau^\rho_\mu = -\varphi_{\mu\sigma}\varphi^{\rho\sigma} + \tfrac{1}{4}\delta^\rho_\mu\varphi_{\lambda\sigma}\varphi^{\lambda\sigma} .$$

The equation $\nabla_\rho T^\rho_\mu = 0$ is then expressed by

(XI-95) $$\nabla_\rho (M^\rho_\mu + \tau^\rho_\mu) = 0$$

that is, according to (XI-93) and (XI-94), by

(XI-96) $$u^{\rho}\nabla_{\rho}u_{\mu} = -\frac{1}{\mu_0 c^2}\varphi_{\mu\rho}j^{\rho} = -\frac{4\pi\rho_0}{\mu_0 c^2}\varphi_{\mu\rho}u^{\rho} = \frac{4\pi}{\mu_0 c^2}f_{\mu}$$

putting

(XI-97) $$j^{\rho} = \nabla_{\sigma}\varphi^{\rho\sigma}, \qquad f_{\mu} = \frac{1}{4\pi}\varphi_{\rho\mu}j^{\rho}.$$

The trajectory of a neutral particle ($\rho_0 = 0$) then reduces to

(XI-98) $$u^{\rho}\nabla_{\rho}u_{\mu} = 0$$

and may also be written, taking into account the definition

$$u^{\mu} = \frac{dy^{\mu}}{ds}$$

(XI-99) $$\boxed{\frac{d^2y^{\rho}}{ds^2} + \left\{ {\rho \atop \mu\nu} \right\} \frac{dy^{\mu}}{ds}\frac{dy^{\nu}}{ds} = 0}$$

This trajectory (Cf. XV-154) is further obtainable from a variational principle

(XI-100) $$\boxed{\delta \int ds = 0}$$

applied to the elementary interval

(XI-101) $$ds^2 = g_{\mu\nu}\,dy^{\mu}\,dy^{\nu}.$$

The trajectories of a neutral particle defined by (XI-98) coincide thus with the shortest lines – or geodesics – of a Riemannian space whose elementary interval is defined by (XI-101). Thus, the Riemannian space which includes the gravitational field in its structure, possesses geodesics that are the trajectories of neutral particles. Here we can consider these particles as free particles describing geodesics (XI-99).

The trajectories of free particles can always be represented by (XI-99) in any chosen system of curvilinear coordinates. This conclusion is valid for every space whose affine connection is represented by the symbols $\left\{ {\rho \atop \mu\nu} \right\}$, that is, for Riemannian as well as Euclidean space. But if we are dealing with an Euclidean space, we can always choose a system of orthonormalized coordinates ($\left\{ {\rho \atop \mu\nu} \right\} = 0$) and relate all the phenomena of an extended region of this space to that system. In this special system $(d^2x^\mu/ds^2) = (du^\mu/ds) = 0$, the components of the world-velocity are constant and the free material point describes a straight line with a uniform motion.

On the other hand, in a Riemannian space, the equation of the geodesics (XI-99) cannot be reduced to u^μ = constant, by a choice of coordinates valid throughout the space. Equation (XI-99) describes the motion of a "free" material point in Riemannian space, that is, of a material point subject only to gravitational forces (neutral particle) in the equivalent Euclidean description. In contrast, the motion of a charged particle ($\rho \neq 0$) is obtained from (XI-96) and differs, by virtue of the right-hand member of (XI-96), from the expression of a Riemannian geodesic.[27]

NOTES

1. Cf. J. Chazy [19] t. 1, p. 140.
2. These values have been confirmed to within very close approximations in the work of Doolittle (1912) and Ross.
   Poincaré considered that, of the three discrepancies noted between the motion of the larger planets and the Newtonian predictions, the discrepancy relative to the perihelion of Mercury is certain, the one concerning the motion of Venus probable, and the one having to do with the perihelion of Mars, doubtful.
3. The advance of the perihelion is $\pi(N - 2)$ for all planets.
4. Of course, it is also possible to choose
   $$C = \sqrt{G}, \qquad M = \sqrt{G}\, m, \qquad K = 1 \qquad \text{whence} \qquad F = -\frac{MM'}{r^2}.$$
5. This result is, indeed, very different from that obtained, for example, in electrostatics. In that case one has
   $$F = -q \operatorname{grad} V \text{ and } F = m\gamma$$
   Whence
   $$\gamma = -\frac{q}{m} \operatorname{\mathbf{grad}} V.$$
   The acceleration depends on the ratio q/m, which is different for every type of particle.
6. L. Southerns: *Proc. Roy. Soc. Londen* A **84** (1910) 325.
7. R. V. Eötvös: *Math. und Naturwiss. Ber. aus Ungarn* **8** (1890) 65; *Ann. d. Phys.* **59** (1896) 354;
   R. V. Eötvös, D. Pekar and E. Fekete: *Ann. d. Phys.* **68** (1922) 11.
8. P. Zeeman: *Proc. Roy. Amsterdam* **20** (1917) 542.
9. More detailed calculations relative to this experiment will be found in Von Laue [24], vol. II.
10. Gauss' theorem, in the form (XI-18) or (XI-19), is here deduced from (XI-13) and (XI-15), that is, from Newton's law. By contrast, if one refuses to establish

Maxwell's theory on the basis of an action-at-a-distance principle, Gauss's theorem (experimentally verifiable) is postulated in electrostatics, and from its consequences Coulomb's law is obtained.

11. On the optical effects of accelerated rectilinear motions, cf. the study by E. DURAND: *Ann. Phys.* **20** (1945) 535–544; **21** (1946) 216–231.
12. F. HARRESS: *Thesis*. Jena, 1912.
13. G. SAGNAC: *Compt. Rend.* **157** (1913) 708 and 1410; *J. Phys.* **4** (1914) 177.
14. A. EINSTEIN: *Jb.f. Rad. u. El.* **4** (1907) 411; *Ann. d. Phys.* **35** (1911) 898; **38** (1912) 443; *Phys. Z.* **14** (1913) 1249.
15. We are here using an argument of Cartan's. Cf. 'Les variétés à connexion affine et la Relativité générale', *Ann. Ec. Norm.* **40** (1923).
16. Naturally, this conclusion assumes the following hypothesis: Observer S admits that the measurements made on S and S_0 by means of the Galilean standards of S_0 lead to a Euclidean geometry. This hypothesis itself depends on the special character of Galilean observations and, consequently, on the possibility of detecting the "absolute" motion of the system S. This possibility (experimentally realizable) obviously contradicts, by the very principle involved, the equivalence (likewise, experimentally noted) of Galilean systems and the reciprocity of the conclusions arising therefrom.
17. Cf. P. LANGEVIN: *Compt. Read.* **173** (1921) 831; **200** (1935) 48; **205** (1937) 304. Cf. also: O. COSTA DE BEAUREGARD [11], p. 45; H. ARZELIÈS [8], p. 153; C. MØLLER [16], p. 241; A. S. EDDINGTON [22], p. 112; B. KURSONOGLU: 'Space-time on the Rotating Disk', *Proc. Camb. Phil. Soc.* **47** (1951) 177.
In particular, the method here used is the one adopted by C. Møller (the constant K becoming α) and H. Arzeliès (K becoming A).
18. Cf., for example, C. MØLLER [16], p. 226, which gives a demonstration of formula (XI-53) that runs as follows:
A two-dimensional surface immersed in a three-dimensional Euclidean space-related to a Cartesian system (x, y, z) allows the following parametric representation

$$(1)\ x^1 = f(y^1y^2)\,,\qquad x^2 = g(y^1y^2)\,,\qquad x^3 = h(y^1y^2)\,,$$

f, g, h being functions of two parameters y^1 and y^2
The distance between two points of the surface corresponding to the values y^a and $y^\alpha + dy^\alpha$ of the parameters $(a = 1, 2)$ is then

$$(2)\ ds^2 = \Sigma_p (dx^p)^2\,,\qquad (p = 1, 2, 3)\,.$$

which may also be written

$$(3)\ ds^2 = g_{ab}\, dy^a\, dy^b$$

with

$$(4)\ g_{ab} = \frac{\partial f}{\partial y^a}\frac{\partial f}{\partial y^b} + \frac{\partial g}{\partial y^a}\frac{\partial g}{\partial y^b} + \frac{\partial h}{\partial y^a}\frac{\partial h}{\partial y^b}\,.$$

Moreover, the cosine formed by the two directions of the components dx^p and δx^p is:

$$(5)\ \cos\theta = \frac{dx^p\,\delta x^p}{ds\,\delta s}\,,\quad ds = \sqrt{\Sigma (dx^p)^2}\,,\quad \delta s = \sqrt{(\Sigma \delta x^p)^2}\,,\quad (p = 1, 2, 3)\,.$$

By differentiating (1), we then have

$$(6)\ \cos\theta = \frac{g_{ab}\,dy^a\,\delta y^b}{ds\,\delta s} = \frac{g_{ab}\,dy^a\,\delta y^b}{\sqrt{g_{cd}\,dy^c\,dy^d}\,\sqrt{g_{ef}\,\delta y^e\,\delta y^f}}, \ (a, b... = 1, 2)\,.$$

as functions of the increments dy^a and δy^a.

19. In particular, all the angles of the geodesic triangle $AA'B$, constructed on the diameter, are zero. Whence $\varphi_{AA'B} = 0$.
20. We have here adopted the terminology of H. ARZELIÈS [8], p. 166.
Arzeliès replaces the definition of local time by a definition of central time t_e, which would be the time indicated by clocks running

$$\sqrt{1 - \frac{r^2\omega^2}{c^2}}$$

times slower. The central time t_c is identical with the time indicated by the clocks of S_0:

$$t_c = \frac{t}{\sqrt{1 - \dfrac{\omega^2 r^2}{c^2}}} = t_0\,.$$

21. Of course, if a clock regulated by H_1 moves along a circle S slowly enough so that its rate is not affected by that motion, its readings on going back to H_1 would coincide with those of H_1 but would differ by $q\tau_0$ from the data given by H.
22. Let us consider a moving body M starting out from M_0 and describing the circle of radius r with the uniform velocity V_0 with respect to S_0, and V with respect to S. At the end of the time dt_0 the axis OY of S moves to OY_1, and the moving body is at M' or M'', depending upon whether it is moving in the same or opposite direction as the rotating disc.
In system S_0, it will have covered the distance

$$(1) \qquad d\sigma_0 = V_0\,dt_0 = dl_0 \pm r_0\omega\,dt_0$$

dl_0 being the length M_1M' (or M_1M'') in S_0. One will then have

$$(2) \qquad V_0 = \frac{dl_0}{dt_0} \pm r_0\omega\,.$$

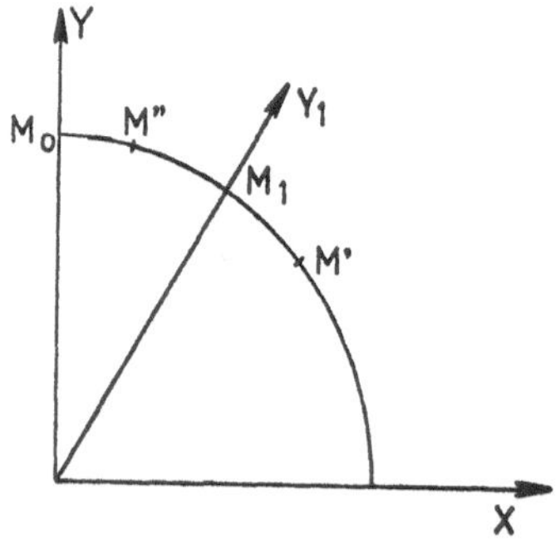

Fig. 41. The moving body M and the rotating disc.

In system S,

$$V = \frac{dl}{d\tau}. \tag{3}$$

By using the natural time and the natural geometry of the disc

$$dl = \frac{dl_0}{\sqrt{1 - \frac{r^2\omega^2}{c^2}}}, \quad d\tau = \sqrt{1 - \frac{r^2\omega^2}{c^2}}\left(dt_0 \mp \frac{\omega r^2 d\theta}{c^2\left(1 - \frac{r^2\omega^2}{c^2}\right)}\right) \tag{4}$$

and

$$dt_1 \sqrt{1 - \frac{r^2\omega^2}{c^2}} = r\, d\theta. \tag{5}$$

By comparing (2), (3), (4) and (5) we find

$$V = \frac{V_0 \mp r\omega}{1 \mp \frac{\omega r}{c^2} V_0}, \quad V_0 = \frac{V \pm r\omega}{1 \pm \frac{\omega r V}{c^2}}. \tag{6}$$

If two moving bodies leave together from M_0 with uniform velocities $+V$ and $-V$ with respect to S, the natural time required by each of them to cover dl_0 will be, according to (6) and (1):

$$(dt_0)_1 = \frac{dl_0}{(V_0)_1 - r\omega} = \frac{dl_0}{V} \frac{1 + \frac{\omega r}{c^2} V}{1 - \frac{\omega^2 r^2}{c^2}},$$

$$(dt_0)_2 = \frac{dl_0}{(V_0)_2 - r\omega} = \frac{-dl_0}{V} \frac{1 - \frac{\omega r}{c^2} V}{1 - \frac{\omega^2 r^2}{c^2}}.$$

The time-difference in covering the distance:

$$\Delta t_0 = (dt_0)_1 - (dt_0)_2 = \frac{2dl_0}{c^2} \frac{\omega r}{1 - \frac{\omega^2 r^2}{c^2}},$$

is independent of V. When the moving bodies have covered one entire circumference ($\int dl_0 = 2\pi r$), it is then:

$$\Delta t_0 = \frac{4\omega \mathscr{S}}{c^2\left(1 - \frac{\omega^2 r^2}{c^2}\right)} \neq \frac{4\omega \mathscr{S}}{c^2}.$$

In the system S attached to the rotating disc, the difference between the triptime for the two moving bodies M_1 and M_2 would be zero if one evaluated it in terms of the natural time of the systems attached to each of them (M_1 and M_2, in fact, cover the same distance with the same velocity). This natural time ought then to be measured by two clocks H_1 and H_2 accompanying each of the moving bodies and, at departure, synchronized with H fastened to S and then regulated with H by an exchange of light-signals. Each of the clocks H_1 and H_2 indicates the same time when the corresponding moving bodies M_1 and M_2 return to the point of departure. But H_1 (moving in the direction of the rotation of the disc) is slow by $2\omega\mathscr{S}/c^2$ and H_2 (moving counter-rotationally) is ahead by $2\omega\mathscr{S}/c^2$ with respect to H [cf. (XI-76)]. M_1, moving in the direction of the rotation of the disc, thus takes a longer time than M_2 to cover a complete circuit. The difference between the duration of the two trips, evaluated by the natural time of the disc and by the same clock H, is then

$$\frac{4\omega\mathscr{S}}{c^2}.$$

23. A detailed study of the motion of bodies moving on a rotating disc has been made especially by H. Arzeliès, who uses various reference systems and various definitions of time (cf. [8] p. 175).
    The comparison of the measurements made in system S_0 and in any accelerated system S is set forth completely in the book by C. Møller [16], p. 233. We shall refer to the examples studied by Møller.
24. A. Einstein [23].
25. D. Hilbert: *Gött. Nachr.* 1915, 395.
26. In fact, the tensor $S_\mu{}^\rho$, which is a function of the $g_{\mu\nu}$, and of their first and second order derivatives and satisfying (XI-87), can be written:

$$S_\mu{}^\sigma = h\left[G_\mu{}^\sigma - \frac{1}{2}\delta_\mu^\sigma(G - 2\lambda)\right].$$

    We suppose here $\lambda = 0$ and $h = 1$.
27. It can be shown that the trajectories of the charged particles are the geodesics of Finsler space-families characterized by the elementary interval

$$ds = \sqrt{g_{\mu\nu}dy^\mu dy^\nu} - \frac{e}{m}\varphi_\mu dy^\mu.$$

    Every Finsler space is determined by the value of e/m relative to every type of particle (cf. for example, A. Lichnerowicz [25], p. 155).

THE DEVELOPMENT OF GENERAL RELATIVITY AND SOME OF ITS CONSEQUENCES

A. THE EQUATIONS IN VARIOUS APPROXIMATIONS

1. *The gravitational potential in the Newtonian approximation*

Let us assume that the gravitational field is sufficiently weak so that the space-time differs very little from a Euclidean space. One can then choose a particular reference system x^μ such that the $g_{\mu\nu}$ differ very little from their Galilean values.

(XII-1) $$\eta_{\mu\nu} = \delta_{\mu\nu}(-1, -1, -1, +1).$$

We will then put

(XII-2) $$g_{\mu\nu} = \eta_{\mu\nu} + h_{\mu\nu}, \qquad g^{\mu\nu} = \eta^{\mu\nu} + h^{\mu\nu}$$

and neglect the squares and products of the $h_{\mu\nu}$ and their derivatives.

The geodesics of Riemannian space defined by the equations

(XII-3) $$\frac{d^2x^\rho}{ds^2} + \begin{Bmatrix} \rho \\ \mu\nu \end{Bmatrix} \frac{dx^\mu}{ds}\frac{dx^\nu}{ds} = 0$$

represent the trajectories described by a neutral particle in a gravitational field. Let us suppose that the velocity of this particle is small relative to that of light:

(XII-4) $$\frac{dx^p}{dt} \ll c \qquad \text{or} \qquad \frac{dx^p}{dx^0} \ll 1 \qquad (x^0 = ct),$$

where the Latin indices p, q, r still vary from one to three and the Greek indices μ, ν, ρ... take on the values 1, 2, 3, 0. One will have

(XII-5) $$\left(\frac{ds}{dx^0}\right)^2 = g_{pq}\frac{dx^p}{dx^0}\frac{dx^q}{dx^0} + 2\,g_{p0}\frac{dx^p}{dx^0} + g_{00} \simeq g_{00}$$

$$(p, q = 1, 2, 3)$$

and, consequently, according to (XII-4) and (XII-2):

$$\begin{aligned} \left\{ {p \atop \mu\nu} \right\} \frac{dx^\mu}{ds}\frac{dx^\nu}{ds} &\simeq \left\{ {p \atop 00} \right\} \left(\frac{dx^0}{ds} \right)^2 \simeq \\ &\simeq \frac{g^{pq}}{2\, g_{00}} (2\, \partial_0\, g_{0q} - \partial_q\, g_{00}) \simeq \\ &\simeq -(\partial^0\, g_{0p} - \tfrac{1}{2} \partial_p\, g_{00}) . \end{aligned} \qquad \text{(XII-6)}$$

If the masses are moving slowly enough, the derivatives involving $\partial_0 = (1/c)(\partial/\partial t)$ are small with respect to the derivatives involving $\partial_p = (\partial/\partial x^p)$ and may be neglected when they occur together with the latter type of derivative. So the three first equations (XII-3) are still written:

$$\frac{d^2 x^p}{ds^2} = -\tfrac{1}{2} \partial_p\, g_{00} . \qquad \text{(XII-7)}$$

Now, taking into account (XII-5)

$$\frac{d^2 x^p}{ds^2} \simeq \frac{d^2 x^p}{(dx^0)^2} = \frac{1}{c^2}\frac{d^2 x^p}{dt^2} . \qquad \text{(XII-8)}$$

By comparing (XII-7) and (XII-8), one can then write the approximate equation of the geodesics in the following form

$$\boxed{\frac{d^2 x^p}{dt^2} = -\frac{c^2}{2} \partial_p\, g_{00} = \frac{\partial U}{\partial x^p}} \qquad \text{(XII-9)}$$

putting

$$\boxed{U = -\frac{c^2}{2} h_{00} + \text{const.}} \qquad \text{(XII-10)}$$

Thus, *in a universe where masses are very small and move rather slowly* ($v \ll c$), *the trajectories of the particles* (geodesics) *are approximately those required by Classical Newtonian mechanics*: they result from a force that is the gradient of a potential U.

In the approximation accepted, the g_{00} component occurs only in the equations of motion, and that is why it is possible to assimilate it to a scalar U and to ignore the real tensor character of the potential of gravitation. In a rigorous theory, the potential of gravitation is correctly represented *by a symmetrical tensor of the second order*. Its ten components $g_{\mu\nu}$ enter into determining the motion of the masses.

2. *The equations of the gravitational field in a system of De Donder and quasi-Galilean coordinates*

Let us consider the equations of the field (XI-90) relative to the "interior" case (matter or electromagnetic field). These are the equations with a right-hand member

(XII-11) $$S_{\mu\nu} \equiv G_{\mu\nu} - \tfrac{1}{2} g_{\mu\nu} G = \chi T_{\mu\nu} .$$

They involve the Ricci tensor

(XII-12) $$G_{\mu\nu} = \partial_\rho \begin{Bmatrix} \rho \\ \mu\nu \end{Bmatrix} - \partial_\nu \begin{Bmatrix} \rho \\ \mu\rho \end{Bmatrix} + \begin{Bmatrix} \lambda \\ \mu\nu \end{Bmatrix} \begin{Bmatrix} \rho \\ \lambda\rho \end{Bmatrix} - \begin{Bmatrix} \lambda \\ \mu\rho \end{Bmatrix} \begin{Bmatrix} \rho \\ \lambda\nu \end{Bmatrix}$$

and the scalar curvature

(XII-13) $$G = g^{\mu\nu} G_{\mu\nu} .$$

One can easily prove that expressions (XII-12) and (XII-13) of $G_{\mu\nu}$ and G can also be written in the following manner:

(XII-14) $$\begin{aligned} G_{\mu\nu} = & - \tfrac{1}{2} g^{\rho\sigma} \partial_\rho \partial_\sigma g_{\mu\nu} - \tfrac{1}{2} \sigma^\rho \partial_\rho g_{\mu\nu} - \\ & - \tfrac{1}{2} (g_{\mu\rho} \partial_\nu \sigma^\rho + g_{\nu\rho} \partial_\mu \sigma^\rho) + \\ & + g^{\lambda\tau} g^{\rho\sigma} (\partial_\rho g_{\mu\lambda} \partial_\sigma g_{\nu\tau} - [\lambda\rho, \mu][\tau\sigma, \nu]) \end{aligned}$$

and

(XII-15)
$$G = - g^{\rho\sigma} \partial_\rho \partial_\sigma \log \sqrt{-g} \quad - \sigma^\rho \partial_\rho \log \sqrt{-g} \quad - \partial_\rho \sigma^\rho - \tfrac{1}{2} \begin{Bmatrix} \lambda \\ \rho\sigma \end{Bmatrix} \partial_\lambda g^{\rho\sigma} .$$

We put

(XII-16) $$[\mu\nu, \rho] = g_{\rho\sigma} \begin{Bmatrix} \sigma \\ \mu\nu \end{Bmatrix} = \tfrac{1}{2} (\partial_\mu g_{\nu\rho} + \partial_\nu g_{\mu\rho} - \partial_\rho g_{\mu\nu}) ,$$

and designate the following expression[1]

(XII-17) $$\sigma^\lambda = - g^{\rho\sigma} \begin{Bmatrix} \lambda \\ \rho\sigma \end{Bmatrix} = \frac{1}{\sqrt{-g}} \partial_\mu (\sqrt{-g}\, g^{\mu\lambda}) .$$

simply as σ^λ.

It will be noted that (XII-14) and (XII-15) may be simplified by choosing a coordinate system such that

(XII-18) $\sigma^{\lambda} = 0.$

These particular coordinates y^{ρ} are known as *isothermal or harmonic coordinates.* They can also be defined in the following manner: expression (XII-17) for the σ^{λ} may also be written:

(XII-19)
$$\Box f^{(\lambda)} = g^{\rho\sigma}\nabla_{\rho}\nabla_{\sigma} f^{(\lambda)} = g^{\rho\sigma}\left(\partial_{\rho}\partial_{\sigma} - \begin{Bmatrix}\tau \\ \rho\sigma\end{Bmatrix}\partial_{\tau}\right) f^{(\lambda)} = = - g^{\rho\sigma}\begin{Bmatrix}\lambda \\ \rho\sigma\end{Bmatrix} = \sigma^{\lambda}$$

with

$$\partial_{\tau} f^{(\lambda)} = \partial_{\tau} y^{\lambda} = \delta^{\lambda}_{\tau}.$$

If the four functions $y^{\lambda} = f^{(\lambda)}$ are solutions of (XII-18), that is, of $\Box f^{(\lambda)} = 0$, they define characteristic isothermal varieties $f^{(\lambda)}$ = constant.

The condition of isothermality thus amounts to adopting a coordinate system such that[2]

(XII-20) $\Box y^{\lambda} = 0 \quad \text{with} \quad \Box = \nabla^{\rho}\nabla_{\rho} = g^{\rho\sigma}\nabla_{\rho}\nabla_{\sigma}.$

With that condition, the gravitational equations (XII-11) have the simple form:

(XII-21)
$$- \tfrac{1}{2} g^{\rho\sigma}\partial_{\rho}\partial_{\sigma} g_{\mu\nu} + g^{\lambda\tau} g^{\rho\sigma}(\partial_{\rho} g_{\mu\lambda}\partial_{\sigma} g_{\nu\tau} - [\lambda\rho, \mu][\tau\sigma, \nu]) + + \tfrac{1}{2} g_{\mu\nu} g^{\rho\sigma}\partial_{\rho}\partial_{\sigma}\log\sqrt{-g} + \tfrac{1}{4} g_{\mu\nu}\begin{Bmatrix}\lambda \\ \rho\sigma\end{Bmatrix}\partial_{\lambda} g^{\rho\sigma} = \chi T_{\mu\nu}.$$

The Quasi-Galilean approximation. Let us now assume that the gravitational field is very weak. It is then possible to select a system of coordinates such that the $g_{\mu\nu}$ differ very little from their Galilean values

(XII-1) $\eta_{\mu\nu} = \delta_{\mu\nu}(-1, -1, -1, +1).$

We shall then put[3]

(XII-22)$_1$ $g_{00} = 1 + h_{00} = 1 + \varepsilon^2 \underset{2}{h}_{00} + 0(\varepsilon^4)$

(XII-22)$_2$ $g_{p0} = h_{p0} = \varepsilon^3 \underset{3}{h}_{p0} + 0(\varepsilon^5) \qquad (p, q = 1, 2, 3)$

(XII-22)$_3$ $g_{pq} = -\delta_{pq} + h_{pq} = -\delta_{pq} + \varepsilon^2 \underset{2}{h}_{pq} + 0(\varepsilon^4)$

with

(XII-23) $\varepsilon^2 = \dfrac{1}{c^2}.$

We shall say that a system of equations is *of the second order* if it is limited to the infinitesimals ε^2 and ε^3. In such a system, then, only the terms in $\underset{2}{h}$ and $\underset{3}{h}$ occur – terms whose products ($\leqslant \varepsilon^4$) are negligible.

It can be shown that *it is always possible to select a quasi-Galilean system that is also an isothermal system.*[4] The choice of a quasi-Galilean system thus results in the vanishing of the σ^λ for every order of approximation. Under these circumstances, the gravitational equations still have the simple form (XII-21) and, in expressing them, one can substitute expansions (XII-22) of the $g_{\mu\nu}$.

These results must be completed by the following, which are easily obtained from (XII-22):

(XII-24) $$g = \text{determinant}\, g_{\mu\nu} = -1 - \varepsilon^2 (\underset{2}{h}_{00} - \Sigma_p \underset{2}{h}_{pp}) + 0(\varepsilon^4)$$

(XII-25)$_1$ $$g^{00} = \frac{1}{g}\,\text{cofactor}\, g_{00} = 1 - \varepsilon^2 \underset{3}{h}_{00} + 0(\varepsilon^4)$$

(XII-25)$_2$ $$g^{0p} = \frac{1}{g}\,\text{cofactor}\, g_{0p} = \varepsilon^3 \underset{3}{h}_{p0} + 0(\varepsilon^5)$$

(XII-25)$_3$ $$g^{pq} = \frac{1}{g}\,\text{cofactor}\, g_{pq} = -\delta_{pq} - \varepsilon^2 \underset{2}{h}_{pq} + 0(\varepsilon^4).$$

By substituting (XII-22), (XII-24) and (XII-25) in (XII-21) we obtain, for the chosen approximation,

(XII-26)$_1$ $$-\tfrac{1}{2}\eta^{\rho\sigma}\partial_\rho\partial_\sigma\left[\underset{2}{h}_{00} - \tfrac{1}{2}(\underset{2}{h}_{00} - \sum_p \underset{2}{h}_{pp})\right] = \frac{\chi}{\varepsilon^2} T_{00}$$

(XII-26)$_2$ $$-\tfrac{1}{2}\eta^{\rho\sigma}\partial_\rho\partial_\sigma \underset{3}{h}_{p0} = \frac{\chi}{\varepsilon^3} T_{p0}$$

(XII-26)$_3$ $$-\tfrac{1}{2}\eta^{\rho\sigma}\partial_\rho\partial_\sigma\left[\underset{2}{h}_{pq} + \tfrac{1}{2}\delta_{pq}(\underset{2}{h}_{00} - \sum_r \underset{2}{h}_{rr})\right] = \frac{\chi}{\varepsilon^2} T_{pq}.$$

To these equations, the condition of isothermality (XII-18) must be added. We write it in the following form:

(XII-27) $$\sigma_\mu = g_{\mu\nu}\sigma^\nu = -g^{\rho\sigma}[\rho\sigma, \mu] = 0$$

that is

(XII-28) $$\sigma_\mu = -g^{\rho\sigma}(\partial_\rho g_{\sigma\mu} - \tfrac{1}{2}\partial_\mu g_{\rho\sigma}) = 0.$$

In a quasi-Galilean system, one then has

$$\text{(XII-29)} \qquad -\underset{2}{\sigma_\mu} = \eta^{\rho\sigma}\partial_\rho \underset{2}{h_{\sigma\mu}} - \tfrac{1}{2}\partial_\mu \underset{2}{h} = 0$$

putting

$$\text{(XII-30)} \qquad \underset{2}{h} = \eta^{\rho\sigma} \underset{2}{h_{\rho\sigma}} .$$

(XII-28) may also be written

$$\text{(XII-31)} \qquad \eta^{\rho\sigma}\partial_\rho [\underset{2}{h_{\sigma\mu}} - \tfrac{1}{2}\eta_{\sigma\mu} \underset{2}{h}] = 0 .$$

If the gravitational field is weak, it satisfies equations (XII-26) and (XII-31), valid to the second order of approximation in an isothermal quasi-Galilean system.

Putting

$$\text{(XII-32)} \qquad \underset{p}{\gamma_{\mu\nu}} = \underset{p}{h_{\mu\nu}} - \tfrac{1}{2}\eta_{\mu\nu} \underset{p}{h} \qquad (\underset{p}{h} = \eta^{\rho\sigma} \underset{p}{h_{\rho\sigma}})$$

and, conversely,

$$\text{(XII-33)} \qquad \underset{p}{h_{\mu\nu}} = \underset{p}{\gamma_{\mu\nu}} - \tfrac{1}{2}\eta_{\mu\nu} \underset{p}{\gamma} \qquad (\underset{p}{\gamma} = \eta^{\rho\sigma} \underset{p}{\gamma_{\rho\sigma}}),$$

one notes that the first-order equations (XII-26) and (XII-31) then have the following form:

$$\text{(XII-34)}_1 \qquad -\tfrac{1}{2}\eta^{\rho\sigma}\partial_\rho\partial_\sigma \underset{2}{\gamma_{00}} = \frac{\chi}{\varepsilon^2} T_{00}$$

$$\text{(XII-34)}_2 \qquad -\tfrac{1}{2}\eta^{\rho\sigma}\partial_\rho\partial_\sigma \underset{3}{\gamma_{p0}} = \frac{\chi}{\varepsilon^3} T_{p0}$$

$$\text{(XII-34)}_3 \qquad -\tfrac{1}{2}\eta^{\rho\sigma}\partial_\rho\partial_\sigma \underset{2}{\gamma_{pq}} = \frac{\chi}{\varepsilon^2} T_{pq}$$

and

$$\text{(XII-35)} \qquad \eta^{\rho\sigma}\partial_\rho \underset{p}{\gamma_{\sigma\mu}} = 0 .$$

These relations may be further written as

$$\text{(XII-36)} \qquad \mathring{\Box}\, \gamma_{\mu\nu} = -2\chi T_{\mu\nu}$$

$$\text{(XII-37)} \qquad \eta^{\rho\sigma}\partial_\rho \gamma_{\sigma\mu} = 0$$

putting

(XII-38) $$\overset{\circ}{\square} = \eta^{\rho\sigma}\partial_\rho\partial_\sigma = \partial_0^2 - \sum_p \partial_p^2$$

and

(XII-39) $$\gamma_{pq} = \varepsilon^2 \underset{2}{\gamma}_{pq}, \qquad \gamma_{p0} = \varepsilon^3 \underset{3}{\gamma}_{p0}, \qquad \gamma_{00} = \varepsilon^2 \underset{2}{\gamma}_{00}.$$

3. *Application to a continuous material medium treated as a perfect gas*

The tensor $T_{\mu\nu}$ represents the contribution of the sources of the field. Let us suppose that these sources are made up of uncharged particles forming a continuous medium of the perfect-gas type. $T_{\mu\nu}$ does not include any term of electromagnetic origin (which would be represented by the Maxwell tensor $\tau_{\mu\nu}$) and reduces to the matter tensor $M_{\mu\nu}$ relative to a perfect fluid (Cf. VIII-166).

(XII-40) $$T_{\mu\nu} = M_{\mu\nu} = (\mu_0 c^2 + p)\, u_\mu u_\nu - p g_{\mu\nu}.$$

The expression for it involves the world four-velocities

(XII-41)$_1$ $$u_p = g_{pq}u^q + g_{p0}u^0 = \left(g_{pq}\frac{v^q}{c} + g_{p0}\right)u^0, \qquad p, q = 1, 2, 3$$

(XII-41)$_2$ $$u_0 = g_{p0}u^p + g_{00}u^0 = \left(g_{p0}\frac{v^p}{c} + g_{00}\right)u^0$$

putting

(XII-42) $$v^p = \frac{dx^p}{dt}, \quad u^p = \frac{dx^p}{ds}, \quad u^0 = \frac{dx^0}{ds} = c\frac{dt}{ds} = \frac{dt}{d\tau} = \frac{1}{\sqrt{1-\beta^2}}.$$

By substituting (XII-41) in (XII-40), and by limiting ourselves to the largest terms, we then obtain

(XII-43)$_1$ $$M_{00} \simeq \mu_0 c^2$$

(XII-43)$_2$ $$M_{p0} \simeq \mu_0 v_p c$$

(XII-43)$_3$ $$M_{pq} \simeq \mu_0 v_p v_q + p\delta_{pq}.$$

Let us insert (XII-43) into the three equations (XII-34); one will then have for the second order

(XII-44)$_1$ $$\overset{\circ}{\square}\, \underset{2}{\gamma}_{00} = -\,2\chi\mu_0 c^4$$

$$\text{(XII-44)}_2 \qquad \overset{\circ}{\Box} \underset{3}{\gamma}_{p0} = -2\chi\mu_0 v_p c^4$$

$$\text{(XII-44)}_3 \qquad \overset{\circ}{\Box} \underset{2}{\gamma}_{pq} = -2\chi c^2 (\mu_0 v_p v_q + p\delta_{pq}).$$

The right-hand members of (XII-44) will be finite if χ is of the order $\varepsilon^4 = 1/c^4$. With

$$\text{(XII-45)} \qquad \chi = \varepsilon^4 \chi_1 = \frac{\chi_1}{c^4}$$

equations (XII-44) will be written

$$\text{(XII-46)}_1 \qquad \overset{\circ}{\Box} \underset{2}{\gamma}_{00} = -2\chi_1\mu_0$$

$$\text{(XII-46)}_2 \qquad \overset{\circ}{\Box} \underset{3}{\gamma}_{p0} = -2\chi_1\mu_0 v_p$$

$$\text{(XII-46)}_3 \qquad \overset{\circ}{\Box} \underset{2}{\gamma}_{pq} = 0.$$

The static solutions. Let us seek the static solutions of equations (XII-46), that is, the $\underset{p}{h}_{\mu\nu}$ solutions not dependent on time. We must solve the following systems:

$$\text{(XII-47)}_1 \qquad \Delta \underset{2}{\gamma}_{00} = 2\chi_1\mu_0$$

$$\text{(XII-47)}_2 \qquad \Delta \underset{3}{\gamma}_{p0} = 2\chi_1\mu_0 v_p$$

$$\text{(XII-47)}_3 \qquad \Delta \underset{2}{\gamma}_{pq} = 0$$

because

$$\text{(XII-48)} \qquad \overset{\circ}{\Box} = \partial_0^2 - \Delta, \qquad \Delta = \sum_p \partial_p^2.$$

Equation $(\text{XII-47})_3$ is satisfied by the solution

$$\text{(XII-49)} \qquad \underset{2}{\gamma}_{pq} = \underset{2}{h}_{pq} - \tfrac{1}{2}\eta_{pq}\underset{2}{h} = \underset{2}{h}_{pq} + \tfrac{1}{2}\delta_{pq}(\underset{2}{h}_{00} - \sum_r \underset{2}{h}_{rr}) = 0$$

which leads, by summation, to

$$\text{(XII-50)} \qquad \sum_p \underset{2}{\gamma}_{pp} = \tfrac{3}{2}\underset{2}{h}_{00} - \tfrac{1}{2}\sum_p \underset{2}{h}_{pp} = 0.$$

Let us put, as in (XII-10)

$$\text{(XII-51)} \qquad U = -\frac{c^2}{2} h_{00} = -\tfrac{1}{2} \underset{2}{h_{00}} + 0(\varepsilon^4).$$

We will then obtain from (XII-50)

$$\text{(XII-52)} \qquad \underset{2}{h_{00}} = -2U$$

$$\text{(XII-53)} \qquad \sum_p \underset{2}{h_{pp}} = -6U$$

and, by substitution in (XII-49) and in (XII-32)

$$\text{(XII-54)} \qquad \underset{2}{h_{pq}} = -2U\delta_{pq}$$

$$\text{(XII-55)} \qquad \underset{2}{\gamma_{00}} = \underset{2}{h_{00}} - \tfrac{1}{2}\eta_{00}(\underset{2}{h_{00}} - \sum_r \underset{2}{h_{rr}}) = \tfrac{1}{2}(\underset{2}{h_{00}} + \sum_r \underset{2}{h_{rr}}) = -4U.$$

Equation $(\text{XII-47})_1$ is then written

$$\text{(XII-56)} \qquad \Delta U = -\frac{\chi_1}{2}\mu_0 .$$

It is identical to Poisson's equation

$$\text{(XII-23)} \qquad \Delta U = -4\pi G \mu_0$$

deduced from the Newtonian law of gravity. One has only to put

$$\text{(XII-57)} \qquad \chi_1 = 8\pi G$$

that is[5]

$$\text{(XII-58)} \qquad \boxed{\chi = \frac{\chi_1}{c^4} = \frac{8\pi G}{c^4}.}$$

Now, according to (XI-14)

$$\text{(XII-59)} \qquad G = 6.664 \cdot 10^{-8}\,\text{cm}^3\,\text{g}^{-1}\,\text{sec}^{-2}.$$

The value of χ will then be

$$\text{(XII-60)} \qquad \chi = \frac{8\pi G}{c^4} = 2.073 \cdot 10^{-48}\,\text{cm}^{-1}\,\text{g}^{-1}\,\text{sec}^{-1}.$$

Under these conditions, the system (XII-47) may also be written:

(XII-61) $$\boxed{\Delta U = -\frac{\chi c^4}{2}\mu_0 = -4\pi G\mu_0}$$

(XII-62) $$\underset{3}{\Delta \gamma_{p0}} = 2\chi c^4 \mu_0 v_p = 16\pi G\mu_0 v_p .$$

Within the allowable approximation, the potential U created at point $P(r)$ by a continuous static distribution in the volume $d\mathscr{V}$ surrounding $M'(r')$ is

(XII-63) $$\boxed{U = \frac{\chi c^4}{8\pi}\int \frac{\mu_0(r')\, d\mathscr{V}'}{|r - r'|} .}$$

The potential U created at $P(r)$ by N bodies $A, B\ldots$ with mass-densities $\mu_A, \mu_B\ldots$ will then be:

(XII-64) $$U = \sum_{A=1}^{N} U_A ,$$

using U_A to designate the solution of

(XII-65) $$\Delta U_A = -\frac{\chi c^4}{2}\mu_A = -4\pi G\mu_A$$

and putting

(XII-66) $$\mu_0 = \sum_{A=1}^{N} \mu_A .$$

On the other hand, if $\overset{A}{v}_p$ represents the components of the velocity of body A, (XII-62) may be further written

(XII-67) $$\underset{3}{\Delta \gamma_{p0}} = 16\pi G \sum_{A=1}^{N} \mu_A \overset{A}{v}_p = -4\sum_{A=1}^{N} \overset{A}{v}_p \Delta U_A$$

taking (XII-65) into account.

Now $\overset{A}{v}_p$ is practically constant over body A. One then has

(XII-68) $$\underset{3}{\Delta \gamma_{p0}} = -4\sum_{A=1}^{N} \Delta(\overset{A}{v}_p U_A)$$

that is

(XII-69) $$\underset{3}{\gamma_{p0}} = \underset{3}{h_{p0}} = -4\sum_{A=1}^{N} \overset{A}{v}_p U_A = 4\sum_{A=1}^{N} \overset{A}{v}{}^p U_A .$$

Let us note that the density μ_0 is introduced into equation (XII-47) by means of the matter tensor $M_{\mu\nu}$. So we are concerned with a *density of an inertial mass.* On the other hand, in the definition of the Newtonian potential (XI-23), μ represents a *density relative to the gravitational mass.* So one of the consequences of the principles of the theory is the identity between gravitational mass and inertial mass.

The identity between gravitational mass and inertial mass. The *inertial mass* of a particle expresses the particle's *reaction to inertial forces.*

The *gravitational mass* of a test-particle represents, in a similar manner, the *reaction* of the particle *to the gravitational forces* exerted on it. In addition, by virtue of the very form of the Newtonian law of gravity, that law also determines *the ability of a body to create a certain gravitational field.*

Now, depending on whether it expresses the behavior of a test-particle subjected to a gravitational field, or the production of a field by one or more sources, the equivalence principle will take on one or the other of the two following forms:

a) A neutral particle subjected to the gravitational field only describes the geodesics of a Riemannian space. The characteristics of this particular Riemannian space and, consequently, the trajectory of the test-particle, are determined by the equations $S_{\mu\nu} = \chi T_{\mu\nu}$, which are independent of the data concerning the test-particle.

So far as the test-particle is concerned, the postulate of geodesics is the expression of the principle of equivalence existing between the inertial and gravitational forces. It implies, by the very terms in which it is stated, an identity between the inertial and gravitational masses, the gravitational mass being defined as a reaction of the particle to gravitational forces. Any distinction between gravitational mass and inertial mass is, by definition, pointless.

b) If a gravitational field appears, for example, inside a continuous medium, the equations

$$\text{(XI-92)} \qquad S_{\mu\nu} \equiv G_{\mu\nu} - \tfrac{1}{2} g_{\mu\nu} G = \chi T_{\mu\nu}$$

determine the behavior of the field, in other words, they tell us what kind of Riemannian space results from the presence of a material medium.

Now, in Equations (XI-92), the energy-momentum tensor $T_{\mu\nu}$ causes the appearance (Cf. XII-40) of the *inertial masses* m_i of the particles (or the *density of the inertial mass* μ_0 of the medium) that create the field.

The approximate equations (XII-47) or (XII-56) have (XII-63) as a possible solution. Assuming that the field is weak and created by bodies $A, B, C\ldots$ sufficiently far apart, we will have

(XII-70) $$g_{\mu\nu} = \eta_{\mu\nu} + \frac{1}{c^2} \underset{2}{h}_{\mu\nu}, \qquad \underset{2}{h}_{\mu\nu} = -2\delta_{\mu\nu} \sum_{A=1}^{N} U_A$$

with

(XII-71) $$U_A = \frac{\chi c^4}{8\pi} \int \frac{(\mu_A)_i \, d\mathscr{V}'}{|r - r'|} \simeq \frac{\chi c^4}{8\pi r} (m_A)_i$$

$(m_A)_i$ here representing the *inertial mass* of the body A creating the field.

On the other hand, if the gravitational law relative to the external case

(XI-83) $$G_{\mu\nu} = 0$$

is applicable, one should substitute for (XII-56) the approximation

(XII-72) $$\Delta U = 0.$$

One then obtains the solution

(XII-73) $$U_A = \frac{G \sum_{A=1}^{N} (m_A)_g}{r}$$

$(m_A)_g$ is a constant occurring in the integration of the Equations (XII-72). This constant of integration depends solely on the properties of the body creating the gravitational field: except for the universal constant (G), this constant of integration thus represents the gravitational mass $(m_A)_g$ of the body.[6]

Now, within the approximation considered (a weak field and sufficiently separated masses), the two solutions (XII-71) and (XII-73) are identical. One will then have

(XII-74) $$Gm_g = \frac{\chi c^4}{8\pi} m_i$$

that is

(XII-75) $$m_g = m_i$$

putting[7]

(XII-58) $$\chi = \frac{8\pi G}{c^4}\,.$$

Hence, these are the conditions for reconciling the solutions of (XI-92) and of (XI-83) corresponding to the interior (pure matter) and the exterior cases, which constitute the expression of the principle of equivalence for fields created by a distribution of matter.[8] These conditions, after an adjustment of universal constants, express the identity between the inertial and the gravitational mass of the particles or the distribution that creates the field.

4. *Equations of the exterior case*

The gravitational equations valid outside matter and in the absence of any electromagnetic field are easily obtained from (XII-47) by setting the tensor $T_{\mu\nu}$ equal to zero.

1) *Static solutions.* If we limit ourselves to the static case and to the Newtonian approximation, we obtain in the exterior case, instead of (XII-47)

(XII-76) $$\Delta \underset{p}{\gamma_{\mu\nu}} = 0\,.$$

But the isothermal conditions (XII-37) remain unchanged

(XII-37) $$\eta^{\sigma\rho}\partial_\rho \gamma_{\sigma\mu} = 0\,.$$

A particular solution of these equations may be sought by arbitrarily combining, by means of constant coefficients, terms in $1/r$, in $\partial_\rho\,(1/r)$ and in $\partial_p\partial_q\,(1/r)$.[9]
Thus:

(XII-77) $$\underset{2}{\gamma_{00}} = -\frac{4a}{r}, \qquad a = \text{const}\,,$$

the other $\underset{2}{\gamma_{\mu\nu}}$ components being zero. By referring to the definitions (XII-33) and (XII-30) one then has

(XII-78) $$\underset{2}{h_{00}} = \underset{2}{\gamma_{00}} - \tfrac{1}{2}\eta_{00}\underset{2}{\gamma} = \tfrac{1}{2}\underset{2}{\gamma_{00}}$$

(XII-79) $$\underset{2}{h_{pq}} = \underset{2}{\gamma_{pq}} - \tfrac{1}{2}\eta_{pq}\underset{2}{\gamma} = \tfrac{1}{2}\delta_{pq}\underset{2}{\gamma_{00}}\,.$$

The only non-zero $\underset{2}{h}_{\mu\nu}$ will thus be:

(XII-80) $$\underset{2}{h}_{11} = \underset{2}{h}_{22} = \underset{2}{h}_{33} = \underset{2}{h}_{00} = -\frac{2a}{r}.$$

Now, according to (XII-10),

(XII-10) $$U = -\frac{c^2}{2}h_{00} = -\tfrac{1}{2}\underset{2}{h}_{00} = -\tfrac{1}{4}\gamma_{00} = \frac{a}{r}.$$

We will then put

(XII-81) $$a = Gm'$$

if we are concerned with the potential created by a mass m'. From this, there results

(XII-82) $$\frac{a}{r} = \frac{Gm'}{r} = U.$$

With this choice of coordinates, and within the chosen approximation

$$g_{\mu\nu} = \eta_{\mu\nu} + \frac{1}{c^2}\underset{2}{h}_{\mu\nu} + 0(\varepsilon^4).$$

The ds^2 corresponding to (XII-80) is written simply

(XII-83) $$ds^2 = -\left(1 + \frac{2U}{c^2}\right)(dx^2 + dy^2 + dz^2) + \left(1 - \frac{2U}{c^2}\right)c^2\,dt^2.$$

2) *Gravitational waves.* If we consider, on the other hand, the non-static solutions of the exterior case, one can predict the existence of gravitational waves. These would arise in a rapidly varying field. They have no equivalent in the Newtonian theory.

If we limit ourselves to the approximate equations

(XII-84) $$\Box\underset{2}{\gamma}_{\mu\nu} = 0,$$

one can determine plane-wave solutions. A plane wave propagated along ox is described by the solutions

(XII-85) $$\underset{2}{\gamma}_{\mu\nu} = \underset{2}{\gamma}_{\mu\nu}(x^1 - x^0).$$

The conditions imposed by coordinates make it possible to introduce other restrictions. It can be shown[10] that a change of coordinates makes it possible to describe plane waves solely by means of the non-zero components

$$\text{(XII-86)} \qquad \underset{2}{\gamma_{22}} = -\underset{2}{\gamma_{33}} \quad \text{or} \quad \underset{2}{\gamma_{23}},$$

the alternative here being reduced, moreover, to the choice $\underset{2}{\gamma_{22}} = -\underset{2}{\gamma_{33}}$ by a rotation of the axes through $\pi/4$ around *ox*.

The plane waves, moreover, disappear the moment one goes back to the rigorous field-equations. The rigorous solutions that are obtainable represent cylindrical waves. These waves have been studied by EINSTEIN and ROSEN[11] and then, more recently, by W. B. BONNOR[12]. P. G. BERGMANN has proposed the following explanation for the absence of plane gravitational waves[13]: these waves transport an energy that produces a stationary gravitational field. Now such a field, according to the principles of the theory, involves a modification of the geometry. A plane wave would transport a constant and finite energy at every point. The infinite deformation of space-time (starting from Euclidean data) in all directions would become possible.

Let us add that there is no experimental evidence for gravitational waves and that their interest – which, however, is considerable – remains purely theoretical.[14]

5. *Equations of the field and motion of the sources*

The equations of the gravitational field and those of the electromagnetic field are of different types and give rise to dissimilar consequences. The former are non-linear equations; they involve the products of the gravitational potentials and of their first-order derivatives. The latter form a linear system in which the products of the fields and their first derivatives do not occur, at least when we limit ourselves to the Maxwellian formalism.

And so the relationship between the field and its sources is very different in the two theories.

Let us consider, for example, an ensemble of charged particles. One could, in principle, separate the total electromagnetic field from the field created by each separate particle. In fact, both these fields, and their difference, are solutions of the linear system formed by the equations related to the electromagnetic field.[15] Moreover, the expression of the force acting on a charged particle (Lorentz force) is totally independent of the postulate on which the Maxwell equations are based. Nor would the occurrence of a non-electromagnetic force modify the field equations.

On the other hand, one cannot rigorously separate the gravitational field acting on a specific particle in the total field produced by several particles. The difference between these two fields would not be a solution of the (non-linear) equations of gravitation. In addition, the field equations are not independent of the force of attraction acting on a particle possessing weight; if a non-gravitational force acts on this particle, it modifies the tensor $T_{\mu\nu}$ and, consequently, the field equations.

The motion of a particle, which, in classical electrodynamics, cannot be deduced from the field equations, is thus intimately connected with them in a non-linear theory such as General Relativity. Indeed, it can be shown, by two quite different methods, that it is possible to deduce the motion of a neutral particle from the postulates upon which the non-linear equations of the gravitational field are based. The equations of motion then appear as conditions that must be verified for every order of approximation if the validity of the corresponding field equations is to be maintained.

1) *Deduction of the equations of motion from the field equations without a right-hand member: the "singularity" method.* a) *The field equations for different orders of approximation.* The singularity method was developed chiefly by EINSTEIN, INFELD and HOFFMANN.[16] It consists in using as a point of departure the field equations written for the *exterior case*; the matter tensor $M_{\mu\nu}$ does not appear. So the sources appear as singularities in the field, and the equations devoid of a right-hand member are valid outside the closed surfaces surrounding these singularities.

It is convenient to substitute for the field equations

(XI-83) $$G_{\mu\nu} = 0$$

the equivalent system

(XII-87) $$S^{*}_{\mu\nu} \equiv G_{\mu\nu} - \tfrac{1}{2}\eta_{\mu\nu}\eta^{\rho\sigma}G_{\rho\sigma} = 0\,.$$

In fact, if we put, as we did in (XII-1) and (XII-2)

$$g_{\mu\nu} = \eta_{\mu\nu} + h_{\mu\nu}, \qquad \eta_{\mu\nu} = \delta_{\mu\nu}(-1, -1, -1, +1), \tag{XII-88}$$

the substitution of (XII-88) in (XII-87) gives us the expressions

$$\gamma_{\mu\nu} = h_{\mu\nu} - \tfrac{1}{2}\eta_{\mu\nu}\eta^{\rho\sigma}h_{\rho\sigma} \tag{XII-89}$$

defined for the second order by (XII-32). By substituting (XII-88) in (XII-87) one obtains, in fact, the following system[17]

$$\partial_p^2\gamma_{00} + 2\Lambda_{00} = 0 \tag{XII-90$_1$}$$

$$\partial_p^2\gamma_{0r} - \partial_p\partial_r\gamma_{0p} + 2\Lambda_{0r} = 0 \tag{XII-90$_2$}$$

$$\partial_p^2\gamma_{rs} - \partial_p\partial_r\gamma_{ps} - \partial_p\partial_s\gamma_{pr} + \delta_{rs}\partial_p\partial_q\gamma_{pq} + 2\Lambda_{rs} = 0 \tag{XII-90$_3$}$$

with

$$2\Lambda_{00} = -\partial_p\partial_q\gamma_{pq} + G'_{pp} + G'_{00} \tag{XII-91$_1$}$$

$$2\Lambda_{0r} = \partial_r\partial_0\gamma_{00} - \partial_0\partial_p\gamma_{pr} + 2G'_{0r} \tag{XII-91$_2$}$$

$$\begin{aligned}2\Lambda_{rs} = {} & \partial_0^2(\delta_{rs}\gamma_{00} - \gamma_{rs}) - 2\delta_{rs}\partial_0\partial_p\gamma_{p0} + \partial_0\partial_r\gamma_{s0} + \\ & + \partial_0\partial_s\gamma_{r0} + 2\left(G'_{rs} + \tfrac{1}{2}\delta_{rs}(G'_{00} - G'_{pp})\right).\end{aligned} \tag{XII-91$_3$}$$

The expressions $G'_{\mu\nu}$ represent the quadratic quantities formed from the $g_{\mu\nu}$ and their first derivatives.

Still assuming that the gravitational field is weak ($h_{\mu\nu} \ll 1$), it is then possible to adopt a quasi-Galilean reference system and to develop the $g_{\mu\nu}$ as we did in (XII-22), according to the increasing powers of an infinitesimally small parameter $\varepsilon = 1/c$.

In keeping with the results of the previous section, one will obtain for the $\gamma_{\mu\nu}$ developments having the following form:

$$\gamma_{00} = \varepsilon^2 \underset{2}{\gamma_{00}} + \dots + \varepsilon^{2l-2} \underset{2l-2}{\gamma_{00}} + 0(\varepsilon^{2l}) \tag{XII-92$_1$}$$

$$\gamma_{0p} = \varepsilon^3 \underset{3}{\gamma_{0p}} + \dots + \varepsilon^{2l-1} \underset{2l-1}{\gamma_{0p}} + 0(\varepsilon^{2l+1}) \tag{XII-92$_2$}$$

$$\gamma_{pq} = \varepsilon^4 \underset{4}{\gamma_{pq}} + \dots + \varepsilon^{2l} \underset{2l}{\gamma_{pq}} + 0(\varepsilon^{2l+2}). \tag{XII-92$_3$}$$

These developments are substituted in (XII-90) by admitting a *quasi-*

static approximation, that is, by assuming that the order of the derivatives $\partial_0 = (1/c)(\partial/\partial t)$ is one degree higher than that of the derivatives $\partial_p = (\partial/\partial x^p)$. With this hypothesis, (XII-90) reduces to the approximate equations of the order l:

$$\text{(XII-93)}_1 \qquad \partial_p^2 \underset{2l-2}{\gamma_{00}} + 2 \underset{2l-2}{\Lambda_{00}} = 0$$

$$\text{(XII-93)}_2 \qquad \partial_p^2 \underset{2l-1}{\gamma_{0r}} - \partial_p \partial_r \underset{2l-1}{\gamma_{0p}} + 2 \underset{2l-1}{\Lambda_{0r}} = 0$$

$$\text{(XII-93)}_3 \qquad \partial_p^2 \underset{2l}{\gamma_{rs}} - \partial_p \partial_r \underset{2l}{\gamma_{ps}} - \partial_p \partial_s \underset{2l}{\gamma_{pr}} + \delta_{rs} \partial_p \partial_q \underset{2l}{\gamma_{pq}} + 2 \underset{2l}{\Lambda_{rs}} = 0 .$$

b) *Integration of the field equations.* Let us suppose that the matter is formed of $1 \ldots k \ldots n$ singularities, and let us integrate equations (XII-93) over a closed surface S surrounding each singularity with the coordinates $\overset{k}{\xi}{}^p$.

Equation (XII-93)_1, in which Λ_{00} is known for every order of approximation, makes it possible to determine γ_{00}, but that determination can only be effected for the order $2l - 2$ to within a harmonic function $\psi = 1/r$. One can thus predict the contribution of k poles $(\bar{\gamma}_{00}^{(p)})$, and of k dipoles $(\bar{\gamma}_{00}^{(d)})$, which will be expressed by solutions of the type

$$\text{(XII-94)} \qquad \bar{\gamma}_{00} = \bar{\gamma}_{00}^{(p)} + \bar{\gamma}_{00}^{(d)} .$$

A solution of (XII-93) will thus be

$$\text{(XII-95)} \qquad \gamma'_{00} = \gamma_{00} + \bar{\gamma}_{00} ,$$

where the barred notations refer to the solutions introduced as a result of the presence of singularities.

In order to find solutions that apply to the presence of particles, we will suppose that the contributions $\bar{\gamma}_{00}^{(p)}$ and $\bar{\gamma}_{00}^{(d)}$ of the poles and dipoles have the following form:

$$\text{(XII-96)} \qquad \underset{2l-2}{\bar{\gamma}_{00}^{(p)}} = - 4 \underset{2l-2}{\overset{k}{m} \overset{k}{\psi}}$$

$$\text{(XII-97)} \qquad \underset{2l-2}{\bar{\gamma}_{00}^{(d)}} = - \underset{2l-2}{\overset{k}{S}_r \partial_r \overset{k}{\psi}}$$

$\overset{k}{m}$ and $\overset{k}{S}_r$ are here functions of time and $\overset{k}{\psi}$ a harmonic function relative to the kth dipole $(\overset{k}{\xi^p})$:

$$\text{(XII-98)} \qquad \overset{k}{\psi} = \frac{1}{\overset{k}{r}} \qquad \text{with} \qquad (\overset{k}{r})^2 = (x^p - \overset{k}{\xi^p})(x^p - \overset{k}{\xi^p}).$$

Let us now integrate $(\text{XII-93})_2$ and $(\text{XII-93})_3$ over the surface S, by changing $\gamma_{\mu\nu}$ and $\Lambda_{\mu\nu}$ to $\gamma'_{\mu\nu}$ and $\Lambda'_{\mu\nu}$, in order to take into account the contribution of the singularities.

We note first of all that the terms of $(\text{XII-93})_2$ and $(\text{XII-93})_3$ other than $\Lambda'_{\mu\nu}$ contribute nothing to the surface integral. In fact, these terms can still be written in the following form:

$$\text{(XII-99)} \qquad \partial_p \{F_{\mu[rp]}\} = \partial_p \{\partial_p \gamma_{\mu r} - \partial_r \gamma_{\mu p} + \delta_{\mu r} \partial_s \gamma_{ps} - \delta_{\mu p} \partial_s \gamma_{rs}\},$$

$F_{\mu[rp]}$ being antisymmetrical in r, p. The surface integral of the divergence of $F_{\mu[rp]}$ is then identically zero.

We then have only to form and then cancel out the surface integrals of the expressions $\Lambda'_{\mu\nu}$. Equations $(\text{XII-93})_2$ and $(\text{XII-93})_3$ are integrable if:

$$\text{(XII-100)} \qquad \frac{1}{4\pi}\int \Lambda'_{\mu r} n_r \, dS \equiv \frac{1}{4\pi}\int \Lambda_{\mu r} n_r \, dS + \frac{1}{4\pi}\int \bar{\Lambda}_{\mu r} n_r \, dS = 0.$$

$\Lambda_{\mu r}$ corresponds to the singularity-free solutions, and $\bar{\Lambda}_{\mu r}$ to the contribution made by the singularities: poles and dipoles. n_r represents the components of the unit vector normal to the surface at the point x^r

$$\text{(XII-101)} \qquad n_r = \cos \widehat{(x^r, \mathbf{n})}.$$

Let us first of all consider the integral of $\Lambda_{\mu r}$. It can be shown that it does not depend on the form of the surface S, but solely on time.[18]

One will then have:

$$\text{(XII-102)} \qquad \frac{1}{4\pi}\int \Lambda_{\mu r} n_r \, ds \equiv c_\mu(\tau)$$

and the condition of integrability (XII-100) takes on the following form:

$$\text{(XII-103)} \qquad c_\mu(\tau) = -\frac{1}{4\pi}\int \bar{\Lambda}_{\mu r} n_r \, dS.$$

$c_\mu(\tau)$ is not, in general, zero. This is why the conditions of integrability of (XII-93) require the effects of poles and dipoles.

With the supplementary condition $S_r = 0$ (which involves, to the order $2l-2$, the elimination of the dipoles), equations (XII-103) assure the integrability of $(\text{XII-93})_3$. To the order $2l$, these three conditions determine the motion of the singularities. The equations of the motion of the singularities thus appear as the conditions of integrability of the field equations (XII-93) to the corresponding order.

c) *Choice of the coordinate system.* One can simplify the expression of the field equations and their integration by referring them to a special system of coordinates. This choice is possible, thanks to the existence of four arbitrary functions which occur in each order of approximation.[19]

Let us consider, for example, a system of isothermal coordinates defined by the following conditions (cf. XII-18):

$$\text{(XII-104)} \qquad g^{\mu\nu}\begin{Bmatrix}\rho\\ \mu\nu\end{Bmatrix} = 0 \qquad \text{or} \qquad \partial_\rho(\sqrt{-g}\, g^{\mu\rho}) = 0 .$$

We note that the equations

$$(\text{XII-105})_1 \qquad \underset{1}{\partial_0}\, \underset{2l-2}{\gamma_{00}} - \partial_p \underset{2l-1}{\gamma_{0p}} = 0$$

$$(\text{XII-105})_2 \qquad \underset{1}{\partial_0}\, \underset{2l-1}{\gamma_{0r}} - \partial_s \underset{2l}{\gamma_{rs}} = 0$$

constitute, for $l = 2$, the first approximation of (XII-104).

So we will adopt a coordinate system such that equations (XII-105) are satisfied.[20]

With this choice, the field equations (XII-93) take on the simpler form:

$$(\text{XII-106})_1 \qquad \partial_p^2 \gamma_{00} + 2\Lambda_{00} = 0$$

$$(\text{XII-106})_2 \qquad \partial_p^2 \gamma_{0r} - \partial_p \partial_r \gamma_{0p} + 2\Lambda_{0r} = 0$$

$$(\text{XII-106})_3 \qquad \partial_p^2 \gamma_{rs} - \partial_r \partial_0 \gamma_{0s} - \partial_s \partial_0 \gamma_{0r} + \delta_{rs} \partial_p \partial_0 \gamma_{0p} + 2\Lambda_{rs} = 0 ,$$

with

$$(\text{XII-107})_1 \qquad 2\underset{2l-2}{\Lambda_{00}} = -\partial_p \underset{1}{\partial_0}\, \underset{2l-3}{\gamma_{0p}} + \underset{2l-2}{G'_{pp}} + \underset{2l-2}{G'_{00}}$$

$$(\text{XII-107})_2 \qquad 2\underset{2l-1}{\Lambda_{0r}} = \underset{1}{\partial_0}\partial_r \underset{2l-2}{\gamma_{00}} - \underset{2}{\partial_0^2}\, \underset{2l-3}{\gamma_{0r}} + 2\underset{2l-1}{G'_{0r}}$$

$$(\text{XII-107})_3 \qquad 2\underset{2l}{\Lambda_{rs}} = -\underset{2}{\partial_0^2}(\delta_{rs}\underset{2l-2}{\gamma_{00}} + \underset{2l-2}{\gamma_{rs}}) + \underset{1}{\partial_0}\partial_r\underset{2l-1}{\gamma_{s0}} + \underset{1}{\partial_0}\partial_s\underset{2l-1}{\gamma_{r0}} + $$
$$+ 2\left[\underset{2l}{G'_{rs}} + \tfrac{1}{2}\delta_{rs}(\underset{2l}{G'_{00}} - \underset{2l}{G'_{pp}})\right].$$

d) *Approximation of the second order.* To the second order of approximation ($l = 2$), one will obtain according to (XII-107) and (XII-92):

$$(\text{XII-108})_1 \qquad 2\underset{2}{\Lambda_{00}} = 0$$

$$(\text{XII-108})_2 \qquad 2\underset{3}{\Lambda_{0r}} = \underset{1}{\partial_0}\partial_r\underset{2}{\gamma_{00}}$$

$$(\text{XII-108})_3 \qquad 2\underset{4}{\Lambda_{rs}} = -\underset{2}{\partial_0^2}\delta_{rs}\underset{2}{\gamma_{00}} + \underset{1}{\partial_0}\partial_r\underset{3}{\gamma_{s0}} + \underset{1}{\partial_0}\partial_s\underset{3}{\gamma_{r0}} + $$
$$+ \tfrac{1}{4}\partial_r\underset{2}{\gamma_{00}}\partial_s\underset{2}{\gamma_{00}} + \tfrac{1}{2}\underset{2}{\gamma_{00}}\partial_{rs}^2\underset{2}{\gamma_{00}} - \tfrac{3}{8}\delta_{rs}\partial_p\underset{2}{\gamma_{00}}\partial_p\underset{2}{\gamma_{00}}$$

where the last three terms stem from the $\underset{4}{G'_{\mu\nu}}$.

By substituting $(\text{XII-108})_1$ and $(\text{XII-108})_2$ in $(\text{XII-106})_1$ and in $(\text{XII-106})_2$ one then obtains, taking (XII-105) into account,

$$(\text{XII-109})_1 \qquad \partial_p^2\underset{2}{\gamma_{00}} = 0$$

$$(\text{XII-109})_2 \qquad \partial_p^2\underset{3}{\gamma_{0r}} = 0$$

equations which, in the absence of singularities, permit the solution

$$(\text{XII-110}) \qquad \underset{2}{\gamma_{00}} = 0, \qquad \underset{3}{\gamma_{0r}} = 0.$$

From this there results, according to (XII-108), the fact that the integrals (XII-102) are zero.

$$(\text{XII-111}) \qquad \underset{3}{c_0}(\tau) \equiv \frac{1}{4\pi}\int\underset{3}{\Lambda_{0r}}n_r\,dS = 0, \qquad \underset{4}{c_r}(\tau) \equiv \frac{1}{4\pi}\int\underset{4}{\Lambda_{rs}}n_s\,dS = 0.$$

To the second order of approximation, the equations of motion are then as follows:

$$(\text{XII-112}) \qquad \underset{3}{c_0}(\tau) \equiv -\frac{1}{4\pi}\int\underset{3}{\bar{\Lambda}_{0r}}n_r\,dS = 0$$

$$(\text{XII-113}) \qquad \underset{4}{c_r}(\tau) \equiv -\frac{1}{4\pi}\int\underset{4}{\bar{\Lambda}_{rs}}n_s\,dS = 0.$$

To this order of approximation, they represent the conditions of integrability for equations $(\text{XII-93})_2$ and $(\text{XII-93})_3$.

α) Let us now consider the solution of $(\text{XII-93})_1$

$$\underset{2}{\bar{\gamma}_{00}} = -4\,\underset{2}{\overset{k}{m}}\,\overset{k}{\psi} \tag{XII-114}$$

corresponding to the existence of k poles and to the exclusion of dipoles. $\overset{k}{\psi}$ has the expression (XII-98), $\overset{k}{m}$ is a function of $\tau = x^0$.

By substituting in $(\text{XII-108})_2$, one finds

$$2\underset{3}{\bar{\Lambda}_{0r}} \simeq -4\,\partial_r(\underset{3}{\overset{k}{\dot{m}}}\,\overset{k}{\psi}) \tag{XII-115}$$

by putting

$$\overset{k}{\dot{m}} = \frac{\partial \overset{k}{m}}{\partial \tau} \tag{XII-116}$$

and by neglecting the terms that are of an order lower than $\left(\overset{k}{r}\right)^{-2}$. These are, according to (XII-98), the expressions

$$\partial_0 \overset{k}{\psi} = \partial_0\left(\frac{1}{\overset{k}{r}}\right) = -\frac{1}{2\left(\overset{k}{r}\right)^3}\,\partial_0\left(\overset{k}{r}\right)^2 = \frac{(x^p - \overset{k}{\xi}{}^p)}{\left(\overset{k}{r}\right)^3}\,\dot{\overset{k}{\xi}}{}^p = -\dot{\overset{k}{\xi}}{}^p\,\partial_p \overset{k}{\psi} \tag{XII-117}$$

with

$$\dot{\xi}^p = \frac{\partial \xi^p}{\partial \tau}. \tag{XII-118}$$

On substituting (XII-115) into (XII-112) one notes immediately that this condition is satisfied if

$$\underset{2}{\overset{k}{\dot{m}}} = 0, \qquad \underset{2}{\overset{k}{m}} = \text{const.} \tag{XII-119}$$

To the approximation $l = 2$, *the masses are independent of time.*

β) Now let us take conditions (XII-113), which constitute the equations of motion. In order to expand $\bar{\Lambda}_{rs}$, according to $(\text{XII-108})_3$, one must know, in addition to $\underset{2}{\bar{\gamma}_{00}}$ the expression for $\underset{3}{\bar{\gamma}_{0r}}$.

According to the approximation $(\text{XII-105})_1$ of the conditions of isothermality

$$\partial_p \underset{3}{\bar{\gamma}_{0p}} = \underset{1}{\partial_0}\underset{2}{\bar{\gamma}_{00}} = -4\,\underset{2}{\overset{k}{m}}\,\underset{1}{\partial_0}\overset{k}{\psi}. \tag{XII-120}$$

Taking (XII-117) into account, one will then have:

$$\partial_p \bar{\gamma}_{0p} = 4 \underset{2}{\overset{k}{m}}\, \underset{1}{\overset{k}{\xi}}{}^p \partial_p \overset{k}{\psi} \tag{XII-121}$$

that is

$$\underset{3}{\bar{\gamma}}_{0p} = 4 \underset{2}{\overset{k}{m}}\, \underset{1}{\overset{k}{\dot{\xi}}}_p \overset{k}{\psi} \tag{XII-122}$$

and $(\text{XII-108})_3$ is written simply

$$\underset{4}{\bar{\Lambda}}_{rs} = 2 \underset{2}{\overset{k}{m}}\, \underset{1}{\partial_0} \partial_s \underset{1}{\overset{k}{\dot{\xi}}}{}^r \overset{k}{\psi} + 2 \partial_r \underset{2}{U} \partial_s \underset{2}{U} + 4 \underset{2}{U} \partial^2_{rs} \underset{2}{U} - 3 \delta_{rs} \partial_p \underset{2}{U} \partial_p \underset{2}{U} \tag{XII-123}$$

with

$$\underset{2}{U} = \tfrac{1}{4} \underset{2}{\bar{\gamma}}_{00} = - G \underset{2}{\overset{k}{m}} \overset{k}{\psi}. \tag{XII-124}$$

In order to evaluate the quadratic terms, let us develop the harmonic function U in the vicinity of the point $x^p = \overset{k}{\xi}{}^p$.

For $l \neq k$, one obtains in this way

$$\overset{l}{U} = \overset{l}{\tilde{U}} + (x^p - \overset{k}{\xi}{}^p) \tilde{\partial}_p \overset{l}{U} + \frac{1}{2!}(x^p - \overset{k}{\xi}{}^p)(x^q - \overset{k}{\xi}{}^q) \tilde{\partial}^2_{pq} \overset{l}{U} + \ldots \tag{XII-125}$$

where the notation $\sim$ designates the values of U, $\partial_p U$, $\partial^2_{pq} U$ at the point $x^p = \overset{k}{\xi}{}^p$. By calculating the different terms of (XII-123) and by substituting the values thus obtained in this equation, one can then write

$$\begin{aligned} \underset{4}{\bar{\Lambda}}_{rs} \simeq{} & 2 \underset{2}{\overset{k}{m}} \partial_s \underset{2}{\overset{k}{\ddot{\xi}}}_r \overset{k}{\psi} + 2 (\partial_r \underset{2}{\bar{U}} \Sigma'_l \tilde{\partial}_s \underset{2}{\overset{l}{U}} + \partial_s \underset{2}{\overset{k}{U}} \Sigma'_l \tilde{\partial}_r \underset{2}{\overset{l}{U}}) + \\ & + 4 \partial_{rs} \underset{2}{\overset{k}{U}} \Sigma'_l (x^p - \overset{k}{\xi}{}^p) \tilde{\partial}_p \underset{2}{\overset{l}{U}} - 6 \delta_{rs} \partial_p \underset{2}{\overset{k}{U}} \Sigma'_l \tilde{\partial}_p \underset{2}{\overset{l}{U}}, \end{aligned} \tag{XII-126}$$

neglecting the lower-order terms and designating by Σ'_l the summation performed for $l \neq k$.

Let us integrate (XII-126) over the surface S with the center $x^p = \overset{k}{\xi}{}^p$. One finally obtains

$$\underset{4}{c}_r(\tau) = - \frac{1}{4\pi} \int \underset{4}{\bar{\Lambda}}_{rs} n_s \, dS = - 2 G \underset{2}{\overset{k}{m}} (\overset{k}{\ddot{\xi}}{}^r + \Sigma'_l \tilde{\partial}_r \underset{2}{\overset{l}{U}}). \tag{XII-127}$$

For $l = 2$, the conditions (XII-113) are then written:

$$\text{(XII-128)} \qquad \boxed{\overset{k}{\underset{2}{\ddot{\xi}}}{}^{r} = - \Sigma_l' \tilde{\partial}_r \overset{l}{\underset{2}{U}}.}$$

And *these are the equations of motion predicted by Newtonian mechanics: a material point with coordinates* $x^p = \overset{k}{\xi}{}^p$ *is subject to a force deriving from the potential* $- \Sigma_l' \overset{l}{U}$. By means of successive approximations involving calculations analogous to those just described, one could deduce from the field equations, the higher-order equations of motion.

2) *Deduction of the equations of motion from the field equations with right-hand member (energy-momentum tensor method).*

This method of obtaining the equations of motion is to a large extent the result of works by G. DARMOIS[21] and TH. DE DONDER[22]. The calculation of the solutions was achieved by widely varying methods in the works of V. A. FOCK[23], A. PAPAPETROU[24], PETROVA[25], and F. HENNEQUIN[26].

We start with the field equations written for the internal case and valid for a continuous distribution of uncharged and unpolarizable matter. We can assume, as in Section 3 of this chapter, that this distribution may be treated as if it were a perfect fluid. Its contribution is given by the matter tensor.

$$\text{(XII-129)} \qquad M_{\mu\nu} = (\mu_0 c^2 + p) u_\mu u_\nu - p g_{\mu\nu}.$$

The field equations are, then,

$$\text{(XII-130)} \qquad S_{\mu\nu} \equiv G_{\mu\nu} - \tfrac{1}{2} g_{\mu\nu} G = \chi M_{\mu\nu}.$$

Let us select the system of isothermal coordinates that is defined by (XII-18). In Section 2 we calculated the expression of $S_{\mu\nu}$ in this particular reference system. The field equations then have the form (XII-21).

By then adopting a quasi-Galilean reference system permitting developments (XII-22) of the $g_{\mu\nu}$, one can write equations (XII-130) for any desired order of approximation. In the second order, these equations reduce to (XII-47). Their solutions, indicated in Section 3, are given by

(XII-52), (XII-54) and by (XII-69), that is, by[27]

(XII-131) $$\underset{2}{h}_{pq} = -2\,\delta_{pq}\Sigma_A U_A, \qquad \underset{2}{h}_{00} = -2\,\Sigma_A U_A$$

(XII-132) $$h_{p0} = 4\,\Sigma_A \overset{A}{v^p} U_A,$$

letting U_A designate the potential created by the body A of density μ_A. This potential satisfies the equation

(XII-133) $$\Delta U_A = -\frac{\chi}{2} c^4 \mu_A = -4\pi G \mu_A .$$

With the $h_{\mu\nu}$ thus determined for the various orders of approximation, one can substitute their values (XII-131) and (XII-132) in the equations for the motion of a perfect fluid. The motion of a perfect fluid, without a charge, in a gravitational field characterized by the potentials $g_{\mu\nu}$, is, in fact, described by equation (VIII-194) in which $F_p = 0$. One thus obtains

(XII-134) $$\frac{d}{dt}\int \mathscr{M}_p^0\, d\mathscr{V} = \frac{c}{2}\int \mathscr{M}^{\rho\sigma}\partial_p g_{\rho\sigma}\, d\mathscr{V} .$$

Now, limiting ourselves to the second-order approximation, one simply has

(XII-135) $$\frac{1}{c}\mathscr{M}_p^0 = \frac{\sqrt{-g}}{c}(\mu_0 c^2 + p)\, u_p u^0 \simeq -\mu_0 v^p$$

and

(XII-136) $$\mathscr{M}^{\rho\sigma}\partial_p g_{\rho\sigma} = \sqrt{-g}\,[(\mu_0 c^2 + p)\, u^\rho u^\sigma \partial_p g_{\rho\sigma} - p g^{\rho\sigma}\partial_p g_{\rho\sigma}] \simeq \\ \simeq -2\,\mu_0 \partial_p U ,$$

taking into account (VIII-191), (XII-52) and the expressions of u_p and u_0 already calculated in (XII-41).

Thus, the motion of the body A, resulting from (XII-134), is described, in the second-order approximation, by the following equations

(XII-137) $$\frac{d}{dt}\int \mu_A \overset{A}{v_p}\, d\mathscr{V}_A = \int \mu_A \partial_p U\, d\mathscr{V}_A .$$

Now, within this approximation[28]

(XII-138) $$m_A = \int \mu_A\, d\mathscr{V}_A$$

and $\overset{A}{v}_p$ is constant over A. (XII-137) is then written:

(XII-139) $$\frac{d}{dt}(m_A \overset{A}{v}_p) = \int \mu_A \partial_p U \, d\mathscr{V}_A .$$

The potential U has the following form:

(XII-140) $$U = U_A + \Sigma' U_B(A).$$

U_A is the potential created at $x(A)$ by the body A itself. $U_B(A)$ is the potential – practically constant over A – created at $x(A)$ by the body B; the sign Σ' indicates a summation over all bodies *other than A*. One thus obtains, starting with (XII-140):

(XII-141) $$\int \mu_A \partial_p U \, d\mathscr{V}_A = m_A \partial_p (\Sigma' U_B(A)) + \int \mu_A \partial_p U_A \, d\mathscr{V}_A .$$

The motion of body A is thus determined by the equation:

(XII-142) $$m_A \ddot{a}^p = m_A \partial_p (\Sigma' U_B(A)) + \int \mu_A \partial_p U_A \, d\mathscr{V}_A$$

where a^p designates the coordinates of the center of gravity of A.

Equation (XII-142) is simplified if the body A possesses spherical symmetry. In that case:

(XII-143) $$\partial_p U_A = \frac{\partial U_A}{\partial r} \frac{\partial r}{\partial x^p} = \frac{\partial U_A}{\partial r} \frac{x^p - a^p}{r}$$

if

(XII-144) $$r^2 = (x^p - a^p)(x^p - a^p).$$

One then obtains:

(XII-145) $$\int \mu_A \partial_p U_A \, d\mathscr{V}_A = \int \mu_A \frac{\partial U_A}{\partial r} \frac{x^p - a^p}{r} \, d\mathscr{V}_A = 0 .$$

Thus, from (XII-142) and to the second order approximation, the equations of motion which it is possible to derive from the General Relativistic field equations are written:

(XII-146) $$\boxed{\ddot{a}^p = \partial_p (\Sigma' U_B(A)).}$$

These coincide with the Newtonian equations of motion.

At higher approximations, calculations have been developed by the authors already cited (see notes 23, 24, 25, 26). The results are analogous to those obtained by the singularity method.

B. STUDY OF A RIGOROUS BUT SPECIAL SOLUTION OF THE FIELD EQUATIONS: SCHWARZSCHILD'S SOLUTION

6. *The gravitational field created in the neighbourhood of a static mass possessing spherical symmetry*

The rigorous determination of a gravitational field created by a spherically symmetric neutral mass becomes extremely important the moment we seek to establish the experimental consequences of General Relativity. The solution of this problem was first determined by SCHWARZSCHILD.[29]

And as a matter of fact, his solution gives, in very approximate terms, the field created in the neighbourhood of most celestial bodies. The structure of the Riemannian space in the vicinity of these bodies is entirely determined by the elementary interval

$$ds^2 = g_{\mu\nu}\, dy^\mu\, dy^\nu . \tag{XII-147}$$

In order to determine the exact form of the $g_{\mu\nu}$, we employ the equations for the gravitational field in the vicinity of attracting masses, namely, the then differential equations:

$$G_{\mu\nu} = G^{\rho}_{\mu\nu\rho} = \partial_\rho \begin{Bmatrix} \rho \\ \mu\nu \end{Bmatrix} - \partial_\nu \begin{Bmatrix} \rho \\ \mu\rho \end{Bmatrix} + \begin{Bmatrix} \lambda \\ \mu\nu \end{Bmatrix} \begin{Bmatrix} \rho \\ \lambda\rho \end{Bmatrix} - \begin{Bmatrix} \lambda \\ \mu\rho \end{Bmatrix} \begin{Bmatrix} \rho \\ \lambda\nu \end{Bmatrix} = 0 . \tag{XII-148}$$

We know that these equations are not independent; there exist between them the four identities (XI-84). Equations (XII-148) then permit us to determine six (i.e., $10 - 4$) potentials of gravitation. The arbitrary character of four of the $g_{\mu\nu}$ is tied in with the arbitrary choice of the reference system.

The calculation of the Christoffel symbols

$$\begin{Bmatrix} \rho \\ \mu\nu \end{Bmatrix} = \tfrac{1}{2}\, g^{\rho\sigma} (\partial_\mu g_{\nu\sigma} + \partial_\nu g_{\mu\sigma} - \partial_\sigma g_{\mu\nu}) \tag{XII-149}$$

becomes much simpler in the case where the static mass creating the field possesses spherical symmetry; the ds^2 in its vicinity then also possesses spherical symmetry, and this means that one can *a priori* say what its form will be and thus reduce the number of $g_{\mu\nu}$ functions that have to be determined.

Let us, in point of fact, select the following variables

(XII-150) $$y^1 = r, \quad y^2 = \theta, \quad y^3 = \varphi, \quad y^0 = ct.$$

In a purely static case, the Riemannian ds^2 with spherical symmetry may be written in the following form[30]:

(XII-151) $$ds^2 = -\alpha\, dr^2 - \beta(d\theta^2 + \sin^2\theta\, d\varphi^2) + \sigma c^2\, dt^2.$$

In the absence of gravitational field, or if r tends to infinity, one must have the following limiting conditions:

(XII-152) $$\alpha \to 1, \quad \beta \to r^2, \quad \sigma \to 1$$

and the ds^2 (XII-151) then describes a Euclidean space referred to a system of polar coordinates.

The determination of the $g_{\mu\nu}$

(XII-153) $$g_{11} = -\alpha, \quad g_{22} = -\beta, \quad g_{33} = -\beta \sin^2\theta, \quad g_{00} = \sigma,$$

is thus reduced to the determining of the three functions α, β and σ.

In a static case these are functions of r.

If, with the help of (XII-153), we form the Christoffel symbols (XII-149), and if we then cancel out, following (XII-148), the components of the Ricci tensor that are written as a result of the use of those symbols, we see that one of the functions α, β or σ remains arbitrary. This property, resulting from the existence of the conservation identities (XI-84), allows us to choose α, β or σ quite arbitrarily. We usually put:

(XII-154) $$\beta = r^2.$$

It is convenient to use the notations:

(XII-155) $$\alpha = e^{2l}, \quad \sigma = e^{2n}.$$

Calculation of the Christoffel symbols, in which we substitute the values

(XII-156) $$g_{11} = -e^{2l}, \quad g_{22} = -r^2, \quad g_{33} = -r^2 \sin^2\theta, \quad g_{00} = e^{2n},$$

then leads us to the following results:

(XII-157)
$$\begin{aligned}
&\left\{\begin{matrix}1\\11\end{matrix}\right\} = l', \qquad \left\{\begin{matrix}1\\33\end{matrix}\right\} = \left\{\begin{matrix}1\\22\end{matrix}\right\}\sin^2\theta = -re^{-2l}\sin^2\theta,\\
&\left\{\begin{matrix}1\\00\end{matrix}\right\} = n'e^{2(n-l)}\\
&\left\{\begin{matrix}2\\33\end{matrix}\right\} = \sin\theta\cos\theta, \qquad \left\{\begin{matrix}2\\12\end{matrix}\right\} = \left\{\begin{matrix}3\\13\end{matrix}\right\} = \frac{1}{r},\\
&\left\{\begin{matrix}3\\23\end{matrix}\right\} = -\frac{1}{\tan\theta}, \qquad \left\{\begin{matrix}0\\10\end{matrix}\right\} = n'.
\end{aligned}$$

By substituting in (XII-148), we note that the only non-identically zero components of the Ricci tensor are thus

(XII-158)$_1$ $$G_{11} \equiv -n'' - n'^2 + l'\left(\frac{2}{r} + n'\right) = 0$$

(XII-158)$_2$ $$G_{22} = \frac{G_{33}}{\cos^2\theta} = 1 + re^{-2l}\left(l' - \frac{1}{r} - n'\right) = 0$$

(XII-158)$_3$ $$G_{00} = e^{2(n-l)}\left(n'' + n'^2 - n'\left[l' - \frac{2}{r}\right]\right) = 0.$$

By forming $e^{2(n-l)}$ (XII-158)$_1$ + (XII-158)$_3$ we then obtain

(XII-159) $$l' + n' = 0 \qquad \text{or} \qquad \log\alpha\sigma = \text{const}.$$

From this we deduce:

(XII-160) $$\alpha\sigma = \text{const} = 1$$

since, according to (XII-152)

$$\alpha\sigma \to 1, \quad \text{if} \quad r \to \infty.$$

So we postulate

(XII-161) $$l + n = 0.$$

By inserting this in (XII-158)$_2$, we obtain two equations that finally reduce to

$$\text{(XII-162)} \qquad e^{2n}(2rn' + 1) = k^2 = 1$$

and to the relation resulting from the derivation (XII-162).

We must select $k^2 = 1$ in order to satisfy the limiting conditions (XII-159) valid for $n' = 0$.

So we have only to integrate (XII-162), which gives us

$$\text{(XII-163)} \qquad e^{2n} r = r - \frac{2a}{c^2}$$

that is

$$\text{(XII-164)} \qquad \sigma = 1 - \frac{2a}{c^2 r} = \frac{1}{\alpha}$$

a/c^2 being a constant of integration.

The Riemannian ds^2 created in the neighbourhood of a static mass with spherical symmetry may thus have the following rigorous solution:

$$\text{(XII-165)} \qquad ds^2 = -\frac{dr^2}{1 - \dfrac{2a}{rc^2}} - r^2(d\theta^2 + \sin^2\theta\, d\varphi^2) + \left(1 - \frac{2a}{rc^2}\right) c^2\, dt^2 .$$

If r tends toward infinity, then this solution satisfies the limiting conditions (XII-152), and the ds^2 is then Euclidean.

Instead of relating the ds^2 to the polar coordinates (XII-150), we can choose the system of coordinates r_1, θ, φ by putting

$$\text{(XII-166)} \qquad r = \left(1 + \frac{a}{2r_1 c^2}\right)^2 r_1 .$$

Starting from that definition, we then arrive at

$$\text{(XII-167)} \qquad dr = \left(1 - \frac{a^2}{4r_1^2 c^4}\right) dr_1 , \qquad \sigma = \frac{\left(1 - \dfrac{a}{2r_1 c^2}\right)^2}{\left(1 + \dfrac{a}{2r_1 c^2}\right)^2} .$$

By substitution in (XII-165) we thus obtain:

$$\text{(XII-168)} \qquad ds^2 = -\left(1 + \frac{a}{2r_1c^2}\right)^4 (dr_1^2 + r_1^2\,d\theta^2 + r_1^2 \sin^2\theta\, d\varphi^2) + \\ + \frac{\left(1 - \frac{a}{2r_1c^2}\right)^2 c^2\,dt^2}{\left(1 + \frac{a}{2r_1c^2}\right)^2}.$$

Let us now define the following coordinates, referred to as *isotropic* coordinates.

$$\text{(XII-169)} \qquad x^1 = r_1 \sin\theta \cos\varphi, \quad y^1 = r_1 \sin\theta \sin\varphi, \quad z^1 = r_1 \cos\theta.$$

They make it possible to write expression (XII-168) of ds^2 in the following form:

$$\text{(XII-170)} \qquad ds^2 = -\left(1 + \frac{a}{2r_1c^2}\right)^4 [(dx^1)^2 + (dy^1)^2 + (dz^1)^2] + \\ + \frac{\left(1 - \frac{a}{2r_1c^2}\right)^2}{\left(1 + \frac{a}{2r_1c^2}\right)^2} c^2\,dt^2.$$

This ds^2 leads to the approximate form:

$$\text{(XII-171)} \qquad ds^2 = -\left(1 + \frac{2U}{c^2}\right)[(dx^1)^2 + (dy^1)^2 + (dz^1)^2] + \\ + \left(1 - \frac{2U}{c^2}\right)c^2\,dt^2,$$

with

$$\text{(XII-172)} \qquad U = \frac{a}{r_1}$$

if r_1 is sufficiently large.

Now (XII-171) is expression (XII-83) of the ds^2 obtained in Section 4 by using an approximate method. In (XII-83), U is the Newtonian potential.

(XII-173) $$U = \frac{KM'}{r}, \qquad \text{whence} \qquad a = KM'.$$

In addition, according to (XII-164), the constant of integration a is connected with the characteristics of the body producing the gravitational field; it represents its inertial mass m'. One will then have

$$a = K_1 m' \qquad \text{that is} \qquad M' = \frac{K_1}{K} m'$$

and Newton's law may then also be written

$$F = -K\frac{MM'}{r^2} = -\frac{K_1^2}{K}\frac{mm'}{r^2} = -G\frac{mm'}{{}^2}.$$

By choosing $K_1 = K$, one will thus obtain

$$M' = m', K_1 = K = G, a = GM' = Gm'$$

that is

(XII-174) $$\sigma \simeq \frac{1}{\alpha} = 1 - \frac{2Gm}{c^2 r}.$$

Remark. Instead of considering a static case, one could seek spherically symmetric solutions depending on time. The $g_{\mu\nu}$ are then functions of r and t. It has been shown that this solution reduces, of necessity, to the solution of the equations for the static case.[31]

7. *The field created in the neighbourhood of a spherically symmetric charged particle*

In this case the equations of the gravitational field have a right-hand member representing the contribution of the electromagnetic field. In the presence of this field, but outside matter (the purely electromagnetic interior case), it is allowed that this contribution be represented by the Maxwell tensor $\tau_{\mu\nu}$:

(XII-175) $$T_{\mu\nu} = \tau_{\mu\nu} = -\varphi_{\mu\rho}\varphi_\nu{}^\rho + \tfrac{1}{4} g_{\mu\nu}\varphi_{\rho\sigma}\varphi^{\rho\sigma}.$$

So, instead of (XII-148), one will write

(XII-176) $$S_{\mu\nu} \equiv G_{\mu\nu} - \tfrac{1}{2} g_{\mu\nu} G = \chi T_{\mu\nu} = \chi\tau_{\mu\nu}.$$

Multiplying (XII-176) by $g^{\mu\nu}$ and summing, one will then have

(XII-177) $$-G = \chi T \qquad \text{with} \qquad T = g^{\mu\nu} T_{\mu\nu}$$

and, by substitution in (XII-176):

(XII-178) $$G_{\mu\nu} = \chi(T_{\mu\nu} - \tfrac{1}{2} g_{\mu\nu} T).$$

But it will be immediately evident that the invariant τ, formed from the Maxwell tensor, is identically zero:

(XII-179) $$\tau = g^{\mu\nu}\tau_{\mu\nu} = -\varphi_{\mu\rho}\varphi^{\mu\rho} + \tfrac{1}{4}\delta^{\mu}_{\mu}\varphi_{\rho\sigma}\varphi^{\rho\sigma} \equiv 0.$$

The equations valid outside matter and in the presence of the electromagnetic field only are thus reduced to:

(XII-180) $$G_{\mu\nu} = \chi\tau_{\mu\nu}.$$

Let us limit ourselves to the static solutions. In a system of polar coordinates, the only non-zero component of the electrostatic field is

(XII-181) $$\varphi_{10} = \frac{e}{r^2}$$

and the only non-zero $g_{\mu\nu}$ have the form (XII-153).

By substitution in (XII-175) one then finds

(XII-182)$_1$ $$\tau_{11} = \tfrac{1}{4} g_{11}(2\varphi_{10}\, g^{11} g^{00} \varphi_{10}) - \varphi_{10}\, g^{00} \varphi_{10} = = -\tfrac{1}{2} g^{00} \varphi_{10}\varphi_{10} = -\frac{1}{2\sigma}\frac{e^2}{r^4}$$

(XII-182)$_2$ $$\tau_{22} = \frac{\tau_{33}}{\cos^2\theta} = \tfrac{1}{4} g_{22}(2\varphi_{10}\, g^{11} g^{00} \varphi_{10}) = \frac{1}{2\alpha\sigma}\frac{e^2}{r^2}$$

(XII-182)$_3$ $$\tau_{00} = \tfrac{1}{4} g_{00}(2\varphi_{10}\, g^{11} g^{00} \varphi_{10}) - \varphi_{01}\, g^{11} \varphi_{01} = = -\tfrac{1}{2} g^{11} \varphi_{01}\varphi_{01} = \frac{1}{2\alpha}\frac{e^2}{r^4},$$

the other $\tau_{\mu\nu}$ being zero. We must put these expressions into the right-hand member of equations (XII-180). A calculation similar to the one developed in the preceding section then leads us to the following solution [32]:

(XII-183) $$\beta = r^2, \qquad \sigma = \frac{1}{\alpha} = 1 - \frac{2Gm}{c^2 r} + \frac{\chi e^2}{2r^2}.$$

The constants e and m characterizing the particle creating the gravitational field then occurs in a very different fashion; m is a constant of integration resulting from the actual solution of the equations; e is introduced through the Maxwell tensor, that is, as a result of data having nothing to do with the actual structure of the field, which the presence of e serves to modify.

8. *The trajectory of a neutral particle in the neighbourhood of a static mass having spherical symmetry*

The trajectories described by a neutral particle in a gravitational field are the geodesics of the Riemannian space:

$$\text{(XII-184)} \qquad \frac{d^2y^\rho}{ds^2} + \begin{Bmatrix} \rho \\ \mu\nu \end{Bmatrix} \frac{dy^\mu}{ds}\frac{dy^\nu}{ds} = 0 .$$

In the special case where the gravitational field is produced by a static mass possessing spherical symmetry, it is convenient to replace the symbols $\{^{\rho}_{\mu\nu}\}$ by their values as calculated from the potentials

$$\text{(XII-185)} \qquad g_{00} = -\frac{1}{g_{11}} = \left(1 - \frac{2mG}{c^2 r}\right), \qquad g_{22} = \frac{g_{33}}{\sin^2\theta} = -r^2 .$$

Let us first of all write (XII-184) for $\rho = 2$. By putting $y^1 = r$, $y^2 = \theta$, $y^3 = \varphi$ we obtain

$$\text{(XII-184)}_2 \qquad \frac{d^2\theta}{ds^2} + \frac{2}{r}\frac{dr}{ds}\frac{d\theta}{ds} - \cos\theta \sin\theta \left(\frac{d\varphi}{ds}\right)^2 = 0 .$$

By choosing a coordinate system such that the initial motion takes place in the plane $\theta = \pi/2$, the initial conditions $d\theta/ds = 0$, $\cos\theta = 0$, impose

$$\frac{d^2\theta}{ds^2} = 0$$

and the motion thus continues to take place in the plane $\theta = \pi/2$ during the following instants.

Equations (XII-184) relative to $\rho = 3$ and $\rho = 0$, are then written

$$\text{(XII-184)}_3 \qquad \frac{d^2\varphi}{ds^2} + \frac{2}{r}\frac{d\varphi}{ds}\frac{dr}{ds} = 0$$

$$\text{(XII-184)}_0 \qquad \frac{d^2 t}{ds^2} + \frac{\sigma'}{\sigma}\frac{dr}{ds}\frac{dt}{ds} = 0, \qquad \left(\sigma' = \frac{d\sigma}{dr}\right)$$

By integrating we then obtain

$$\text{(XII-186)} \qquad r^2 \frac{d\varphi}{ds} = \frac{h}{c}, \qquad \frac{dt}{ds} = \frac{k}{\sigma c^2},$$

h and k being two constants of integration.

One can then integrate (XII-184) for $\rho = 1$. It makes no difference if we use the expression (XII-165) for ds^2

$$\text{(XII-187)} \qquad ds^2 = -\frac{1}{\sigma} dr^2 - r^2 (d\theta^2 + \sin^2\theta \, d\varphi^2) + \sigma c^2 \, dt^2$$

and eliminate dt and ds from this relation by taking (XII-186) into account. In that way we obtain

$$\text{(XII-188)} \qquad \frac{1}{\sigma}\left(\frac{h}{r^2 c}\frac{\partial r}{\partial \varphi}\right)^2 + \frac{h^2}{r^2 c^2} - \frac{k^2}{\sigma c^2} = -1 .$$

Let us now put

$$\text{(XII-189)} \qquad \frac{1}{r} = u .$$

Taking (XII-174) into account, (XII-188) may further be written

$$\text{(XII-190)} \qquad \left(\frac{\partial u}{\partial \varphi}\right)^2 + u^2 = -\frac{c^2}{h^2}\left(1 - \frac{k^2}{c^4}\right) + \frac{2Gm}{h^2} u + \frac{2Gm}{c^2} u^3 .$$

Differentiating with respect to φ and comparing this to (XII-188), we obtain the equations for the trajectories:

$$\text{(XII-191)} \qquad \boxed{\frac{d^2 u}{d\varphi^2} + u = \frac{Gm}{h^2} + \frac{3Gmu^2}{c^2}}$$

$$\text{(XII-192)} \qquad \boxed{r^2 \frac{d\varphi}{ds} = \frac{h}{c},}$$

or, according to (XII-186),

$$\text{(XII-193)} \qquad r^2 \frac{d\varphi}{dt} = \frac{h\sigma c}{k} = \frac{hc}{k}\left(1 - \frac{2Gm}{rc^2}\right).$$

These equations, to a very good order of approximation, are applicable in Celestial Mechanics. Nevertheless, in order that these equations provide a test in favour of the validity of General Relativity, they would have to make possible predictions that are quite different from those arising from the Newtonian theory; in that case, experimental evidence would be decisive in favour of one or the other of the two theories.

Now, in Newtonian Mechanics, the equations determining the trajectory of a test-body in a gravitational field created by a static mass having spherical symmetry, are as follows:

$$\frac{d^2u}{d\varphi^2} + u = \frac{Gm}{h^2} \tag{XII-194}$$

$$r^2 \frac{d\varphi}{dt} = h\,. \tag{XII-195}$$

They differ from equations (XII-191) and (XII-192), resulting from General Relativity, by the term $(3Gm/c^2)\,u^2$. If the velocity of a test-body is small with respect to c, this term is then generally too small for its effects to be detected. One has indeed

$$\frac{\dfrac{3Gmu^2}{c^2}}{\dfrac{Gm}{h^2}} = \frac{3h^2u^2}{c^2} = 3\left(r\frac{d\varphi}{ds}\right)^2 = 3\left(\frac{r\,d\varphi}{c\,d\tau}\right)^2, \tag{XII-196}$$

and $r\,(d\varphi/d\tau)$, the tangential velocity of the planet, is small with respect to c.

One can, however, prove the influence of the supplementary term of (XII-191) in a few special cases that we now consider.

9. *The experimental verifications of Schwarzschild's solution*

1) *The advance in the perihelions of the planets.* We have pointed out that one of the principle disagreements between Newtonian Mechanics and experience concerns the motion of the planet Mercury, whose perihelion shows a secular advance of 43″ of an arc per century. The predictions involved are deduced from the Newtonian equation (XII-194) having the solution

$$u_0 = \frac{mG}{h^2}\left[1 + A\cos(\varphi - \varpi)\right]. \tag{XII-197}$$

A and $\bar{\omega}$ are constants of integration. They represent the eccentricity of the planet and the longitude of its perihelion.

Let us, in fact, consider once more the classical equation of an elliptical trajectory with axes a and b, expressed in terms of a system of polar coordinates having as its center one of the foci:

(XII-198) $$\frac{1}{r} = \frac{1 + e\cos\varphi}{a(1 - e^2)} = \frac{1}{p}(1 + e\cos\varphi)$$

with

(XII-199) $$e = \frac{r_{\max} - r_{\min}}{r_{\max} + r_{\min}}, \qquad \text{excentricity of the ellipse}$$

(XII-200) $$p = \frac{b^2}{a} = a(1 - e^2), \qquad \text{parameter of the ellipse.}$$

Comparing (XII-197) and (XII-198), we will then put:

(XII-201) $$\frac{Gm}{h^2} = \frac{1}{p}, \qquad A = e$$

which may also be written in the form

(XII-202) $$h^2 = (Gm)\,p = Gma(1 - e^2).$$

Let us now go back to equation (XII-191) deduced from General Relativity. Let us calculate the solutions of this equation by successive approximations, substituting solution (XII-197) – which is an approximate solution of (XII-191) – in the supplementary term that is involved in this equation. One thus obtains:

(XII-203) $$\frac{d^2u}{d\varphi^2} + u \sim + \frac{6\,G^3m^3}{c^2h^4}\,e\cos(\varphi - \bar{\omega}).$$

The other terms actually furnish only a negligible contribution. On the other hand, the term in $\cos(\varphi - \bar{\omega})$ of (XII-203) involves the appearance of the true period and may give rise to resonance effects. So equation (XII-203) allows the special solution

(XII-204) $$u_1 = \frac{3\,G^3m^3}{c^2h^4}\,e\varphi\sin(\varphi - \bar{\omega}).$$

A solution to the second order of approximation is formed by the superposition of (XII-197) and (XII-204)

$$u = u_0 + u_1 = \frac{mG}{h^2}[1 + e\cos(\varphi - \bar{\omega} - \delta\bar{\omega})] \tag{XII-205}$$

putting

$$\delta\bar{\omega} = \frac{3m^2G^2}{c^2h^2}\varphi. \tag{XII-206}$$

We deduce from this, by comparison with (XII-202),

$$\frac{\delta\bar{\omega}}{\varphi} = \frac{3\,m^2G^2}{c^2h^2} = \frac{3\,mG}{ac^2(1-e^2)}. \tag{XII-207}$$

For one complete revolution of the planet, that is, for $\varphi = 2\pi$, the advance of the perihelion is, thus:

$$\delta\bar{\omega} = \frac{6\pi mG}{ac^2(1-e^2)}. \tag{XII-208}$$

In the special case of a revolution around the sun

$$m_{\text{sun}} = 1.983 \cdot 10^{33}\,\text{g}. \tag{XII-209}$$

$$\frac{2\,mG}{c^2} = \frac{2 \times 1.983 \times 10^{33} \times 6.66 \times 10^{-8}}{9 \cdot 10^{20}} = 2.95 \cdot 10^5\,\text{cm} \tag{XII-210}$$

and

$$\begin{aligned} \delta\bar{\omega} &= \frac{3\pi \cdot 2.95 \cdot 10^5}{a(1-e^2)}\,\text{rad} = \\ &= \frac{360 \times 3\,600}{2\pi} \cdot \frac{3\pi \times 2.95 \cdot 10^5}{a(1-e^2)}\,\text{sec of arc} \\ &= \frac{57.348 \cdot 10^{10}}{a(1-e^2)}\,\text{sec of arc}. \end{aligned} \tag{XII-211}$$

If T is the duration of the planet's revolution expressed in sidereal days, the advance of the perihelion in the course of a century will be

$$d\Omega = \frac{100\,T_{\text{earth}}}{T_{\text{planet}}}\delta\bar{\omega} = \frac{36.525\,\delta\bar{\omega}}{T} \tag{XII-212}$$

that is

$$(\text{XII-213}) \qquad d\Omega = \frac{20\,946{,}357 \cdot 10^{12}}{a(1-e^2)T} \text{ sec of arc.}$$

The correction $\delta\bar{\omega}$ may be perceptible for the small planets (a being small) having great eccentricity. The eccentricity is maximum in the case of the planet Mercury, for which we have the following data:

$$(\text{XII-214}) \qquad \begin{cases} a = 5.8 \cdot 10^{12}\,\text{cm} \\ e = 0.2056 \\ T = 87.97\,\text{days}. \end{cases}$$

Thus

$$(\text{XII-215}) \qquad a(1-e^2) = 5.555 \cdot 10^{12}$$

and, substituting these values in (XII-213), we obtain the following value for the secular advance of the perihelion of Mercury:

$$(\text{XII-216}) \qquad d\Omega = \frac{20 \cdot 946.36 \cdot 10^{12}}{5.55 \times 87.97 \cdot 10^{12}} = 42.9''.$$

This value is an exceedingly close approximation of the experimental results.

2) *The deflexion of light-rays by a gravitational field.* Schwarzschild's solution is further valid for the special case in which we consider, not the motion of a test-body of mass m, but the motion of light in a gravitational field produced by a static spherical body. In this case, the trajectories of the light-rays are still geodesics, but geodesics of zero length. The condition

$$(\text{XII-217}) \qquad ds = 0$$

thus entailing, according to (XII-192)

$$(\text{XII-218}) \qquad h \to \infty.$$

The equations representing the trajectories of light-rays then have, according to (XII-191) and (XII-218), the following form:

$$(\text{XII-219}) \qquad \frac{d^2u}{d\varphi^2} + u = \frac{3\,Gmu^2}{c^2}.$$

We again integrate (XII-219) by successive approximations. The solution of a homogeneous equation of the type (XII-219) is

$$u_0 = \frac{\cos\varphi}{R}, \tag{XII-220}$$

R being a constant of integration.

Let us substitute this solution in the term $3\,Gmu^2/c^2 \ll 1$.

The equation with right-hand member thus obtained has the particular solution

$$u_1 = \frac{G}{c^2}\frac{m}{R^2}(\cos^2\varphi + 2\sin^2\varphi). \tag{XII-221}$$

And in the second approximation, the solution of (XII-219) thus has the following expression

$$u = u_0 + u_1 = \frac{\cos\varphi}{R} + \frac{G}{c^2}\frac{m}{R^2}(\cos^2\varphi + 2\sin^2\varphi), \qquad \left(u = \frac{1}{r}\right). \tag{XII-222}$$

Using Cartesian coordinates ($x = r\cos\varphi$, $y = r\sin\varphi$) one obtains the equation for the trajectories:

$$x = R - \frac{mG(x^2 + 2y^2)}{c^2R\sqrt{x^2+y^2}}. \tag{XII-223}$$

The last term expresses a deviation of the light-rays from the straight line $x = R$. The angle α measuring this deviation has the value

$$\alpha = \frac{2mG}{c^2R}\left(\frac{x^2+2y^2}{\sqrt{x^2+y^2}}\right)_{y \gg x} = \frac{4mGy}{c^2R}. \tag{XII-224}$$

This is double the value one would obtain by using the Newtonian theory. One would, in fact, have:

$$\gamma = \mathbf{grad}\,U \qquad \text{with} \qquad U = \frac{Gm}{r}. \tag{XII-225}$$

If the particle moves along a parallel to Oy, passing at a distance $x = R$ from the body of mass m:

$$\frac{d^2x}{dt^2} = Gm\frac{\partial}{\partial x}\left(\frac{1}{r}\right) = -\frac{Gm}{r^2}\frac{\partial r}{\partial x} = -\frac{Gmx}{r^3} \tag{XII-226}$$

and if the particle has the velocity of light:

(XII-227) $$\frac{d^2y}{dt^2} = 0, \qquad \frac{dy}{dt} = c.$$

For $x = R$, the acceleration will then have the following value

(XII-228) $$\frac{d^2x}{dt^2} = c^2 \frac{d^2x}{dy^2} = -\frac{GmR}{(R^2 + y^2)^{\frac{3}{2}}}$$

and, approximately

(XII-229) $$x = R - \frac{Gmy}{c^2R}.$$

The angle measuring the deviation of the light-rays is then

(XII-230) $$\alpha' = \frac{2mG}{c^2R}.$$

Experimentally, the deviation of light-rays may be measured in a sufficiently intense gravitational field by observing the fixed stars, for example, in the apparent vicinity of the sun. Such a measurement is possible during an eclipse, since at that moment the brilliance of the sun is sufficiently dimmed. The stars situated in its apparent vicinity are then visible, and the light-rays that proceed from them could be deflected by the sun's gravitational field close to their trajectories. If such is indeed the case, one should observe an apparent displacement of these stars on the celestial sphere. Thus, stars that are normally hidden by the sun, according to the hypothesis of rectilinear propagation, should become visible by virtue of the bending of the light rays.

The deviations observed at the time of the 1919 eclipse were connected with stars in the Hyades constellation, which was at that time in the apparent vicinity of the sun. The values obtained agree very well with those predicted by Einstein's theory. But, of course, we are here concerned with an order of magnitude, since the deviations that may arise from the solar field are slight (in this instance, $\alpha \cong 1.75''$).

Let us likewise cite the results obtained by CAMPBELL and TRUMPLER[33], who obtain, by the use of two types of photographic tests

(XII-231) $$\alpha_1 = 1.72'' \pm 0.11'' \qquad \alpha_2 = 1.82'' \pm 0.15'',$$

Nevertheless, the phenomena observed are on the very limit of experimental errors.

3) *Another consequence of General Relativity: The red shift.* The experiments that may be performed in connection with the perihelion of Mercury or the deviation of light-rays in the vicinity of a heavy body, constitute tests that are especially adapted to Schwarzschild's solution. On the other hand, if the red shift *may* be explained by recourse to certain properties of this solution, it does not seem to be *necessarily* connected with the description of trajectories in the neighborhood of a spherical mass, as predicted by General Relativity.

First of all, it is necessary to distinguish between the red shift that we are studying in this section, and the red shift discovered by HUBBLE in 1929 in the spectrum of the extra-galactic nebulae. This latter phenomenon, of a considerably high order of magnitude, still remains very difficult to explain. Two kinds of theory have been tried in seeking to account for it:

a) Cosmological theories presenting different models of an expanding universe;

b) Theories concerning the "aging" of light as it traverses interstellar spaces.

Here, we shall not study the Hubble effect; we will limit ourselves to the red shift of the spectral lines of a body situated in the gravitational field of a *single* static mass with spherical symmetry – in the gravitational field of the sun, for example.

The atoms making up the gaseous atmosphere of a fixed star constitute oscillators situated in the star's gravitational field. Each of them forms a kind of clock whose vibrations measure the proper time of the atom. The proper frequency ν_0 of this atom is the number of vibrations per unit of proper time.

$$\text{(XII-232)} \qquad \nu_0 = \frac{dN}{d\tau} \qquad \text{with} \qquad ds^2 = c^2\, d\tau^2 .$$

Let us adopt a coordinate system in which the star is at rest. In such a system the average velocity of an atom is zero and the ds^2 at the corresponding point has the following value:

$$\text{(XII-233)} \qquad ds^2 = g_{00}\, c^2\, dt^2 .$$

In a reference system defined in this way, an observer attributes to the atom under consideration the frequency

(XII-234) $$\nu = \frac{dN}{dt} = \nu_0 \sqrt{g_{00}}\,.$$

Now, in the neighborhood of the star, the ds^2 is the one resulting from the Schwarzschild solution reduced to the term in dt^2. According to (XII-164)

(XII-235) $$g_{00} = \sigma = 1 - \frac{2m}{rc^2} G\,.$$

Substituting in (XII-234), one will then have

(XII-236) $$\nu = \sqrt{1 - \frac{2mG}{rc^2}}\, \nu_0 \sim \nu_0 \left(1 - \frac{mG}{rc^2}\right)$$

and, consequently,

(XII-237) $$\frac{\Delta \nu}{\nu_0} = - \frac{mG}{rc^2}\,.$$

An atom on the surface of the sun will then have, for a terrestrial observer, a frequency lower by $\Delta \nu$ than its proper frequency. If M and R represent the mass and the radius of the sun,

(XII-238) $$\frac{\Delta \nu}{\nu_0} = - \frac{MG}{c^2 R}\,.$$

Now the proper frequency ν_0 is approximately that emitted, on earth, by an atom of the same substance. In fact, the distance from this terrestrial atom to the star is great enough for g_{00} to be approximately equal to unity and for $\Delta \nu$ to be practically zero.

An atom situated on a star thus has the spectral lines that it emits displaced towards red with respect to the lines of the terrestrial spectrum arising from an atom of the same substance.

The displacement thus produced will be perceptible if the gravitational field resulting from the star is sufficiently intense. For example, if we are concerned with the sun

(XII-239) $$M = 1.983 \cdot 10^{33}\,\text{g}, \qquad R = 6.95 \cdot 10^{10}\,\text{cm}.$$

Whence

(XII-240) $$\frac{\Delta \nu}{\nu_0} = -2.10 \cdot 10^{-6}.$$

An effect thirty times more powerful would be obtained by observing the spectrum of an atom situated on the companion of Sirius, whose density is 10^6 times that of water. The measurements that have actually been made show satisfactory agreement with the predictions of the theory.[34]

Let us note, however, that the red shift assumes solely the form (XII-235) as a function of the potential g_{00}, that is

(XII-241) $$g_{00} = 1 + U.$$

It is perfectly compatible with a theory predicting a modification of frequency of the form (XII-236). Such is the case of the Euclidean theories of gravitation (BIRKHOFF-MOSHINSKY).[35] Thus, although it *may* be interpreted by use of the consequences of General Relativity, the spectral shift towards the red cannot be used as proof of a geometrical interpretation of gravitation. It can likewise be explained within the framework of a Minkowski universe and by the principles of Special Relativity.

Let us finally add that, if the three effects that have been examined here lead to correct verifications, the two latter ones remain on the very limits of the possibilities of experimental verifications.

It is the internal coherence of General Relativity, the simplicity of its principles, the natural generalization of the relativistic principle that it presupposes that remain the strongest supports of a theory that experiment favours without imposing. This theory proposes, indeed, the only natural explanation of the identity noted between gravitational mass and inertial mass. It constitutes the most remarkable example of a true field theory, that is, of a theory such that the motion of the sources is subsumed by the laws of the field.

It is, therefore, very tempting to seek a way to incorporate into this synthesis the electromagnetic field, which still remains profoundly alien to all geometrization. After serving as a guide in formulating a principle of Special Relativity, the electromagnetic theory seems to resist all attempts at applying any results of a principle of General Relativity. The contribution of a Maxwellian field expressed by the tensor $\tau_{\mu\nu}$ is purely phenomenological and results from data foreign to the theory. The motion of the charges is deduced from a force that remains entirely independent of the behavior of the field.

One can, on the other hand, entertain the hope that the geometrization of a generalized field may make it possible to extend to the motion of charged particles the result achieved in the case of neutral particles; their motion might be determined by the very compatibility of the field equations. Finally, this primacy of the notion of a field may appear as the prelude to a pure field theory in which the properties of the particles would be entirely reduced to those of the field. These vast ambitions are as yet far from being realized, even in the purely classical domain. Of course, the concepts introduced by quantum theories remain foreign, to a large degree, to the usual domain of General Relativity. The quantization of non-linear equations, as of the equations of the gravitational field, poses difficult problems, which have not yet found solutions, and whose interpretation is far from being clear.

NOTES

1. Indeed

$$
\begin{aligned}
- g^{\rho\sigma}\left\{ {\lambda \atop \rho\sigma} \right\} &= -\frac{1}{2} g^{\rho\sigma} g^{\lambda\tau}(2\partial_\rho g_{\sigma\tau} - \partial_\tau g_{\rho\sigma}) \\
&= -\partial_\rho(g^{\rho\sigma} g^{\lambda\tau} g_{\sigma\tau}) + g_{\sigma\tau}\partial_\rho(g^{\rho\sigma} g^{\lambda\tau}) + \frac{1}{2} g^{\lambda\rho}\frac{\partial_\rho g}{g} \\
&= -\partial_\rho g^{\rho\lambda} + \delta_\tau{}^\rho \partial_\rho g^{\lambda\tau} + \delta_\sigma{}^\lambda \partial_\rho g^{\rho\sigma} + \frac{1}{2} g^{\lambda\rho}\frac{\partial_\rho g}{g} \\
&= \partial_\rho g^{\rho\lambda} + g^{\lambda\rho}\frac{\partial_\rho\sqrt{-g}}{\sqrt{-g}} = \frac{1}{\sqrt{-g}}\,\partial_\rho(\sqrt{-g}\, g^{\lambda\rho}
\end{aligned}
$$

taking into account (cf. Chap. XV, p. 447);

$$g^{\rho\sigma} g_{\sigma\tau} = \delta_\tau{}^\rho \quad \text{and from} \quad dg = g g^{\mu\nu}\, dg_{\mu\nu}$$

2. On the choice of isothermal coordinates, see: G. Darmois [20], Chapter III.
   Th. de Donder [21], p. 40.
   J. Chazy [19], t. II, p. 143.
3. The development of the $h_{\mu\nu}$ with respect to the infinitesimals ε, ε^2, ε^3 is here arbitrary. We shall see that the choice is imposed by the order of magnitude of the terms characterizing the energy tensor (cf. XII-43).

4. See J. Chazy [19], t. II, p. 147.
5. One often puts $\chi = (8\pi G/c^2)$. This value refers to the expression $M_{\mu\nu} = \mu_0 \mu_\mu u_\nu$ of the matter tensor, which is then substituted for (XII-40).
On the other hand, one can also adopt a new system of units such that $c = 1$ and $G = 1$. In order to do this, one has only to modify the units of time and mass, leaving the unit of length unchanged.
In this new system

$$[L]' = [L] = 1 \text{ cm}$$

$$[T]' = [T] \frac{c'}{c} = \frac{1}{3 \cdot 10^{10}} = 3.33 \cdot 10^{-11} \text{ sec}$$

$$[M]' = [M] \frac{G'}{G} \frac{[T]^2}{[T]'^2} = \frac{(3 \cdot 10^{10})^2}{6.66 \cdot 10^{-2}} = 1.35 \cdot 10^{28} \text{ g}.$$

In this new system of units, (XI-90) is written simply

$$S_{\mu\nu} \equiv G_{\mu\nu} - \frac{1}{2} g_{\mu\nu} G = 8\pi T_{\mu\nu} .$$

6. The same result is also deduced from Schwarzschild's solution expressed in *isotropic* coordinates and then subjected to the same approximation.
7. Or $\chi = 8\pi$ in the system of units $G = 1$, $c = 1$.
8. Naturally, agreement of (XII-73) with (XII-71) amounts to identifying (XII-71) with the solution of Poisson's equation (XI-23) in which μ is the density of the gravitational mass μ_g. But Poisson's equation is deduced from the Newtonian law of gravitation by already postulating the identity of gravitational mass and inertial mass.
9. Cf. for example, P. G. Bergmann [9], p. 186. One there finds a development of the solution of (XII-76) and (XII-37).
10. See also a study of plane waves in P. G. Bergmann [9], p. 188.
11. A. Einstein and Rosen: 'On Gravitational Waves', *Journ. Franklin Inst.* **223** (1937) 43.
12. W. B. Bonnor: *Ann. Inst. H. Poincaré* XVI, (1957) III, 146; *Nature* **181** (1958) 1196.
13. P. G. Bergmann [9], p. 189.
14. Mention must be made of results recently obtained by Pirani (cf. for example, *Actes du Congrès sur la gravitation, Chapel Hill, 1957*) and A. Lichnerowicz: *Compt. Rend.*, July 1958.
15. In point of fact, this separation would not be unequivocally determined.
16. A. Einstein, L. Infeld, B. Hoffmann: *Ann. Math.* **39** (1938) 65.
A. Einstein, L. Infeld: *Ann. Math.* **41** (1940) 455; *Canad. J. Math.* **1** (1949) 209.
L. Infeld and P. R. Wallace: *Phys. Rev.* **57** (1940) 797.
L. Infeld and A. Schild: *Rev. Mod. Phys.* **21** (1949) 408.
L. Infeld: *Acta Phys. Polo.* **13** (1954) 187.
Pham Tan Hoang: Thesis. *La méthode des singularités pour les équations du mouvement en Rélativité Générale et en théorie du champ unifié.* Paris, 1957.
17. Throughout this section the summation sign is assumed, even when the indices are repeated below *or* above.
18. Indeed, let us assume that the field equations are satisfied for the order $2l - 2$

(1) $$\underset{2l-2}{S_{\mu\nu}} = 0.$$

In that case, the conservational identities (XI-84) valid for $\underset{2l-1}{S_{0\mu}}$ reduce to:

(2) $$\partial_r \underset{2l-1}{S_{0r}} = 0$$

since the supplementary terms that would occur in these identities are all expressed as functions of the $\underset{2l-2}{S_{\mu\nu}}$, which are zero according to (1).

But (2) is also expressible in the order $2l-2$ by $\partial_r S_{0r} = 0$ by virtue of (1) and the definition (XII-87) of $S_{\mu\nu}{}^*$. The expression of $S_{0r}{}^*$ is exactly given by (XII-90)$_2$, so that (2) necessarily imposes

(3) $$\partial_r \underset{2l-1}{\Lambda_{0r}} = 0.$$

In a similar manner if, besides (1), the field equations also include

(4) $$\underset{2l-1}{S_{0m}} = 0.$$

the conservational identities applied to $\underset{2l}{S_{rs}}$ reduce to

(5) $$\partial_r \underset{2l}{S_{rs}} = 0.$$

since the other terms disappear by virtue of field equations (1) and (4).
Finally, (5) is equivalent to $\partial_r S_{rs} = 0$, and, according to expression (XII-90)$_3$ of $S_{rs}{}^*$, to the condition

(6) $$\partial_r \underset{2l}{\Lambda_{rs}} = 0.$$

Thus, by virtue of the existence of four identities of conservation, the validity of the field equations up to the order $2l-2$ and $2l-1$ lead, for the order $2l-1$ and $2l$, to conditions (3) and (6).
These conditions, always fulfilled according to the field equations, express the fact that the surface integrals of $\Lambda_{\mu\nu}$ *do* not depend on the form of the surface, but solely on time. One will then have (XII-102) by virtue of the identities of conservation (XI-84).

19. If $\gamma_{00}{}^*$, $\gamma_{0m}{}^*$, $\gamma_{mn}{}^*$ represent solutions of (XII-93)

$$\underset{2l-2}{\gamma_{00}} = \underset{2l-2}{\gamma_{00}{}^*} \qquad \text{with } a_m, a_0 \text{ arbitraires}$$

$$\underset{2l-1}{\gamma_{0m}} = \underset{2l-1}{\gamma_{0m}{}^*} + \partial_m \underset{2l-1}{a_0}$$

$$\underset{2l}{\gamma_{mn}} = \underset{2l}{\gamma_{mn}{}^*} + \partial_n \underset{2l}{a_m} - \partial_m \underset{2l}{a_n} - \underset{1}{\gamma_{mn}}\partial_r \underset{2l}{a_r} + \delta_{mn}\partial_0 \underset{2l-1}{a_0}$$

are also solutions of (XII-93).

20. A. Einstein and L. Infeld choose the following conditions

$$\partial_0\gamma_{00} - \partial_r\gamma_{0r} = 0, \qquad \partial_r\underset{2l}{\gamma_{rs}} = 0.$$

Pham Tan Hoang has developed the calculations based on the use of the condition of isothermy (XII-104) admitting the approximation (XII-105). (Cf. *Thèse*, Paris, 1957).

21. G. Darmois [20] and *Les équations de la gravitation.* Paris, Gauthier-Villars, 1927.
22. Th. de Donder [21] and *La gravifique einsteinienne.* Paris, Gauthier-Villars, 1921.
23. G. A. Fock: *J. Phys. Ac. Sc. U.R.S.S.* **1** (1939) 81.
24. A. Papapetrou: *Proc. Phys. Soc.* **64** (1951) 37.
25. Petrova: *Zh. Eksper. Teor. Fiz.* **19** (1949) 989.
26. F. Hennequin: *Thèse.* Paris, 1956.
27. We here designate by a^p the coordinates of the center of gravity of body A. If one refers to the explanation of the Singularity Method, this notation corresponds to the $\overset{k}{\xi}{}^p$, which designate the coordinates of the kth body. It follows that $\ddot{a}^p$ corresponds to $\ddot{\overset{k}{\xi}}{}^p$ and that $\Sigma' U_B(A)$ corresponds to $G\Sigma' \overset{k}{m}\overset{k}{\psi}$. We retain both notations (each of which has its advantages) in order to facilitate the reading of the original papers.
28. It would be natural to adopt definition (VIII-187) of the mass, which is a definition that arises in the conservation equations. However, to the approximation required in this paragraph, it is equivalent to choosing (XII-138).
29. Schwarzschild: *Berl. Ber.* 1916, 189.
30. More generally one can seek the spherically symmetric solution such that

$$ds^2 = g_{00}\,(dy^0)^2 + 2g_{p0}\,dy^p\,dy^0 + g_{pq}\,dy^p\,dy^q\,, \qquad (1)$$

where the coordinates y^p satisfy

$$r^2 = \sum_{p=1}^{3} (y^p)^2.$$

One can reduce (1) to the following expression:

$$ds^2 = g_{00}\,(dy^0)^2 - g_{pq}\,dy^p\,dy^q \qquad (2)$$

putting

$$g_{pq} = \delta_{pq} - D(r)\chi_p\chi_q \qquad \text{with} \qquad \chi_p = \frac{y^p}{r}.$$

In this way one obtains the following static solution:

$$g_{00}(r) = 1 - \frac{2Gm}{rc^2}, \quad g_{pq} = -\,\delta_{pq} - \frac{\dfrac{2Gm}{rc^2}}{1 - \dfrac{2Gm}{rc^2}}\,\chi_p\chi_q, \quad g_{p0} = 0.$$

Cf. P. G. Bergmann [9], p. 198.

31. Birkhoff: *Relativity and Modern Physics.* Harvard University Press, 1923, p. 253.
H. Mineur: *Bulletin de la Société Math. de France* **56** (1928) 50.
This work had as its aim the representation of the gravitational field created by pulsating Cepheids by means of non-static spherically symmetric solutions.
32. G. B. Jeffery: *Proc. Roy. Soc.* **99A,** 123.
H. Rassner; *Ann. d. Phys.* **50** (1916) 106.
H. Weyl: *Ann. d. Phys.* **54** (1917) 117; (27).

32a. The ten $G_{\mu\nu}$ components of the Ricci tensor are first reduced to three independent components in the special case of ds^2 with spherical symmetry. Further, the ten identities (XI-87) are reduced to a single identity in the same special case. So finally there exist $3 - 1 = 2$ independent $G_{\mu\nu}$ components. One can thus write only two Einstein equations between the three variables α, β and σ.

33. W. W. CAMPBELL and R. TRUMPLER: *Lick Observatory Bull.* **11** (1923) 41 and **13** (1928) 130.
W. M. OVENDEN: *Sc. Progr.* **40** (1952) 645.
S. A. MITCHELL: *Eclipses of the Sun.* New York, 1951.

34. SAINT-JOHN: *Astrophys. Journ.* **67** (1928) 195.
ADAMS: *Proc. Nat. Acad.* **11** (1925) 383.
M. G. ADAM: *Mon. Not. Roy. Astr. Soc.* **108** (1948) 446.
D. M. POPPER: *Astrophys. Journ.* **120** (1954) 316.
G. P. KUIPER: *Act. Scient. Industr.* 1941.

35. BIRKHOFF: *Proc. Nat. Acad.* **29** (1943) 231.

CHAPTER XIII

UNIFIED THEORIES OF ELECTROMAGNETISM AND GRAVITATION

CHARACTERISTICS OF A PURE FIELD THEORY UNIFIED THEORIES AND NON-DUALISTIC THEORIES

By "unified theory" one usually indicates a similar and, most frequently, geometric interpretation of electromagnetic and of gravitational phenomena. The field equations thus obtained would constitute the structural conditions of a non-Euclidean universe.

But the term "unified" is also used to qualify a theory that seeks to reduce the notions of field and particle to a single concept.

Generally speaking, these two kinds of unified theories have nothing in common but a regrettable similarity in name. Attempts have been made, however, at constructing theories that would make possible a double synthesis, namely: the geometrization of both the electromagnetic and gravitational fields; and, on the other hand, the reduction of the characteristics of the sources to the single basic datum of a generalized field. At the present time, no such theory has yet been achieved, even in the purely classical domain. Nevertheless, even without constructing an actual theory of the pure field, one can think that a unified field theory (gravitation-electromagnetism) would make it possible to deduce the motion of the singularities – the sources of the field – from the basic data of the equations of a generalized field. The existence of the singularities would not be made dependent on the existence of the field, but the motion of the singularities would arise from the conditions to which the field is subjected.

So there is a certain relationship between the two synthetical operations – that is, between a common interpretation of fields and the reduction of the sources of the field to the field itself or – what is really very different – the deduction of the motion of the sources from the conditions imposed upon a generalized field.

In order to avoid all confusion, we reserve the name *unified theories* to attempted *syntheses of electromagnetism and gravitation*. We shall call *non-dualistic* those attempts at reducing the sources of the field to the field itself.

A. UNIFIED THEORIES

1. *Unified theories until the advent of General Relativity*

One can first of all ask to what extent a connection between the domains of electromagnetism and gravitation seems forced upon us. Charges and masses both obey a law of action at a distance involving $1/r^2$. Nevertheless, in spite of this purely formal analogy, it has never been possible to construct a unified theory in which the charge q would play a role analogous to that of gravitational mass.

There has, however, been an attempt (Föppl-Wien) made at describing gravitational actions by using electrical actions at a distance as a model. In this instance it was a matter of reducing the gravitational force to a sort of compensation of interactions between charges and thus achieve a gravitational theory along the lines of Lorentz's electronic theory. Unfortunately, the opposite signs occurring in the expression of the Newton and Coulomb forces (for two charges having the same sign) give rise to considerable differences between the two theories – differences of a sort that seem quite irreducible.

After the development of Special Relativity, the need for constructing a unified theory, or at the very least, of finding a relativistic law of gravity along the lines of the laws of Optics, became more and more imperative.

As a matter of fact, it was Optics that had triumphed over Mechanics in the conflict between them. It was the Maxwell equations, which are invariant in a Lorentz transformation, that were to serve as a model for the new relativistic physics. Thus the theories of action at a distance, which had dominated all of physics at the outset of the 19th century yielded to the relativistic field theories their role as a guide. By a reversal of the situation, it seemed that gravity would now have to be patterned after Optics.

Yet, in spite of some very ingenious attempts, no relativistic theory of gravitation seemed acceptable. Attempts by Minkowski, Winiewski and Poincaré came up, especially against the obvious antagonism which, in Special Relativity, exists between the conservative character of the charge and the fact of a mass varying with the velocity. So, until 1916, that is until the completion of the formulation of General Relativity, the balance-sheet in the construction of unified theories showed total failure, in spite of some very interesting attempts.

2. *General Relativity and the construction of unified theories*

In achieving General Relativity, Einstein put forward an interpretation of gravitational phenomena as profound as it was original. The law of gravitation, adapted to structural conditions that a Riemannian space must satisfy, becomes relativistic. Moreover, it makes possible the recovery of the results of the Newtonian theory as a first approximation and the practical elimination of certain experimental discrepancies, the most celebrated of which was the unexplained advance in the perihelion of Mercury.

But the problem of the relationship with electromagnetism is thereby transported onto an entirely different plane. A geometrical interpretation of gravitation profoundly isolates gravitational phenomena from the rest of physics, and from electromagnetism in particular. A unified theory thus no longer consists of bringing together two more or less dissimilar phenomena, but rather of extending the geometrical properties characteristic of gravitation to electromagnetism. Otherwise it becomes necessary to allow the privileged character of the phenomena of gravity and of the very special way they intervene in determining the motion of a charged particle of matter.

3. *Interpretation of the electromagnetic and gravitational fields proposed by unified theories*

After General Relativity, the aim of most of the unified theories will thus be to extend to the electromagnetic field the geometrization that had proved so satisfactory in the case of the gravitational field.

General Relativity reduces the laws of gravitation to ten structural conditions that the curvature of a four-dimensional Riemannian space manifold must satisfy. This curvature is, indeed, the only characteristic of such a manifold. Now, if it is possible, on this basis, to interpret the laws of gravitation as geometrical conditions, there is no room left for a possible analogous attempt with reference to electromagnetism. In fact, the possibilities left open by a Riemannian structure do not permit the existence of supplementary conditions that might correspond to Maxwell's equations. In order to deduce electromagnetism from the geometrical properties of the universe, it is necessary to enlarge the Riemannian and four-dimensional framework of General Relativity. In this case it is possible to proceed along two very different lines.

1) *Riemannian theories of more than four dimensions.* Though still retaining the Riemannian character of the manifold, one can assume that it has five or six dimensions. This approach was first made by KALUZA[1] (1921) and has subsequently been followed, with some variations in the EINSTEIN-MAYER theory[2] and in the so-called projective theories.[3] We may also mention the more recent theories using fifteen field variables (JORDAN[4], THIRY[5]) and the theory involving a six-dimensional manifold developed by PODOLANSKI[6].

The auxiliary role of the five-dimensional continuum employed comes out quite clearly in the projective theories. This synthesis takes on real physical meaning only in this auxiliary manifold which, however, is not the actual physical space. In this regard it is interesting to contrast this interpretation with the one proposed, for example, by the EINSTEIN-BERGMANN-BARGMANN theory[7] or even by PODOLANSKI's theory[8]. Stated in rather more complicated terms, these theories seek to interpret the five- or six-dimensional continuum as real physical space, endowed, it is true, with a rather special structure. For example, the five-dimensional space of the Einstein-Bergmann-Bargmann theory is closed with respect to the fifth dimension; Podolanski's physical space does indeed have six dimensions, but it is endowed with a laminated structure such that all the points in a given layer correspond to a single point in the four-dimensional space-time. All these attempts are in accordance with the spirit of General Relativity; they seek to introduce a Riemannian space that not only provides a satisfactory unified formalism, but that also constitutes the real physical space.

The five-dimensional theories have the great advantage of providing an easy interpretation of the trajectories of charged material particles as families of geodesics (families which correspond to the various values of e/m) of five-dimensional space. This represents an extension of the result achieved by General Relativity for the case of neutral particles.

Yet, many physicists still consider, in spite of everything, that the success of this sort of formalism is rather artificial. The success of a language adapted to a five-dimensional manifold is, according to them, only a way of concealing the lack of developments truly adaptable to the four-dimensional universe, which remains the true physical universe. The cylindrical condition constitutes the weak point in the five-dimensional transposition. That condition expresses the fact that everything observable depends physically on four (and not five) parameters. And this condition is expressed by a limitation on the covariance of the equations, since the coordinate x^5 must play a special role. The condition thus makes it impossible to achieve a complete unified synthesis in the way that, for example, Maxwell's theory achieved a synthesis of the electric and magnetic fields.

2) *Non-Riemannian four-dimensional theories.* One can, on the contrary, retain the essentially four-dimensional character of physical space. But then it is necessary, in order to find supplementary geometrical conditions, to give up the Riemannian structure.

It thus becomes necessary to make the geometry of the four-dimensional manifold much more complex. Now a manifold with a very general affine connection may possess torsion and two sorts of curvature:

a) the usual *Riemannian* curvature, or the *rotational* curvature defined by the Riemann-Christoffel tensor

$$\Omega^\rho_\mu = R^\rho_{\mu\nu\sigma}[dy^\nu \delta y^\sigma]$$

b) the *homothetic* curvature defined by the invariant

$$\Omega = \Omega^\mu_\mu = R^\mu_{\mu\nu\sigma}[dy^\nu \delta y^\sigma]$$

c) finally, *torsion*, introduced by Elie Cartan, is linked with the non-symmetrical character of an affine connection $\Gamma^\rho_{\mu\nu}$, which generalizes the Christoffel symbols $\{^{\ \rho}_{\mu\nu}\}$.

$$\Omega^\rho = \Gamma^\rho_{\mu\nu}[dy^\mu \delta y^\nu].$$

In point of fact, most authors did not immediately notice the wealth (curvatures and torsion) of structural elements at their disposal. And that is why the unified theories appeared, for quite a long time, as a rather arbitrary ensemble based on a geometry that could always be quite artificially modified.

But, in actual fact, the possibility open to unified theories within the framework of a manifold with affine connection are limited to the judicious coordinating of the structural elements (curvatures and torsion) of the manifold. Such has been the case of theories without torsion but having two curvatures (WEYL's theory[8]), and of theories using torsion but leaving out curvature (EINSTEIN's 1929 theory), allowing the existence of an absolute parallelism.[9] And finally, one can assume the existence of a manifold with any affine connection endowed *a priori* with torsion and two sorts of curvature. It is this latter hypothesis that made it possible for Einstein to develop a very general unified theory which is, at the same time, the most natural extension of the theory of gravitation.[10]

4. *Classical unified theories and the possibility of further predictions*

Within the classical domain, the interest presented by the unified theories, in terms of systematization, is undeniable. They permit a vast geometrical synthesis that leads quite naturally to the equations of gravitation and electromagnetism.

Nevertheless, this sort of theory has often been criticized as contenting itself too easily with such a synthesis without seeking to arrive, in its own terms, at original conclusions. Now, if it is interesting to associate Maxwell's equations with Einstein's formalism, it may nevertheless seem rather annoying not to be able to go beyond that synthesis. In fact, it is quite natural to think that a synthesis as important as that of electromagnetism and gravitation ought necessarily to furnish new theoretical extensions. Unfortunately, the unified theories have too often appeared refractory to anything but pure systematization.

The fifteen-field-variable theories and the recent Einstein-Schrödinger theory seem, on the contrary, to escape this sort of criticism. These theories no longer seek to establish a pure and simple synthesis between already established theories; instead, they produce new laws, of which Einstein's and Maxwell's theories are a first approximation.

For example, one can interpret the conclusions of the fifteen-field-variable theory by assuming that the gravitational factor χ is no longer constant[11], or that there exists a polarization of empty space[12], or that it is convenient to incorporate a conformal space.[13] The field equations resulting from this theory, lead us to equations of motion valid for a charged particle. This motion can be determined as precisely as we wish by proceeding through successive approximations.

In addition, the Einstein-Schrödinger theory leads to non-linear electromagnetic equations in which terms related to gravitation occur. In this way there exists a much closer interdependence between electromagnetism and gravitation. Finally, the theory makes it possible to avoid certain difficulties inherent in linear electromagnetism.

In practical terms, these theories could be harmonized with experimental proof if the effects predicted did not remain beyond the threshold of possible investigations. Far from making them unassailable, this circumstance constitutes the prime weakness of these theories.

Let us consider, for example, the variations of the gravitational factor χ – variations that are associated with those of the factor e/m in certain interpretations of the fifteen-field-variable theory. This prediction might seem to be in harmony with the existence of a rotational magnetism of neutral masses ($\mu \neq 0$, $\rho = 0$). That is precisely what the Blackett effect was, for which an empirical expression had been proposed. Unfortunately, it would seem that a rotational magnetism, if it exists, would exert far weaker effects than those predicted by Blackett's formula. Between certain limits, the disappearance or modification of such an effect cannot invalidate the interpretations of the Jordan-Thiry theory. But of course, neither can it confirm these interpretations.

The same situation holds for the influence existing between the gravitational field and the electromagnetic field as foreseen by the Einstein-Schrödinger theory. For example, the existence of a current due exclusively to the presence of a curvature of the universe remains well beyond the limits of verifiable manifestations. The specific effects predicted by this unified theory are far too weak to be detectable experimentally, and especially to serve as a guide in the physical interpretation of geometrical magnitudes resulting from the theory itself. These difficulties arise from the experimental deficiencies inherent in the domain exploited by this sort of theory.

5. *Unified theories and quantum theories*

The very existence of quantum theory is in itself the most serious objection to the construction of unified theories. In fact, every attempt thus far made at a unification of a gravitational field and a quantized electromagnetic field has remained in an embryonic stage. One may even seriously doubt that the gravitational field is quantizable or even whether it is desirable to attempt such a quantization.

The various attempts at quantization of the gravitational field vary considerably, depending on the classical description adopted as the basis for each particular attempt:

1) *The linear theories* of the gravitational field assume that gravitational phenomena are *rigorously* described by a system of linear equations. These linear equations can be deduced from the quasi-Galilean approximation of Einstein's law, but this involves a purely heuristic procedure which, in principle, is not connected with any rigorous non-Euclidean interpretation. One can equally well deduce the linear equations of gravitation from the equations of a field created by a particle of spin two or from the corpuscular wave-equations connected with the particle in question.

Under these circumstances, it is much easier to conceive a *rapprochement* between this type of equations and the electromagnetic equations, which are also linear. Moreover, the quantization of the linear equations of gravitation, in principle, presents no difficulty, since quantization requires a covariance valid in Minkowski space, which is to say, related exclusively to the Lorentz transformations.

Unfortunately, when deprived of the geometrical interpretation that remains the pivot of General Relativity, the equations to be quantized remain largely arbitrary. They give rise to a deceptive simplicity that proceeds from no available evidence and that is in no way imposed by experience. Quite to the contrary – to cite but one example – the at least problematic existence of gravitational waves seems to divert the theory from any obvious linear formalism.

2) *The non-linear theories* of the gravitational field postulate that the gravitational field obeys non-linear differential equations arising from Einstein's theory, but these nonlinear theories exploit the conclusions of General Relativity in a very different spirit.

Either they assume that the Einstein equations make up an acceptable formalism, but one relative to a *strictly Euclidean space* – in which case we confine ourselves to postulating the existence of a covariance limited to the Lorentz transformations[14] – or else they grant to the non-linear equations the *non-Euclidean interpretation* proposed by Einstein. In that case it is necessary to allow a general covariance of the equations – a covariance valid for any transformation of coordinates. The difficulties which the quantization methods then run into are numerous. It does seem, as of the present moment, that the achieving of a quantized law of gravity cannot be unequivocally determined.

Thus, even assuming that the quantization of the gravitational field is an unrestrictedly desirable objective, it is certainly not being attained in any satisfying way. This quantization applies, in fact, to linear equations that remain to a great extent artificial, or else it runs headlong into complications that make it arbitrary. On the other hand, the necessity of applying a quantum treatment to other fields seems, in spite of everything, inescapable. So one can subscribe to the following conclusion: So long as one remains in the classical realm, the construction of unified theories is a natural objective. In many respects it would seem that such a construction ought to be a source of progress permitting the elaboration of a non-linear electromagnetic theory with all the advantages (and disadvantages, as well) arising from non-linearity. Such a construction makes possible a more satisfying interpretation of the motion of the sources and could reveal an interdependence between the fields. Finally, it permits the elimination of the arbitrariness that, in General Relativity, governs the choice of the electromagnetic tensor $T_{\mu\nu}$, which is of a purely phenomenological origin. Thus, it constitutes the natural extension of field theories and requires only a more precise confrontation with the experience.

In the quantum domain, the construction of unified theories, or even of a gravitational theory, has not been achieved in a satisfactory manner. It is possible that such a construction is not realizable and that such a goal will turn out to be fruitless.

During recent years it has seemed as if quantum field theory has blithely taken over the legitimate ambitions of theoretical physics. Quantum field theory has in fact succeeded in agreeing in a very precise way with experiment, and the success of a theory depends above all on its possibilities of accurate prediction. Nevertheless, it is beyond question that predictions cannot be resolved by the formulation of a series – often a very limited series – of experiments, thus allowing the very notion of an explanation to drop out of sight.

The presence of deductive theories proceeding from a simple geometrical principle and capable of being amply and logically elaborated reminds us that there are ways, often unworkable, but none the less tempting. These ways are the natural outcome of classical field theories. One can thus hope that classical unified theories will ultimately succeed in coming up with new and verifiable predictions and with which quantum electrodynamics will succeed in establishing a closer coordination of its own developments. Perhaps it is the very principle of a *rapprochement* that it is almost impossible or unreasonable to undertake for systematically; but it nevertheless remains indispensable for the harmonious extension of field theories.

B. NON-DUALISTIC THEORIES

6. *The field and its sources*

The connection between the field and its sources has been and continues to be one of the most difficult problems that have to be resolved by electrodynamics, both in the classical and quantum realms. Putting the question in the simplest possible terms, one may say that the idea of a point source runs into the well known difficulties of infinite self-energy; but on the other hand, the notion of an extended source runs counter to relativistic requirements and is accompanied by hypotheses that are very often arbitrary. Among the many theories that have been put forward concerning this problem, one can distinguish *dualistic* and, contrariwise, *non-dualistic* theories.

a) The dualistic theories assume that the particles, which are the sources of the field, constitute entities that, along with their various characteristics, mass and charge, remain fundamentally different from the field itself. Along this line of thought one can cite, for example, those studies that circumvent the difficulties of infinite energy by recourse to two compensating fields. Theories based on rather different principles have subsequently been developed, and these latter theories form the bases of present-day classical electrodynamics.

b) On the other hand, non-dualistic theories assume that the sources of the field are not essentially different in nature from the field itself.

In order to avoid the structural hypotheses of the older electronic theories, which are unfit for a relativistic formulation, one is led to construct theories using multiple times, or even non-linear theories. In any case, both sorts appear to resist all attempts at quantization.

We have, for example, examined in Chapter IX the principles of Mie's theory and of the Born-Infeld theory. The latter theory is based on the existence of non-linear relations between the field and the electromagnetic induction. This non-linearity makes it possible to define a finite electromagnetic field at every point, even at the origin. Instead of the Maxwell electrostatic field, which is infinite at the point-source, we shall have a field that remains finite at the origin.

One can then define specific quantities attached to the particle (charge and, possibly, even mass) as functions of the quantities that are characteristic of the field. One will thus render the free densities of charge, current and mass, entirely reducible to the field quantities.

This assimilation is possible insofar as the electric field (characterizing the density of charge) remains finite at the origin. And that property, as we have seen, itself results from the non-linear character of the equations of the theory.

Of course, Born's theory is purely electromagnetic and purely Euclidean. It has no direct connection whatever with the construction of unified theories.

7. *Non-linearity and the characteristics of a pure field theory*

Nevertheless, there is one property common to both Born's electrodynamic theory and to gravitational theories: the property of non-linearity that we have previously discussed. Now, according to the results of General Relativity, this non-linearity is a necessary, but not a sufficient[15], condition for deducing the motion of the singularities of the very equations of the field.

1) *The deduction of the equations of motion.* Classical electrodynamics rests on the basic acceptance of Maxwell's equations, which are partial differential equations of the first order. These equations are linear, since they do not involve either the product of the fields or of their derivatives. As a direct consequence the sum (or difference) of two solutions of the field equations still constitute a solution of those equations. If two charged particles create, independent of each other, two electromagnetic fields that are solutions of Maxwell's equations, the field resulting from the superposition of these two elementary fields still satisfies Maxwell's equations. But in this situation it is impossible, by the exclusive use of Maxwell's equations, to find a supplementary condition that will characterize the interaction of the two particles – that is to say Coulomb's law. In order to do that, it would be necessary, in fact, for the total field formed by the two particles to be a solution of Maxwell's equations if, *and only if*, such a condition was fulfilled. This condition of compatibility imposed on the very equations of the field would constitute precisely the equations of

motion. Now, in a linear theory, such a condition is not necessary, since the sum of the partial fields is always an acceptable solution.

So we cannot derive this supplementary condition from the field equations. It will have to be independently postulated: this will be the expression of the Lorentz force.

If, on the contrary, the field equations were non-linear, the superposition of two fields – each of which is, taken separately, a solution – no longer satisfies the basic equations. The total field of the two particles can be accepted only when conditions of compatibility have been satisfactorily established. These conditions may eventually define the equations of motion of the particles. The motion of these particles is then determined by the field equations themselves; it is the particular motion that makes possible the compatibility of the fundamental equations.

Now the gravitational equations put forward by Einstein are not linear. The left-hand member $S_{\mu\nu}$ is a function of the second derivatives of the gravitational potential $g_{\rho\sigma}$ and of the products of the first-order derivatives $S_{\mu\nu}(\partial_\lambda^2 g_{\rho\sigma}, \partial_\alpha g_{\rho\sigma}, g_{\rho\sigma})$. The right-hand member $T_{\mu\nu}$ represents the contribution of the matter creating the gravitational field. It will thus be possible to obtain the equations of motion – that is, as a first approximation, Newton's law – from the law of the gravitational field. And that is what leads us to believe that General Relativity is a more perfect field theory than is electrodynamics.

2) *The reduction of the sources of the field to the field itself.* In Einstein's last efforts, the particle was to be integrated into the very structure of the field. "We could regard matter as being made up of regions of space in which the field is extremely intense... A rock thrown is, from this point of view, a varying field, a field in which the states of greatest intensity move forward with the velocity of the rock. There would be no room in this new physics for both field and matter, for the field would be the only reality."

As a matter of fact, Einstein tries to explain the contribution of matter (the right-hand member $T_{\mu\nu}$ of the field equations) in terms of the purely geometrical contribution ($S_{\mu\nu}$ of the left-hand member) of the generalized field. For this to be possible, it must also be possible, if need be, for $S_{\mu\nu}$ to express the intervention of an electromagnetic field. Thus, in the view expressed by General Relativity, a truly non-dualistic theory (the elimination of the sources of the field) is likewise of necessity a unified theory (electromagnetic-gravitational field).

So it seems that a unified theory can achieve any sort of internal cohesion in physics only if it is doubly unified, assuring the synthesis

a) of the electrodynamic field and the gravitational field, so that classical electrodynamics is subsumed,

b) of the generalized field thus defined and of the particles – the motion of the particles being deducible from the properties of the generalized field.

"A coherent field theory", writes Einstein, "requires that all its elements be continuous... And from this requirement arises the fact that the material particle has no place as a basic concept in a field theory. Thus, even apart from the fact that it does not include gravitation, Maxwell's theory cannot be considered a complete theory."

C. UNIFIED AND NON-DUALISTIC THEORIES

So it becomes interesting to compare Einstein's conclusions with those that could be drawn from, for example, the electrodynamics of Born-Infeld.

In both cases there is an attempt made to reduce the sources of the field to the field itself (that is, they are both non-dualistic theories).

In both cases non-linear equations are employed.

The non-linearity of the field equations is a necessary condition for obtaining the equations of motion. It does not constitute, however, a sufficient condition for achieving that objective. The electrodynamics of Born-Infeld, for example, cannot, on its own, come up with Coulomb's law. On the other hand, the law motion for charged particles will be obtained if the equations of the gravitational field contain, as expressed in the right-hand member, the contribution of the electromagnetic field.

It is, thus, by means of a unified theory of the Einstein-Schrödinger type that we may be able to move closer to the coherent development of a pure field theory.

Like Born's theory, it is a non-dualistic and non-linear theory. But it *also* leads to a unified interpretation of the electromagnetic and gravitational fields, whereas the Born-Infeld theory could contain only electrodynamics. The geometrical interpretation that is the very basis of the Einstein-Schrödinger theory makes possible, moreover, an intuitive justification of the theory and, above all, the avoidance of the arbitrariness that appears in the new definitions of a function of action and a generalized energy tensor.

Naturally, the difficulties in the interpretation arising from the introduction of a doubly unified theory (that is to say both unified and non-dualistic) are very numerous. The Einstein-Schrödinger theory presents us with a great many such difficulties, and they may perhaps remain as insurmountable obstacles. Nevertheless, that theory points the way which, possibly in a modified form, may lead to interesting results.

This search for unity in terms of classical field theories does not manifest Einstein's attachment for a well-determined form of theory. The search cannot be explained by invoking a "rigid adherence to classical theory". What really is, Einstein asks, a "classical" theory? Newton's theory introduced forces that were replaced by the theory of the continuous field of Hertz and Maxwell – a theory that was likewise, but in a very different way, classical. General Relativity proposes yet a different version, which moves toward being, though not altogether achieving, a pure field theory. Nevertheless, Einstein continues, the classical theory exists, but "as a program". It presents no decisive argument to the physicist who doubts the very notion of continuity. "Such a doubt is entirely legitimate, but where shall we seek some other way out?"

Is geometry, or rather, the physical theory considered in its totality and thus incorporating the geometry that it presupposes – a verifiable construction? In point of fact, the question is put with the same degree of arbitrariness as in the rest of physics. The force of this need for unification causes us to seek satisfaction in the data we find at our disposal in any given epoch; at the present time this means in certain mathematical constructions that sustain and guide our quest. This formalism is often far ahead of the intuitive interpretation of the results to which the formalism leads. The difficulties then reside, first, in the ambiguity of verifications, which tend to appeal to experimental data that are often uncertain (cosmological theories), and secondly in the not entirely reassuring adaptation of techniques (suggested, for example, by quantum theory) to a very different framework.

The way opened by the unified theories seems to require to be deepened rather than further extension. It might then become most useful as an introduction to an indispensable but difficult synthesis.

NOTES

1. KALUZA: *Sitzungsber. d. Preuss. Acad. d. Wiss.* 1921, 966.
2. A. EINSTEIN and MAYER: *Berl. Ber.* 1931, 541 and 1932, 130.
3. O. VEBLEN: *Projektive Relativitätstheorie.* Berlin, 1933.
   W. PAULI: *Ann. d. Phys.* **18** (1933) 305.
4. JORDAN: *Ann. Phys.* 1947, 219.
5. Y. R. THIRY: *C. R. Ac. Sc.* **226** (1948) 216 and 1881; *Thesis.* Paris, 1950.
6. PODOLANSKI: *Proc. Roy. Soc.* **201** (1950) 234.
7. A. EINSTEIN, V. BARGMANN, P. G. BERGMANN: *Theodore von Kármán Anniversary Volume.* Pasadena, 1941, p. 212.
8. H. WEYL: *Sitzungsber. d. Preuss. Acad. d. Wiss.* 1918, 465; *Ann. d. Phys.* **59** (1919) 101; [27], Chapter XI.
9. A. EINSTEIN: 'Théorie unitaire du champ physique', *Ann. Inst. H. Poincaré* 1931.
10. A. EINSTEIN: *The Meaning of Relativity.* Appendix II.
    M. A. TONNELAT: *La théorie du champ unifié d'Einstein.* Paris, 1955. (Cf. Bibliography.)
11. Y. THIRY: *Thesis.* Paris, 1950.
12. A. LICHNEROWICZ [25], p. 201.
13. F. HENNEQUIN: *Thesis.* Paris, 1955.
14. Cf. for example GUPTA: 'Quantification of Einstein's gravitational field. Linear approximation'. *Proc. Phys. Soc.* **65** (1962) 161.
15. The field equations must also satisfy at least four identities – which is not true for electrodynamics, even when it is non-linear. Cf. BERGMANN [9], p. 241.

PART FOUR

MATHEMATICAL SUPPLEMENT

CHAPTER XIV

TENSOR CALCULUS IN AN EUCLIDEAN VECTOR SPACE[1]

We shall call an n-dimensional space a *vector space* when the set of its elements **A**, **B**, ... up to n components possesses the usual commutative, associative and distributive properties of vectors. The elements **A**, **B**... are the *vectors* of this space.

A vector space is Euclidean if the operation (**A**, **B**) corresponds to a number and possesses the usual properties of *scalar multiplication.*

The summation convention. We shall always use the following summation convention: *the sign Σ relative to an index will be taken for granted when the index in question is repeated above and below within the same monomial*:

$$A_\mu B^\mu = A_1 B^1 + A_2 B^2 + \dots + A_n B^n$$

This index, which merely indicates a summation, has no determined value; it is a *dummy index*, which may be represented by any letter ($A_\mu B^\mu = A_\lambda B^\lambda$).

The results of the present chapter are valid for an n-dimensional Euclidean space. But the application that will be made will be limited to four-dimensional Euclidean space. Following accepted usage, the Greek indices μ, ν, ρ, σ... then take on the values 1, 2, 3, 0.

A. RECTILINEAR AXES

1. *Covariance and contravariance*

Let there be a reference system S defined by the rectilinear axes x^μ. On each of these axes let us locate a unit vector $\mathbf{e}_\mu$.

Any arbitrary vector **A** will be represented by:

(XIV-1) $$\mathbf{A} = A^\mu \cdot \mathbf{e}_\mu .$$

The quantities A^μ represent the projections of the vector **A** drawn *parallel to the coordinate axes.* By definition, we call them *contravariant components of the vector* **A**.

On the other hand, one can define the quantities such that

(XIV-2) $$A_\mu = \mathbf{A}\cdot\mathbf{e}_\mu .$$

These are the orthogonal projections of vector **A** along the axes; we call them the *covariant components of vector* **A**.

We shall designate by $g_{\mu\nu}$ the *scalar products of the base vectors*:

(XIV-3) $$(\mathbf{e}_\mu\cdot\mathbf{e}_\nu) = g_{\mu\nu} .$$

The $g_{\mu\nu}$, as the base vectors, *are constants* in the system of rectilinear axes we have chosen. According to their definition, we still have

(XIV-4) $$g_{\mu\nu} = g_{\nu\mu} .$$

The covariant components of a vector are easily obtained from its contra-

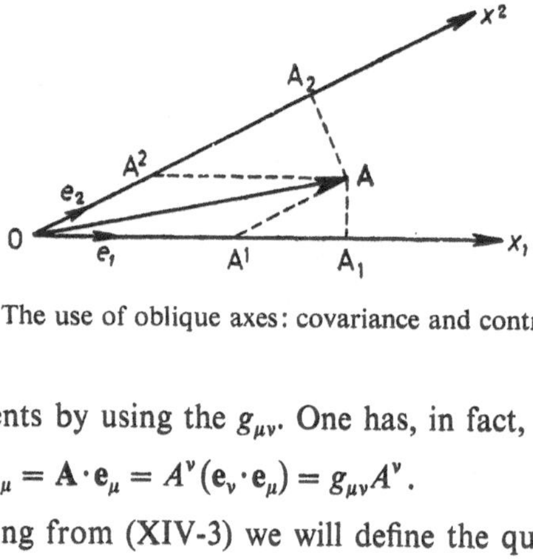

Fig. 42. The use of oblique axes: covariance and contravariance.

variant components by using the $g_{\mu\nu}$. One has, in fact,

(XIV-5) $$A_\mu = \mathbf{A}\cdot\mathbf{e}_\mu = A^\nu(\mathbf{e}_\nu\cdot\mathbf{e}_\mu) = g_{\mu\nu}A^\nu .$$

Finally, proceeding from (XIV-3) we will define the quantities $g^{\mu\nu}$ such that

(XIV-6) $$g_{\mu\rho}g^{\nu\rho} = \delta^\nu_\mu .$$

We then deduce from (XIV-5) and (XIV-6):

(XIV-7) $$g^{\mu\rho}A_\rho = g^{\mu\rho}g_{\rho\sigma}A^\sigma = \delta^\mu_\sigma A^\sigma = A^\mu .$$

And thus we pass from the covariant components of a vector to its contravariant components, and, conversely, by means of the relations

(XIV-8) $$A_\mu = g_{\mu\nu}A^\nu , \qquad A^\mu = g^{\mu\nu}A_\nu .$$

Let us designate by g the determinant formed by the $g_{\mu\nu}$:

(XIV-9) $$g = \det g_{\mu\nu} .$$

By taking (XIV-6) into account, we can easily verify

(XIV-10) $\text{cofactor } g_{\mu\nu} = g g^{\mu\nu}.$

Special case: The use of orthogonal systems. An Euclidean space may be referred to a system of orthogonal coordinates. According to the preceding definitions, we will them have

(XIV-11) $\mathbf{e}_\mu \cdot \mathbf{e}_\nu = \delta_{\mu\nu}$

with

(XIV-12) $\delta_{\mu\nu} = \begin{cases} 0 & \text{if } \mu \neq \nu \\ 1 & \text{if } \mu = \nu . \end{cases}$

And thus

(XIV-13) $g_{\mu\nu} = \delta_{\mu\nu}$

and

(XIV-14) $A^\mu = A_\mu .$

In a Euclidean space, the use of *a system of orthogonal coordinates* ($g_{\mu\nu} = \delta_{\mu\nu}$) *makes it possible to identify the contravariant components of a vector with its covariant components.* So there is no point, in this case, in distinguishing between covariance and contravariance. And that is why all the problems involving a three-dimensional Euclidean space dispense with this distinction, provided that – and it is always possible – a system of rectangular axes is employed.

2. *The norm of a vector. The scalar product of two vectors*

In a vector space, the *scalar product* of two vectors **A** and **B** has the following expression, by definition:

(XIV-15) $\mathbf{A} \cdot \mathbf{B} = (A^\mu \mathbf{e}_\mu) \cdot (B^\nu \mathbf{e}_\nu) = A^\mu B^\nu (\mathbf{e}_\mu \cdot \mathbf{e}_\nu)$

that is, according to (XIV-3)

(XIV-16) $\mathbf{A} \cdot \mathbf{B} = g_{\mu\nu} A^\mu B^\nu .$

Two vectors will be *orthogonal* if their product is zero.

(XIV-17) $\mathbf{A} \cdot \mathbf{B} \equiv g_{\mu\nu} A^\mu B^\nu = 0 .$

The *norm of a vector* is the scalar product of a vector with itself. This norm likewise represents the square of the length of the vector:

(XIV-18) $$|A|^2 = g_{\mu\nu} A^\mu A^\nu .$$

If the vector-norm is equal to one, the vector is said to be a *normalized* or *unit* vector.

We call a space a *true Euclidean* space when the norm of all non-zero vectors is positive.

More especially, the square of the vector **ds**, whose components are dx^μ, may be written in the form

(XIV-19) $$ds^2 = g_{\mu\nu}\, dx^\mu\, dx^\nu .$$

A special case: The use of orthogonal coordinates. If four-dimensional space is related to a system of rectangular axes

(XIV-13) $$g_{\mu\nu} = \delta_{\mu\nu} ,$$

the scalar product of two vectors **A** and **B** is simply written:

(XIV-20) $$\mathbf{A}\cdot\mathbf{B} = g_{\mu\nu} A^\mu B^\nu = \Sigma_\mu A^\mu B^\mu$$

and the "norm", that is, the square of the length of a vector, becomes, according to (XIV-20)

(XIV-21) $$|A|^2 = g_{\mu\nu} A^\mu A^\nu = \Sigma_\mu (A^\mu)^2 .$$

In particular:

(XIV-22) $$ds^2 = \Sigma_\mu (dx^\mu)^2 .$$

The expressions (XIV-20), (XIV-21) and (XIV-22) generalize the definitions, in three-dimensional space, of the scalar product, the length of a vector and the distance of two infinitesimally separated points $M\,(x^\mu)$ and $M'(x^\mu + dx^\mu)$ related to trirectangular axes.

3. *Transformation of rectilinear axes*

Let us make a change of the basic system in four-dimensional Euclidean space. In doing so, we transform from the unit vectors $\mathbf{e}_\mu$ associated with the rectilinear axes x^μ to the new vectors $\mathbf{e}'_\mu$ associated with the rectilinear axes x'^μ.

(XIV-23) $$\mathbf{e}'_\mu = a^\nu_{\mu'} \mathbf{e}_\nu .$$

The $a^\nu_{\mu'}$ are here the constant coefficients characterizing the transformation of oblique axes.

The new base vectors $\mathbf{e}'_\mu$ are independent if:

(XIV-24) $\quad a = |a^{\nu}_{\mu'}| \neq 0\,.$

Conversely, the $\mathbf{e}_\mu$ are expressed as functions of the $\mathbf{e}'_\mu$ by the formulas

(XIV-25) $\quad \mathbf{e}_\mu = a^{\nu'}_{\mu}\, \mathbf{e}'_\nu$

with

(XIV-26) $\quad a' = |a^{\nu'}_{\mu}| \neq 0\,.$

One then has, by combining (XIV-23) and (XIV-25)

(XIV-27) $\quad \mathbf{e}'_\mu = a^{\nu}_{\mu'}\, \mathbf{e}_\nu = a^{\nu}_{\mu'}\, a^{\rho'}_{\nu}\, \mathbf{e}'_\rho\,, \qquad \mathbf{e}_\mu = a^{\nu'}_{\mu}\, \mathbf{e}'_\nu = a^{\nu'}_{\mu}\, a^{\rho}_{\nu'}\, \mathbf{e}_\rho$

that is, the so-called condition of orthogonality:

(XIV-28) $\quad a^{\nu}_{\mu'}\, a^{\rho'}_{\nu} = a^{\rho}_{\nu'}\, a^{\nu'}_{\mu} = \delta^{\rho}_{\mu}\,,$

with

(XIV-29) $\quad \delta^{\rho}_{\mu} = \begin{cases} 0 & \text{if} \quad \mu \neq \rho \\ 1 & \text{if} \quad \mu = \rho\,. \end{cases}$

The conditions (XIV-28) are equivalent to the definition of the $a^{\nu}_{\mu'}$ as a function of the $a^{\sigma'}_{\rho}$, which results from the linear relationship (XIV-23) and their converse. One has, in fact,

(XIV-30) $\quad a^{\nu'}_{\mu} = \dfrac{\text{cofactor}\, a^{\mu}_{\nu'}}{[a]}\,, \qquad a^{\nu}_{\mu'} = \dfrac{\text{cofactor}\, a^{\mu'}_{\nu}}{[a']}\,,$

which does indeed produce conditions (XIV-28).

From the properties of the determinants one obtains, in particular

(XIV-31) $\quad [a]\,[a'] = 1\,.$

Let us now consider the vector

(XIV-32) $\quad \mathbf{x} = x^{\mu}\mathbf{e}_\mu$

whose contravariant components are x^μ in the system of rectilinear axes with the base $\mathbf{e}_\mu$, and x'^μ in the system with base $\mathbf{e}'_\mu$. One has

(XIV-33) $\quad \mathbf{x} = x^{\mu}\mathbf{e}_\mu = x'^{\mu}\mathbf{e}'_\mu$

that is

(XIV-34) $\quad x^{\mu} a^{\nu'}_{\mu}\, \mathbf{e}'_\nu = x'^{\mu}\mathbf{e}'_\mu\,, \qquad x'^{\mu} a^{\nu}_{\mu'}\, \mathbf{e}_\nu = x^{\mu}\mathbf{e}_\mu\,,$

whatever the bases $\mathbf{e}_\mu$ and $\mathbf{e}'_\mu$ may be.

Whence the transformation formulas:

(XIV-35) $x^{\mu} = a^{\mu}_{\nu'} x'^{\nu}$

and conversely

(XIV-36) $x'^{\mu} = a^{\mu'}_{\nu} x^{\nu}$.

4. *Invariants, four-vectors and tensors*

1) *An invariant* is a quantity that retains the same form when a change of the basic system (XIV-23) is made. In this chapter, equation (XIV-23) is limited to a change of rectilinear axes, $a^{\nu'}_{\mu}$ and $a^{\nu}_{\mu'}$ being constants. The expressions

(XIV-37) $ds^2 = dx_{\mu}\, dx^{\mu}, \qquad \Box = \dfrac{\partial^2}{\partial x_{\mu}\, \partial x^{\mu}}$

are invariants.

2) *A vector* **A** is a quantity whose covariant components are transformed, for a change of reference system at the same point, like the base vectors $\mathbf{e}_{\mu}$. Its contravariant components are transformed like the coordinates x^{μ}.

One has, in fact, for the covariant components:

(XIV-38) $A_{\mu} = \mathbf{A} \cdot \mathbf{e}_{\mu} = \mathbf{A} a^{\nu'}_{\mu} \mathbf{e}'_{\nu} = a^{\nu'}_{\mu} A'_{\nu}$

and conversely

(XIV-39) $A'_{\mu} = \mathbf{A}\mathbf{e}'_{\mu} = \mathbf{A} a^{\nu}_{\mu'} \mathbf{e}_{\nu} = a^{\nu}_{\mu'} A_{\nu}$.

On the other hand, by using the contravariant components, we can write

(XIV-40) $\mathbf{A} = A^{\mu} \mathbf{e}_{\mu} = A'^{\mu} \mathbf{e}'_{\mu}$

that is:

(XIV-41) $A^{\mu} a^{\nu'}_{\mu} \mathbf{e}'_{\nu} = A'^{\mu} \mathbf{e}'_{\mu}, \qquad A'^{\mu} \cdot a^{\nu}_{\mu'} \mathbf{e}_{\nu} = A^{\mu} \mathbf{e}_{\mu}$.

Whence the transformation formulas, analogous to (XIV-38), for the contravariant components:

(XIV-42) $A^{\mu} = a^{\mu}_{\nu'} A'^{\nu}, \qquad A'^{\mu} = a^{\mu'}_{\nu} A^{\nu}$.

3) By starting with either the covariant or contravariant components of

two vectors **A** and **B**, one can form the expressions:

(XIV-43) $C_{\mu\nu} = A_\mu B_\nu, \quad C^{\mu\nu} = A^\mu B^\nu, \quad C^\nu_\mu = A_\mu B^\nu.$

They are transformed as follows:

(XIV-44) $C'_{\mu\nu} = A'_\mu B'_\nu = a^\rho_{\mu'} a^\sigma_{\nu'} A_\rho B_\sigma = a^\rho_{\mu'} a^\sigma_{\nu'} C_{\rho\sigma}$

(XIV-45) $C'^{\mu\nu} = A'^\mu B'^\nu = a^{\mu'}_\rho a^{\nu'}_\sigma A^\rho B^\sigma = a^{\mu'}_\rho a^{\nu'}_\sigma C^{\rho\sigma}$

(XIV-46) $C'^\nu_\mu = A'_\mu B'^\nu = a^\rho_{\mu'} a^{\nu'}_\sigma A_\rho B^\sigma = a^\rho_{\mu'} a^{\nu'}_\sigma C^\sigma_\rho.$

Any quantity that obeys this law of transformation in change of base-systems is called a *tensor.*

The name *tensor of the nth order* is usually applied to a quantity whose components are characterized by n indices and obey the following law of transformation:

(XIV-47) $A'_{\mu\nu..\rho\sigma} = a^\alpha_{\mu'} a^\beta_{\nu'} .. a^\gamma_{\rho'} a^\lambda_{\sigma'} A_{\alpha\beta..\gamma\lambda}$

(XIV-48) $A'^{\mu\nu..\rho\sigma} = a^{\mu'}_\alpha a^{\nu'}_\beta .. a^{\rho'}_\gamma a^{\sigma'}_\lambda A^{\alpha\beta..\gamma\lambda}$

(XIV-49) $A'^{\rho\sigma}_{\mu\nu} = a^\alpha_{\mu'} a^\beta_{\nu'} .. a^{\rho'}_\gamma a^{\sigma'}_\lambda A_{\alpha\beta}^{..\gamma\lambda},$

and inversely

(XIV-50) $A_{\mu\nu..\rho\sigma} = a^{\alpha'}_\mu a^{\beta'}_\nu .. a^{\gamma'}_\rho a^{\lambda'}_\sigma A'_{\alpha\beta..\gamma\lambda}$

(XIV-51) $A^{\mu\nu..\rho\sigma} = a^\mu_{\alpha'} a^\nu_{\beta'} .. a^\rho_{\gamma'} a^\sigma_{\lambda'} A'^{\alpha\beta..\gamma\lambda}$

(XIV-52) $A_{\mu\nu}{}^{\rho\sigma} = a^{\alpha'}_\mu a^{\beta'}_\nu .. a^\rho_{\gamma'} a^\sigma_{\lambda'} A'^{\gamma\lambda}_{\alpha\beta}.$

5. *Symmetry and antisymmetry*

A tensor is *symmetrical* with respect to the μ and ν indices if its $A_{\mu\nu..}$ (or $A^{\mu\nu..}$) components are such that:

(XIV-53) $A_{\mu\nu..} = A_{\nu\mu..}$

In that case they will be represented by $A_{(\mu\nu)}$ (and occasionally by $A_{\underline{\mu\nu}}$).

A tensor is *antisymmetrical* with respect to the μ and ν indices if its components $A_{\mu\nu..}$ are such that:

(XIV-54) $A_{\mu\nu} = - A_{\nu\mu}$

They will then be represented by $A_{[\mu\nu]}$ (and occasionally by $A_{\underset{\vee}{\mu\nu}..}$).

Any tensor may be decomposed into a sum of symmetrical and antisymmetrical tensors

(XIV-55) $$A_{\mu\nu..} = \tfrac{1}{2}(A_{\mu\nu} + A_{\nu\mu}) + \tfrac{1}{2}(A_{\mu\nu} - A_{\nu\mu}) = A_{(\mu\nu)} + A_{[\mu\nu]}.$$

One can easily verify the fact that the symmetrical or antisymmetrical character of a tensor is not altered by a change of base-system.

Special remark. Every tensor constitutes an autonomous geometrical entity. By this is meant that a change of oblique axes (XIV-23) provides it with new components that may be completely deduced from the old components and the coefficients $a^{\nu}_{\mu'}$ or $a^{\nu'}_{\mu}$.

6. *Transformation of the metric tensor. A special case: Utilization of orthogonal frames of reference*

The law of transformation for the symmetrical tensor $g_{\mu\nu}$ is immediately apparent from its definition (XIV-3):

(XIV-56) $$g'_{\mu\nu} = (\mathbf{e}'_{\mu} \cdot \mathbf{e}'_{\nu}) = a^{\rho}_{\mu'} a^{\sigma}_{\nu'} (\mathbf{e}_{\rho} \cdot \mathbf{e}_{\sigma}) = a^{\rho}_{\mu'} a^{\sigma}_{\nu'} g_{\rho\sigma}$$

Conversely:

(XIV-57) $$g_{\mu\nu} = (\mathbf{e}_{\mu} \cdot \mathbf{e}_{\nu}) = a^{\rho'}_{\mu} a^{\sigma'}_{\nu} (\mathbf{e}'_{\rho} \cdot \mathbf{e}'_{\sigma}) = a^{\rho'}_{\mu} a^{\sigma'}_{\nu} g'_{\rho\sigma}.$$

The transformation of the contravariant components $g'^{\mu\nu}$ and $g^{\mu\nu}$ is immediately deduced from (XIV-56) and (XIV-57) when (XIV-6) is taken into account. One thus obtains[2]:

(XIV-58) $$g'^{\mu\nu} = a^{\mu'}_{\rho} a^{\nu'}_{\sigma} g^{\rho\sigma}$$

and conversely

(XIV-59) $$g^{\mu\nu} = a^{\mu}_{\rho'} a^{\nu}_{\sigma'} g'^{\rho\sigma}.$$

The invariance of the ds^2

(XIV-60) $$ds^2 = dx_{\mu}\, dx^{\mu}$$

is also verified by proceeding from (XIV-56)

(XIV-61) $$ds'^2 = g'_{\mu\nu}\, dx'^{\mu}\, dx'^{\nu} = g_{\mu\nu}\, dx^{\mu}\, dx^{\nu} = ds^2.$$

Special case: Utilization of orthogonal frames of reference. If the rectilinear reference systems defined by the $\mathbf{e}_{\mu}$ and the $\mathbf{e}'_{\mu}$ are orthogonal, one will have:

(XIV-62) $$g_{\mu\nu} = (\mathbf{e}_{\mu} \cdot \mathbf{e}_{\nu}) = \delta_{\mu\nu}, \qquad g'_{\mu\nu} = (\mathbf{e}'_{\mu} \cdot \mathbf{e}'_{\nu}) = \delta_{\mu\nu}$$

that is, according to (XIV-62)

(XIV-63) $\delta_{\mu\nu} = a^{\rho}_{\mu'} a^{\sigma}_{\nu'} \delta_{\rho\sigma} = \Sigma_{\rho} a^{\rho}_{\mu'} a^{\rho}_{\nu'}$

Multiplying by $a^{\nu'}_{\lambda}$ and summing over ν, there will result, if we take into account the conditions for orthogonality (XIV-28),

(XIV-64) $a^{\nu'}_{\lambda} \delta_{\mu\nu} = a^{\nu'}_{\lambda} \Sigma_{\rho} a^{\rho}_{\mu'} a^{\rho}_{\nu'} = a^{\lambda}_{\mu'}$

that is:

(XIV-65) $a^{\mu'}_{\lambda} = a^{\lambda}_{\mu'}$.

This relation results in the equality of the determinants $| a |$ and $| a' |$

(XIV-66) $[a'] = [a]$.

Moreover, according to (XIV-28), the determinants $| a |$ and $| a' |$ related to the change of base system still satisfy the relation:

(XIV-31) $[a][a'] = 1$.

By comparing (XIV-31) and (XIV-66) we then obtain:

(XIV-67) $[a] = [a'] = \pm 1$.

7. *The rotations of axes in a four-dimensional Euclidean space*

Let us consider the change of axes

(XIV-23) $\mathbf{e}'_{\mu} = a^{\nu}_{\mu'} \mathbf{e}_{\nu}, \qquad \mathbf{e}_{\mu} = a^{\nu'}_{\mu} \mathbf{e}'_{\nu}$

subjected to the conditions

(XIV-28) $a^{\rho'}_{\mu} a^{\nu}_{\rho'} = \delta^{\nu}_{\mu}$

which insure the invariance of the length of a vector.

The changes of the oblique axes will define the rotations if the two following conditions are satisfied:

1) The scalar products $g_{\mu\nu}$ and $g'_{\mu\nu}$ must be equal:

(XIV-68) $g'_{\mu\nu} = (\mathbf{e}'_{\mu} \cdot \mathbf{e}'_{\nu}) = (\mathbf{e}_{\mu} \cdot \mathbf{e}_{\nu}) = g_{\mu\nu}$.

One will then have:

(XIV-69) $a^{\rho}_{\mu'} a^{\sigma}_{\nu'} g_{\rho\sigma} = g_{\mu\nu}$

that is, multiplying by $a^{\nu'}_{\lambda}$ and summing, while taking (XIV-28) into account

(XIV-70) $$a^{\rho}_{\mu'} g_{\rho\lambda} = a^{\nu'}_{\lambda} g_{\mu\nu}.$$

Usually one is considering the case of orthogonal systems (Cartesian axes), so that condition (XIV-70) with $g_{\mu\nu} = g'_{\mu\nu} = \delta_{\mu\nu}$ reduces to (XIV-65)

(XIV-65) $$a^{\lambda}_{\mu'} = a^{\mu'}_{\lambda}$$

2) One must assume

(XIV-71) $$[a] = [a'] = +1.$$

We know, as a matter of fact, from the preceding section, that condition (XIV-65) together with (XIV-31) gives $|a| = |a'| = \pm 1$.

By choosing hypothesis (XIV-71), the orthogonal systems defined by the $\mathbf{e}_\mu$ and by the $\mathbf{e}'_\mu$ *are in the same sense*, and one passes from the one to the other *by means of a rotation.*

On the other hand, by putting $|a| = |a'| = -1$, the two systems are by definition oriented *in opposite senses*, and one passes from the one to the other *by means of a reflexion.*

Summing up: a *rotation of orthogonal axes* in a four-dimensional Euclidean space is defined by the following group of relations:

(XIV-23) $$\boxed{\mathbf{e}'_\mu = a^{\nu}_{\mu'}\mathbf{e}_\nu, \qquad \mathbf{e}_\mu = a^{\nu'}_{\mu}\mathbf{e}'_\nu}$$

with

(XIV-28) $$\boxed{a^{\rho}_{\mu'} a^{\nu'}_{\rho} = \delta^{\nu}_{\mu}}$$

and also by:

(XIV-65) $$a^{\lambda}_{\mu'} = a^{\mu'}_{\lambda}$$

(XIV-71) $$[a] = [a'] = 1$$

Relations (XIV-23) and (XIV-28) are valid for any linear transformation of oblique axes.

Condition (XIV-65) assures the preservation of the orthogonality of the axes.

Finally (XIV-71) insures the rotational character of the performed transformation.

B. USE OF ARBITRARY CURVILINEAR COORDINATES

8. *Passage from one system of curvilinear coordinates to another in an Euclidean vector space*

We have studied the transformations of one rectilinear reference system (x^μ) into another rectilinear system (x'^μ). In an Euclidean vector space it is always possible to relate the phenomena involved in any extended region of that space to a rectilinear system that may be orthogonal. In that case the ds^2 takes on the reduced form (XIV-22).

Nevertheless, it is also possible, and in certain cases advantageous, to describe the phenomena in terms of arbitrary curvilinear coordinates (y^μ). In order to pass from the components $A^{\rho\sigma}_{\mu\nu}$ of a tensor, which have been related to the system (M, y^μ), to the components $A'^{\rho\sigma}_{\mu\nu}$ of this tensor related to system (M', y'^μ), it is first necessary to find the connection between the two reference systems (M, y^μ) and (M', y'^μ). In other words, we must know how to determine (M', y'^μ) as a function of (M, y^μ). It is the laws of this sort of determination that we shall now derive.

1) *Curvilinear coordinates and their associated natural reference system.* Let us consider at the point M of an Euclidean vector space a system of curvilinear coordinates (y^μ) and an associated system formed by the unit vectors $\mathbf{e}_\mu$ tangent to the y^μ.[3] By definition [4]

(XIV-72) $$dM = \mathbf{e}_\mu \, dy^\mu .$$

The system of rectilinear axes $\mathbf{e}_\mu$ constitutes, at the point M, the *natural reference system associated with the curvilinear coordinates* y^μ.

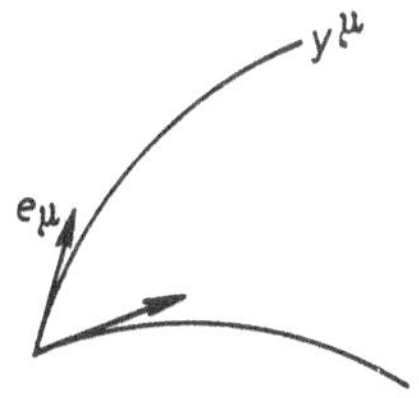

Fig. 43. Associated natural reference system.

If we choose, at the point M, another system of coordinates (y'^μ), we must associate with this system a new natural system formed by the unit vectors $\mathbf{e}'_\mu$ tangent to y'^μ. We shall thus have:

(XIV-73) $$dM = \mathbf{e}'_\mu \, dy'^\mu$$

that is

(XIV-74) $$\mathbf{e}'_\mu = \frac{\partial \mathbf{M}}{\partial y'^\mu} = \frac{\partial \mathbf{M}}{\partial y^\nu}\frac{\partial y^\nu}{\partial y'^\mu} = a^\nu_{\mu'}\mathbf{e}_\nu$$

putting

(XIV-75) $$a^\nu_{\mu'} = \frac{\partial y^\nu}{\partial y'^\mu}.$$

Inversely

(XIV-76) $$\mathbf{e}_\mu = a^{\nu'}_\mu \mathbf{e}'_\nu$$

with

(XIV-77) $$a^{\nu'}_\mu = \frac{\partial y'^\nu}{\partial y^\mu}.$$

In either curvilinear system (y^μ) or (y'^μ), the expression of the ds^2 is then:

(XIV-78) $$ds^2 = g_{\mu\nu}\, dy^\mu\, dy^\nu = g'_{\mu\nu}\, dy'^\mu\, dy'^\nu$$

and one has, as in (XIV-3):

(XIV-79) $$g_{\mu\nu} = (\mathbf{e}_\mu \cdot \mathbf{e}_\nu), \qquad g'_{\mu\nu} = (\mathbf{e}'_\mu \cdot \mathbf{e}'_\nu).$$

The $g_{\mu\nu}$ that reduce to $\delta_{\mu\nu}$ in an orthonormalized system have, at point M, a value determined by the natural system formed by the oblique axes $\mathbf{e}_\mu$ associated with the y^μ. One still has:

(XIV-80) $$g_{\mu\nu} = (\mathbf{e}_\mu \cdot \mathbf{e}_\nu) = a^{\rho'}_\mu a^{\sigma'}_\nu (\mathbf{e}'_\rho \mathbf{e}'_\sigma) = a^{\rho'}_\mu a^{\sigma'}_\nu g'_{\rho\sigma}.$$

2) Let us consider two infinitesimally separated points M and M' in an Euclidean vector space such that $\mathbf{OM}' = \mathbf{OM} + d\mathbf{M}$. To simplify things, let us represent $\mathbf{OM}'$ and $\mathbf{OM}$ by $\mathbf{M}'$ and $\mathbf{M}$.[4] The natural reference system $\mathbf{e}_\mu$ is associated with the curvilinear coordinate stysem y^μ at $\mathbf{M}$. At $\mathbf{M} + d\mathbf{M}$ there corresponds to $\mathbf{e}_\mu$ the natural reference system $\mathbf{e}_\mu + d\mathbf{e}_\mu$. We seek to connect the reference system obtained from ($\mathbf{M} + d\mathbf{M}$, $\mathbf{e}_\mu + d\mathbf{e}_\mu$) with the frame of reference using system ($\mathbf{M}$, $\mathbf{e}_\mu$). In order to do this we must express $d\mathbf{M}$ and $d\mathbf{e}_\mu$ in the natural reference system at point M:

(XIV-81) $$\boxed{d\mathbf{M} = \omega^\mu \mathbf{e}_\mu}$$

(XIV-82) $$\boxed{d\mathbf{e}_\mu = \omega^\nu_\mu \mathbf{e}_\nu}$$

with

(XIV-83) $\omega^\mu = dy^\mu, \qquad \omega_\mu^\nu = \Gamma^\nu_{\mu\rho}\, dy^\rho,$

ω_μ^ν being a linear function of the differentials dy^ρ. One shall then have:

(XIV-84) $d\mathbf{M} = dy^\mu \mathbf{e}_\mu$

(XIV-85) $d\mathbf{e}_\mu = \Gamma^\nu_{\mu\rho}\mathbf{e}_\nu\, dy^\rho.$

From this we deduce

(XIV-86) $dg_{\mu\nu} = d(\mathbf{e}_\mu\cdot\mathbf{e}_\nu) = \Gamma^\lambda_{\mu\rho}(\mathbf{e}_\lambda\cdot\mathbf{e}_\nu)\, dy^\rho + \Gamma^\lambda_{\nu\rho}(\mathbf{e}_\mu\cdot\mathbf{e}_\lambda)\, dy^\rho = (\Gamma_{\mu\rho,\nu} + \Gamma_{\nu\rho,\mu})\, dy^\rho$

putting

(XIV-87) $\Gamma^\lambda_{\mu\rho} g_{\lambda\nu} = \Gamma_{\mu\rho,\nu}.$

In addition, according to (XIV-84),

(XIV-88) $\partial_\mu \mathbf{M} = \mathbf{e}_\mu.$

The conditions

(XIV-89) $\partial_\nu(\partial_\mu \mathbf{M}) = \partial_\mu(\partial_\nu \mathbf{M}) \qquad \text{or} \qquad \partial_\nu \mathbf{e}_\mu = \partial_\mu \mathbf{e}_\nu$

are then expressed, taking (XIV-85) into account, by:

(XIV-90) $\Gamma^\lambda_{\mu\nu}\cdot\mathbf{e}_\lambda = \Gamma^\lambda_{\nu\mu}\cdot\mathbf{e}_\lambda$

or:

(XIV-91) $\Gamma^\lambda_{\mu\nu} = \Gamma^\lambda_{\nu\mu}, \qquad \Gamma_{\mu\nu,\rho} = \Gamma_{\nu\mu,\rho}.$

The coefficients $\Gamma^\alpha_{\mu\nu}$ are symmetrical in μ and ν.
If we now write (XIV-86) in the following form:

(XIV-92)$_1$ $\Gamma_{\mu\rho,\nu} + \Gamma_{\nu\rho,\mu} = \partial_\rho g_{\mu\nu},$

and if we change μ to ρ, and then ν to ρ, in (XIV-92)$_1$

(XIV-92)$_2$ $\Gamma_{\rho\mu,\nu} + \Gamma_{\nu\mu,\rho} = \partial_\mu g_{\rho\nu}$

(XIV-92)$_3$ $\Gamma_{\mu\nu,\rho} + \Gamma_{\rho\nu,\mu} = \partial_\nu g_{\mu\rho},$

equation (XIV-92)$_2$ + (XIV-92)$_3$ − (XIV-92)$_1$ may further be written

(XIV-93) $\Gamma_{\mu\nu,\rho} = \tfrac{1}{2}(\partial_\mu g_{\nu\rho} + \partial_\nu g_{\mu\rho} - \partial_\rho g_{\mu\nu}),$

taking into account the symmetry of the $\Gamma_{\mu\nu,\rho}$ expressed by (XIV-91).

Finally, putting:

(XIV-94) $$[\mu\nu,\rho] = \tfrac{1}{2}(\partial_\mu g_{\nu\rho} + \partial_\nu g_{\mu\rho} - \partial_\rho g_{\mu\nu})$$

(XIV-95) $$\left\{\begin{matrix}\rho \\ \mu\nu\end{matrix}\right\} = \tfrac{1}{2} g^{\rho\sigma}(\partial_\mu g_{\nu\sigma} + \partial_\nu g_{\mu\sigma} - \partial_\sigma g_{\mu\nu})$$

the coefficients $\Gamma^\rho_{\mu\nu}$ will be reduced to the following expressions:

(XIV-96) $$\Gamma_{\mu\nu,\rho} = [\mu\nu,\rho], \qquad \Gamma^\rho_{\mu\nu} = \left\{\begin{matrix}\rho \\ \mu\nu\end{matrix}\right\}.$$

Expressions (XIV-94) and (XIV-95) are called Christoffel symbols of the first and second kind. The coefficients $\Gamma^\rho_{\mu\nu}$, identical to the Christoffel symbols, are then completely expressed as a function of the $g_{\mu\nu}$ – that is, of the natural reference system at M – and of their derivatives. The increments $d\mathbf{M}$ and $d\mathbf{e}_\mu$ are thus related to the reference system defined at M. The relation between the two reference systems ($M + d\mathbf{M}, \mathbf{e}_\mu + d\mathbf{e}_\mu$) and ($\mathbf{M}, \mathbf{e}_\mu$) is then clearly determined.

Remark I. The coefficients $\Gamma^\rho_{\mu\nu}$, which make it possible to determine the increments $d\mathbf{e}_\mu$, that is, to express the relation of the natural reference systems at two infinitesimally separated points in space, are called *affine connexion coefficients*. The $\Gamma^\rho_{\mu\nu}$ do not form the components of a tensor.[5]

Remark II. a) If an Euclidean vector space is related to any *system of curvilinear coordinates*, the associated natural reference system varies from one point to another. The $g_{\mu\nu}$, by their very definition, $g_{\mu\nu} = (\mathbf{e}_\mu \cdot \mathbf{e}_\nu)$, *vary from one point to another and are functions of the* y^μ.

The affine connexion that determines the relationship of the natural systems is then *expressed by the Christoffel symbols.*

b) If we relate an Euclidean vector space to a *system of rectilinear coordinates*, the natural systems are equally applicable at every point. In a system of rectilinear axes, the $g_{\mu\nu}$ *are constants.* The coefficients $\Gamma^\rho_{\mu\nu}$, which express the variation of one natural system with respect to another infinitesimally separated natural system, then disappear identically.

In the even more special case where the rectilinear coordinates are also *orthogonal*, the $g_{\mu\nu}$ constants are reduced to $\delta_{\mu\nu}$.

9. *Differential relations between the components of the metric tensor*

Let us call g the determinant formed by the covariant components of the $g_{\mu\nu}$.

(XIV-9) $$g = \det g_{\mu\nu}.$$

We verify, as in (XIV-10)

(XIV-10) $\qquad \text{cofactor } g_{\mu\nu} = g g^{\mu\nu}$,

using the definition

$$\sum_{\mu} g_{\mu\rho} \text{ cofactor } g_{\mu\nu} = g \delta_{\rho\nu}$$

and the relation (XIV-6)

(XIV-6) $\qquad g_{\mu\rho} g^{\mu\sigma} = \delta_{\rho}^{\sigma}$.

Now, by definition

(XIV-97) $\qquad dg = \sum_{\mu\nu} \text{cofactor } g_{\mu\nu} \cdot dg_{\mu\nu} = g g^{\mu\nu} dg_{\mu\nu}$

and, taking (XIV-6) into account:

(XIV-98) $\qquad dg = -g g_{\mu\nu} dg^{\mu\nu}$.

Then let us form,

(XIV-99)
$$\begin{aligned} dg_{\mu\nu} &= d(g_{\mu\rho} g_{\nu\sigma} g^{\rho\sigma}) \\ &= g_{\mu\rho} g_{\nu\sigma} dg^{\rho\sigma} + \delta_{\nu}^{\rho} dg_{\mu\rho} + \delta_{\mu}^{\sigma} dg_{\nu\sigma} \\ &= g_{\mu\rho} g_{\nu\sigma} dg^{\rho\sigma} + 2 dg_{\mu\nu}. \end{aligned}$$

One then obtains

(XIV-100) $\qquad dg_{\mu\nu} = -g_{\mu\rho} g_{\nu\sigma} dg^{\rho\sigma}$.

Inversely

(XIV-101) $\qquad dg^{\mu\nu} = -g^{\mu\rho} g^{\nu\sigma} dg_{\rho\sigma}$.

Let us now go back to definition (XIV-95) of the Christoffel symbols. If $\rho = \nu$

(XIV-102) $\qquad \begin{Bmatrix} \rho \\ \mu\rho \end{Bmatrix} = \frac{1}{2} g^{\rho\sigma} \partial_{\mu} g_{\rho\sigma}$

and, taking (XIV-97) into account[6]:

(XIV-103) $\qquad \begin{Bmatrix} \rho \\ \mu\rho \end{Bmatrix} = \frac{1}{2g} \partial_{\mu} g = \partial_{\mu} \log \sqrt{|g|}$.

10. *Covariant differentation*

a) In the change of systems $(\mathbf{M}, \mathbf{e}_\mu) - (\mathbf{M} + d\mathbf{M}, \mathbf{e}_\mu + d\mathbf{e}_\mu)$ the *contravariant components* A^μ of a vector are modified, and the natural system defined by the curvilinear coordinates at M is likewise changed. One still has

(XIV-1) $\mathbf{A} = A^\mu \mathbf{e}_\mu$

that is:

(XIV-104) $d\mathbf{A} = dA^\mu \cdot \mathbf{e}_\mu + A^\mu d\mathbf{e}_\mu = (dA^\mu + \omega^\mu_\rho A^\rho)\,\mathbf{e}_\mu = \nabla A^\mu \mathbf{e}_\mu .$

In the natural system at M, the contravariant components of the increment $\mathbf{dA}$ of the vector $\mathbf{A}$ are then:

(XIV-105) $\nabla A^\mu = dA^\mu + \omega^\mu_\sigma A^\sigma = dA^\mu + \Gamma^\mu_{\sigma\rho} A^\sigma\, dy^\rho .$

By definition, these increments ∇A^μ are the components of a vector. The expression

$$\nabla_\rho A^\mu = \frac{\nabla A^\mu}{dy^\rho}$$

thus appears as a tensor. One will have:

(XIV-106) $$\boxed{\nabla_\rho A^\mu \equiv \frac{\nabla A^\mu}{dy^\rho} = \partial_\rho A^\mu + \Gamma^\mu_{\sigma\rho} A^\sigma .}$$

On the other hand, the term $\Gamma^\mu_{\sigma\rho} A^\sigma$ does not represent the components of a tensor; and, consequently, the same is true of the ordinary derivative $\partial_\rho A^\mu$.

The quantity $\nabla_\rho A^\mu$ is called the *covariant* or *absolute derivative* of the contravariant components A^μ. The expression ∇A^μ is the *absolute differential of* A^μ. It represents the *real increment of vector* A *related to the reference system* (y^μ) *defined at* M. In other words, this "real" increment results on one hand from the variation of the components A^μ, and on the other from the modification of the natural system when passing from point M (y^μ) to the infinitesimally separated point $\mathrm{M} + d\mathrm{M}$ $(y^\mu + dy^\mu)$. Both the absolute differential and the covariant derivative possess a tensor character.

b) According to (XIV-104), the absolute differential ∇A^μ of the contravariant components of a vector A represent the contravariant components $(dA)^\mu$ of the increment dA.

In a similar manner, the absolute differential ∇A_μ *of the covariant components of* **A** defines the *covariant components* $(dA)_\mu$ *of this same increment d***A**. One will then have:

(XIV-107) $\quad \nabla A_\mu = (dA)_\mu = d\mathbf{A}\cdot\mathbf{e}_\mu$

that is, taking (XIV-2) and (XIV-82) into account

(XIV-108) $\quad \nabla A_\mu = d(\mathbf{A}\cdot\mathbf{e}_\mu) - \mathbf{A}\,d\mathbf{e}_\mu = dA_\mu - \mathbf{A}\omega^\sigma_\mu\mathbf{e}_\sigma$

or

(XIV-109) $\quad \nabla A_\mu = dA_\mu - \omega^\sigma_\mu A_\sigma = dA_\mu - \Gamma^\sigma_{\mu\rho}A_\sigma\,dy^\rho\,.$

The covariant derivative of the components A_μ then has the form:

(XIV-110) $$\boxed{\nabla_\rho A_\mu \equiv \frac{\nabla A_\mu}{dy_\rho} = \partial_\rho A_\mu - \Gamma^\sigma_{\mu\rho}A_\sigma\,.}$$

Let us note that (XIV-109) also gives us

(XIV-111) $\quad \nabla\mathbf{e}_\mu \equiv d\mathbf{e}_\mu - \omega^\sigma_\mu\mathbf{e}_\sigma \equiv 0$

or, according to (XIV-79), which constitutes the *definition* of the $g_{\mu\nu}$ in the same Euclidean space[7]

(XIV-112) $\quad \nabla g_{\mu\nu} \equiv 0\,.$

Of course, with the help of the metric tensor, one can establish a bond between definitions (XIV-104) and (XIV-107). According to (XIV-107) one does indeed have:

(XIV-113) $\quad (dA)_\mu = d\mathbf{A}\cdot\mathbf{e}_\mu = \nabla A^\rho(\mathbf{e}_\rho\cdot\mathbf{e}_\mu) = g_{\mu\rho}\nabla A^\rho\,.$

Thus, by comparing (XIV-107) and (XIV-113)[8]

(XIV-114) $\quad \nabla A_\mu = g_{\mu\rho}\nabla A^\rho\,.$

c) The covariant or absolute derivative of a tensor is obtained by generalizing expressions (XIV-106) and (XIV-110), the plus-sign being adopted for every contravariant index and the minus-sign for every covariant index. One will have:

(XIV-115) $$\boxed{\begin{aligned}\nabla_\rho A^{\sigma\tau}_{\mu\nu} = \partial_\rho A^{\sigma\tau}_{\mu\nu} + \Gamma^\sigma_{\lambda\rho}A^{\lambda\tau}_{\mu\nu} + \Gamma^\tau_{\lambda\rho}A^{\sigma\lambda}_{\mu\nu} + \dots\\ - \Gamma^\lambda_{\mu\rho}A^{\sigma\tau}_{\lambda\nu} - \Gamma^\lambda_{\nu\rho}A^{\sigma\tau}_{\mu\lambda} - \dots\end{aligned}}$$

d) *The covariant derivative of the metric tensor.* By applying (XIV-115) to the covariant components $g_{\mu\nu} = \mathbf{e}_\mu \cdot \mathbf{e}_\nu$ of the metric tensor, one obtains:

(XIV-116) $$\nabla_\rho g_{\mu\nu} = \partial_\rho g_{\mu\nu} - \Gamma^\sigma_{\mu\rho} g_{\sigma\nu} - \Gamma^\sigma_{\nu\rho} g_{\mu\sigma}.$$

In fact, by definition, the affine connection of an Euclidean vector space reduces to the Christoffel symbols. One will then have:

(XIV-117) $$\nabla_\rho g_{\mu\nu} = \partial_\rho g_{\mu\nu} - \begin{Bmatrix}\sigma\\ \mu\rho\end{Bmatrix} g_{\sigma\nu} - \begin{Bmatrix}\sigma\\ \nu\rho\end{Bmatrix} g_{\mu\sigma}.$$

According to definition (XIV-95) for the symbols $\{{}^{\rho}_{\mu\nu}\}$, the expression (XIV-117) is identically zero. In an Euclidean vector space, one always has:

(XIV-118) $$\boxed{\nabla_\rho g_{\mu\nu} \equiv 0.}$$

And in a like manner for the contravariant components:

(XIV-119) $$\boxed{\nabla_\rho g^{\mu\nu} \equiv 0.}$$

11. *Tensor densities*

A *tensor density* is the product of a tensor multiplied by $\sqrt{-g}$, g being the determinant of the metric tensor. We shall put:

(XIV-120) $$\boxed{\mathscr{A}^{\mu\nu}_{\rho\sigma} = \sqrt{-g}\, A^{\mu\nu}_{\rho\sigma}.}$$

In particular, a scalar density is defined by beginning with the scalar A:

(XIV-121) $$\mathscr{A} = \sqrt{-g}\, A.$$

It is transformed like $\sqrt{-g}$ in a change of reference system. One then obtains:

(XIV-122) $$g'_{\mu\nu} = a^\rho_{\mu'} a^\sigma_{\nu'} g_{\rho\sigma}$$

that is

(XIV-123) $$g' = a^2 g, \qquad a = \det a^\rho_{\mu'} = \frac{1}{a'}, \qquad (a' = \det a^{\rho'}_\mu).$$

The scalar density $\mathscr{A}$ is then transformed in the following manner:

(XIV-124) $$\mathscr{A}' = \sqrt{-g'}\, A' = a\sqrt{-g}\, A = a\mathscr{A}.$$

Thus, one is led to substitute for the element of volume

(XIV-125) $$d\tau = dy^1 \wedge dy^2 \ldots \wedge dy^n.$$

the invariant

(XIV-126) $$\sqrt{-g}\, d\tau = \sqrt{-g}\, dy^1 \wedge dy^2 \ldots \wedge dy^n.$$

In fact, according to Jacobi's theorem, (XIV-125) is transformed as follows:

(XIV-127) $$d\tau' = d\tau \quad \text{determ} \quad \frac{dy'^{\rho}}{dy^{\sigma}} = a' d\tau = \frac{1}{a} d\tau.$$

Consequently, (XIV-126) is indeed an invariant:

(XIV-128) $$\sqrt{-g'}\, d\tau' = a\sqrt{-g}\,\frac{1}{a}\, d\tau = \sqrt{-g}\, d\tau.$$

The introduction of tensor densities frequently makes it possible to simplify the formalism of the field equations when they are written for a Riemannian space or for an Euclidean space related to arbitrary curvilinear coordinates.

Indeed, the covariant derivation of a density $\mathscr{A}^{\mu\nu\ldots} = \sqrt{-g}\, A^{\mu\nu\ldots}$ leads to the following result

(XIV-129) $$\begin{aligned} \nabla_\rho \mathscr{A}^{\mu\nu\ldots} &= \sqrt{-g}\, \nabla_\rho A^{\mu\nu\ldots} = \\ &= \sqrt{-g}\left(\partial_\rho A^{\mu\nu\ldots} + \begin{Bmatrix}\mu \\ \sigma\rho\end{Bmatrix} A^{\sigma\nu\ldots} + \begin{Bmatrix}\nu \\ \sigma\rho\end{Bmatrix} A^{\mu\sigma\ldots}\right) \end{aligned}$$

or

(XIV-130) $$\begin{aligned} \nabla_\rho \mathscr{A}^{\mu\nu} &= \partial_\rho \mathscr{A}^{\mu\nu\ldots} + \begin{Bmatrix}\mu \\ \sigma\rho\end{Bmatrix} \mathscr{A}^{\sigma\nu} + \\ &+ \begin{Bmatrix}\nu \\ \sigma\rho\end{Bmatrix} \mathscr{A}^{\mu\sigma} - \mathscr{A}^{\mu\nu} \frac{\partial_\rho \sqrt{-g}}{\sqrt{-g}} \\ &= \partial_\rho \mathscr{A}^{\mu\nu} + \begin{Bmatrix}\mu \\ \sigma\rho\end{Bmatrix} \mathscr{A}^{\sigma\nu} + \begin{Bmatrix}\nu \\ \sigma\rho\end{Bmatrix} \mathscr{A}^{\mu\sigma} - \begin{Bmatrix}\sigma \\ \sigma\rho\end{Bmatrix} \mathscr{A}^{\mu\nu} \end{aligned}$$

Let us consider in particular an antisymmetrical tensor of the second order whose contravariant components are $A^{\mu\nu}_{\vee}$. According to (XIV-130), we shall obtain, by taking the divergence of the corresponding density

(XIV-131) $$\nabla_\rho \mathscr{A}^{\mu\rho}_{\vee} = \partial_\rho \mathscr{A}^{\mu\rho}_{\vee}$$

or

(XIV-132) $$\boxed{\nabla_\rho A^{\mu\rho}_{\vee} = \frac{1}{\sqrt{-g}} \partial_\rho (\sqrt{-g}\, A^{\mu\rho}_{\vee}).}$$

The introduction of tensor densities thus reduces, in this particular case, the formalism of covariant differentiation to a formalism of ordinary differentiation.

Similarly, if we are concerned with a vector A^ρ corresponding to the density $\mathscr{A}^\rho = \sqrt{-g}\, A^\rho$, one immediately obtains, according to (XIV-130),

(XIV-133) $$\nabla_\rho \mathscr{A}^\rho = \partial_\rho \mathscr{A}^\rho$$

that is

(XIV-134) $$\boxed{\nabla_\rho A^\rho = \frac{1}{\sqrt{-g}} \partial_\rho \mathscr{A}^\rho}$$

PROBLEMS

1. Show that the properties of symmetry or antisymmetry of a tensor are left unchanged by a general transformation of coordinates.

2. Prove the validity of transformation law (6) (in footnote 5, p. 439) for the expression

$$\tfrac{1}{2} g^{\lambda\rho} (\partial_\mu g_{\nu\rho} + \partial_\nu g_{\mu\rho} - \partial_\rho g_{\mu\nu})$$

which is explicitly of the form of a Christoffel symbol.

3. Show that the expressions

$$\varphi_{\mu\nu} = \partial_\mu \varphi_\nu - \partial_\nu \varphi_\mu$$

$$\varphi_{\mu\nu\rho} = \partial_\mu \varphi_{\nu\rho} + \partial_\rho \varphi_{\mu\nu} + \partial_\nu \varphi_{\rho\mu}$$

are transformed like tensors.

4. In any system of coordinates, the Laplacian has the following expression

$$\Delta f = g^{\mu\nu}\nabla_\mu \nabla_\nu f$$

a) Expand this expression by use of the Christoffel symbols.
b) Show that it can be written in the form of a divergence

$$\frac{1}{\sqrt{-g}}\partial_\rho(\sqrt{-g}\,V^\rho)$$

c) Give the form for Δ in a system of orthogonal coordinates such that

$$ds^2 = \pm \sum_{\mu=1}^{n} \omega_\mu^2$$

where $\omega_\mu = \alpha^{(\mu)}(y^\rho)\,dy^\mu$ (without summation).

NOTES

1. The results of the A part of this chapter are used in Chapter VI on "The Four-dimensional Formalism of Special Relativity". However, the formalism of Special Relativity introduces restrictions in the hypotheses of part A by limiting them exclusively to orthonormalized systems. Nevertheless, it is useful, as we explained at the outset of Chapter VI, to introduce a general method of localization in Euclidean space. This latter topic we have confined to the supplementary mathematical considerations taken up in the present chapter.
2. In fact, according to (XIV-6) and (XIV-56)

$$g_{\rho\sigma}{}'g'^{\rho\tau} = a_{\rho'}{}^{\lambda}a_{\sigma'}{}^{\delta}g_{\lambda\delta}g'^{\rho\tau} = \delta_\sigma{}^\tau\,.$$

Let us multiply by $a_{\tau'}{}^{\nu}\,a_\pi{}^{\sigma'}$ and sum, taking (XIV-6) into account

$$a_{\tau'}{}^{\nu}a_\pi{}^{\sigma'}a_{\rho'}{}^{\lambda}a_{\sigma'}{}^{\delta}g_{\lambda\delta}g'^{\rho\tau} = \delta_\sigma{}^\tau a_{\tau'}{}^{\nu}a_\pi{}^{\sigma'} = a_{\sigma'}{}^{\nu}a_\pi{}^{\sigma'} = \delta_\pi{}^\nu$$
$$\delta_\pi{}^\delta a_{\tau'}{}^{\nu}a_{\rho'}{}^{\lambda}g_{\lambda\delta}g'^{\rho\tau} = a_{\tau'}{}^{\nu}a_{\rho'}{}^{\lambda}g_{\lambda'\pi}g'^{\rho\tau} = \delta_\pi{}^\nu$$

and then, multiplying by $g^{\mu\pi}$, we obtain (XIV-59)

$$g^{\mu\pi}a_{\tau'}{}^{\nu}a_{\rho'}{}^{\lambda}g_{\lambda\pi}g'^{\rho\tau} = g^{\mu\pi}\delta_\pi{}^\nu$$
$$\delta_\lambda{}^\mu a_{\tau'}{}^{\nu}a_{\rho'}{}^{\lambda}g'^{\rho\tau} = a_{\rho'}{}^{\mu}a_{\tau'}{}^{\nu}g'^{\rho\tau} = g^{\mu\nu}.$$

3. More precisely, the $\mathbf{e}_\mu$ are tangent to the "coordinate curves" y^μ. A "coordinate curve" is the curve obtained by allowing only one of the coordinates to vary, the others remaining constant.
If we are concerned with a curvilinear system, the y^μ are the coordinate curves.
If one considers, for example, a system of polar coordinates (r, θ, φ), the coordinate curves are formed by the radius vector, the meridian and the parallel passing through point M. These curves are orthogonal at M and the associated natural system formed by the unit vectors $\mathbf{e}_\mu$ tangent to these curves is an orthogonal system. Nevertheless, at the point M in the system $y^\mu(r, \theta, \varphi)$, the $g_{\mu\nu}$ are not $\delta_{\mu\nu}$ but depend on the coordinates of M. In fact, by unit vectors we mean the vectors $\mathbf{e}_\mu$, which serve as *standards*

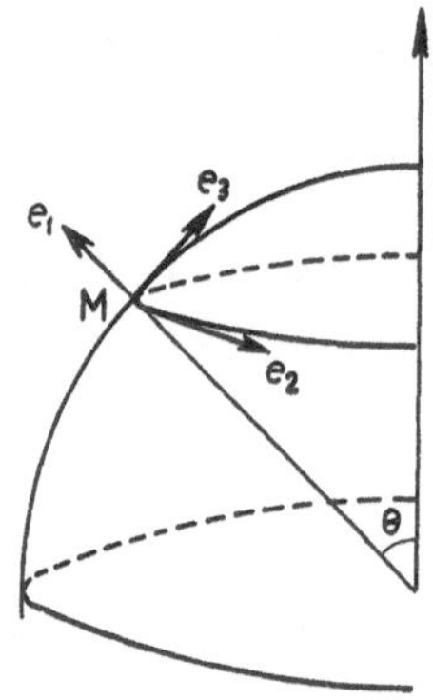

Fig. 44. Coordinate curves and the associated natural reference system.

of length. Let us consider, for example, the Cartesian system x', y', z' with the origin M, equally applicable with the rectangular system $O\,x\,y\,z$. One will have

(1) $$\mathbf{e}_1' = 1\,, \qquad \mathbf{e}_2' = 1\,, \qquad \mathbf{e}_3' = 1\,,$$

and

$$g_{pq}' = \mathbf{e}_p'\cdot\mathbf{e}_q' = \delta_{pq}.$$

It is immediately apparent that the change of the $\mathbf{e}_\mu'$ to $\mathbf{e}_\mu$, previously defined as vectors tangent to the coordinate curves y_r, y_θ, y_φ, does not allow us to put the standards of length, defined by each $\mathbf{e}_\mu$, equal to unity. One has, as a matter of fact according to (XIV-25)

(2) $$\mathbf{e}_p = a_p{}^{q'}\mathbf{e}_q'$$

and, since the axes $\mathbf{e}_q'$ are rectangular and equal to one

(3) $$(e_p{}^2) = (a_p{}^1)^2 + (a_p{}^2)^2 + (a_p{}^3)^2\,.$$

We deduce the values of the

$$a_p{}^{q'} = \frac{\partial x'^q}{\partial x^p}$$

with

$$\begin{cases} x'^1 = x\,, & x'^2 = y\,, & x'^3 = z\,, \\ x^1 = r\,, & x^2 = \theta\,, & x^3 = \varphi\,, \end{cases}$$

from the relations

(4) $$x = x'^1 = r\sin\theta\cos\varphi, \qquad y = x'^2 = r\sin\theta\sin\varphi, \qquad z = x'^3 = r\cos\theta.$$

By substituting in (3) we obtain

(5) $$\mathbf{e}_1 = 1,\ \mathbf{e}_2 = r,\ \mathbf{e}_3 = r\sin\theta\,.$$

Whence the value of the $g_{\mu\nu} = (\mathbf{e}_\mu\cdot\mathbf{e}_\nu)$ at M in the system r, θ, φ. Cf. LICHNEROWICZ [35], p. 81.

4. If, to every value of a parameter ξ, there corresponds one, and only one point M, the notation $d\mathbf{M}$ represents, as an abbreviation, the differential

$$d\,\mathbf{OM} = d\,\mathbf{O'M} = \xi' d\xi$$

which does not depend on the arbitrary origin O or O', but solely on the point M.

5. In fact, a change of curvilinear system at the point M allows us to write, according to (XIV-25)

(1) $$\mathbf{e}_\mu = a_\mu{}^{\nu'}\,\mathbf{e}_{\nu}'$$

Whence

(2) $$d\mathbf{e}_\mu = a_\mu{}^{\nu'}\,d\mathbf{e}_\nu' + (da_\mu{}^{\nu'})\,\mathbf{e}_\nu'$$

and according to (XIV-82)

(3) $$d\mathbf{e}_\mu = a_\mu{}^{\nu'}\omega_{\nu'}{}^{\rho'}\mathbf{e}_\rho' + (da_\mu{}^{\rho'})\mathbf{e}_\rho' = (da_\mu{}^{\rho'} + a_\mu{}^{\nu'}\omega_{\nu'}{}^{\rho'})\mathbf{e}_\rho' = (da_\mu{}^{\rho'} + a_\mu{}^{\nu'}\omega_{\nu'}{}^{\rho'})a_{\rho'}{}^{\sigma}\mathbf{e}_\sigma .$$

But one also has at point M

(4) $$d\mathbf{e}_\mu = \omega_\mu{}^\sigma \mathbf{e}_\sigma .$$

So one will have, by comparing (3) and (4):

(5) $$\omega_\mu{}^\sigma = \Gamma_{\mu\lambda}{}^\sigma\, dy^\lambda = (da_\mu{}^{\rho'} + a_\mu{}^{\nu'}\omega_{\nu'}{}^{\rho'})\, a_{\rho'}{}^\sigma$$

that is

(6) $$\Gamma_{\mu\lambda}{}^\sigma = (\partial_\lambda a_\mu{}^{\rho'} + a_\mu{}^{\nu'}\Gamma_{\nu\tau}{}'^{\rho} a_\lambda{}^{\tau'})\, a_{\rho'}{}^\sigma ,$$

since

$$\omega_{\nu'}{}^{\rho'} = \Gamma_{\nu\tau}{}'^{\rho} dy'^{\tau} = \Gamma_{\nu\tau}{}'^{\rho} a_\lambda{}^{\tau'}\, dy^\lambda .$$

Besides the term $a_\mu{}^{\nu'}\, a_\lambda{}^{\tau'}\, a_{\rho'}{}^\sigma\, \Gamma_{\nu\tau}{}'^{\rho}$ which would occur if $\Gamma_{\mu\lambda}{}^\sigma$ were transformed like a tensor, the change of reference system also introduces the appearance of $a_{\rho'}{}^\sigma \partial_\lambda\, a_\mu{}^{\rho'} = a_{\rho'}{}^\sigma\, \partial_\lambda \partial_\mu y'^{\rho}$. The change of curvilinear coordinates thus shows that the affine connection coefficients are not transformed in the same way as the components of a tensor. This demonstration involves no hypothesis concerning the symmetry of the $\Gamma_{\mu\nu}{}^\rho$. If we supposed that the $\Gamma_{\mu\nu}{}^\rho$ were non-symmetrical, the antisymmetrical part of the connection $\Gamma_{\mu\nu}{}^\rho = \frac{1}{2}(\Gamma_{\mu\nu}{}^\rho - \Gamma_{\nu\mu}{}^\rho)$ would be transformed like the components of a tensor, since the supplementary term in $\partial_\mu\partial_\nu - \partial_\nu\partial_\mu$ would then disappear.

6. In a Euclidean space with orthonormalized axes (reduced form)

$$g = g_{11}g_{22}g_{33}g_{00} = -1$$

by choosing the real coordinates ($x^0 = ct$).

7. In a non-Euclidean manifold, the relation (XIV-79), which is no longer a simple definition of the $g_{\mu\nu}$, is not always differentiable, and (XIV-12) is not necessarily valid. If $\nabla g_{\mu\nu} \neq 0$, one cannot define a *single* unit of length valid for every point of the manifold (cf. XV-76, p. 452).

8. Beginning with (XIV-114) and (XIV-105), it would be possible to establish the expression for the absolute differential ∇A_μ directly. As a matter of fact:

$$\begin{aligned}\nabla A_\mu &= g_{\mu\rho}(dA^\rho + \Gamma_{\sigma\lambda}{}^\rho A^\sigma\, dy^\lambda) = dA_\mu - A^\rho(dg_{\mu\rho}) + g_{\mu\rho}\Gamma_{\sigma\lambda}{}^\rho A^\sigma\, dy^\lambda \\ &= dA_\mu - A^\rho(\Gamma_{\mu\lambda}{}^\sigma g_{\sigma\rho} + \Gamma_{\rho\lambda}{}^\sigma g_{\mu\sigma})\, dy^\lambda + g_{\mu\sigma}\Gamma_{\rho\lambda}{}^\sigma A^\rho\, dy^\lambda \\ &= dA_\mu - \Gamma_{\mu\lambda}{}^\sigma A_\sigma\, dy^\lambda\end{aligned}$$

if

$$\begin{aligned}dg_{\mu\rho} &= d(e_\mu \cdot e_\rho) = e_\mu\omega_\rho{}^\sigma e_\sigma + e_\rho\omega_\mu{}^\sigma e_\sigma \\ &= (\Gamma_{\rho\lambda}{}^\sigma g_{\mu\sigma} + \Gamma_{\mu\lambda}{}^\sigma g_{\rho\sigma})\, dy^\lambda .\end{aligned}$$

But this method assumes that $g_{\mu\rho} = \mathbf{e}_\mu \mathbf{e}_\rho$ is differentiable or that $\nabla g_{\mu\rho} = 0$, which amounts to the same thing. Though valid in the case of Euclidean space (and consequently throughout the present chapter), this deduction cannot be unrestrictedly extended to any and all non-Euclidean manifolds.

CHAPTER XV

TENSOR CALCULUS IN A NON-EUCLIDEAN METRIC MANIFOLD. APPLICATION TO RIEMANNIAN SPACE

1. *Metric space and tangent Euclidean space*

A metric space is characterized by the following property: One can arbitrarily define, at every point in the space, a standard of length. This standard generally varies from one point to another.

Let y^μ be a system of coordinates with origin at point M_0 of the manifold. We shall say that the differential quadratic expression

(XV-1) $$ds^2 = g_{\mu\nu}\, dy^\mu\, dy^\nu$$

represents, by definition, the elementary interval between two infinitesimally separated points $M_0\,(y^\mu)$ and $M'\,(y^\mu + dy^\mu)$ of the manifold:

(XV-2) $$d\mathbf{M}^2 = g_{\mu\nu}\, dy^\mu\, dy^\nu .$$

The $g_{\mu\nu}$ occurring in (XV-1) form ten continuously differentiable functions of the y^μ.

Let us now consider an Euclidean space associated with the point $M_0 \equiv m_0$, and let us relate this space to a system of rectilinear axes extending from m_0. Let us then apply an identical unit of length to each one of these axes. This operation is possible, since we have defined a unit of length for every point. We shall choose our system of unit vectors $(\mathbf{e}_\mu)_0$ such that the following relation

(XV-3) $$(\mathbf{e}_\mu)_0\,(\mathbf{e}_\nu)_0 = (g_{\mu\nu})_0$$

is satisfied at the point m_0. In this relation, $(g_{\mu\nu})_0$ represents the values taken on by the ten $g_{\mu\nu}$ functions at the point $M_0 \equiv m_0$.

In the neighborhood of point M_0, we then have, according to (XV-2) and (XV-3)

(XV-4) $$(d\mathbf{M})_0^2 = (g_{\mu\nu})_0\, dy^\mu\, dy^\nu = (\mathbf{e}_\mu)_0\,(\mathbf{e}_\nu)_0\, dy^\mu\, dy^\nu = (\mathbf{e}_\mu\, dy^\mu)_0^2$$

that is

(XV-5) $$\left(\frac{\partial \mathbf{M}}{\partial y^\mu}\right)_0 = (\mathbf{e}_\mu)_0 .$$

The $(\mathbf{e}_\mu)_0$ then define axes tangent at M_0 to every curve y^μ, that is, the natural system for point M_0. The system of the $(\mathbf{e}_\mu)_0$ is thus related to the space tangent at M to the non-Euclidean manifold.

The space tangent to every point of a metric manifold is an Euclidean space.

The elementary interval of the associated Euclidean space

(XV-6) $$ds^2 = \overline{g}_{\mu\nu}\, dy^\mu\, dy^\nu$$

corresponds, in the neighborhood of the point $M_0 \equiv m_0$, to the square of the distance between points m_0 and $m = m_0 + dm_0$ of that space. Expression (XV-6) involves the coefficients

(XV-7) $$\overline{g}_{\mu\nu} = \overline{\mathbf{e}}_\mu \cdot \overline{\mathbf{e}}_\nu$$

which, at the point m_0 and in the system defined by the $(\mathbf{e}_\mu)_0$, will have the following values:

(XV-8) $$(\overline{g_{\mu\nu}})_0 = (e_\mu)_0 (e_\nu)_0 .$$

Thus, by comparing (XV-3) and (XV-8)

(XV-9) $$(g_{\mu\nu})_0 = (\overline{g_{\mu\nu}})_0 .$$

One can, thus, choose a coordinate system such that the two metrics – that of the manifold and that of the associated Euclidean space – be identical at $M = m_0$; and we then say that *the metrics are tangent at this point.*

2. *Affine connection*

At two infinitesimally separated points $M_0(y_0^\rho)$ and $M(y_0^\rho + dy^\rho)$ of a metric manifold, let us define the unit-vectors $(\mathbf{e}_\mu)_0$ and $\mathbf{e}_\mu$, with origins M_0

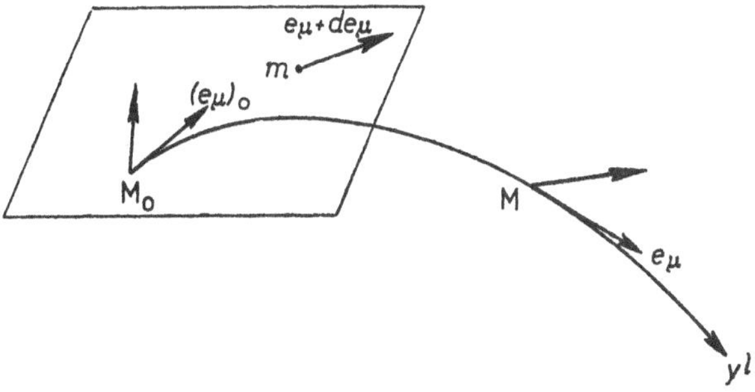

Fig. 45. Localization in a space tangent to infinitesimally separated points.

and M, tangent respectively to the y_0^ρ and the y^ρ. The systems of axes $(\mathbf{e}_\mu)_0$ and $\mathbf{e}_\mu$ constitute the natural systems at M_0 and at M and determine the Euclidean spaces tangent to these points of the manifold.

We wish to establish a correspondence between the point M, the vectors $\mathbf{e}_\mu$ (belonging to the Euclidean space tangent at M) and the elements m, $\mathbf{e}_\mu$ of the space tangent to M_0. This correspondence makes it possible to coordinate the local reference systems that successively define the Euclidean spaces tangent to the manifold.

When the variables y^ρ become $y'^\rho = y^\rho + dy^\rho$, the point $M\ (y^\rho)$ and the vectors $\mathbf{e}_\mu\ (y^\rho)$ undergo the corresponding variations

(XV-10) $$d\mathbf{M} = dy^\mu \mathbf{e}_\mu$$

(XV-11) $$d\mathbf{e}_\mu = \omega_\mu^\nu \mathbf{e}_\nu = \Gamma_{\mu\rho}^\nu \mathbf{e}_\nu\, dy^\rho$$

in the Euclidean space tangent at M.

We shall make "correspond" to the point M of the manifold the "image-point" $m = m_0 + d\mathbf{m}$ of the Euclidean space tangent at M_0; and to the vector $\mathbf{e}_\mu(M)$ of the manifold, the "image-vector" $\bar{\mathbf{e}}_\mu + d\bar{\mathbf{e}}_\mu$ of this tangent space. We thus define, in the Euclidean space tangent to M_0, a representation of the manifold in the neighborhood of the point M_0.

The coefficients $\Gamma_{\mu\rho}^\nu$, which allow us to define at every point a correspondence between the space tangent to the point and the space tangent to the point infinitesimally separated from it, are called the *affine connection coefficients of the manifold.*

3. *Representation of the first order*

Let us write relations (XV-10) and (XV-11) at the point $M_0 \equiv m_0$ of the Euclidean space tangent to M_0:

(XV-12) $$(d\mathbf{M})_0 = dy^\mu (\mathbf{e}_\mu)_0$$

(XV-13) $$(d\mathbf{e}_\mu)_0 = (\Gamma_{\mu\rho}^\nu)_0 (e_\nu)_0\, dy^\rho .$$

If $dy^\nu = y^\nu - y_0^\nu$ are infinitesimals of the first order, the correspondence defined by (XV-12) and by (XV-13) is likewise of the first order.

Moreover, the vectors $\bar{\mathbf{e}}'_\mu(m')$ of an Euclidean space, when they are defined at the point $m + d\mathbf{m}$ infinitesimally separated from point m, are expressed by $\bar{\mathbf{e}}_\mu + d\bar{\mathbf{e}}_\mu$ in the natural system at m. In this latter system one has:

(XV-14) $$d\mathbf{m} = dy^{\mu}\,\overline{\mathbf{e}_{\mu}}$$

(XV-15) $$d\overline{\mathbf{e}_{\mu}} = \overline{\omega_{\mu}^{\nu}}\,\overline{\mathbf{e}_{\nu}} = \overline{\begin{Bmatrix} \nu \\ \mu\rho \end{Bmatrix}}\,\overline{\mathbf{e}_{\nu}}\,dy^{\rho},$$

the barred notations still referring here to the symbols formed from the $g_{\mu\nu}$ of Euclidean space.

In the Euclidean space tangent to the manifold at M_0, one will have, in particular,

(XV-16) $$(d\mathbf{m})_0 = dy^{\mu}(\mathbf{e}_{\mu})_0$$

(XV-17) $$d(\mathbf{e}_{\mu})_0 = \overline{\begin{Bmatrix} \nu \\ \mu\rho \end{Bmatrix}}(\mathbf{e}_{\nu})_0\,dy^{\rho}.$$

In fact, by hypothesis,

(XV-18) $$(\overline{\mathbf{e}_{\mu}})_0 = (\mathbf{e}_{\mu})_0\,.$$

By comparing (XV-12) and (XV-16) one then has

(XV-19) $$(d\mathbf{M})_0 = (d\mathbf{m})_0$$

or

(XV-20) $$\left(\frac{\partial \mathbf{M}}{\partial y^{\mu}}\right)_0 = \left(\frac{\partial \mathbf{m}}{\partial y^{\mu}}\right)_0 = (\mathbf{e}_{\mu})_0\,.$$

The results obtained in Section 1 arise from the representation of the first order of the manifold in the associated Euclidean space.

The Euclidean space is tangent to the manifold at the point $M_0 \equiv m_0$. Furthermore, one has generally

(XV-21) $$(d\mathbf{e}_{\mu})_0 \neq d(\mathbf{e}_{\mu})_0\,.$$

Indeed, the vectors $\mathbf{e}_{\mu}$ tangent to the curves y^{μ} of the manifold coincide with the natural Euclidean system $\bar{\mathbf{e}}_{\mu}$ at the point $M_0 \equiv m_0$, but such is no longer the case for the infinitesimally separated point: in the tangent space of M_0, the vectors $(\mathbf{e}_{\mu})_0 + (d\mathbf{e}_{\mu})_0$ corresponding to the displacement dy^{ρ} in the manifold and the vectors $(\mathbf{e}_{\mu})_0 + d(\mathbf{e}_{\mu})_0$ corresponding to the displacement dy^{ρ} in the tangent space are not, generally, the same.

In addition, according to (XV-13) and (XV-17)

(XV-22) $$(\Gamma^{\rho}_{\mu\nu})_0 \neq \overline{\begin{Bmatrix}\rho\\ \mu\nu\end{Bmatrix}}_0 .$$

On the other hand, with the hypothesis

(XV-23) $$(d\mathbf{e}_\mu)_0 = d(\mathbf{e}_\mu)_0$$

one would obtain, through scalar multiplication by $(\mathbf{e}_\mu)_0 = (\overline{\mathbf{e}_\mu})_0$ and by symmetrizing

(XV-24) $$(\mathbf{e}_\nu)_0(d\mathbf{e}_\mu)_0 + (\mathbf{e}_\mu)_0(d\mathbf{e}_\nu)_0 = (\mathbf{e}_\nu)_0\, d(\mathbf{e}_\mu)_0 + (\mathbf{e}_\mu)_0\, d(\mathbf{e}_\nu)_0$$

that is

(XV-25) $$d(\mathbf{e}_\mu)_0(\mathbf{e}_\nu)_0 = (d\mathbf{e}_\mu\mathbf{e}_\nu)_0$$

or

(XV-26) $$(dg_{\mu\nu})_0 = d(\mathbf{e}_\mu)_0(\mathbf{e}_\nu)_0 = d(g_{\mu\nu})_0 .$$

By suppressing the zero indices, one would then have in the neighborhood of every point M:

(XV-27) $$dg_{\mu\nu} = d(\mathbf{e}_\mu\cdot\mathbf{e}_\nu) .$$

4. *Representation of the second order*

We are obliged to have recourse to a representation of the second order of the metric manifold in the Euclidean space tangent at a point M as soon as we have to relate to this space the resultant of two infinitesimal displacements in that space.

Let us suppose, for example, that the point M describes in the manifold an infinitesimal closed contour. In order to give a representation of the successive increments $\mathbf{d}M$ and $\mathbf{de}_\mu$, we must transport them into one and the same tangent Euclidean space. It is convenient to select a point $m_0 \equiv M_0$ inside the contour infinitesimally separated from the points on the contour.

To every position of the point M in the manifold there will then correspond a point m of the Euclidean space tangent at m_0.

The unit-vectors, tangent to the curves y^μ at the point M, are expressed in the following manner in the Euclidean space tangent at M_0:

(XV-28) $$\mathbf{e}_\mu = (\mathbf{e}_\mu)_0 + (\omega^\nu_\mu)_0(\mathbf{e}_\nu)_0 = (\mathbf{e}_\mu)_0 + (\Gamma^\nu_{\mu\rho})_0(y^\rho - y^\rho_0)(\mathbf{e}_\nu)_0$$

taking (XV-13) into account.

The shift from point M and from the $\mathbf{e}_\mu(M)$ to the infinitesimally separated point M' and to the natural system $\mathbf{e}'_\mu(M')$ is expressed by the increments (XV-10)

(XV-10) $$d\mathbf{m} = \omega^\mu \mathbf{e}_\mu = dy^\mu \mathbf{e}_\mu$$

(XV-11) $$d\mathbf{e}_\mu = \omega_\mu^\nu \mathbf{e}_\nu = \Gamma_{\mu\rho}^\nu \mathbf{e}_\nu \, dy^\rho$$

both plotted in the plane tangent to the point $M \equiv m$.

By relating both these increments to the plane tangent to the fixed point $M_0 \equiv m_0$, one obtains by substituting (XV-28) into (XV-10) and (XV-11):

(XV-29) $$(d\mathbf{m})_0 = [dy^\mu + (\Gamma_{\nu\rho}^\mu)_0 (y^\rho - y_0^\rho)\, dy^\nu](\mathbf{e}_\mu)_0$$

(XV-30) $$(d\mathbf{e}_\mu)_0 = [\omega_\mu^\nu + (\Gamma_{\sigma\rho}^\nu)_0 (y^\rho - y_0^\rho)\, \omega_\mu^\sigma](\mathbf{e}_\nu)_0 .$$

This correspondence, in which the infinitesimal products $(y^\rho - y_0^\rho)\, dy^\nu$ appear, is called the *representation of the second order.*

According to (XV-29)

(XV-31) $$\left(\frac{\partial^2 \mathbf{m}}{\partial y^\mu \partial y^\nu}\right)_0 = (\Gamma_{\nu\mu}^\rho)_0 (\mathbf{e}_\rho)_0 .$$

Whence

(XV-32) $$\left(\frac{\partial^2 \mathbf{m}}{\partial y^\mu \partial y^\nu}\right)_0 \neq \left(\frac{\partial^2 \mathbf{m}}{\partial y^\nu \partial y^\mu}\right)_0 .$$

The displacement $d\mathbf{m}$ is generally not integrable.

Moreover, if we compare (XV-29) and (XV-30) with the expression for the Euclidean increments written to second order:

(XV-33) $$d\mathbf{m}_0 = \left[dy^\mu + \begin{Bmatrix}\mu \\ \nu\rho\end{Bmatrix}_0 (y^\rho - y_0^\rho)\, dy^\nu\right](\mathbf{e}_\mu)_0$$

(XV-34) $$d(\mathbf{e}_\mu)_0 = \left[(\overline{\omega}_\mu^\nu)_0 + \begin{Bmatrix}\nu \\ \sigma\rho\end{Bmatrix}_0 (y^\rho - y_0^\rho)(\omega_\mu^\sigma)_0\right](\mathbf{e}_\nu)_0 ,$$

one obtains

(XV-35) $$\frac{\partial^2 \mathbf{m}}{\partial y^\mu \partial y^\nu} \neq \left(\frac{\partial^2 \mathbf{m}}{\partial y^\mu \partial y^\nu}\right)_0, \qquad \frac{\partial (\mathbf{e}_\nu)_0}{\partial y^\mu} \neq \left(\frac{\partial \mathbf{e}_\nu}{\partial y^\mu}\right)_0 .$$

Thus, any metric manifold is tangent at every point to a Euclidean space

$$\left(\frac{\partial \mathbf{M}}{\partial y^\mu}\right)_0 = \left(\frac{\partial \mathbf{m}}{\partial y^\mu}\right)_0$$

but, in general, it is a non-osculating tangency.

5. *Vectors and tensors associated with a metric manifold*

Let us consider the natural system at every point M of a metric manifold. In this way we define the tangent Euclidean spaces successively associated with the manifold. Each of these tangent Euclidean spaces is characterized by a system which introduces the coefficients $g_{\mu\nu}$ in such a way that

(XV-9) $$(\bar{g}_{\mu\nu})_0 = (g_{\mu\nu})_0$$

In each of these spaces, let us give the components of a vector or tensor in terms of the system of axes thus defined. In this way we obtain for the manifold a *field of vectors* or a *field of tensors.*

If we bring about a change of coordinates, and consequently a change of natural systems at this same point M, the components of a vector or tensor are transformed as indicated in Chapter XIV [cf. (XIV-44) through (XIV-52)]. Naturally, the $a^\nu_{\mu'}$ and $a^{\nu'}_\mu$ coefficients of the transformation vary from one point to another of the manifold.

At every point M of the manifold, the definitions of Chapter XIV apply to the tangent Euclidean space. The *contravariant components* A^μ such that

(XV-36) $$\mathbf{A} = A^\mu \mathbf{e}_\mu$$

are defined along each of the axes $\mathbf{e}_\mu$.

Moreover, the scalar product

(XV-37) $$\mathbf{A}\mathbf{e}_\mu = A_\mu$$

represents the *covariant components* of the vector A.

One then has, as in (XIV-5)

(XV-38) $$A_\mu = (A^\rho \mathbf{e}_\rho)\,\mathbf{e}_\mu = g_{\mu\rho} A^\rho\,,$$

and one will define, as in (XIV-6), the contravariant components of the metric tensor in such a way that

(XV-39) $$g_{\mu\rho}\, g^{\nu\rho} = \delta^\nu_\mu\,.$$

From this we deduce, as in Chapter XIV, the reciprocal relation (XIV-38)

(XV-40) $$g^{\mu\rho}A_\rho = g^{\mu\rho}g_{\rho\sigma}A^\sigma = \delta^\mu_\sigma A^\sigma = A^\mu$$

and we generalize, as in Chapter XIV, relations (XV-38) and (XV-40) to a tensor p-fold covariant and q-fold contravariant:

(XV-41) $$A_{\mu\nu} = g_{\mu\rho}g_{\nu\sigma}A^{\rho\sigma}, \qquad A^{\mu\nu} = g^{\mu\rho}g^{\nu\sigma}A_{\rho\sigma}.$$

(XV-42) $$A^{\mu\nu\dots}_{\rho\sigma\dots} = g^{\mu\lambda}g^{\nu\tau}\dots g_{\rho\delta}g_{\sigma\varepsilon\dots}A^{\delta\varepsilon\dots}_{\lambda\tau\dots}.$$

Finally, the *scalar product* of two vectors is then written as in (XIV-16)

(XV-43) $$\mathbf{A}\cdot\mathbf{B} = A^\mu B_\mu = g_{\mu\nu}A^\mu B^\nu.$$

The *norm* or *square of the length* of a vector **A** thus has the following expression

(XV-44) $$|A|^2 = A_\mu A^\mu = g_{\mu\nu}A^\mu A^\nu.$$

In particular, the square of the length of the vector $d\mathbf{M}$ (dy^ρ) is, in the Euclidean space tangent to M:

(XV-45) $$|d\mathbf{M}|^2 = dy_\mu dy^\mu = g_{\mu\nu}\,dy^\mu dy^\nu.$$

Thus, the metric tensor $g_{\mu\nu}$, at the point M of the manifold, has the properties which are summed up in formula (XIV-97) and those following it in the preceding chapter.

If g is the determinant formed by the $g_{\mu\nu}$

(XV-46) $$g = \det g_{\mu\nu}$$

one verifies, as in (XIV-10) – that is, taking (XV-39) into account – that

(XV-47) $$\text{cofactor}\, g_{\mu\nu} = g g^{\mu\nu}.$$

According to the rule for differentiating a determinant, one then obtains relations (XIV-97), (XIV-98) and (XIV-100), that is:

(XV-48) $$dg = g g^{\mu\nu}\,dg_{\mu\nu} = -\,g g_{\mu\nu}\,dg^{\mu\nu}$$

and

(XV-49) $$dg_{\mu\nu} = -\,g_{\mu\rho}g_{\nu\sigma}\,dg^{\rho\sigma}$$

(XV-50) $$dg^{\mu\nu} = -\,g^{\mu\rho}g^{\nu\sigma}\,dg_{\rho\sigma}$$

valid at every point M of the manifold.

We now seek to localize the vectors and the tensors in one of two infinitesimally separated points of the manifold with respect to those of the other point. This operation is possible if we know how to relate, by infinitesimal steps, the Euclidean tangent spaces; this is to say, if we can determine the affine connection of the manifold.

6. *Covariant derivation*

By definition, we say that the vector $\mathbf{e}'_\mu\,(M')$ "corresponding" to the vector $\mathbf{e}_\mu + d\mathbf{e}_\mu$ of the space tangent to M is *equipollent* to the vector $\mathbf{e}_\mu\,(M)$, or else that it is deduced from $\mathbf{e}_\mu\,(M)$ by *parallel transport*.

We shall define a vector **A** by listing its contravariant components at every point. We shall then establish a correspondence between the components of the vector $\mathbf{A}\,(y^\rho)$ with origin M and those of vector $\mathbf{A}'\,(y'^\rho)$ whose origin is at a point M' infinitesimally separated from M. In order to do this, we make the vector $\mathbf{A} + d\mathbf{A}$ of the space tangent to M correspond to A′.

The increment $d\mathbf{A}$ is called the *real or absolute increment of the vector* **A** when we pass from M to M'. This increment arises:

1) from the variation of the contravariant components when we pass from M to M',

2) from the change of local systems inherent to this displacement.

One has, as a matter of fact,

$$\text{(XV-36)} \qquad \mathbf{A} = A^\mu \mathbf{e}_\mu$$

and, in the space tangent to M:

$$\text{(XV-51)} \qquad d\mathbf{A} = dA^\mu \cdot \mathbf{e}_\mu + A^\mu\, d\mathbf{e}_\mu = (dA^\mu + \omega^\mu_\rho A^\rho)\,\mathbf{e}_\mu = DA^\mu \cdot \mathbf{e}_\mu\,.$$

In the space tangent to M, the real increment of the contravariant components of **A** is then:

$$\text{(XV-52)} \qquad \boxed{DA^\mu = dA^\mu + \omega^\mu_\sigma A^\sigma = dA^\mu + \Gamma^\mu_{\sigma\rho} A^\sigma\, dy^\rho\,.}$$

According to definition (XV-51), this increment forms the components of a tensor (whereas the increment dA^μ does not, according to (XV-52), have a tensor meaning). DA^μ represents the *absolute differential* of the contravariant components A^μ, that is, the absolute variation, independent of the representation adopted, when one passes from M to M'.

We call the ratio between the absolute increment DA^μ and the increment dy^ρ of the variables *the covariant derivative* ($D_\rho A^\mu$ or $A^\mu{}_{;\rho}$) of the components A^μ.

(XV-53) $$\boxed{D_\rho A^\mu \equiv A^\mu{}_{;\rho} \equiv \frac{DA^\mu}{dy^\rho} \equiv \partial_\rho A^\mu + \Gamma^\mu_{\sigma\rho} A^\sigma .}$$

Like the DA^μ, the covariant derivatives have a clearly defined tensor character.

The absolute differential DA^μ represents the contravariant components $(dA)^\mu$ of the increment $d\mathbf{A}$ in the system with base $\mathbf{e}_\mu$ of the space tangent at M.

Similarly, the absolute differential DA_μ of the covariant components defines the components $(dA)_\mu$ of the same increment $d\mathbf{A}$ with respect to system $\mathbf{e}_\mu$

(XV-54) $$DA_\mu = (dA)_\mu = d\mathbf{A} \cdot \mathbf{e}_\mu .$$

This relation may also be written

(XV-55) $$DA_\mu = d(\mathbf{A} \cdot \mathbf{e}_\mu) - \mathbf{A} \cdot d\mathbf{e}_\mu$$

and, taking (XV-37) and (XV-11) into account

(XV-56) $$DA_\mu = dA_\mu - \omega^\sigma_\mu \mathbf{A} \cdot \mathbf{e}_\sigma .$$

One then obtains:

(XV-57) $$DA_\mu = dA_\mu - \omega^\sigma_\mu A_\sigma = dA_\mu - \Gamma^\sigma_{\mu\rho} A_\sigma \, dy^\rho .$$

The covariant derivative ($D_\rho A_\mu$ or $A_{\mu;\rho}$) follows immediately

(XV-58) $$\boxed{D_\rho A_\mu \equiv A_{\mu;\rho} = \frac{DA_\mu}{dy^\rho} = \partial_\rho A_\mu - \Gamma^\sigma_{\mu\rho} A_\sigma .}$$

It is, of course, understood that definition (XV-57) imposes

(XV-59) $$D\mathbf{e}_\mu \equiv d\mathbf{e}_\mu - \omega^\sigma_\mu \mathbf{e}_\sigma \equiv 0 .$$

One will also have, according to (XV-54) and (XV-51)

(XV-60) $$(dA)_\mu = d\mathbf{A} \cdot \mathbf{e}_\mu = (DA^\rho)\, e_\rho \cdot e_\mu = g_{\mu\rho} DA^\rho$$

and, by comparison with definition (XV-54)

(XV-61) $$DA_\mu = g_{\mu\rho} DA^\rho .$$

The condition

(XV-62) $$Dg_{\mu\nu} = 0$$

is necessary to assure the compatability of (XV-53) and (XV-58) in a metric variety where $A_\mu = g_{\mu\rho} A^\rho$.

If $Dg_{\mu\nu} = 0$, that is if $g_{\mu\nu} = (\mathbf{e}_\mu \cdot \mathbf{e}_\nu)$ is differentiable – but only in this case – then the substitution of (XV-52) in (XV-62) leads to the definition (XV-58).

One can very easily generalize definitions (XV-53) and (XV-58) so as to make them applicable to the tensor $A^{\lambda\sigma\cdot\cdot}_{\mu\nu\cdot\cdot}$. The covariant derivative of this tensor is then

(XV-63) $$\boxed{\begin{aligned} D_\rho A^{\lambda\sigma}_{\mu\nu\cdot\cdot} \equiv A^{\lambda\sigma\cdot\cdot}_{\mu\nu\cdot\cdot;\,\rho} = \partial_\rho A^{\lambda\sigma\cdot\cdot}_{\mu\nu\cdot\cdot} - \Gamma^\tau_{\mu\rho} A^{\lambda\sigma}_{\tau\nu} - \Gamma^\tau_{\nu\rho} A^{\lambda\sigma\cdot\cdot}_{\mu\tau\cdot\cdot} \\ + \Gamma^\lambda_{\tau\rho} A^{\tau\sigma\cdot\cdot}_{\mu\nu\cdot\cdot} + \Gamma^\sigma_{\tau\rho} A^{\lambda\tau\cdot}_{\mu\nu\cdot\cdot}\,. \end{aligned}}$$

In particular, if we are concerned with the components of the metric tensor $g_{\mu\nu}$, one will have

(XV-64) $$D_\rho g_{\mu\nu} = \partial_\rho g_{\mu\nu} - \Gamma^\sigma_{\mu\rho} g_{\sigma\nu} - \Gamma^\sigma_{\nu\rho} g_{\mu\sigma}\,.$$

Condition (XV-60) may then be written

(XV-65) $$\partial_\rho g_{\mu\nu} = \Gamma^\sigma_{\mu\rho}(\mathbf{e}_\sigma \mathbf{e}_\nu) + \Gamma^\sigma_{\nu\rho}(\mathbf{e}_\mu \mathbf{e}_\sigma)\,,$$

and, according to (XV-11):

(XV-66) $$\partial_\rho g_{\mu\nu} = (\partial_\rho \mathbf{e}_\mu)\,\mathbf{e}_\nu + (\partial_\rho \mathbf{e}_\nu)\,\mathbf{e}_\mu = \partial_\rho(\mathbf{e}_\mu \cdot \mathbf{e}_\nu)\,.$$

Equation (XV-60) therefore has the following meaning: The condition $g_{\mu\nu} = \mathbf{e}_\mu \cdot \mathbf{e}_\nu$, valid at the point M, is differentiable. We shall see in the following section that this property expresses the possibility of defining an absolute unit of length, the same for every point of the manifold.[1]

7. *The parallel transport of a vector*

Let us suppose that condition (XV-60) or (XV-66) has been satisfied. According to (XV-52) and (XV-57), the real increment acquired by the components of a vector for an increment dy^μ of the variables, may be put in the following form:

(XV-67) $$DA^\mu = dA^\mu + \Gamma^\mu_{\nu\rho} A^\nu dy^\rho = (A^\mu + dA^\mu) - (A^\mu - \Gamma^\mu_{\nu\rho} A^\nu dy^\rho)$$

(XV-68) $$DA_\mu = dA_\mu - \Gamma^\sigma_{\mu\rho} A_\sigma dy^\rho = (A_\mu + dA_\mu) - (A_\mu + \Gamma^\sigma_{\mu\rho} A_\sigma dy^\rho)\,.$$

So this real increment appears as a difference between two terms:
The first $A^\mu + dA^\mu$ (or $A_\mu + dA_\mu$) is the value that the components A^μ (or A_μ) would take on if the vector **A** were transported in *any way whatsoever* from M to M'.

The second term $A^\mu - \Gamma^\mu_{\nu\rho} A^\nu \, dy^\rho$ (or $A_\mu + \Gamma^\sigma_{\mu\rho} A_\sigma dy^\rho$) corresponds, by definition, to the values that the components A^μ (or A_μ) would have if the displacement of the vector from M to M' were effected by *parallel transport*. The increase undergone by the components of a vector transported parallel to itself is then:

(XV-69) $\qquad (dA^\mu)_\| = -\Gamma^\mu_{\nu\rho} A^\nu \, dy^\rho$

(XV-70) $\qquad (dA_\mu)_\| = \Gamma^\sigma_{\mu\rho} A_\sigma \, dy^\rho$

With this convention, the absolute increment of the components A^μ and A_μ takes on the following form:

(XV-71) $\qquad DA^\mu = dA^\mu - (dA^\mu)_\|$

(XV-72) $\qquad DA_\mu = dA_\mu - (dA_\mu)_\|$

It is equal to the difference between two increments that would occur in the components of a vector transported from M to M' in any way whatsoever and the increments which would result from a parallel transport.

If the vector **A**′ (M') is indeed deduced from **A** (M) *by a parallel transport*, then $dA^\mu = (dA^\mu)_\|$. According to (XV-71) and (XV-72)

(XV-73) $\qquad DA^\mu \equiv 0$

(XV-74) $\qquad DA_\mu \equiv 0.$

If *a vector is transported parallel to itself, the absolute differential* – and consequently the *covariant derivative – of its components is zero.*

We deduce from definitions (XV-51) that the vector **A** + d**A**, which in the flat space tangent to M corresponds to the vector **A**′(M'), reduces to vector **A** when the change **A** → **A**′ is effected by parallel transport. The parallelism at two infinitesimally separated points is thus expressed by a *parallelism in the usual sense* (equality of the components of the vector A and of the vector **A** + d**A** corresponding to **A**′) *in the tangent Euclidean space.*

Finally, it must be noted that if vector **A** is transported parallel to itself, we must take into account the condition

(XV-74) $\qquad DA^\mu = 0 \qquad$ or $\qquad dA^\mu = -\Gamma^\mu_{\sigma\rho} A^\sigma \, dy^\rho$

in order to express the variation of the square of the length of this vector:

(XV-75) $$d(l^2) = d(g_{\mu\nu}A^\mu A^\nu) = dg_{\mu\nu} \cdot A^\mu A^\nu + 2g_{\mu\nu}A^\nu dA^\mu .$$

By substituting (XV-74) in (XV-75):

(XV-76) $$\begin{aligned} d(l^2) &= dg_{\mu\nu} \cdot A^\mu A^\nu - 2g_{\mu\nu}A^\nu \Gamma^\mu_{\sigma\rho} A^\sigma dy^\rho \\ &= (dg_{\mu\nu} - \Gamma^\lambda_{\mu\rho} g_{\lambda\nu} dy^\rho - \Gamma^\lambda_{\nu\rho} g_{\lambda\mu} dy^\rho) A^\mu A^\nu \\ &= (Dg_{\mu\nu}) A^\mu A^\nu , \end{aligned}$$

taking (XV-64) into account.

Condition (XV-60), $Dg_{\mu\nu} = 0$, *then expresses the conservation of the length of a vector when it is transported parallel to itself.*

In particular, the unit-vectors $\mathbf{e}_\mu$ are always, by definition, transported parallel to themselves

They thus retain the same length at every point of the manifold, and it is possible, by means of (XV-60), to define an absolute unit of length.

8. *The conditions of integrability and the structure of space*

1) *Euclidean space.* A space is *Euclidean* when the conditions of integrability relative to the increments

(XV-10) $$\mathbf{dm} = \mathbf{e}_\mu dy^\mu$$

(XV-11) $$\mathbf{de}_\mu = \omega^\nu_\mu \mathbf{e}_\nu = \Gamma^\nu_{\mu\rho} \mathbf{e}_\nu dy^\rho$$

are satisfied. One thus obtains, by integrating over an infinitesimal closed contour

(XV-77) $$\boxed{\int \mathbf{dm} = 0}$$

(XV-78) $$\boxed{\int \mathbf{de}_\mu = 0}$$

a) *The increment* **dm** *is integrable.* This condition is expressed by the symmetry of the affine connection coefficients.

In fact, every increment **dm** is independent of the path followed in going from $\mathbf{m}_0$ to $\mathbf{m}$ in the tangent Euclidean space. One then has:

(XV-79) $$\partial_\mu \partial_\nu \mathbf{m} = \partial_\nu \partial_\mu \mathbf{m}$$

or according to (XV-10),

(XV-80) $$\partial_\mu \mathbf{e}_\nu = \partial_\nu \mathbf{e}_\mu .$$

Taking (XV-11) into account, one will then have, as in (XIV-90),

(XV-81) $$\Gamma^{\rho}_{\mu\nu}\mathbf{e}_{\rho} = \Gamma^{\rho}_{\nu\mu}\mathbf{e}_{\rho}$$

that is[2]

(XV-82) $$\boxed{\Gamma^{\rho}_{\mu\nu} = \Gamma^{\rho}_{\nu\mu}\,.}$$

b) *The increment* $d\mathbf{e}_{\mu}$ *is integrable.* This condition is expressed by restrictions that must be satisfied by the $g_{\mu\nu}$ and their derivatives of the first and second order.[3] If it is possible to define a system of coordinates such that the corresponding $g_{\mu\nu}$ satisfy these restrictions, the space is Euclidean.

These restrictions were not set down in Chapter XIV, but we did take into account the following consequence, which arises from the integrability of $d\mathbf{e}_{\mu}$:

b′) *The unit length* l, which a certain standard of length defines in the tangent Euclidean space, *is not modified when that standard* A, *transported parallel to itself, has described an infinitesimal closed contour.*[4] Now, according to (XV-76)

(XV-76) $$d(l^2) = (Dg_{\mu\nu})\,A^{\mu}A^{\nu}\,,$$

putting

(XV-83) $$Dg_{\mu\nu} = dg_{\mu\nu} - \Gamma^{\lambda}_{\mu\rho}g_{\lambda\nu}\,dy^{\rho} - \Gamma^{\lambda}_{\nu\rho}g_{\mu\lambda}\,dy^{\rho}\,.$$

$d(l^2)$ is zero if $Dg_{\mu\nu} = 0$, that is (Cf. XV-66) if

(XV-66) $$d_{\rho}g_{\mu\nu} = d_{\rho}(\mathbf{e}_{\mu}\mathbf{e}_{\nu})\,.$$

The relation $g_{\mu\nu} = (\mathbf{e}_{\mu}\cdot\mathbf{e}_{\nu})$ is then differentiable.

We have implicitly used in Chapter XIV (cf. XIV-86) the two relations (XV-82) and (XV-66), valid in a Euclidean space, in determining the expression of the affine connection as a function of the $g_{\mu\nu}$ and their derivatives. Conditions (XV-82) and (XV-66) are actually sufficient for the attainment of this end. By expanding (XV-66), one in fact obtains:

(XV-84) $$d_{\rho}g_{\mu\nu} = \Gamma^{\sigma}_{\mu\rho}g_{\sigma\nu} + \Gamma^{\sigma}_{\nu\rho}g_{\mu\sigma} \qquad \text{that is} \qquad D_{\rho}\,g_{\mu\nu} = 0$$

and, by proceeding as we did in (XIV-92), taking (XV-82) into account, one obtains the previously established results.

(XV-85) $$\Gamma^{\rho}_{\mu\nu} = \left\{ {\rho \atop \mu\nu} \right\}$$

with

$$\text{(XV-86)}\qquad \left\{ \begin{matrix} \rho \\ \mu\nu \end{matrix} \right\} = \tfrac{1}{2} g^{\rho\sigma} (\partial_\mu g_{\sigma\nu} + \partial_\nu g_{\mu\sigma} - \partial_\sigma g_{\mu\nu}).$$

The affine connection of a Euclidean space is equal at every point, to the Christoffel symbols.

And yet, even if this condition is *necessarily fulfilled* in a Euclidean space, *it is not sufficient* to express the structure of such a space. It expresses, as a matter of fact, (XV-82) and (XV-66), that is (XV-77) and a consequence of (XV-78). The condition (XV-78) will be fulfilled and the space will be Euclidean if the affine connection, defined by (XV-85), satisfies, in addition to certain other restrictions, those specific restrictions that express the disappearance of curvature. If that is not the case, that is, if it is impossible to choose a reference system such that the conditions resulting from (XV-78) are satisfied by the connection $\{{}^{\rho}_{\mu\nu}\}$, the space is then Riemannian.

2) *Riemannian space.* If a space is Riemannian,

a) *The increment* $d\mathbf{m}$ *is integrable.* This condition

$$\text{(XV-77)}\qquad \int d\mathbf{m} = 0$$

is expressed by

$$\text{(XV-82)}\qquad \Gamma^{\rho}_{\mu\nu} = \Gamma^{\rho}_{\nu\mu}$$

b) *It is possible to define an absolute unit of length.*

Going back to (XV-66), this condition is expressed by

$$\text{(XV-87)}\qquad D_\rho g_{\mu\nu} = 0 \qquad \text{or} \qquad d_\rho g_{\mu\nu} = d_\rho(\mathbf{e}_\mu \cdot \mathbf{e}_\nu).$$

So, beginning with (XV-82) and (XV-87), one then obtains, as in the case of a Euclidean space

$$\text{(XV-85)}\qquad \boxed{\Gamma^{\rho}_{\mu\nu} = \left\{ \begin{matrix} \rho \\ \mu\nu \end{matrix} \right\}}$$

c) *The increment* $d\mathbf{e}_\mu$*, however, is not integrable.* So we cannot choose a system of coordinates such that the $g_{\mu\nu}$ fulfil the conditions resulting from (XV-78). *So there exists a world-curvature* whose expression and properties will be studied in Section 9.

3) *Arbitrary metric manifold.* Before limiting ourselves to the special case of Riemannian space, let us situate that case in the general framework of all possible metric manifolds.

In such a manifold, neither the increments $d\mathbf{m}$ nor $d\mathbf{e}_\mu$ is integrable[5]:

$$\boxed{\begin{aligned} &\int \mathbf{dm} \neq 0 \qquad \text{(XV-88)} \\ &\int \mathbf{de}_\mu \neq 0 \qquad \text{(XV-89)} \end{aligned}}$$

According to (XV-88), an infinitesimal closed contour does not have, in the tangent Euclidean space, a closed "corresponding element": *the space is endowed with a torsion.*

On the other hand, according to (XV-89), the standard of length is not orientable and, in general, not transportable; there exists a *rotational curvature of the manifold* and if the standard of length is not transportable, there also exists an invariant form: *the segmental curvature.* If it is not zero, one cannot define an absolute unit of length, that is, one retaining the same value at every point of the manifold. Nevertheless, this segmental curvature may disappear (and, consequently, an absolute unit of length can be defined), even though $d\mathbf{e}_\mu$ is not integrable. We then say that there exists *an invariance of gauge* (H. Weyl). If, in addition, there is no torsion, one obtains the structure of the Riemannian space.

9. *The curvature of Riemannian space. The Riemann-Christoffel tensor*

In a Riemannian space, just as in a Euclidean space, the integrability of $d\mathbf{m}$ and the conditions of gauge invariance impose

$$\Gamma^\rho_{\mu\nu} = \begin{Bmatrix} \rho \\ \mu\nu \end{Bmatrix}. \qquad \text{(XV-85)}$$

The difference between Euclidean space and Riemannian space appears when we form the integral

$$\int d\mathbf{e}_\mu \qquad \text{(XV-90)}$$

along an infinitesimal closed contour. If it is possible to choose a coordinate system such that (XV-90) is zero, then the space is Euclidean. If such a choice is impossible, the condition $\int \mathbf{de}_\mu \neq 0$ makes it possible to define a curvature. And this curvature characterizes Riemannian space.

For an increment dy^ρ of the variables, the vector $\mathbf{e}_\mu(M)$ becomes $\mathbf{e}'_\mu(y^\rho + dy^\rho)$. Corresponding to it in the plane tangent to M is a vector $\mathbf{e}_\mu + \mathbf{de}_\mu$ such that:

$$\text{(XV-91)} \qquad d\mathbf{e}_\mu = \omega_\mu^\nu \mathbf{e}_\nu = \begin{Bmatrix} \nu \\ \mu\sigma \end{Bmatrix} \mathbf{e}_\nu \, dy^\sigma$$

since the affine connection of the Riemannian manifold has the expression (XV-85). As we did in Section 4, let us relate the successive increments of $\mathbf{e}_\mu$ to the Euclidean space tangent to a point M_0 infinitesimally separated from all the points of the contour. One will then have, as in (XV-30):

$$\text{(XV-92)} \qquad (d\mathbf{e}_\mu)_0 = \left[\omega_\mu^\nu + \begin{Bmatrix} \nu \\ \sigma\rho \end{Bmatrix}_0 (y^\rho - y_0^\rho)\, \omega_\mu^\sigma \right] (\mathbf{e}_\nu)_0 \,.$$

Let us integrate (XV-92) along the contour C, assuming that the contour is formed by an infinitesimal curvilinear parallelogram with sides d and δ. The application of Stokes's theorem then makes it possible to pass from the curvilinear integral (XV-90) to a surface integral

$$\text{(XV-93)} \qquad \int d\mathbf{e}_\mu = \iint d\delta \mathbf{e}_\mu - \delta d\mathbf{e}_\mu = \iint (d\mathbf{e}_\mu)'$$

putting

$$\text{(XV-94)} \qquad (d\mathbf{e}_\mu)' = d\delta \mathbf{e}_\mu - \delta d\mathbf{e}_\mu \,.$$

By substituting (XV-92), this expression may also be written:

$$\text{(XV-95)} \qquad (d\mathbf{e}_\mu)' = d\left[\omega_\mu^\nu(\delta) + \begin{Bmatrix} \nu \\ \sigma\rho \end{Bmatrix}_0 (y^\rho - y_0^\rho)\, \omega_\mu^\sigma(\delta) \right] (\mathbf{e}_\nu)_0 - \delta\left[\omega_\mu^\nu(d) + \begin{Bmatrix} \nu \\ \sigma\rho \end{Bmatrix}_0 (y^\rho - y_0^\rho)\, \omega_\mu^\sigma(d) \right] (\mathbf{e}_\nu)_0$$

that is

$$\text{(XV-96)} \qquad (d\mathbf{e}_\mu)' = \left[(\omega_\mu^\nu)' + \begin{Bmatrix} \nu \\ \sigma\rho \end{Bmatrix}_0 [\mathrm{d}y^\rho \omega_\mu^\sigma(\delta) - \delta y^\rho \omega_\mu^\sigma(d)] + \begin{Bmatrix} \nu \\ \sigma\rho \end{Bmatrix}_0 (y^\rho - y_0^\rho)\, \omega_\mu^{\sigma\prime} \right] (\mathbf{e}_\nu)_0$$

putting

$$\text{(XV-97)} \qquad (\omega_\mu^\nu)' = d\omega_\mu^\nu(\delta) - \delta\omega_\mu^\nu(d) \,.$$

The first two terms of the right-hand member of (XV-96) are of the second order for d and δ. Neglecting the third-order terms, that is, the third term of (XV-96) and the difference between $\{^{\mu}_{\nu\rho}\}$ and $\{^{\mu}_{\nu\rho}\}_0$, we obtain:

(XV-98) $$(d\mathbf{e}_\mu)' = \left[(\omega^\nu_\mu)' + \begin{Bmatrix}\nu\\ \sigma\rho\end{Bmatrix}[dy^\rho\omega^\sigma_\mu(\delta) - \delta y^\rho\omega^\sigma_\mu(d)]\right](\mathbf{e}_\nu)_0 .$$

Taking (XV-91) into account, which may also be written in the following form:

(XV-99) $$\omega^\nu_\mu(d) = \begin{Bmatrix}\nu\\ \mu\rho\end{Bmatrix}dy^\rho, \qquad \omega^\nu_\mu(\delta) = \begin{Bmatrix}\nu\\ \mu\rho\end{Bmatrix}\delta y^\rho,$$

(XV-98) then reduces to

(XV-100) $$(d\mathbf{e}_\mu)' = \{(\omega^\nu_\mu)' + \omega^\nu_\sigma(d)\,\omega^\sigma_\mu(\delta) - \omega^\nu_\sigma(\delta)\,\omega^\sigma_\mu(d)\}\,(\mathbf{e}_\nu)_0 .$$

or

(XV-101) $$(d\mathbf{e}_\mu)' = \{(\omega^\nu_\mu)' - [\omega^\sigma_\mu\omega^\nu_\sigma]\}\,(\mathbf{e}_\nu)_0$$

putting

(XV-102) $$[\omega^\sigma_\mu\omega^\nu_\sigma] = \omega^\sigma_\mu(d)\,\omega^\nu_\sigma(\delta) - \omega^\sigma_\mu(\delta)\,\omega^\nu_\sigma(d).$$

Expression (XV-100) of the increment $(d\mathbf{e}_\mu)'$ then leads us to define the tensor of the second order whose components are

(XV-103) $$\Omega^\nu{}_\mu = (\omega^\nu_\mu)' - [\omega^\sigma_\mu\omega^\nu_\sigma].$$

The sum of the increments of the $d\mathbf{e}_\mu$ components is then

(XV-104) $$\int d\mathbf{e}_\mu = \int\int \Omega^\nu{}_\mu\mathbf{e}_\nu .$$

The tensor $\Omega^\nu{}_\mu$ defines the curvature of Riemannian space.

By expanding (XV-103), one can put its components in the following form

(XV-105) $$\Omega^\nu{}_\mu = d\left(\begin{Bmatrix}\nu\\ \mu\rho\end{Bmatrix}\delta y^\rho\right) - \delta\left(\begin{Bmatrix}\nu\\ \mu\rho\end{Bmatrix}dy^\rho\right) - \begin{Bmatrix}\sigma\\ \mu\rho\end{Bmatrix}dy^\rho\begin{Bmatrix}\nu\\ \sigma\lambda\end{Bmatrix}\delta y^\lambda + \begin{Bmatrix}\sigma\\ \mu\rho\end{Bmatrix}\delta y^\rho\begin{Bmatrix}\nu\\ \sigma\lambda\end{Bmatrix}dy^\lambda,$$

taking (XV-99) into account.

It should be noted that (XV-105) may further be written

(XV-106) $$\Omega^{\nu}{}_{\mu} = \left(\partial_{\lambda}\begin{Bmatrix}\nu\\ \mu\rho\end{Bmatrix} + \begin{Bmatrix}\sigma\\ \mu\rho\end{Bmatrix}\begin{Bmatrix}\nu\\ \sigma\lambda\end{Bmatrix}\right)dy^{\lambda}\delta y^{\rho} - \left(\partial_{\lambda}\begin{Bmatrix}\nu\\ \mu\rho\end{Bmatrix} + \begin{Bmatrix}\sigma\\ \mu\rho\end{Bmatrix}\begin{Bmatrix}\nu\\ \sigma\lambda\end{Bmatrix}\right)dy^{\rho}\delta y^{\lambda}$$

or, by permutation of the dummy indices ρ and λ in the last two terms

(XV-107) $$\Omega^{\nu}{}_{\mu} = \left(\partial_{\lambda}\begin{Bmatrix}\nu\\ \mu\rho\end{Bmatrix} - \partial_{\rho}\begin{Bmatrix}\nu\\ \mu\lambda\end{Bmatrix} + \begin{Bmatrix}\sigma\\ \mu\rho\end{Bmatrix}\begin{Bmatrix}\nu\\ \sigma\lambda\end{Bmatrix} - \begin{Bmatrix}\sigma\\ \mu\lambda\end{Bmatrix}\begin{Bmatrix}\nu\\ \sigma\rho\end{Bmatrix}\right)dy^{\lambda}\delta y^{\rho}.$$

The expression in parentheses is antisymmetrical in λ and ρ. One can define the area of the infinitesimal parallelogram constructed by d and δ:

(XV-108) $$ds^{\lambda\rho} = \tfrac{1}{2}(dy^{\lambda}\delta y^{\rho} - dy^{\rho}\delta y^{\lambda}).$$

One will thus obtain

(XV-109) $$\Omega^{\nu}{}_{\mu} = \left(\partial_{\lambda}\begin{Bmatrix}\nu\\ \mu\rho\end{Bmatrix} - \partial_{\rho}\begin{Bmatrix}\nu\\ \mu\lambda\end{Bmatrix} + \begin{Bmatrix}\sigma\\ \mu\rho\end{Bmatrix}\begin{Bmatrix}\nu\\ \sigma\lambda\end{Bmatrix} - \begin{Bmatrix}\sigma\\ \mu\lambda\end{Bmatrix}\begin{Bmatrix}\nu\\ \sigma\rho\end{Bmatrix}\right)ds^{\lambda\rho}.$$

Let us put

(XV-110) $$G^{\nu}{}_{\mu\rho\lambda} = \partial_{\lambda}\begin{Bmatrix}\nu\\ \mu\rho\end{Bmatrix} - \partial_{\rho}\begin{Bmatrix}\nu\\ \mu\lambda\end{Bmatrix} + \begin{Bmatrix}\sigma\\ \mu\rho\end{Bmatrix}\begin{Bmatrix}\nu\\ \sigma\lambda\end{Bmatrix} - \begin{Bmatrix}\sigma\\ \mu\lambda\end{Bmatrix}\begin{Bmatrix}\nu\\ \sigma\rho\end{Bmatrix}.$$

The components $\Omega^{\nu}{}_{\mu}$ of the curvature tensor finally take on the following expression [6]:

(XV-111) $$\boxed{\Omega^{\nu}{}_{\mu} = -\,G^{\nu}{}_{\mu\rho\lambda}\,ds^{\rho\lambda}.}$$

The curvature tensor whose components are $G^{\nu}{}_{\mu\rho\lambda}$ is called the *Riemann-Christoffel tensor.*

One still has, according to (XV-110)

(XV-112) $$G^{\mu}{}_{\mu\rho\lambda} = \partial_{\lambda}\begin{Bmatrix}\mu\\ \mu\rho\end{Bmatrix} - \partial_{\rho}\begin{Bmatrix}\mu\\ \mu\lambda\end{Bmatrix}$$

and, taking the following expressions into consideration:

$$(\text{XV-113})\qquad \begin{Bmatrix}\mu\\ \mu\rho\end{Bmatrix} = \tfrac{1}{2}\, g^{\mu\sigma}\,\partial_\rho g_{\mu\sigma} = \frac{1}{2g}\,\partial_\rho g = \tfrac{1}{2}\,\partial_\rho \log g\,,$$

by substitution in (XV-112) one obtains:

$$(\text{XV-114})\qquad G^{\mu}{}_{\mu\rho\lambda} \equiv \tfrac{1}{2}(\partial_\lambda\partial_\rho \log g - \partial_\rho\partial_\lambda \log g) \equiv 0\,,$$

Thus

$$(\text{XV-115})\qquad \Omega = \Omega^{\mu}{}_{\mu} = -\, G^{\mu}{}_{\mu\rho\lambda}\, ds^{\rho\lambda} = 0\,.$$

The invariant Ω formed from the Riemann-Christoffel tensor (cf. Equation (10) in footnote 5) represents the *homothetic curvature*. In a Riemannian space this invariant is identically zero.

Summing up: the connection of a Riemannian space is defined by (XV-85), just as in the case of Euclidean space. And like Euclidean space, Riemannian space also possesses: zero torsion $\Omega^\mu = 0$ $(\Gamma^\rho_{\nu\mu} = \Gamma^\rho_{\nu\mu})$ and zero homothetic curvature $\Omega = 0$.
But, on the other hand, one can define a non-zero curvature

$$\Omega^{\nu}{}_{\mu} = -\, G^{\nu}{}_{\mu\rho\sigma}\, ds^{\rho\sigma} \neq 0\,.$$

Thus, when a vector **A** describes, when transported parallel to itself, an infinitesimal contour in a Riemannian space, the variation

$$(\text{XV-116})\qquad dA_\mu = \begin{Bmatrix}\nu\\ \mu\rho\end{Bmatrix} A_\nu\, dy^\rho\,, \qquad (DA_\mu)_{\|} = 0$$

of the covariant components produces a total increment

$$(\text{XV-117})\qquad \int dA_\mu = \iint (dA)'_\mu\,.$$

According to (XV-103)

$$(\text{XV-118})\qquad \int dA_\mu = \iint \Omega^{\nu}{}_{\mu} A_\nu = -\iint G^{\nu}{}_{\mu\rho\sigma} A_\nu\, ds^{\rho\sigma}\,.$$

On the other hand, the variation

$$(\text{XV-119})\qquad dA^\mu = -\begin{Bmatrix}\mu\\ \sigma\rho\end{Bmatrix} A^\sigma\, dy^\rho$$

of the contravariant components produces an increment

(XV-120) $\int dA^{\mu} = \int\int (dA^{\mu})'$

along the contour. One also has[7]:

(XV-121) $\int dA^{\mu} = \int\int \Omega^{\mu}{}_{\nu} A^{\nu} = -\int\int G^{\mu}{}_{\nu\rho\sigma} A^{\nu}\, ds^{\rho\sigma}.$

To sum up: the structural conditions of a Riemannian space impose its various properties in the following way:

a) *The torsion is zero* ($\Omega^{\mu} = 0$), and consequently, the affine connection coefficients are symmetrical:

(XV-82) $\Gamma^{\rho}_{\mu\nu} = \Gamma^{\rho}_{\nu\mu}$

b) *The homothetic curvature is zero* ($\Omega = \Omega^{\mu}_{\mu} = 0$). Therefore, the length of a vector describing, when transported parallel to itself, an infinitesimal closed contour does not vary. It is then possible to choose a system of coordinates such that the unit of length is the same for every point in the space. Thus, one still has, according to (XV-76)

(XV-122) $dl^2 \equiv (Dg_{\mu\nu})\, A^{\mu} A^{\nu} = 0$

calling $Dg_{\mu\nu}$ expression (XV-64). Condition (XV-122) thus presupposes the possibility of choosing a system of coordinates such that

(XV-123) $D_{\rho} g_{\mu\nu} = 0.$

From (XV-82) and (XV-123) one deduces the expression for the affine connection as a function of the $g_{\mu\nu}$ and their first derivatives:

(XV-124) $\Gamma^{\rho}_{\mu\nu} = \left\{ \begin{matrix} \rho \\ \mu\nu \end{matrix} \right\}.$

c) *The rotational curvature is not zero* ($\Omega^{\nu}{}_{\mu} \neq 0$). So one cannot find a coordinate system such that the $g_{\mu\nu}$ and their derivatives of the first two orders will satisfy the conditions

(XV-125) $\Omega^{\nu}{}_{\mu} = 0$ or $G^{\nu}{}_{\mu\rho\sigma} = 0$

as it is possible to do for Euclidean space.

The structure of Riemannian space is entirely defined by the properties a), b), c). It should be noted that these characteristics – which really reduce to the single property of curvature – depend solely on the $g_{\mu\nu}$ and their derivatives of the first two orders. We may say that the properties of Riemannian space are entirely deduced from the character of the basic interval:

(XV-126) $$ds^2 = g_{\mu\nu}\, dy^\mu\, dy^\nu .$$

10. *Properties of the Riemann-Christoffel tensor*

The curvature of Riemannian space is expressed by means of the Riemann-Christoffel tensor

(XV-110) $$\boxed{G^\rho{}_{\mu\nu\sigma} = \partial_\sigma \begin{Bmatrix} \rho \\ \mu\nu \end{Bmatrix} - \partial_\nu \begin{Bmatrix} \rho \\ \mu\sigma \end{Bmatrix} + \begin{Bmatrix} \lambda \\ \mu\nu \end{Bmatrix}\begin{Bmatrix} \rho \\ \lambda\sigma \end{Bmatrix} - \begin{Bmatrix} \lambda \\ \mu\sigma \end{Bmatrix}\begin{Bmatrix} \rho \\ \lambda\nu \end{Bmatrix}.}$$

By lowering the ρ index, we form the tensor

(XV-127) $$G_{\mu\nu\rho\sigma} = g_{\mu\lambda}\, G^\lambda{}_{\nu\rho\sigma} .$$

According to (XV-110), this tensor has the following expression

(XV-128) $$\boxed{\begin{aligned} G_{\mu\nu\rho\sigma} = {} & \tfrac{1}{2}(\partial_\mu \partial_\rho g_{\nu\sigma} + \partial_\nu \partial_\sigma g_{\mu\rho} - \partial_\mu \partial_\sigma g_{\nu\rho} - \partial_\nu \partial_\rho g_{\mu\sigma}) + \\ & + \begin{Bmatrix} \lambda \\ \nu\sigma \end{Bmatrix} [\rho\mu, \lambda] - \begin{Bmatrix} \lambda \\ \sigma\mu \end{Bmatrix} [\nu\rho, \lambda] . \end{aligned}}$$

It is antisymmetrical in ρ, σ, in μ, ν, and symmetrical in the pairs of indices $\mu\nu$ and $\rho\sigma$:

(XV-129) $$G_{\mu\nu\rho\sigma} = - G_{\mu\nu\sigma\rho} , \qquad G_{\mu\nu\rho\sigma} = - G_{\nu\mu\rho\sigma} ,$$

(XV-130) $$G_{\mu\nu\rho\sigma} = G_{\rho\sigma\mu\nu} .$$

The Ricci tensor. The name *Ricci tensor* has been given to the second-order tensor formed by contraction of $G^\rho{}_{\mu\nu\sigma}$ with respect to ρ and σ.

(XV-131) $$\boxed{\begin{aligned} G_{\mu\nu} = G^\rho{}_{\mu\nu\rho} = {} & \partial_\rho \begin{Bmatrix} \rho \\ \mu\nu \end{Bmatrix} - \partial_\nu \begin{Bmatrix} \rho \\ \mu\rho \end{Bmatrix} + \\ & + \begin{Bmatrix} \lambda \\ \mu\nu \end{Bmatrix}\begin{Bmatrix} \rho \\ \lambda\rho \end{Bmatrix} - \begin{Bmatrix} \lambda \\ \mu\rho \end{Bmatrix}\begin{Bmatrix} \rho \\ \lambda\nu \end{Bmatrix} \end{aligned}}$$

[the contracted tensor $G^{\rho}{}_{\rho\mu\nu}$ is always zero in a Riemannian space, cf. (XV-114)].

The tensor $G_{\mu\nu}$ is symmetrical

(XV-132) $$G_{\mu\nu} = G_{\nu\mu}.$$

Starting with this tensor, one can finally define a *scalar Riemannian curvature*

(XV-133) $$G = g^{\mu\nu} G_{\mu\nu} = g^{\mu\nu} G^{\rho}{}_{\mu\nu\rho} = G^{\rho\nu}{}_{\nu\rho}.$$

Noteworthy identities. The presence of a curvature has the following consequence: the difference between the second derivatives $\nabla_\rho \nabla_\sigma$ and $\nabla_\sigma \nabla_\rho$ of a tensor is not zero. By referring to the definitions (XV-63) one can show that[8]

(XV-134) $$(\nabla_\rho \nabla_\sigma - \nabla_\sigma \nabla_\rho) A^\nu = - G^{\nu}{}_{\tau\rho\sigma} A^\tau$$

(XV-135) $$(\nabla_\rho \nabla_\sigma - \nabla_\sigma \nabla_\rho) A_\mu = G^{\tau}{}_{\mu\rho\sigma} A_\tau$$

and, consequently,

(XV-136) $$(\nabla_\rho \nabla_\sigma - \nabla_\sigma \nabla_\rho) A^{\nu\cdot\cdot}_{\mu\cdot\cdot} = - G^{\nu}{}_{\tau\rho\sigma} A^{\tau\cdot\cdot}_{\mu\cdot\cdot} + G^{\tau}{}_{\mu\rho\sigma} A^{\nu\cdot\cdot}_{\tau\cdot\cdot}.$$

Let us apply the formula thus obtained to the tensor $A_{\mu\nu} = \nabla_\mu A_\nu$. One obtains:

(XV-137) $$(\nabla_\rho \nabla_\sigma - \nabla_\sigma \nabla_\rho) \nabla_\mu A_\nu = G^{\tau}{}_{\mu\rho\sigma} \nabla_\tau A_\nu + G^{\tau}{}_{\nu\rho\sigma} \nabla_\mu A_\tau.$$

By forming two analogous equations through a cyclic permutation of ρ, σ, ν and by then adding them to (XV-137), we obtain a relation in which one must separately cancel out the coefficients of $\nabla_\mu A_\nu$ and of A_ν. One has, thus, the two identities

(XV-138) $$G^{\tau}{}_{\nu\rho\sigma} + G^{\tau}{}_{\sigma\nu\rho} + G^{\tau}{}_{\rho\sigma\nu} \equiv 0$$

(XV-139) $$\nabla_\rho G^{\tau}{}_{\mu\sigma\nu} + \nabla_\nu G^{\tau}{}_{\mu\rho\sigma} + \nabla_\sigma G^{\tau}{}_{\mu\nu\rho} \equiv 0.$$

Identity (XV-139) is usually called the *Bianchi identity.*

If we raise the index μ (through multiplication by $g^{\mu\lambda}$ whose covariant derivative is zero) and then sum over μ, we obtain:

(XV-140) $$\nabla_\rho G^{\tau\lambda}{}_{\sigma\nu} + \nabla_\nu G^{\tau\lambda}{}_{\rho\sigma} + \nabla_\sigma G^{\tau\lambda}{}_{\nu\rho} \equiv 0.$$

Let us contract in λ, σ and in τ, ν. There results:

(XV-141) $$\nabla_\rho G^{\nu\sigma}{}_{\sigma\nu} + \nabla_\nu G^{\nu\sigma}{}_{\rho\sigma} + \nabla_\sigma G^{\nu\sigma}{}_{\nu\rho} \equiv 0.$$

According to the symmetrical properties (XV-129) of the $G^{\mu\nu}{}_{\rho\sigma}$ and keeping in mind definitions (XV-131) and (XV-133), one thus has:

(XV-142) $$\nabla_\rho G - 2\nabla_\nu G^\nu_\rho \equiv 0$$

or

(XV-143) $$\nabla_\nu (G^\nu_\rho - \tfrac{1}{2}\delta^\nu_\rho G) \equiv 0 .$$

This identity is fundamental in General Relativity. The tensor

(XV-144) $$\boxed{S^\nu_\rho = G^\nu_\rho - \tfrac{1}{2}\delta^\nu_\rho G}$$

is called Einstein's tensor. Its divergence is identically zero:

(XV-145) $$\boxed{\nabla_\nu S^\nu_\rho \equiv 0 .}$$

11. *The geodesics of Riemannian space as analogues of the straight lines of Euclidean space*

The elementary interval of a four-dimensional metric space has the following expression in any system of coordinates whatever

(XV-146) $$ds^2 = g_{\mu\nu}\, dy^\mu\, dy^\nu .$$

If the properties of the space are entirely deducible from the given characteristics of the ds^2, that space is Euclidean (if the Riemann-Christoffel tensor is zero) or Riemannian.

We shall call the "shortest" lines – that is, the extremals that make ds stationary – the geodesics of the Riemannian space. If we are dealing with a Euclidean space, the geodesics will be the straight lines of that space.

A geodesic is thus defined by the condition

(XV-147) $$\delta \int ds = 0 .$$

Now, according to (XV-146)

(XV-148) $$2ds\,\delta\, ds = \delta g_{\mu\nu}\, dy^\mu\, dy^\nu + g_{\mu\nu}\, dy^\nu\, \delta\, dy^\mu + g_{\mu\nu}\, dy^\mu\, \delta\, dy^\nu .$$

(XV-147) is then written as follows:

(XV-149) $$\delta \int ds = \frac{1}{2}\int \left[\frac{\partial g_{\mu\nu}}{\partial y^\rho}\delta y^\rho \frac{dy^\mu}{ds}\frac{dy^\nu}{ds} + \right. \\ \left. + g_{\rho\nu}\frac{dy^\nu}{ds}\frac{d\delta y^\rho}{ds} + g_{\mu\rho}\frac{dy^\mu}{ds}\frac{d\delta y^\rho}{ds}\right] ds$$

and, integrating by parts

$$\int \delta\, ds = \frac{1}{2}\int\left[\frac{\partial g_{\mu\nu}}{\partial y^{\rho}}\frac{dy^{\mu}}{ds}\frac{dy^{\nu}}{ds} - \frac{d}{ds}\left(g_{\rho\nu}\frac{dy^{\nu}}{ds} + g_{\mu\rho}\frac{dy^{\mu}}{ds}\right)\right] \delta y^{\rho}\, ds + \frac{d}{ds}\left(g_{\rho\nu}\frac{dy^{\nu}}{ds}\delta y^{\rho} + g_{\mu\rho}\frac{dy^{\mu}}{ds}\delta y^{\rho}\right) ds. \tag{XV-150}$$

The final term of (XV-150) disappears if we assume that the δy^{ρ} cancel out at the limit of the domain of integration. Condition (XV-147) is then expressed by:

$$\frac{\partial g_{\mu\nu}}{\partial y^{\rho}}\frac{dy^{\mu}}{ds}\frac{dy^{\nu}}{ds} - \frac{d}{ds}\left(g_{\rho\nu}\frac{dy^{\nu}}{ds} + g_{\mu\rho}\frac{dy^{\mu}}{ds}\right) = 0 \tag{XV-151}$$

for (XV-150) must be satisfied identically, no matter what the variation δy^{ρ} may be.

Beginning with (XV-151) one easily obtains

$$\left(\frac{\partial g_{\mu\nu}}{\partial y^{\rho}} - \frac{\partial g_{\rho\nu}}{\partial y^{\mu}} - \frac{\partial g_{\mu\rho}}{\partial y^{\nu}}\right)\frac{dy^{\mu}}{ds}\frac{dy^{\nu}}{ds} - g_{\rho\nu}\frac{d^2 y^{\nu}}{ds^2} - g_{\mu\rho}\frac{d^2 y^{\mu}}{ds^2} = 0. \tag{XV-152}$$

The two final terms are identical. Multiplying by $g^{\rho\sigma}$ one has, at last,

$$\frac{d^2 y^{\sigma}}{ds^2} + \frac{1}{2}g^{\sigma\rho}\left(\frac{\partial g_{\mu\rho}}{\partial y^{\nu}} + \frac{\partial g_{\nu\rho}}{\partial y^{\mu}} - \frac{\partial g_{\mu\nu}}{\partial y^{\rho}_{\rho}}\right)\frac{dy^{\mu}}{ds}\frac{dy^{\nu}}{ds} = 0 \tag{XV-153}$$

or

$$\boxed{\frac{d^2 y^{\sigma}}{ds^2} + \begin{Bmatrix}\sigma\\ \mu\nu\end{Bmatrix}\frac{dy^{\mu}}{ds}\frac{dy^{\nu}}{ds} = 0.} \tag{XV-154}$$

This equation is often written in a slightly different way. Let us put

$$u^{\mu} = \frac{dy^{\mu}}{ds}. \tag{XV-155}$$

(XV-154) then is expressed as a function of the u^{μ}:

$$\frac{du^{\sigma}}{ds} + \begin{Bmatrix}\sigma\\ \mu\nu\end{Bmatrix}u^{\mu}u^{\nu} = 0. \tag{XV-156}$$

Taking into account the definition of an absolute derivative

$$\nabla u^{\sigma} = du^{\sigma} + \begin{Bmatrix}\sigma\\ \mu\nu\end{Bmatrix}u^{\mu}\, dy^{\nu}, \tag{XV-157}$$

(XV-156) may then be written as follows:

$$\text{(XV-158)} \qquad \frac{\nabla u^\sigma}{ds} = \frac{du^\sigma}{ds} + \begin{Bmatrix} \sigma \\ \mu\nu \end{Bmatrix} u^\mu u^\nu = 0, \qquad \frac{dy^\rho}{ds}\frac{\nabla u^\sigma}{dy^\rho} = 0,$$

that is

$$\text{(XV-159)} \qquad \boxed{u^\rho \nabla_\rho u^\sigma = 0.}$$

Equations (XV-154) or (XV-159) define the geodesics of Riemannian space or of Euclidean space. If we are concerned with an Euclidean space, it is always possible to relate the equation of a geodesic to a system of Galilean coordinates such that:

$$\text{(XV-160)} \qquad g_{\mu\nu} = \eta_{\mu\nu}.$$

In this system of special coordinates, (XV-158) reduces to

$$\text{(XV-161)} \qquad \frac{du^p}{ds} = 0, \qquad \frac{du^0}{ds} = 0$$

and according to (VII-12)

$$\text{(XV-162)} \qquad u^p = \frac{v^p}{c} u^0, \qquad u^0 = c\frac{dt}{ds}.$$

In a Galilean system, equations (XV-154) or (XV-159) are then written simply as:

$$\text{(XV-163)} \qquad v^p = \frac{dy^p}{dt} = \text{const}, \qquad y^p = v^p t + a^p.$$

These are, of course, the equations representing a straight line in an orthonormal system.

PROBLEMS

1. We consider a manifold having an arbitrary affine connection ($\Gamma^\rho_{\mu\nu}$ non-symmetrical). Show that

$$D_\rho \log g = 2\left[\begin{Bmatrix} \mu \\ \mu\rho \end{Bmatrix} - \Gamma^\mu_{\mu\rho}\right]$$

$$D_\rho \mathscr{A} = \partial_\rho \mathscr{A} - \mathscr{A}\Gamma^\mu_{\mu\rho}$$

$$D_\rho A^{\mu\rho} = \frac{1}{\sqrt{-g}}\partial_\rho \mathscr{A}^{\mu\rho} + \Gamma^\mu_{\sigma\rho} A^{\sigma\rho} + 2A^{\mu\sigma}\Gamma_\sigma - A^{\mu\sigma} D_\sigma \log\sqrt{-g}.$$

We have denoted

$$\mathscr{A} = \sqrt{-g}\,A, \qquad \mathscr{A}^{\mu\rho} = \sqrt{-g}\,A^{\mu\rho}, \qquad \Gamma_\rho = \Gamma^{\mu}_{\underset{\vee}{\rho\mu}}.$$

2. Show that the foregoing relations may be simplified

a) if $\Gamma^{\rho}_{\mu\nu} = \{^{\rho}_{\mu\nu}\}$ (Riemannian or Euclidean space).

b) if, in addition, the tensor $A^{\mu\rho}$ is antisymmetrical.

3. Solve problem no. 3 at the end of Chapter XIV, assuming that the affine connection is not symmetrical.

Calculate

$$\Phi_{\mu\nu} = D_\mu \varphi_\nu - D_\nu \varphi_\mu$$
$$\Phi_{\mu\nu\rho} = D_\mu \varphi_{\nu\rho} + D_\rho \varphi_{\mu\nu} + D_\nu \varphi_{\rho\mu}$$

as a function of $\varphi_{\mu\nu}$ and of $\varphi_{\mu\nu\rho}$.

4. We consider the two-dimensional manifold formed by the surface of a sphere of radius R. Using a system of coordinates such that

$$ds^2 = f(\rho^2)(dx^2 + dy^2), \qquad f(0) = 1, \qquad \rho^2 = x^2 + y^2$$

calculate x and y as functions of R, θ and φ. (With $X = R\cos\varphi\cos\theta$, $Y = R\cos\varphi\sin\theta$.)

Solution:

$$x = \frac{2X}{1+\sin\varphi}, \qquad \eta = \frac{2Y}{1+\sin\varphi}.$$

5. Write out (XV-134) and (XV-135) and find the identities which generalize (XV-138) and (XV-139) in the case where a space has, in addition to curvature, a torsion ($\Gamma^{\rho}_{\mu\nu} \neq \Gamma^{\rho}_{\nu\mu}$).

NOTES

1. This assumes [cf. footnote 5, equations (19) and (20)] that the manifold possesses no homothetic curvature.
2. This property highlights the absence of torsion in Euclidean space. [Cf. equation (5) in footnote 5.]
3. These restrictions assume the vanishing of all curvature, that is, the vanishing of the tensor $R^{\nu}{}_{\mu\rho\sigma}$ [cf. footnote 5, equation (8)].
4. This requirement assumes the following structural condition: the space is devoid of homothetic curvature. The tensor defined in Note 5, equation (13) disappears.
5. One can treat integrals (XV-88) and (XV-89) in a fashion similar to that developed in Section 9 for the special case of a Riemannian space.

Let us consider the closed contour formed by an infinitesimal parallelogram constructed by the increments d and δ. Its area is then

(1)
$$ds^{\sigma\rho} = \frac{1}{2}(dy^{\sigma}\delta y^{\rho} - dy^{\rho}\delta y^{\sigma})$$

One can transform the curvilinear integrals (XV-88) and (XV-89), relative to this contour, into surface integrals by applying Stokes's theorem. One then obtains

(2)
$$\int \mathbf{dm} = \iint (d\mathbf{m})', \qquad \int d\mathbf{e}_{\mu} = \iint (d\mathbf{e}_{\mu})'$$

putting:

(3)
$$(d\mathbf{m})' = d\delta\mathbf{m} - \delta d\mathbf{m}, \qquad (\mathbf{de}_{\mu})' = d\delta\mathbf{e}_{\mu} - \delta d\mathbf{e}_{\mu}.$$

The calculations of $(\mathbf{dm})'$ and of $(\mathbf{de}_{\mu})'$ defined by (3) is effected without difficulty by using relations (XV-29) and (XV-30), which refer to the Euclidean space tangent to a fixed point M_0 infinitesimally separated from all the points of the contour. By substituting (XV-29) and (XV-30) in (3), one will obtain, neglecting third-order terms:

a) On one hand

(4)
$$(d\mathbf{m})' = \Omega^{\rho}\mathbf{e}_{\rho}$$

with

(5)
$$\Omega^{\rho} = -(\Gamma_{\mu\nu}{}^{\rho} - \Gamma_{\nu\mu}{}^{\rho})\, ds^{\mu\nu}.$$

Ω^{ρ} forms the components of a vector that represents the variety's *torsion.* Condition (XV-82) results in the vanishing of this torsion. Conversely, this vanishing is expressed, according to (5), by the symmetry of the affine connection coefficients.

b) On the other hand

(6)
$$(d\mathbf{e}_{\mu})' = \Omega^{\nu}{}_{\mu}\,\mathbf{e}_{\nu}$$

with

(7)
$$\Omega^{\nu}{}_{\mu} = -R^{\nu}{}_{\mu\rho\sigma}\, ds^{\rho\sigma}$$

and

(8)
$$R^{\nu}{}_{\mu\rho\sigma} = \partial_{\sigma}\Gamma_{\mu\rho}{}^{\nu} - \partial_{\rho}\Gamma_{\mu\sigma}{}^{\nu} + \Gamma_{\mu\rho}{}^{\lambda}\Gamma_{\lambda\sigma}{}^{\nu} - \Gamma_{\mu\sigma}{}^{\lambda}\Gamma_{\lambda\rho}{}^{\nu}.$$

The tensor $\Omega^{\nu}{}_{\mu}$ (or $R^{\nu}{}_{\mu\rho\sigma}$) represents the curvature of the manifold.

In particular,

(9)
$$\int \mathbf{dl} = \mathbf{l}\iint \Omega$$

with

(10)
$$\Omega = \Omega^{\mu}{}_{\mu} = -R^{\mu}{}_{\mu\rho\sigma}\, ds^{\rho\sigma}.$$

The invariant Ω is called the *homothetic* or *segmental curvature* of the manifold. The tensor $\Omega^{\nu}{}_{\mu}$, from which it is derived, is then referred to as the *rotational curvature.* According to (10) and (8)

(11)
$$\Omega = -(\partial_{\sigma}\Gamma_{\mu\rho}{}^{\mu} - \partial_{\rho}\Gamma_{\mu\sigma}{}^{\mu})\, ds^{\rho\sigma}.$$

The standard of length is then not modified ($\int dl = 0$) if

$$\Gamma_{\mu\rho}{}^{\mu} = \partial_\rho \psi \,, \tag{12}$$

ψ being an arbitrary function.
And such is the case in both a Euclidean space and a Riemannian space where (12) is always realized. In fact (cf. XIV-103)

$$\Gamma_{\mu\rho}{}^{\mu} = \left\{ \begin{matrix} \mu \\ \mu\rho \end{matrix} \right\} = \frac{1}{2} g^{\mu\sigma} \partial_\rho g_{\mu\sigma} = \frac{1}{2g} \partial_\rho g = \partial_\rho \log \sqrt{-g} \,. \tag{13}$$

a) Thus, *an arbitrary metric manifold* is endowed with a torsion and two curvatures.

$$\boxed{\Omega^\mu \neq 0, \qquad \Omega^\mu{}_\nu \neq 0, \qquad \Omega \neq 0} \tag{14}$$

so that

$$\boxed{\Gamma_{\mu\nu}{}^{\rho} \neq \Gamma_{\nu\mu}{}^{\rho} \,, \qquad R^\mu{}_{\nu\rho\sigma} \neq 0, \qquad d_\rho g_{\mu\nu} \neq d_\rho(\mathbf{e}_\mu \cdot \mathbf{e}_\nu) \,.} \tag{15}$$

b) A *Riemannian manifold* has no torsion and is endowed with only one sort of curvature. Its structural conditions are then

$$\boxed{\Omega^\mu = 0, \qquad \Omega^\mu{}_\nu \neq 0, \qquad \Omega = 0} \tag{16}$$

and are expressed by

$$\boxed{\Gamma_{\mu\nu}{}^{\rho} = \Gamma_{\nu\mu}{}^{\rho}, \qquad R^\nu{}_{\mu\rho\sigma} \neq 0, \qquad d_\rho g_{\mu\nu} = d_\rho \,(\mathbf{e}_\mu \cdot \mathbf{e}_\nu) \,.} \tag{17}$$

c) Finally, an *Euclidean manifold,* which presupposes the integrability of $d\mathbf{m}$ and of $d\mathbf{e}_\mu$, has neither curvature nor torsion

$$\boxed{\Omega^\mu = 0, \qquad \Omega^\mu{}_\nu = 0, \qquad \Omega = 0} \tag{18}$$

conditions equivalent to

$$\boxed{\Gamma_{\mu\nu}{}^{\rho} = \Gamma_{\nu\mu}{}^{\rho}, \qquad R^\mu{}_{\nu\rho\sigma} = 0, \qquad d_\rho g_{\mu\nu} = d_\rho \,(\mathbf{e}_\mu \cdot \mathbf{e}_\nu) \,.} \tag{19}$$

The first and third of conditions (17) or (19), which are common to both Riemannian and Euclidean manifolds, are expressed, as we have seen by

$$\Gamma_{\mu\nu}{}^{\rho} = \left\{ \begin{matrix} \rho \\ \mu\nu \end{matrix} \right\} \tag{20}$$

On this subject, one may consult E. Cartan [30] and M. A. Tonnelat [26].

6. Tensor $G^\nu{}_{\mu\rho\lambda}$ is the tensor $R^\nu{}_{\mu\rho\lambda}$ ({ }), that is, the curvature tensor of any manifold in which the special expression

$$\left\{ \begin{matrix} \rho \\ \mu\nu \end{matrix} \right\}$$

is substituted for the connection $\Gamma_{\mu\nu}{}^{\rho}$. (Cf. Equation (8) of Note 5.)

7. In fact

(1) $$(dA^{\mu})' = d\delta A^{\mu} - \delta dA^{\mu} = -d\left(\begin{Bmatrix} \mu \\ \sigma\rho \end{Bmatrix} A^{\sigma}\delta y^{\rho}\right) + \delta\left(\begin{Bmatrix} \mu \\ \sigma\rho \end{Bmatrix} A^{\sigma}\, dy^{\rho}\right)$$

according to (XV-52). One then has:

(2) $$(dA^{\mu})' = -\left[\partial_{\tau}\begin{Bmatrix} \mu \\ \sigma\rho \end{Bmatrix} - \begin{Bmatrix} \mu \\ \lambda\rho \end{Bmatrix}\begin{Bmatrix} \lambda \\ \sigma\tau \end{Bmatrix}\right| A^{\sigma}\, dy^{\tau}\delta y^{\rho} +$$

$$\left[\partial_{\tau}\begin{Bmatrix} \mu \\ \lambda\rho \end{Bmatrix} - \begin{Bmatrix} \mu \\ \sigma\rho \end{Bmatrix}\begin{Bmatrix} \sigma \\ \lambda\tau \end{Bmatrix}\right] A^{\lambda}\delta y^{\tau}\, dy^{\rho}$$

$$= \left[\partial_{\rho}\begin{Bmatrix} \mu \\ \lambda\tau \end{Bmatrix} - \partial_{\tau}\begin{Bmatrix} \mu \\ \lambda\rho \end{Bmatrix} + \begin{Bmatrix} \mu \\ \sigma\rho \end{Bmatrix}\begin{Bmatrix} \sigma \\ \lambda\tau \end{Bmatrix} - \begin{Bmatrix} \mu \\ \sigma\tau \end{Bmatrix}\begin{Bmatrix} \sigma \\ \lambda\rho \end{Bmatrix}\right] A^{\lambda}\delta y^{\varrho}\, dy^{\tau}$$

$$= \tfrac{1}{2} G^{\mu}{}_{\lambda\tau\rho}A^{\lambda}(dy^{\tau}\delta y^{\rho} - \delta y^{\tau}\, dy^{\rho}) = G^{\mu}{}_{\lambda\tau\rho}A^{\lambda}\, ds^{\tau\rho} = -\,\Omega^{\mu}{}_{\lambda}A^{\lambda}\,.$$

8. In fact,

$$(\nabla_{\rho}\nabla_{\sigma} - \nabla_{\sigma}\nabla_{\rho})A^{\nu} = \left[\partial_{\rho}(\nabla_{\sigma}A^{\nu}) - \begin{Bmatrix} \lambda \\ \sigma\rho \end{Bmatrix}\nabla_{\lambda}A^{\nu} + \begin{Bmatrix} \nu \\ \lambda\rho \end{Bmatrix}\nabla_{\sigma}A^{\lambda}\right]$$

$$-\left[\partial_{\sigma}(\nabla_{\rho}A^{\nu}) - \begin{Bmatrix} \lambda \\ \rho\sigma \end{Bmatrix}\nabla_{\lambda}A^{\nu} + \begin{Bmatrix} \nu \\ \lambda\sigma \end{Bmatrix}\nabla_{\rho}A^{\lambda}\right]$$

$$= \left[\partial_{\rho}\partial_{\sigma}A^{\nu} + \partial_{\rho}\left(\begin{Bmatrix} \nu \\ \lambda\sigma \end{Bmatrix} A^{\lambda}\right) - \begin{Bmatrix} \lambda \\ \sigma\rho \end{Bmatrix}\nabla_{\lambda}A^{\nu} + \begin{Bmatrix} \nu \\ \lambda\rho \end{Bmatrix}\partial_{\sigma}A^{\lambda} + \begin{Bmatrix} \nu \\ \lambda\rho \end{Bmatrix}\begin{Bmatrix} \lambda \\ \tau\sigma \end{Bmatrix} A^{\tau}\right]$$

$$-\left[\partial_{\sigma}\partial_{\rho}A^{\nu} + \partial_{\sigma}\left(\begin{Bmatrix} \nu \\ \lambda\rho \end{Bmatrix} A^{\lambda}\right) - \begin{Bmatrix} \lambda \\ \rho\sigma \end{Bmatrix}\nabla_{\lambda}A^{\nu} + \begin{Bmatrix} \nu \\ \lambda\sigma \end{Bmatrix}\partial_{\rho}A^{\lambda} + \begin{Bmatrix} \nu \\ \lambda\sigma \end{Bmatrix}\begin{Bmatrix} \lambda \\ \tau\rho \end{Bmatrix} A^{\tau}\right]$$

$$= \left[\partial_{\rho}\begin{Bmatrix} \nu \\ \lambda\sigma \end{Bmatrix} - \partial_{\sigma}\begin{Bmatrix} \nu \\ \lambda\rho \end{Bmatrix} + \begin{Bmatrix} \nu \\ \tau\rho \end{Bmatrix}\begin{Bmatrix} \tau \\ \lambda\sigma \end{Bmatrix} - \begin{Bmatrix} \nu \\ \tau\sigma \end{Bmatrix}\begin{Bmatrix} \tau \\ \lambda\rho \end{Bmatrix}\right] A^{\lambda} = G^{\nu}{}_{\lambda\sigma\rho}A^{\lambda} = -\,G^{\nu}{}_{\tau\rho\sigma}A^{\tau}$$

BIBLIOGRAPHY

PART ONE / ELECTROMAGNETIC THEORY

BECKER, R. [1]: *Théorie des électrons*. Paris, Alcan, 1938.

BLOCH, L. [2]: *Précis d'électricité théorique*. Paris, Gauthier-Villars, 1919.

DURAND, E. [3]: *Électrostatique et magnétostatique*. Paris, Masson, 1953.

JOUGUET, M. [4]: *Le champ électromagnétique*. Paris, Armand Colin, 1954. *Électrostatique*. Paris, Gauthier-Villars, 1952. *Électrocinétique et magnétostatique*. Paris, Gauthier-Villars, 1955.

LORENTZ, H. A. [5]: *The Theory of Electrons*. Leipzig, 1916.

SLATER, C. and N. H. FRANK [6]: *Electromagnetism*. New York, McGraw-Hill Book Company Inc., 1947.

STRATTON, J. A. [7]: *Electromagnetic Theory*. New York, McGraw-Hill Book Company Inc., 1947.

PART TWO / SPECIAL RELATIVITY

ARZELIÈS, H. [8]: *Cinématique relativiste*. Paris, Gauthier-Villars, 1955.

BERGMANN, P. G. [9]: *Introduction to the Theory of Relativity*. New York, Prentice Hall, 1942.

BORN, M. [10]: *La théorie de la relativité d'Einstein*. Paris, Gauthier-Villars, 1923.

COSTA DE BEAUREGARD, O. [11]: *La théorie de la relativité restreinte*. Paris, Masson, 1949.

EINSTEIN, A. [12]: *Ann. d. Phys.* **17** (1905) 891.

EINSTEIN, A. [13]: *Über die spezielle und die allgemeine Relativitätstheorie*. Braunschweig, 1917.

LORENTZ, H. A., A. EINSTEIN and H. MINKOWSKI [14]: *Das Relativitätsprinzip*. 4th Edition, Leipzig, 1922.

LAUE, M. VON [15]: *La théorie de la relativité*, t. I. Paris, Gauthier-Villars, 1924.

MØLLER, C. [16]: *The Theory of Relativity*. Oxford, Clarendon Press, 1952.

PAPAPETROU, A. [17]: *Spezielle Relativitätstheorie.* Berlin, Deutscher Verlag der Wissenschaften, 1955.

SYNGE, J. L. [18]: *Relativity: The Special Theory.* Amsterdam, North-Holland Publishing Company, 1956.

PART THREE / GENERAL RELATIVITY

BERGMANN, P. G. [9]: *Introduction to the Theory of Relativity.* Englewood Cliffs, N.J., Prentice Hall International Inc., 1942.

CHAZY, J. [19]: *La théorie de la relativité et la mécanique céleste*, t. I and II. Paris, Gauthier-Villars, 1930.

DARMOIS, G. [20]: 'Les équations de la gravifique Einsteinienne', *Mém. des Sc. Math.* **25** (1926).

DONDER, TH. DE [21]: 'Introduction à la gravifique Einsteinnienne. Théorie du champ gravifique', *Mém. des Sc. Math.* **8** (1925).

EDDINGTON, A. S. [22]: *The Mathematical Theory of Relativity.* London, Cambridge University Press, 1954.

EINSTEIN, A. [23]: *Berl. Ber.* 1915, 778, 799 and 844;
Ann. d. Phys. **49** (1916) 769;
Les fondements de la relativité générale. Paris, Hermann, 1933.

LAUE, M. VON [24]: *La théorie de la relativité*, t. II. Paris, Gauthier-Villars, 1924.

LICHNEROWICZ, A. [25]: *Théories relativistes de la gravitation et de l'électromagnétisme.* Paris, Masson, 1955.

MØLLER, C. [16]: *The Theory of Relativity.* Oxford, Clarendon Press, 1952.

TONNELAT, M. A. [26]: *Les théories unitaires de l'électromagnetisme et de la gravitation.* Paris, Gauthier-Villars, 1965.

WEYL, H. [27]: *Raum, Zeit, Materie.* Berlin, 1920.

PART FOUR / MATHEMATICAL SUPPLEMENT

BRILLOUIN, L. [28]: *Les tenseurs en mécanique et en élasticité.* Paris, Masson, 1938.

CARTAN, E. [29]: *Leçons sur la géométrie des espaces de Riemann.* Paris, Gauthier-Villars, 1963.

CARTAN, E. [30]: 'Les espaces à connexion affine et la théorie de la Relativité Générale', *Ann. Ec. Norm.* **40**, 23.

CARTAN, E. [31]: *La méthode du repère mobile*. Paris, Hermann, 1935.

EDDINGTON, A. S. [22]: *The Mathematical Theory of Relativity*, Chapter II. London, Cambridge University Press, 1954.

EISENHART, L. P. [32]: *Riemannian Geometry*. Amer. Math. Soc., 1927.

EISENHART, L. P. [33]: *Non Riemannian Geometry*. Amer. Math. Soc.

LEVI CIVITA, T. [34]: 'Nozione di parallelismo in una varietà qualunque'. *Rend. del Circ. Math.*, Palermo, 1917, p. 173.

LICHNEROWICZ, A. [35]: *Élements de calcul tensoriel*. Paris, Armand Colin, 1957.

MAVRIDÈS, S. [36]: *Algèbre et analyse tensorielle. Structure des variétés à connexion affine*. Publications du C.E.M.A., Ch. 4.

TONNELAT, M.-A. [26]: *Les théories unitaires de l'électromagnetisme et de la gravitation*. Paris, Gauthier-Villars, 1965, Ch. I.

SUBJECT INDEX

Aberration 114
Addition of velocities 184, 187
Advance of the perihelions of the planets 386
Affine connection 441
Approximations of the gravitational field equations 351

Base vectors 418
Bending of light-rays by a gravitational field 389
Bianchi identities 462
Biot and Savart's law 34
Born-Infeld theory 271

Characteristic cone 166
Christoffel symbols 453
Clock paradox 146
Coefficients in the general Lorentz transformation 171
Collision between systems of particles 220
Compton effect 301
Conductor in an electric field 74
Conservation laws 237
Continuous media 226
Contraction of lengths 132
Coordinates, curvilinear 232, 427
–, pseudo euclidean 162
–, rectilinear 417
Coulomb's law 12
Covariance and contravariance 417
Covariant differentiation 432, 449
Covariant form of Maxwell's theory 244
Current four-vector 254
Curvature of a Riemannian space 457

Deflexion of light in a gravitational field 389
Deviation of charged particles 294
Dielectrics 26
Dilatation of time-intervals 133
Dipole field 24
Displacement current 51
Doppler effect 115, 150, 283
Dual tensor 249
Dynamics of continuous media 226
– of a point-mass (General Relativity) 344, 351, 366
– (Special Relativity) 209

Eichenwald experiment 75
Einstein's gravitational law 341
– law inside matter 343
– with electromagnetic field 343
Elastic collision 298
Electric charge 83
Electrodynamics in a gravitational field 382
Electromagnetic energy 58
– induction 49
– mass 88
– stress-energy tensors 252, 260
– stresses 85
– waves 61
Electrons (theory of) 83
Electrostatic and electromagnetic units 65
– potential 17
– laws 13
Energy-Momentum method (equation of motion) 374
– tensors 260
Eötvös experiment 319
Equations of Motion 465
Equivalence of mass and energy 213
Equivalence Principle 326
Ether drift 116
Euclidean vector space 417
Experimental verifications of special relativity 282

Faraday's law 49
First order effects 114
Fitzgerald-Lorentz hypothesis 124
Fizeau's experiment 116
Force, electromagnetic 251
–, fictitious 113, 323
–, four-dimensional or Minkowski's 211
Four-dimensional formalism 159
Four-vector (transformation of) 176
Fresnel's dragging coefficient 119, 206

Galilean transformation 112
Gauge transformation 248
Gauss' theorem 14
General Relativity 313
Geodesic lines 234
Geometrical representation of Lorentz formulas 136
Geometry, non-euclidean136
Gravitational field equations 346, 348, 341, 343
– linear approximation 357
Gravitational potential 317, 351
– quasi-static field 351
– shift (spectral lines) 392
– waves 364
Group-velocity 69
Guye and Lavanchy experiments 296

Hertz' hypothesis 78
Hoek's experiment 116
Huygens' Principle 71, 198

Index of a dispersive medium 69
Induction and field 100
– (relativistic relations) 257
Insulator in an electric field 74
– in a magnetic field 76
Isothermal coordinates 354
Isotropic coordinates 381
– vectors 165
Ives and Stillwell's experiments 286

Kinematics 184
Kronecker Symbol 162

Laplace's equation 18
Levi-Civita's symbol 249, 279

Liénard-Wiechert potentials 94
Limiting velocity 193
Linear and othogonal substitutions 167
Lorentz condition 56
– contraction 89, 132
– equations (covariant form) 250
– force 85
– theory 83
– transformation (general) 140
– (special) 131, 139

Mach's Principle 325
Magnetic Media 42
– moment 97
Magnetostatics laws 34
Mass defect 225, 304
Material energy-momentum tensor 230
Maxwell's equations 53f
– (covariant form) 244
Maxwell-Hertz potentials 92
Maxwell tensor 252
Mean values 95
Meson life-time 289
Metric tensor 428, 430
Michelson's experiment 121
Microscopic fields 84
Mie's theory 267
Minkowski (space and formalism) 159
– force 211
Møller's method 140
Momentum energy of a particle 209
Motion laws 237
Motion of a charged particle in an electromagnetic field 292

Newton's law 313, 321
Newtonian gravitational constant 318
Newtonian potential 360
Non-dualistic theories 409
Non-euclidean space 316, 222, 327
Non-linear electromagnetism 266, 272
Norm of a vector 419

Observational data (Newton's law 313
Orthogonal frame 424

Parallel displacement of vectors 450
Perfect fluids 231
Perihelion, advance 386

Phase velocity 203
Plane waves 67
Poisson's law 18, 320
Polarization 97
Potentials (Liénard-Wiechert) 94
–, electromagnetic 55
–, gravitational 351
–, retarded 70, 91
Poynting's theorem 60
Pressure in a continuous media 228
Principle of equivalence 322
– of special relativity 126
Propagation of electromagnetic field 61
– of spherical waves 200
– velocity 203
Proper time 135

Radius of the electron 87
Red shift 392
Reduced forms of the ds^2 164
Refractive medium 205
Relativistic dynamics 209
– electromagnetism 244
– kinematics 184
Relativity principle in electrodynamics 113
– in mechanics 111
Rest mass 210
Retarded potentials 55, 58
Riemannian space 454
Römer's experiment 151
Rotating disc 324, 330
Rotation of axes 425
Rowland's experiment 74

Sagnac's experiment 324
Scalar product 431
Schwarzschild's solution 377
Simultaneity of events 128
Singularity method (Einstein, Infeld, Hoffmann) 366
Slow motion (electromagnetic equation) 73
Slowing-down of clocks 283
Space like vectors 165
Special relativity 111
Spherical conductor 23
– symmetry 377, 382
– waves 70
Structure of the electron 87
Structure relations (non euclidean manifold) 467
Summation convention 161, 417
System, inertial 323
–, uniformly rotating 339
Systems of bound particles 223
– of free particles 218
– of units 53

Tensor algebra 417
– analysis 427
Tensors (transformation) 432
Thomas precession 169
Time-like vectors 165
Trajectories of a neutral particle in a Schwarzschild field 384
Trajectory of a charged material point in an electromagnetic field 263
Transformation of rectilinear axes 420

Unified field theories 400

Variation of mass with velocity 292
Variational principle 265
Velocity of light measurements 65

Wave propagation and Relativistic Kinematics 199
Waves, electromagnetic 61
Wave-trains 67
Wilson's experiment 76
World velocity 84